VICTORIAN GHOST STORIES

VICTORIAN GHOST STORIES

An Oxford Anthology

Selected and introduced by
MICHAEL COX
and
R. A. GILBERT

Oxford New York
OXFORD UNIVERSITY PRESS
1991

Oxford University Press, Walton Street. Oxford OX2 6DP

Oxford New York Toronto
Delhi Bombay Calcutta Madras Karachi
Petaling Jaya Singapore Hong Kong Tokyo
Nairobi Dar es Salaam Cape Town
Melbourne Auckland

and associated companies in
Berlin Ibadan

Oxford is a trade mark of Oxford University Press

British Library Cataloguing in Publication Data
Data available

Library of Congress Cataloging in Publication Data
Victorian ghost stories: an Oxford anthology/selected and introduced by Michael Cox and R. A. Gilbert.
p. cm.
Includes bibliographical references (p.).
1. Ghost stories, English. 2. English fiction—19th century.
I. Cox, Michael, 1948– . II. Gilbert, R. A.
PR1309.G5V54 1991 823'.0873308'09034—dc20 91–2748
ISBN 0-19-214202-X

Typeset by Best-set Typesetters Ltd.
Printed in Great Britain by
Courier International Ltd, East Kilbride

ACKNOWLEDGEMENTS

OUR thanks are due to the following individuals and institutions for advice and assistance received during the compilation of this anthology: Richard Dalby and Raphael Shaberman for their bibliographical help; Dr Glen Cavaliero and Deirdre Toomey; the staff of the Cambridge University Library and the Bodleian Library, Oxford.

CONTENTS

INTRODUCTION

FOR those who look back on it, the Victorian age seems to be invested with a peculiar quality of difference—heightened by its relative proximity in time—that is reflected in its ghosts. It was an age shaped, perhaps more than any other previous period, by the forces of transition. The predominantly feudal and agrarian past had disintegrated under the action of democracy and industrialism; and yet the final consequences of these truly revolutionary processes remained unclear. All that people knew was that a gulf was opening up with the past. As Thackeray noted in 1860: 'Your railroad starts the new era, and we of a certain age belong to the new time and the old one.'[1] With the shadow of change falling across virtually every area of life and thought, the receding past became a focus for anxiety, and in literature the ghost story offered a way of anchoring the past to an unsettled present by operating in a continuum of life and death. In the ghost story, obligations do not cease with death, and the past is never a closed book. What has been can be again, though often terribly transformed. For a progressive age (progressing to what?), the idea of a vindictive past held an especial potential for terror.

In personal terms, ghosts were obvious, though still potent, images of the lost past—past sins, past promises, past attachments, past regrets—and could be used to confront, and exorcize, the demons of guilt and fear. For almost the whole of the Victorian period the ghost story is of a piece: traditional in its forms and intentions, but energetically inventive and infused with a relish of the supernatural that parallels the more general Victorian fascination with the trappings of death—the dark, extravagant splendour of the funeral, the baroque richness of the cemetery, the guilt-laden luxury of mourning. And then, at its most basic, the function of most Victorian ghost stories, like all similar fictions, was simply to produce what Michael Sadleir called 'the pleasurable shudder'—'a horror which we know does not—but none the less conceivably might—threaten ourselves'.[2]

Our aim in this anthology is to map out the development of the Victorian ghost story from *c.*1850. Although there are earlier examples of the form (notably the early stories of J. S. Le Fanu), it is in the 1850s

[1] 'De Juventute' (*Cornhill Magazine*, Aug. 1860).

[2] 'A Master of the Ghost Story', review of M. R. James's *Ghost Stories of an Antiquary*, *The Listener* (29 Apr. 1931).

that the distinct, anti-Gothic character of the Victorian ghost story begins to emerge. Where the Gothic tale of terror had been indulgently heroic and ostentatiously fictitious, the Victorian ghost story was typically domestic in tone and inclined to blur the boundaries between fact and fiction. Our first example, Elizabeth Gaskell's 'The Old Nurse's Story', which dates from 1852, exhibits essential Victorian qualities—not least in its homely detail and disciplined treatment of supernatural events—that are not often apparent in ghost stories of the 1830s and 1840s.

We have not felt obliged to terminate the collection strictly at 1901. An exchange in M. R. James's story 'A Neighbour's Landmark' (published in 1925) both emphasizes the 'otherness' of the Victorian period and justifies our somewhat elastic definition of 'Victorian':

> 'You begin in a deeply Victorian manner,' I said; 'is this to continue?'
>
> 'Remember, if you please,' said my friend, looking at me over his spectacles, 'that I am a Victorian by birth and education, and that the Victorian tree may not unreasonably be expected to bear Victorian fruit.'

'Victorian by birth and education', then, will be the basic qualification for inclusion; publication dates are consequently less important if it is clear that this broad condition can be met. All the authors represented here are self-evidently fruits of the Victorian tree.

As for what we understand by a ghost story, the five criteria followed for *The Oxford Book of English Ghost Stories* remain applicable on the whole to the principles of selection followed in the present volume:

> Each story should reveal to the reader a spectacle of the returning dead, or their agents, and their actions; there must be a dramatic interaction between the living and the dead, more often than not with the intention of frightening or unsettling the reader; the story must exhibit clear literary quality . . . there must be a definable Englishness about the story . . . and finally . . . the story must be relatively short.

The last requirement has unfortunately eliminated several otherwise excellent stories (for instance, we have been unable to justify including anything by the prolix Margaret Oliphant). We have also tried to balance well-known landmarks of the genre with lesser-known examples and one or two genuine rarities.

Ghost stories were something at which the Victorians excelled. They were as typically part of the cultural and literary fabric of the age as imperial confidence or the novel of social realism. Much of the output was hack commercial fiction of the weakest sort; but the Victorians also scored an impressive number of successes and effectively defined the possibilities of the short ghost story, to the extent that all subsequent practitioners have been indebted to the Victorian achievement in some degree or other.

Ghost stories—particularly Victorian ghost stories—are inherently limited in form and dynamic potential. But their constraining qualities can often be a source of strength as well as a weakness. The difficulty of achieving the right balance of elements is one reason why complete success is often elusive. Montague Summers—scholar, occultist, and the compiler of two classic anthologies of ghost stories—pointed out that, 'setting aside the highest masterpieces of literature, there is nothing more difficult to achieve than a first-class ghost story'.[3] Summers was a sturdy apologist for Victorianism, especially for the more arcane literary productions of the age, but he does not overstate the case. The successful ghost story, like the successful detective story, depends on using conventions creatively. The ghost story's basic dynamics are settled in the reader's expectations at the outset. We know that we are to be shown a climactic interaction between the living and the dead, and usually expect to be unsettled by the experience. The skill comes when an author is able to work closely within the limited conventions of the form whilst at the same time reassembling familiar components into something that can still engage and surprise. To some extent this requires a certain complicity between author and reader, whereby the latter becomes a willing accomplice in the whole design. But it also calls for fluent invention, well-developed dramatic instincts, and story-telling capability of a high order on the part of the author. One might add that the author's commitment to the fiction is also a fundamental requisite. While it is not necessary for an author to believe unconditionally in the supernatural for a ghost story to come off (H. G. Wells's 'The Red Room' shows that), it is essential that he or she engages fully in the *pretence* of believing.

The art of the literary ghost story was perfected in the middle decades of the nineteenth century through the medium of magazines. The repeal of the newspaper tax in 1855, the development of new technology, and the accelerating spread of literacy helped to create an unprecedented boom in periodical publishing. In 1859 the publisher George Smith conceived the idea of a low-priced monthly magazine that would combine the attractions of a general review with a wide range of quality fiction. Though Smith's *Cornhill Magazine* was anticipated by the launch of another shilling monthly along similar lines,

[3] Introduction to *Victorian Ghost Stories* (1936), xvii.

Macmillan's Magazine in November 1859, he had a trump card in Thackeray, whose *Lovel the Widower* and *Roundabout Papers* began in the first number (January 1860), together with the commencement of Trollope's *Framley Parsonage*. Thackeray, who was also the *Cornhill*'s editor until May 1862, was proud of the magazine's suitability for a family audience ('Our magazine is written not only for men and women, but for boys, girls, infants'),[4] and initially the *Cornhill* was a considerable success; but by the end of the decade its circulation had fallen off dramatically in the face of competition from a succession of popular monthlies whose aim was to supply easily assimilated entertainment and information. They included *Temple Bar* in December 1860; the *St James's Magazine* in April 1861; *London Society* in February 1862; *The Argosy* in December 1865; *Belgravia* in November 1866; and *Tinsleys' Magazine* in August 1867.

It was in magazines like these that the ghost story came to maturity. Fiction—short stories as well as serializations of full-length novels—was the prime reason for the popularity of these periodicals, reflecting the fact that they catered for a burgeoning middle-class readership that was educated but relatively unsophisticated in its literary tastes. As a prospectus for the launch of the weekly *Cassell's Magazine* in 1868 announced: 'Fiction of powerful interest will form the prominent feature of its pages.' There was, indeed, a sense that fiction now had an essential social function. This was the message of an article by Walter Besant entitled 'The Value of Fiction' in the November 1871 number of *Belgravia*: 'It is interesting to mark', wrote Besant, 'the sudden rush with which the old Puritanical dislike for novels has collapsed, at last, in the present generation.' He even saw an educative function in sensational fiction: 'Ladies who read *Belgravia* do not often penetrate into the slums of the East-end. Fagin and his tribe are as unknown to them as the Esquimaux. It is not, however, bad for ladies to know that such things exist.' Nor, perhaps, for ladies to be pleasantly frightened by a ghost story.

The high point of the periodical trade was the special Christmas Numbers, in which ghost stories came into their own. Richard Altick notes that 'The sales of some of these annual supplements were tremendous . . . They were admirably fitted to the tastes of those whose pocketbooks were opened a little wider than usual under the mellowing influence of the Christmas season, and their presence in English homes had a powerful effect on the spread of reading interest.'[5]

[4] Quoted by Anne Thackeray Ritchie, *Cornhill Magazine* (Jan. 1910).

[5] *The English Common Reader* (1957), 363.

A typical production was *A Stable for Nightmares*, the Christmas Number of *Tinsleys' Magazine* for 1868, which contained twelve anonymous stories (one of them, 'Pichon & Sons, of the Croix Rousse', is reprinted here), three poems, and six illustrations—all for a shilling.

Christmas, which had always had an association with the marvellous and the supernatural ('A sad tale's best for winter: I have one / Of sprites and goblins', says Mamillius in *The Winter's Tale*), now became indelibly identified with the reading of ghost stories. The key figure here was Dickens. Though his own forays into short supernatural fiction were comparatively few, Dickens fostered the ghost story's seasonal association through the Christmas Numbers of the magazines he edited—in particular those of *All the Year Round* (launched in April 1859), which sold on average between 185,000 and 250,000 copies. Stories published by Dickens in *All the Year Round* included: 'How the Third Floor Knew the Potteries' (1863), 'The Phantom Coach' (1864), and 'The Engineer' (1866) by Amelia B. Edwards; 'Not to be Taken at Bed-time' by Rosa Mulholland, and Dickens's own companion piece 'To be Taken with a Grain of Salt' (both 1865); 'The Compensation House' by Charles Collins, and Dickens's 'The Signal-man' (both 1866); 'The Botathen Ghost' (1867) by R. S. Hawker; and several stories by J. S. Le Fanu, amongst them 'Green Tea' (1869), with its famous spectral monkey, 'The White Cat of Drumgunniol', and 'Madam Crowl's Ghost' (both 1870).

At the beginning of his story 'An Account of Some Strange Disturbances in Aungier Street' (1853), Le Fanu had evoked the picture of 'a circle of intelligent and eager faces, lighted up by a good after-dinner fire on a winter's evening, with a cold wind rising and wailing outside, and all snug and cosy within'. But it was Dickens, more than anyone else, who established and exploited the Christmas market for supernatural fiction and embedded the images of 'Winter Stories—Ghost Stories . . . round the Christmas fire'[6] firmly in the national consciousness. By the 1890s the convention had become a national institution and provided Jerome K. Jerome (who none the less could still turn out an excellent conventionally told tale) with an easy target for a deflationary excursus in the introduction to *Told After Supper* (1891):

> There must be something ghostly in the air of Christmas—something about the close, muggy atmosphere that draws up the ghosts, like the dampness of the summer rains brings out the frogs and snails.
>
> And not only do the ghosts themselves always walk on Christmas Eve, but live people always sit and talk about them on Christmas Eve. Whenever five or

[6] 'A Christmas Tree', Extra Christmas Number of *Household Words* (1850).

six English-speaking people meet round a fire on Christmas Eve, they start telling each other ghost stories. Nothing satisfies us on Christmas Eve but to hear each other tell authentic anecdotes about spectres. It is a genial, festive season, and we love to muse upon graves, and dead bodies, and murders, and blood . . . For ghost stories to be told on any other evening than the evening of the twenty-fourth of December would be impossible in English society as at present regulated.

A great many of those who provided this seasonal fare were women, both as writers and as editors of magazines. Mary Elizabeth Braddon, of *Lady Audley's Secret* fame, for example, wrote a number of fine ghost stories—'The Cold Embrace' (1860) and 'Eveline's Visitant' (1867) being her most famous—and from 1866 to 1893 edited *Belgravia*; Mrs Henry Wood, author of the best-selling *East Lynne* (1861), owned and edited *The Argosy* from 1865 to 1887 and wrote several ghost stories that appeared in her *Johnny Ludlow* series; and Mrs J. H. [Charlotte] Riddell edited the *St James's Magazine* for a time and produced some excellent ghost stories, the best of which were collected in 1882 as *Weird Stories*. The reasons why women took to the ghost story so successfully is one of the great unasked critical questions; but one might guess that it was due less to an inherent susceptibility to the supernatural (though doubtless some psycho-cultural thesis could be advanced to explain their achievements) than to the practical—often pressing—need of a certain type of educated woman to earn a living. The monthly magazines required an endless supply of fiction, short and long, and authorship was often the only means some middle-class women had to meet their financial needs. As ghost stories were consistently in demand it was natural that women, who provided so much fiction for the magazines, should provide these too. Charlotte Riddell is an often-quoted example of a not uncommon situation: the woman who had to write constantly to make up the financial deficiencies of her husband. Margaret Oliphant, who wrote long, emotionally charged 'Stories of the Seen and the Unseen', is another. Her diary entry for Christmas night 1887 testifies to the relentless drudgery that went with the regular provision of magazine copy: 'All the things I seem to want are material things. I want money. I want work, work that will pay, enough to keep this house going which there is no-one to provide for but me.'[7] Amelia Edwards, Miss Braddon, Mrs Riddell, and Rhoda Broughton were the most prominent women ghost-story writers of the 1860s and 1870s, but there were many others. In the last two decades of the century the list includes Louisa Molesworth,

[7] *The Autobiography of Margaret Oliphant*, ed. Elisabeth Jay (Oxford, 1990), 155.

Vernon Lee (pseudonym of Violet Paget), Rosa Mulholland, Bithia Mary Croker, Edith Nesbit, Louisa Baldwin (Rudyard Kipling's aunt), and Violet Hunt. In the twentieth century women have been equally productive.

The explosion of periodical publishing from 1860 meant that the rise of the short ghost story was a rapid one. In 1842 the American novelist William Gilmore Simms had prefaced his story '"Murder Will Out"' with the complaint that: 'We can no longer get a good ghost story, either for love or money. The materialists have it all their own way . . . That cold-blooded demon called Science has taken the place of all the other demons.' But only forty years later F. Anstey (the humorist Anstey Guthrie) could reflect with tongue-in-cheek regret that the British Ghost—by which he meant the real thing—was 'fast becoming as extinct as the Great Bustard', thanks to the irresistible advance of the 'Magazine Ghost'.[8]

These two comments frame four decades during which the ghost story proliferated to such an extent that not even the implacable demon of science could snuff out a seemingly insatiable popular taste for these irrational entertainments. Indeed, the ghost story seemed to thrive precisely because it dealt in possibilities that were in fundamental opposition to the explicatory march of science; it was certainly true that 'great as may be the popularity of any clever work that undertakes to explain portents and apparitions on grounds that are called "natural", the vogue of such a work never yet equalled the vogue of a right-down book of ghost-stories'.[9]

From the late 1840s the parallel craze for spiritualism and mesmerism fed popular credulity, on the one hand, and, on the other, stimulated worthy efforts to prove the objective reality of supernatural phenomena. Whilst fiction echoed the veridical literature in its use of such recurring themes as the haunted house and the warning dream, the ghosts of fiction bore only occasional resemblance to the often aimless visitations recorded in the dreary annals of psychical research. As the narrator of Dinah Mulock's story 'The Last House in C—— Street' (1853) puts it: 'They [Ghosts] come—that is, they are reported to come—so irrelevantly, purposelessly—so ridiculously, in short—that one's common sense as regards this world, one's supernatural sense of the other, are alike revolted.' In contrast to the typically spasmodic and mute appearances of veridical apparitions, the ghosts of Victorian fiction, more like their folkloric counterparts, hardly ever

[8] 'The Decay of the British Ghost', *Longman's Magazine* (Jan. 1884).

[9] 'A Physician's Ghosts', *All the Year Round* (6 Aug. 1859).

lacked motivation—even though it might sometimes be fuelled by an anarchic and baffling logic: they revealed secrets, avenged wrongs, re-enacted ancient tragedies, in some cases proffered help and comfort to the living, or bore witness to the workings of divine providence. Most disquieting of all, they could pursue blameless living victims with a relentless and unfathomable malignity.

The relationship between veridical phenomena and imagined ghosts was a complex one. Fiction, for example, often posed as fact, and a range of narrative strategies was deployed to reinforce the masquerade. Although Amelia Edwards suggested that 'nothing, perhaps, is more calculated to throw discredit upon a ghost-story than the least pretension to authenticity',[10] the notion of 'authenticity' was often used by writers to bridge the worlds of fiction and supposed fact. In an early example, Edgar Allan Poe's 'The Facts in the Case of M. Valdemar' (1845), the way the theme of mesmerism 'in articulo mortis' (echoed fifty years later in E. Nesbit's 'Hurst of Hurstcote') is presented anticipates the kind of spurious factuality used by later writers. Whilst many ghost stories embodied a reaction against fact and empirical logic, writers frequently made use of an appearance of fact to enforce an illusion of authenticity. The technique can be seen in R. S. Hawker's 'The Botathen Ghost' (1867), in which the story is partly told through extracts from the diary of Parson Rudall—the '"diurnal" which fell by chance into the hands of the present writer'. But the greatest exponent of the factualizing narrative was M. R. James, whose antiquarian stories set in train a vigorous sub-category of English ghost stories that still continues. In James's 'Canon Alberic's Scrap-book' (1895), for instance, the reader is bombarded with factual detail—or rather a subtle blending of actual fact and invention based on James's formidable learning: bibliographical and historical references, Latin quotations (duly translated for the ignorant), architectural and topographical detail—all delivered in a reticent style that heightens the sense of actuality by distancing the narrator from the events he is reporting.

Contemporary settings, or at least settings with only a slight haze of distance, also gave the Victorian ghost story a sense of solidity lacking in its literary predecessors. The Gothic tale of terror—whose improbable fantasies continued well into the nineteenth century through the work of such authors as G. W. M. Reynolds and J. F. Smith—had revelled in pseudo-historical settings; the Victorian ghost story turned to the prosaic detail of modernity to establish a credible context for

[10] Prefatory note to *Monsieur Maurice* (1873).

supernatural violation. M. R. James, an avid reader of magazine ghost stories as a boy in the 1870s, concluded that:

> On the whole (though not a few instances might be quoted against me) I think that a setting so modern that the ordinary reader can judge of its naturalness for himself is preferable to anything antique. For some degree of actuality is the charm of the best ghost stories; not a very insistent actuality, but one strong enough to allow the reader to identify himself with the patient; while it is almost inevitable that the reader of an antique story should fall into the position of the mere spectator.[11]

Everyday detail abounds in the Victorian ghost story: details of decor and dress, food and drink, furniture and transport, landscape and architecture, as well as the realities of social and sexual relationships. Despite the pace of change, there were still plenty of apparently settled social structures: marriage, the law, landed and aristocratic society, the Church, the universities, the colonial experience. Any one of these could provide an ordered microcosm into which the supernatural could intrude.

Though Sir Walter Scott wrote two of the earliest fictional ghost stories worthy of the name—'Wandering Willie's Tale' (from *Redgauntlet*, 1824) and 'The Tapestried Chamber' (1829)—the ghost story's potential was first revealed by the Irish writer J. S. Le Fanu, who was to dominate Victorian supernatural fiction. M. R. James—the equivalent figure in the twentieth century and who, with S. M. Ellis, was responsible for Le Fanu's rehabilitation after a long period of neglect—placed him 'absolutely in the first rank as a writer of ghost stories . . . nobody sets the scene better than he, nobody touches in the effective detail more deftly'.[12] Le Fanu's power and originality are clearly displayed in one of his early stories, 'Schalken the Painter' (1839), which develops the startling theme of supernatural abduction (and, by implication, rape). 'Schalken' is a rare instance of a successful ghost story with a historical setting (seventeenth-century Holland). Le Fanu's 'An Account of Some Strange Disturbances in Aungier Street', reprinted here, is no less successful, though in a different way. It is perhaps the best of all Victorian haunted-house stories—far more subtle in its effects than Sir Edward Bulwer Lytton's much-vaunted 'The Haunted and the Haunters', published in *Blackwood's Magazine* in August 1859, six years after Le Fanu's story first appeared in the *Dublin University Magazine*. Le Fanu's sureness of touch and keen eye for the disquieting detail are apparent throughout, as in this passage:

[11] Introduction to V. H. Collins (ed.), *Ghosts and Marvels* (Oxford, 1924), vii.
[12] Prologue to J. S. Le Fanu, *Madam Crowl's Ghost* (1923), vii.

It was two o'clock, and the streets were as silent as a churchyard—the sounds were, therefore, perfectly distinct. There was a slow, heavy tread, characterized by the emphasis and deliberation of age, descending by the narrow staircase from above; and, what made the sound more singular, it was plain that the feet which produced it were perfectly bare, measuring the descent with something between a pound and a flop, very ugly to hear.

Le Fanu also wrote tales and novels of mystery (most famously, *Uncle Silas*, 1864), and elements of the mystery story frequently invade his overtly supernatural fiction (and vice versa). These two forms of sensational literature shared several common qualities, and it was not uncommon for elements of the mystery story and tale of detection—the sowing of clues, criminous motivation, final explication—to be combined with a supernatural denouement—as in Wilkie Collins's 'Mrs Zant and the Ghost' (1885) or M. R. James's 'The Treasure of Abbot Thomas' (1904), in which the unravelling of clues in true Holmesian style leads the curious Mr Somerton to the hidden treasure and its terrifying guardian. Such fusion produced another sub-genre, the story of psychic detection, with sleuths such as E. and H. Heron's Flaxman Low, Algernon Blackwood's John Silence, and W. Hope Hodgson's Carnacki pitting their wits against a variety of supernatural opponents. The close relationship between the ghost story and tales of mystery and detection is emphasized by the satisfying fact that Sir Arthur Conan Doyle, creator of the most famous detective of them all, also wrote supernatural stories (we reprint one of his earliest, and best, 'The Captain of the "Pole-star"', 1883).

Le Fanu's first book, the extremely rare *Ghost Stories and Tales of Mystery*, with four illustrations by Dickens's illustrator 'Phiz' (Hablot Knight Browne), appeared in 1851. Though now recognized as a landmark collection, it had far less impact on popular taste for the supernatural than two celebrated volumes by Mrs Catherine Crowe: *The Night-side of Nature* (1848) and *Ghosts and Family Legends* (1859 for 1858), which presented highly embellished versions of what were claimed to be veridical experiences. But the resulting hybrid has not worn well, and Mrs Crowe's writings, despite their considerable contemporary popularity, had negligible influence on the way supernatural fiction developed over the course of the century. In terms of influence, and in the quality of his best work, Le Fanu stands head and shoulders above his contemporaries. One reason for this is that he hardly ever strayed beyond the boundaries of supernatural and mystery fiction: he was a supreme specialist. He appeared to recognize his limitations, but at the same time there was clearly a deep inner compulsion to write the kind of fiction that could accommodate the themes that engaged him so obsessively—the implacability of evil, the demoniacal potential of sexual desire, and, above all, the consequences of guilt. The stark conviction of Captain Barton in 'The Watcher' (1851) is surely Le Fanu's own, and stands as a motto to the darker aspects of the Victorian ghost story:

> There does exist beyond this a spiritual world—a system whose workings are generally in mercy hidden from us—a system which may be, and which is sometimes, partially and terribly revealed. I am sure—I know . . . that there is a God—a dreadful God—and that retribution follows guilt, in ways the most mysterious and stupendous—by agencies the most inexplicable and terrible.

Few Victorian writers of ghost stories wrote with this degree of personal conviction. The majority of stories were written in direct response to popular taste: but the best are none the worse for this, and for those who like a tale well told there are pleasures aplenty to be savoured.

As the century drew to a close the ghost story proved to be remarkably resistant to mainstream literary influences. In the wider sphere of supernatural fantasy, the Decadence produced a strain of lyrical, atavistic horror in the work of Arthur Machen; but Machen's stories belong to a different tradition from the ghost story proper, and apart from a handful of examples—such as Vincent O'Sullivan's 'The Business of Madame Jahn', Ella d'Arcy's 'The Villa Lucienne', or the highly wrought tales of Vernon Lee—there is no discernible *fin de siècle* mood in the stories that continued to be turned out for popular consumption. Nor was the taste for spooks of the old-fashioned kind confined to Britain. A literary tradition independent of, but clearly part of, the British root-stock flourished in New England. The chief figure was Henry James (represented here by an early story, 'The Romance of Certain Old Clothes', 1868), who produced ghost stories, increasingly oblique in style, throughout his career; his friend and admirer Edith Wharton wrote a well-crafted body of ghost stories in the traditional style, whilst Mary E. Wilkins, whose 'The Shadows on the Wall' is reprinted here, fused traditional elements with strong local colour.

If its central characteristics remained immune to change, the ghost story did keep pace with the times through progressive modernization of settings and language. In Barry Pain's 'The Case of Vincent Pyrwhit' (1901), a message from beyond the grave is communicated via the telephone; elsewhere the motor car and other features of modern life are used to vivify stock situations. The magazines remained the chief purveyors of supernatural thrills to a mass public. In the 1860s and 1870s, the work of artists like J. A. Pasquier had occasionally been used to accompany magazine ghost stories; by the 1890s popular monthlies such as the *Pall Mall Magazine* (a particularly rich source of ghost stories), the *Windsor Magazine*, and *The Strand* had become profusely illustrated, an adjunct of doubtful use to a form so dependent upon the individual imagination. The secondary market for collections of ghost stories in book form was equally buoyant—as our Select Conspectus indicates (see p. 493). Ghost stories were being written by mainstream literary figures, such as Kipling and H. G. Wells; by specialists like Algernon Blackwood; and by large numbers of professional writers who combined writing ghost stories with other forms of fiction—E. F. Benson, W. W. Jacobs, Barry Pain, Robert Barr, amongst many others.

The real change in the traditional ghost story came with the upheaval of the First World War, making 1914 an appropriately symbolic termination-point for the Victorian ghost story. The stories written during the first decade of the new century, despite hints of growing uncertainties, remain undeniably fruit from the Victorian tree. In their tone, style, and thematic simplicity, Algernon Blackwood's story 'The Kit-bag' and Perceval Landon's 'Thurnley Abbey' (both 1908) could have been written fifty years earlier; and yet the sense of physical horror—intense in 'Thurnley Abbey'—strikes a new note and signals the end of the true Victorian style. After the Deluge of 1914 the ghost story withered for a time in the face of greater nightmares; but it was quickly to revive, and indeed achieved a second great flowering, with new themes, new modes of expression, and new images of supernatural violation. But that is itself another story.

Michael Cox
R. A. Gilbert

All Souls' Day, 1990

O, tell us a tale of a ghost! now do!
It's a capital time, for the fire burns blue.

ANON, 'The Vicarage Ghost',
Tinsleys' Magazine (Christmas Number, 1868)

And she harbours a silent wrath against Providence for allowing the dead to walk and to molest the living.

SABINE BARING-GOULD, 'The Leaden Ring',
from *A Book of Ghosts* (1904)

The Old Nurse's Story

ELIZABETH GASKELL

You know, my dears, that your mother was an orphan, and an only child; and I dare say you have heard that your grandfather was a clergyman up in Westmorland, where I come from. I was just a girl in the village school, when, one day, your grandmother came in to ask the mistress if there was any scholar there who would do for a nurse-maid; and mighty proud I was, I can tell ye, when the mistress called me up, and spoke to my being a good girl at my needle, and a steady honest girl, and one whose parents were very respectable, though they might be poor. I thought I should like nothing better than to serve the pretty young lady, who was blushing as deep as I was, as she spoke of the coming baby, and what I should have to do with it. However, I see you don't care so much for this part of my story, as for what you think is to come, so I'll tell you at once. I was engaged and settled at the parsonage before Miss Rosamond (that was the baby, who is now your mother) was born. To be sure, I had little enough to do with her when she came, for she was never out of her mother's arms, and slept by her all night long; and proud enough was I sometimes when missis trusted her to me. There never was such a baby before or since, though you've all of you been fine enough in your turns; but for sweet, winning ways, you've none of you come up to your mother. She took after her mother, who was a real lady born; a Miss Furnivall, a granddaughter of Lord Furnivall's, in Northumberland. I believe she had neither brother nor sister, and had been brought up in my lord's family till she had married your grandfather, who was just a curate, son to a shopkeeper in Carlisle—but a clever, fine gentleman as ever was—and one who was a right-down hard worker in his parish, which was very wide, and scattered all abroad over the Westmorland Fells. When your mother, little Miss Rosamond, was about four or five years old, both her parents died in a fortnight—one after the other. Ah! that was a sad time. My pretty young mistress and me was looking for another baby, when my master came home from one of his long rides, wet, and tired, and took the fever he died of; and then she never held up her head again, but just lived to see her dead baby, and have it laid on her breast before she sighed away her life. My mistress had asked me, on her death-bed, never to leave Miss Rosamond; but if she had never spoken a word, I would have gone with the little child to the end of the world.

The next thing, and before we had well stilled our sobs, the executors and guardians came to settle the affairs. They were my poor young mistress's own cousin, Lord Furnivall, and Mr Esthwaite, my master's brother, a shopkeeper in Manchester; not so well-to-do then as he was afterwards, and with a large family rising about him. Well! I don't know if it were their settling, or because of a letter my mistress wrote on her death-bed to her cousin, my lord; but somehow it was settled that Miss Rosamond and me were to go to Furnivall Manor House, in Northumberland, and my lord spoke as if it had been her mother's wish that she should live with his family, and as if he had no objections, for that one or two more or less could make no difference in so grand a household. So though that was not the way in which I should have wished the coming of my bright and pretty pet to have been looked at—who was like a sunbeam in any family, be it never so grand—I was well pleased that all the folks in the Dale should stare and admire, when they heard I was going to be young lady's maid at my Lord Furnivall's at Furnivall Manor.

But I made a mistake in thinking we were to go and live where my lord did. It turned out that the family had left Furnivall Manor House fifty years or more. I could not hear that my poor young mistress had ever been there, though she had been brought up in the family; and I was sorry for that, for I should have liked Miss Rosamond's youth to have passed where her mother's had been.

My lord's gentleman, from whom I asked so many questions as I durst, said that the Manor House was at the foot of the Cumberland Fells, and a very grand place; that an old Miss Furnivall, a great-aunt of my lord's, lived there, with only a few servants; but that it was a very healthy place, and my lord had thought that it would suit Miss Rosamond very well for a few years, and that her being there might perhaps amuse his old aunt.

I was bidden by my lord to have Miss Rosamond's things ready by a certain day. He was a stern proud man, as they say all the Lords Furnivall were; and he never spoke a word more than was necessary. Folk did say he had loved my young mistress; but that, because she knew that his father would object, she would never listen to him, and married Mr Esthwaite; but I don't know. He never married, at any rate. But he never took much notice of Miss Rosamond; which I thought he might have done if he had cared for her dead mother. He sent his gentleman with us to the Manor House, telling him to join him at Newcastle that same evening; so there was no great length of time for him to make us known to all the strangers before he, too, shook us off; and we were left, two lonely young things (I was not eighteen), in the great old Manor House. It seems like yesterday that we drove there. We had left our own dear parsonage very early, and we had both cried as if our hearts would break, though we were travelling in my lord's carriage, which I thought so much of once. And now it was long past noon on a September day, and we stopped to change horses for the last time at a little smoky town, all full of colliers and miners. Miss Rosamond had fallen asleep, but Mr Henry told me to waken her, that she might see the park and the Manor House as we drove up. I thought it rather a pity; but I did what he bade me, for fear he should complain of me to my lord. We had left all signs of a town, or even a village, and were then inside the gates of a large wild park—not like the parks here in the north, but with rocks, and the noise of running water, and gnarled thorn-trees, and old oaks, all white and peeled with age.

The road went up about two miles, and then we saw a great and stately house, with many trees close around it, so close that in some places their branches dragged against the walls when the wind blew; and some hung broken down; for no one seemed to take much charge of the place—to lop the wood, or to keep the moss-covered carriage-way in order. Only in front of the house all was clear. The great oval drive was without a weed; and neither tree nor creeper was allowed to grow over the long, many-windowed front; at both sides of which a wing projected, which were each the ends of other side fronts; for the house, although it was so desolate, was even grander than I expected. Behind it rose the Fells, which seemed unenclosed and bare enough; and on the left hand of the house, as you stood facing it, was a little, old-fashioned flower-garden, as I found out afterwards. A door opened out upon it from the west front; it had been scooped out of the thick dark wood for some old Lady Furnivall; but the branches of the great forest trees had grown and overshadowed it again, and there were very few flowers that would live there at that time.

When we drove up to the great front entrance, and went into the hall I thought we should be lost—it was so large, and vast, and grand. There was a chandelier all of bronze, hung down from the middle of the ceiling; and I had never seen one before, and looked at it all in amaze. Then, at one end of the hall, was a great fire-place, as large as the sides of the houses in my country, with massy andirons and dogs to hold the wood; and by it were heavy old-fashioned sofas. At the opposite end of the hall, to the left as you went in—on the western side—was an organ built into the wall, and so large that it filled up the best part of that end. Beyond it, on the same side, was a door; and opposite, on each side of the fire-place, were also doors leading to the east front; but those I never went through as long as I stayed in the house, so I can't tell you what lay beyond.

The afternoon was closing in, and the hall, which had no fire lighted in it, looked dark and gloomy, but we did not stay there a moment. The old servant, who had opened the door for us, bowed to Mr Henry, and took us in through the door at the further side of the great organ, and led us through several smaller halls and passages into the west drawing-room, where he said that Miss Furnivall was sitting. Poor little Miss Rosamond held very tight to me, as if she were scared and lost in that great place, and as for myself, I was not much better. The west drawing-room was very cheerful-looking, with a warm fire in it, and plenty of good, comfortable furniture about. Miss Furnivall was an old lady not far from eighty, I should think, but I do not know. She was thin and tall, and had a face as full of fine wrinkles as if they had been drawn all over it with a needle's point. Her eyes were very watchful, to make up, I suppose, for her being so deaf as to be obliged to use a trumpet. Sitting with her, working at the same great piece of tapestry, was Mrs Stark, her maid and companion, and almost as old as she was. She had lived with Miss Furnivall ever since they were both young, and now she seemed more like a friend than a servant; she looked so cold and grey, and stony as if she had never loved or cared for any one; and I don't suppose she did care for any one, except her mistress; and, owing to the great deafness of the latter, Mrs Stark treated her very much as if she were a child. Mr Henry gave some message from my lord, and then he bowed goodbye to us all—taking no notice of my sweet little Miss Rosamond's outstretched hand—and left us standing there, being looked at by the two old ladies through their spectacles.

I was right glad when they rung for the old footman who had shown us in at first, and told him to take us to our rooms. So we went out of that great drawing-room, and into another sitting-room, and out of that, and then up a great flight of stairs, and along a broad gallery—which was something like a library, having books all down one side, and windows and writing-tables all down the other—till we came to our rooms, which I was not sorry to hear were just over the kitchens; for I began to think I should be lost in that wilderness of a house. There was an old nursery that had been used for all the little lords and ladies long ago, with a pleasant fire burning in the grate, and the kettle boiling on the hob, and tea-things spread out on the table; and out of that room was the night-nursery, with a little crib for Miss Rosamond close to my bed. And old James called up Dorothy, his wife, to bid us welcome; and both he and she were so hospitable and kind, that by and by Miss Rosamond and me felt quite at home; and by the time tea was over, she was sitting on Dorothy's knee, and chattering away as fast as her little tongue could go. I soon found out that Dorothy was from Westmorland, and that bound her and me together, as it were; and I would never wish to meet with kinder people than were old James and his wife. James had lived pretty nearly all his life in my lord's family, and thought there was no one so grand as they. He even looked down a little on his wife; because, till he had married her, she had never lived in any but a farmer's household. But he was very fond of her, as well he might be. They had one servant under them, to do all the rough work. Agnes they called her; and she and me, and James and Dorothy, with Miss Furnivall and Mrs Stark, made up the family; always remembering my sweet little Miss Rosamond! I used to wonder what they had done before she came, they thought so much of her now. Kitchen and drawing-room, it was all the same. The hard, sad Miss Furnivall, and the cold Mrs Stark, looked pleased when she came fluttering in like a bird, playing and pranking hither and thither, with a continual murmur, and pretty prattle of gladness. I am sure, they were sorry many a time when she flitted away into the kitchen, though they were too proud to ask her to stay with them, and were a little surprised at her taste; though to be sure, as Mrs Stark said, it was not to be wondered at, remembering what stock her father had come of. The great, old rambling house was a famous place for little Miss Rosamond. She made expeditions all over it, with me at her heels; all, except the east wing, which was never opened, and whither we never thought of going. But in the western and northern part was many a pleasant room; full of things that were curiosities to us, though they might not have been to people who had seen more. The windows were darkened by the sweeping boughs of the trees, and the ivy which had overgrown them: but, in the green gloom, we could manage to see old China jars and carved ivory boxes, and great heavy books, and, above all, the old pictures!

Once, I remember, my darling would have Dorothy go with us to tell us who they all were; for they were all portraits of some of my lord's family, though Dorothy could not tell us the names of every one. We had gone through most of the rooms, when we came to the old state drawing-room over the hall, and there was a picture of Miss Furnivall; or, as she was called in those days, Miss Grace, for she was the younger sister. Such a beauty she must have been! but with such a set, proud look, and such scorn looking out of her handsome eyes, with her eyebrows just a little raised, as if she were wondering how any one could have the impertinence to look at her; and her lip curled at us, as we stood there gazing. She had a dress on, the like of which I had never seen before, but it was all the fashion when she was young: a hat of some soft white stuff like beaver, pulled a little over her brows, and a beautiful plume of feathers sweeping round it on one side; and her gown of blue satin was open in front to a quilted white stomacher.

'Well, to be sure!' said I, when I had gazed my fill. 'Flesh is grass, they do say; but who would have thought that Miss Furnivall had been such an out-and-out beauty, to see her now?'

'Yes,' said Dorothy. 'Folks change sadly. But if what my master's father used to say was true, Miss Furnivall, the elder sister, was handsomer than Miss Grace. Her picture is here somewhere; but, if I show it you, you must never let on, even to James, that you have seen it. Can the little lady hold her tongue, think you?' asked she.

I was not so sure, for she was such a little sweet, bold, open-spoken child, so I set her to hide herself; and then I helped Dorothy to turn a great picture, that leaned with its face towards the wall, and was not hung up as the others were. To be sure, it beat Miss Grace for beauty; and, I think, for scornful pride, too, though in that matter it might be hard to choose. I could have looked at it an hour, but Dorothy seemed half frightened at having shown it to me, and hurried it back again, and bade me run and find Miss Rosamond, for that there were some ugly places about the house, where she should like ill for the child to go. I was a brave, high-spirited girl, and thought little of what the old woman said, for I liked hide-and-seek as well as any child in the parish; so off I ran to find my little one.

As winter drew on, and the days grew shorter, I was sometimes almost certain that I heard a noise as if some one was playing on the great organ in the hall. I did not hear it every evening; but, certainly, I did very often; usually when I was sitting with Miss Rosamond, after I had put her to bed, and keeping quite still and silent in the bedroom. Then I used to hear it booming and swelling away in the distance. The first night, when I went down to my supper, I asked Dorothy who had been playing music, and James said very shortly that I was a gowk to take the wind soughing among the trees for music: but I saw Dorothy look at him very fearfully, and Bessy, the kitchen-maid, said something beneath her breath, and went quite white. I saw they did not like my question, so I held my peace till I was with Dorothy alone, when I knew I could get a good deal out of her. So, the next day, I watched my time, and I coaxed and asked her who it was that played the organ: for I knew that it was the organ and not the wind well enough, for all I had kept silence before James. But Dorothy had had her lesson, I'll warrant, and never a word could I get from her. So then I tried Bessy, though I had always held my head rather above her, as I was evened to James and Dorothy, and she was little better than their servant. So she said I must never, never tell; and if I ever told, I was never to say *she* had told me; but it was a very strange noise, and she had heard it many a time, but most of all on winter nights, and before storms; and folks did say, it was the old lord playing on the great organ in the hall, just as he used to do when he was alive; but who the old lord was, or why he played, and why he played on stormy winter evenings in particular, she either could not or would not tell me. Well! I told you I had a brave heart; and I thought it was rather pleasant to have that grand music rolling about the house, let who would be the player; for now it rose above the great gusts of wind, and wailed and triumphed just like a living creature, and then it fell to a softness most complete; only it was always music and tunes, so it was nonsense to call it the wind. I thought at first that it might be Miss Furnivall who played, unknown to Bessy; but one day when I was in the hall by myself, I opened the organ and peeped all about it and around it, as I had done to the organ in Crosthwaite Church once before, and I saw it was all broken and destroyed inside, though it looked so brave and fine; and then, though it was noonday, my flesh began to creep a little, and I shut it up, and run away pretty quickly to my own bright nursery; and I did not like hearing the music for some time after that, any more than James and Dorothy did. All this time Miss Rosamond was making herself more and more beloved. The old ladies liked her to dine with them at their early dinner; James stood behind Miss Furnivall's chair, and I behind Miss Rosamond's all in state; and, after dinner, she would play about in a corner of the great drawing-room, as still as any mouse, while Miss Furnivall slept, and I had my dinner in the kitchen. But she was glad enough to come to me in the nursery afterwards; for, as she said, Miss Furnivall was so sad, and Mrs Stark so dull; but she and I were merry enough; and, by-and-by, I got not to care for that weird rolling music, which did one no harm, if we did not know where it came from.

That winter was very cold. In the middle of October the frosts began, and lasted many, many weeks. I remember, one day at dinner, Miss Furnivall lifted up her sad, heavy eyes, and said to Mrs Stark, 'I am afraid we shall have a terrible winter,' in a strange kind of meaning way. But Mrs Stark pretended not to hear, and talked very loud of something else. My little lady and I did not care for the frost; not we! As long as it was dry we climbed up the steep brows, behind the house, and went up on the Fells, which were bleak, and bare enough, and there we ran races in the fresh, sharp air; and once we came down by a new path that took us past the two old gnarled holly-trees, which grew about halfway down by the east side of the house. But the days grew shorter and shorter; and the old lord, if it was he, played more and more stormily and sadly on the great organ. One Sunday afternoon—it must have been towards the end of November—I asked Dorothy to take charge of little Missey when she came out of the drawing-room, after Miss Furnivall had had her nap; for it was too cold to take her with me to church, and yet I wanted to go. And Dorothy was glad enough to promise, and was so fond of the child that all seemed well; and Bessy and I set off very briskly, though the sky hung heavy and black over the white earth, as if the night had never fully gone away; and the air, though still, was very biting and keen.

'We shall have a fall of snow,' said Bessy to me. And sure enough, even while we were in church, it came down thick, in great large flakes, so thick it almost darkened the windows. It had stopped snowing before we came out, but it lay soft, thick and deep beneath our feet, as we tramped home. Before we got to the hall the moon rose, and I think it was lighter then—what with the moon, and what with the white dazzling snow—than it had been when we went to church, between two and three o'clock. I have not told you that Miss Furnivall and Mrs Stark never went to church: they used to read the prayers together, in their quiet gloomy way; they seemed to feel the Sunday very long without their tapestry-work to be busy at. So when I went to Dorothy in the kitchen, to fetch Miss Rosamond and take her upstairs with me, I did not much wonder when the old woman told me that the ladies had kept the child with them, and that she had never come to the kitchen, as I had bidden her, when she was tired of behaving pretty in the drawing-room. So I took off my things and went to find her, and bring her to her supper in the nursery. But when I went into the best drawing-room there sat the two old ladies, very still and quiet, dropping out a word now and then but looking as if nothing so bright and merry as Miss Rosamond had ever been near them. Still I thought she might be hiding from me; it was one of her pretty ways; and that she had persuaded them to look as if they knew nothing about her; so I went softly peeping under this sofa, and behind that chair, making believe I was sadly frightened at not finding her.

'What's the matter, Hester?' said Mrs Stark, sharply. I don't know if Miss Furnivall had seen me, for, as I told you, she was very deaf, and she sat quite still, idly staring into the fire, with her hopeless face. 'I'm only looking for my little Rosy-Posy,' replied I, still thinking that the child was there, and near me, though I could not see her.

'Miss Rosamond is not here,' said Mrs Stark. 'She went away more than an hour ago to find Dorothy.' And she too turned and went on looking into the fire.

My heart sank at this, and I began to wish I had never left my darling. I went back to Dorothy and told her. James was gone out for the day, but she and me and Bessy took lights and went up into the nursery first, and then we roamed over the great large house, calling and entreating Miss Rosamond to come out of her hiding-place, and not frighten us to death in that way. But there was no answer; no sound.

'Oh!' said I at last, 'Can she have got into the east wing and hidden there?'

But Dorothy said it was not possible, for that she herself had never been there; that the doors were always locked, and my lord's steward had the keys, she believed; at any rate, neither she nor James had ever seen them: so I said I would go back, and see if, after all, she was not hidden in the drawing-room, unknown to the old ladies; and if I found her there, I said, I would whip her well for the fright she had given me; but I never meant to do it. Well, I went back to the west drawing-room, and I told Mrs Stark we could not find her anywhere, and asked for leave to look all about the furniture there, for I thought now, that she might have fallen asleep in some warm hidden corner; but no! we looked, Miss Furnivall got up and looked, trembling all over, and she was nowhere there; then we set off again, every one in the house, and looked in all the places we had searched before, but we could not find her. Miss Furnivall shivered and shook so much that Mrs Stark took her back into the warm drawing-room; but not before they had made me promise to bring her to them when she was found. Well-a-day! I began to think she never would be found, when I bethought me to look out into the great front court, all covered with snow. I was upstairs when I looked out; but it was such clear moonlight, I could see, quite plain, two little footprints, which might be traced from the hall door, and round the corner of the east wing. I don't know how I got down, but I tugged open the great, stiff hall door; and, throwing the skirt of my gown over my head for a cloak, I ran out. I turned the east corner, and there a black shadow fell on the snow; but when I came again into the moonlight, there were the little footmarks going up—up to the Fells. It was bitter cold; so cold that the air almost took the skin off my face as I ran, but I ran on, crying to think how my poor little darling must be perished, and frightened. I was within sight of the holly-trees when I saw a shepherd coming down the hill, bearing something in his arms wrapped in his maud. He shouted to me, and asked me if I had lost a bairn; and, when I could not speak for crying, he bore towards me, and I saw my wee bairnie lying still, and white, and stiff, in his arms, as if she had been dead. He told me he had been up the Fells to gather in his sheep, before the deep cold of night came on, and that under the holly-trees (black marks on the hill-side, where no other bush was for miles around) he had found my little lady—my lamb—my queen—my darling—stiff and cold, in the terrible sleep which is frost-begotten. Oh! the joy, and the tears of having her in my arms once again! for I would not let him carry her; but took her, maud and all, into my own arms, and held her near my own warm neck and heart, and felt the life stealing slowly back again into her little gentle limbs. But she was still insensible when we reached the hall, and I had no breath for speech. We went in by the kitchen door.

'Bring the warming-pan,' said I; and I carried her upstairs and began undressing her by the nursery fire, which Bessy had kept up. I called my little lammie all the sweet and playful names I could think of—even while my eyes were blinded by my tears; and at last, oh! at length she opened her large blue eyes. Then I put her into her warm bed, and sent Dorothy down to tell Miss Furnivall that all was well; and I made up my mind to sit by my darling's bedside the live-long night. She fell away into a soft sleep as soon as her pretty head had touched the pillow, and I watched by her until morning light; when she wakened up bright and clear—or so I thought at first—and, my dears, so I think now.

She said that she had fancied that she should like to go to Dorothy, for that both the old ladies were asleep, and it was very dull in the drawing-room; and that, as she was going through the west lobby, she saw the snow through the high window falling—falling—soft and steady; but she wanted to see it lying pretty and white on the ground; so she made her way into the great hall; and then, going to the window, she saw it bright and soft upon the drive; but while she stood there, she saw a little girl, not so old as she was, 'but so pretty,' said my darling, 'and this little girl beckoned to me to come out; and oh, she was so pretty and so sweet, I could not choose but go.' And then this other little girl had taken her by the hand, and side by side the two had gone round the east corner.

'Now you are a naughty little girl, and telling stories,' said I. 'What would your good mamma, that is in heaven, and never told a story in her life, say to her little Rosamond, if she heard her—and I dare say she does—telling stories!'

'Indeed, Hester,' sobbed out my child, 'I'm telling you true. Indeed I am.'

'Don't tell me!' said I, very stern. 'I tracked you by your footmarks through the snow; there were only yours to be seen: and if you had had a little girl to go hand-in-hand with you up the hill, don't you think the footprints would have gone along with yours?'

'I can't help it, dear, dear Hester,' said she, crying, 'if they did not; I never looked at her feet, but she held my hand fast and tight in her little one, and it was very, very cold. She took me up the Fell-path, up to the holly-trees; and there I saw a lady weeping and crying; but when she saw me, she hushed her weeping, and smiled very proud and grand, and took me on her knee, and began to lull me to sleep; and that's all, Hester—but that is true; and my dear mamma knows it is,' said she, crying. So I thought the child was in a fever, and pretended to believe her, as she went over her story—over and over again, and always the same. At last Dorothy knocked at the door with Miss Rosamond's breakfast; and she told me the old ladies were down in the eating parlour, and that they wanted to speak to me. They had both been into the night-nursery the evening before, but it was after Miss Rosamond was asleep; so they had only looked at her—not asked me any questions.

'I shall catch it,' thought I to myself, as I went along the north gallery. 'And yet,' I thought, taking courage, 'it was in their charge I left her; and it's they that's to blame for letting her steal away unknown and unwatched.' So I went in boldly, and told my story. I told it all to Miss Furnivall, shouting it close to her ear; but when I came to the mention of the other little girl out in the snow, coaxing and tempting her out, and willing her up to the grand and beautiful lady by the holly-tree, she threw her arms up—her old and withered arms—and cried aloud, 'Oh! Heaven, forgive! Have mercy!'

Mrs Stark took hold of her; roughly enough, I thought; but she was past Mrs Stark's management, and spoke to me, in a kind of wild warning and authority.

'Hester! keep her from that child! It will lure her to her death! That evil child! Tell her it is a wicked, naughty child.' Then Mrs Stark hurried me out of the room; where, indeed, I was glad enough to go; but Miss Furnivall kept shrieking out, 'Oh! have mercy! Wilt Thou never forgive! It is many a long year ago'——

I was very uneasy in my mind after that. I durst never leave Miss Rosamond, night or day, for fear lest she might slip off again, after some fancy or other; and all the more because I thought I could make out that Miss Furnivall was crazy, from their odd ways about her; and I was afraid lest something of the same kind (which might be in the family, you know) hung over my darling. And the great frost never ceased all this time; and whenever it was a more stormy night than usual, between the gusts, and through the wind, we heard the old lord playing on the great organ. But, old lord, or not, wherever Miss Rosamond went, there I followed; for my love for her, pretty helpless orphan, was stronger than my fear for the grand and terrible sound. Besides, it rested with me to keep her cheerful and merry, as beseemed her age. So we played together, and wandered together, here and there, and everywhere; for I never dared to lose sight of her again in that large and rambling house. And so it happened, that one afternoon, not long before Christmas Day, we were playing together on the billiard-table in the great hall (not that we knew the way of playing, but she liked to roll the smooth ivory balls with her pretty hands, and I liked to do whatever she did); and, by-and-by, without our noticing it, it grew dusk indoors, though it was still light in the open air, and I was thinking of taking her back into the nursery, when, all of a sudden, she cried out:

'Look, Hester! look! there is my poor little girl out in the snow!'

I turned towards the long narrow windows, and there, sure enough, I saw a little girl, less than my Miss Rosamond—dressed all unfit to be out-of-doors such a bitter night—crying, and beating against the window-panes, as if she wanted to be let in. She seemed to sob and wail, till Miss Rosamond could bear it no longer, and was flying to the door to open it, when, all of a sudden, and close up upon us, the great organ pealed out so loud and thundering, it fairly made me tremble; and all the more, when I remembered me that, even in the stillness of that dead-cold weather, I had heard no sound of little battering hands upon the window-glass, although the Phantom Child had seemed to put forth all its force; and, although I had seen it wail and cry, no faintest touch of sound had fallen upon my ears. Whether I remembered all this at the very moment, I do not know; the great organ sound had so stunned me into terror; but this I know, I caught up Miss Rosamond before she got the hall-door opened, and clutched her, and carried her away, kicking and screaming, into the large bright kitchen, where Dorothy and Agnes were busy with their mince-pies.

'What is the matter with my sweet one?' cried Dorothy, as I bore in Miss Rosamond, who was sobbing as if her heart would break.

'She won't let me open the door for my little girl to come in; and she'll die if she is out on the Fells all night. Cruel, naughty Hester,' she said, slapping me; but she might have struck harder, for I had seen a look of ghastly terror on Dorothy's face, which made my very blood run cold.

'Shut the back-kitchen door fast, and bolt it well,' said she to Agnes. She said no more; she gave me raisins and almonds to quiet Miss

Rosamond: but she sobbed about the little girl in the snow, and would not touch any of the good things. I was thankful when she cried herself to sleep in bed. Then I stole down to the kitchen, and told Dorothy I had made up my mind. I would carry my darling back to my father's house in Applethwaite: where, if we lived humbly, we lived at peace. I said I had been frightened enough with the old lord's organ-playing; but now that I had seen for myself this little moaning child, all decked out as no child in the neighbourhood could be, beating and battering to get in, yet always without any sound or noise—with the dark wound on its right shoulder; and that Miss Rosamond had known it again for the phantom that had nearly lured her to her death (which Dorothy knew was true); I would stand it no longer.

I saw Dorothy change colour once or twice. When I had done, she told me she did not think I could take Miss Rosamond with me, for that she was my lord's ward, and I had no right over her; and she asked me, would I leave the child that I was so fond of, just for sounds and sights that could do me no harm; and that they had all had to get used to in their turns? I was all in a hot, trembling passion; and I said it was very well for her to talk, that knew what these sights and noises betokened, and that had, perhaps, had something to do with the Spectre-Child while it was alive. And I taunted her so, that she told me all she knew, at last; and then I wished I had never been told, for it only made me afraid more than ever.

She said she had heard the tale from old neighbours, that were alive when she was first married; when folks used to come to the hall sometimes, before it had got such a bad name on the country side: it might not be true, or it might, what she had been told.

The old lord was Miss Furnivall's father—Miss Grace as Dorothy called her, for Miss Maude was the elder, and Miss Furnivall by rights. The old lord was eaten up with pride. Such a proud man was never seen or heard of; and his daughters were like him. No one was good enough to wed them, although they had choice enough; for they were the great beauties of their day, as I had seen by their portraits, where they hung in the state drawing-room. But, as the old saying is, 'Pride will have a fall'; and these two haughty beauties fell in love with the same man, and he no better than a foreign musician, whom their father had down from London to play music with him at the Manor House. For, above all things, next to his pride, the old lord loved music. He could play on nearly every instrument that ever was heard of: and it was a strange thing it did not soften him; but he was a fierce dour old man, and had broken his poor wife's heart with his cruelty, they said. He was mad after music, and would pay any money for it. So he got this foreigner to come; who made such beautiful music, that they said the very birds on the trees stopped their singing to listen. And, by degrees, this foreign gentleman got such a hold over the old lord, that nothing would serve him but that he must come every year; and it was he that had the great organ brought from Holland, and built up in the hall, where it stood now. He taught the old lord to play on it; but many and many a time, when Lord Furnivall was thinking of nothing but his fine organ, and his finer music, the dark foreigner was walking abroad in the woods with one of the young ladies; now Miss Maude, and then Miss Grace.

Miss Maude won the day and carried off the prize, such as it was; and he and she were married, all unknown to any one; and before he made his next yearly visit, she had been confined of a little girl at a farm-house on the Moors, while her father and Miss Grace thought she was away at Doncaster Races. But though she was a wife and a mother, she was not a bit softened, but as haughty and as passionate as ever; and perhaps more so, for she was jealous of Miss Grace, to whom her foreign husband paid a deal of court—by way of blinding her—as he told his wife. But Miss Grace triumphed over Miss Maude, and Miss Maude grew fiercer and fiercer, both with her husband and with her sister; and the former—who could easily shake off what was disagreeable, and hide himself in foreign countries—went away a month before his usual time that summer, and half-threatened that he would never come back again. Meanwhile, the little girl was left at the farm-house, and her mother used to have her horse saddled and gallop wildly over the hills to see her once every week, at the very least—for where she loved, she loved; and where she hated, she hated. And the old lord went on playing—playing on his organ; and the servants thought the sweet music he made had soothed down his awful temper, of which (Dorothy said) some terrible tales could be told. He grew infirm too, and had to walk with a crutch; and his son—that was the present Lord Furnivall's father—was with the army in America, and the other son at sea; so Miss Maude had it pretty much her own way, and she and Miss Grace grew colder and bitterer to each other every day; till at last they hardly ever spoke, except when the old lord was by. The foreign musician came again the next summer, but it was for the last time; for they led him such a life with their jealousy and their passions, that he grew weary, and went away, and never was heard of again. And Miss Maude, who had always meant to have her marriage acknowledged when her father should be dead, was left now a deserted wife—whom nobody knew to have been married—with a child that she dared not own, although she loved it to distraction; living with a father whom she feared, and a sister whom she hated. When the next summer passed over and the dark foreigner never came, both Miss Maude and Miss Grace grew gloomy and sad; they had a haggard look about them, though they looked handsome as ever. But by-and-by Miss Maude brightened; for her father grew more and more infirm, and more than ever carried away by his music; and she and Miss Grace lived almost entirely apart, having separate rooms, the one on the west side, Miss Maude on the east—those very rooms which were now shut up. So she thought she might have her little girl with her, and no one need ever know except those who dared not speak about it, and were bound to believe that it was, as she said, a cottager's child she had taken a fancy too. All this, Dorothy said, was pretty well known; but what came afterwards no one knew, except Miss Grace, and Mrs Stark, who was even then her maid, and much more of a friend to her than ever her sister had been. But the servants supposed, from words that were dropped, that Miss Maude had triumphed over Miss Grace, and told her that all the time the dark foreigner had been mocking her with pretended love—he was her own husband; the colour left Miss Grace's cheek and lips that very day for ever, and she was heard to say many a time that sooner or later she would have her revenge; and Mrs Stark was for ever spying about the east rooms.

One fearful night, just after the New Year had come in, when the snow was lying thick and deep, and the flakes were still falling—fast enough to blind any one who might be out and abroad—there was a great and violent noise heard, and the old lord's voice above all, cursing and swearing awfully—and the cries of a little child—and the proud defiance of a fierce woman—and the sound of a blow—and a dead stillness—and moans and wailings dying away on the hill-side! Then the old lord summoned all his servants, and told them, with terrible oaths, and words more terrible, that his daughter had disgraced herself, and that he had turned her out of doors—her, and her child—and that if ever they gave her help—or food—or shelter—he prayed that they might never enter Heaven. And, all the while, Miss Grace stood by him, white and still as any stone; and when he had ended she heaved a great sigh, as much as to say her work was done, and her end was accomplished. But the old lord never touched his organ again, and died within the year; and no wonder! for, on the morrow of that wild and fearful night, the shepherds, coming down the Fell side, found Miss Maude sitting, all crazy and smiling, under the holly-trees, nursing a dead child—with a terrible mark on its right shoulder. 'But that was not what killed it,' said Dorothy; 'it was the frost and the cold; every wild creature was in its hole, and every beast in its fold—while the child and its mother were turned out to wander on the Fells! And now you know all! and I wonder if you are less frightened now?'

I was more frightened than ever; but I said I was not. I wished Miss Rosamond and myself well out of that dreadful house for ever; but I would not leave her, and I dared not take her away. But oh! how I watched her, and guarded her! We bolted the doors and shut the window-shutters fast, an hour or more before dark, rather than leave them open five minutes too late. But my little lady still heard the weird child crying and mourning; and not all we could do or say could keep her from wanting to go to her, and let her in from the cruel wind and the snow. All this time, I kept away from Miss Furnivall and Mrs Stark, as much as ever I could; for I feared them—I knew no good could be about them, with their grey hard faces, and their dreamy eyes, looking back into the ghastly years that were gone. But, even in my fear, I had a kind of pity—for Miss Furnivall, at least. Those gone down to the pit can hardly have a more hopeless look than that which was ever on her face. At last I even got so sorry for her—who never said a word but what was quite forced from her—that I prayed for her; and I taught Miss Rosamond to pray for one who had done a deadly sin; but often when she came to those words, she would listen, and start up from her knees, and say, 'I hear my little girl plaining and crying very sad—Oh! let her in, or she will die!'

One night—just after New Year's Day had come at last, and the long winter had taken a turn, as I hoped—I heard the west drawing-room bell ring three times, which was a signal for me. I would not leave Miss Rosamond alone, for all she was asleep—for the old lord had been playing wilder than ever—and I feared lest my darling should waken to hear the spectre child; see her I knew she could not. I had fastened the windows too well for that. So I took her out of her bed and wrapped her up in such outer clothes as were most handy, and carried her down to the drawing-room, where the old ladies sat at their tapestry work as usual. They looked up when I came in, and Mrs Stark asked, quite astounded, 'Why did I bring Miss Rosamond there, out of her warm bed?' I had begun to whisper, 'Because I was afraid of her being tempted out while I was away, by the wild child in the snow,' when she stopped me short (with a glance at Miss Furnivall), and said Miss Furnivall wanted me to undo some work she had done wrong, and which neither of them could see to unpick. So I laid my pretty dear on the sofa, and sat down on a stool by them, and hardened my heart against them, as I heard the wind rising and howling.

Miss Rosamond slept on sound, for all the wind blew so; and Miss

Furnivall said never a word, nor looked round when the gusts shook the windows. All at once she started up to her full height, and put up one hand, as if to bid us listen.

'I hear voices!' said she, 'I hear terrible screams—I hear my father's voice!'

Just at that moment my darling wakened with a sudden start: 'My little girl is crying, oh, how she is crying!' and she tried to get up and go to her, but she got her feet entangled in the blanket, and I caught her up; for my flesh had begun to creep at these noises, which they heard while we could catch no sound. In a minute or two the noises came, and gathered fast, and filled our ears; we, too, heard voices and screams, and no longer heard the winter's wind that raged abroad. Mrs Stark looked at me, and I at her, but we dared not speak. Suddenly Miss Furnivall went towards the door, out into the ante-room, through the west lobby, and opened the door into the great hall. Mrs Stark followed, and I durst not be left, though my heart almost stopped beating for fear. I wrapped my darling tight in my arms, and went out with them. In the hall the screams were louder than ever; they sounded to come from the east wing—nearer and nearer—close on the other side of the locked-up doors—close behind them. Then I noticed that the great bronze chandelier seemed all alight, though the hall was dim, and that a fire was blazing in the vast hearth-place, though it gave no heat; and I shuddered up with terror, and folded my darling closer to me. But as I did so, the east door shook, and she, suddenly struggling to get free from me, cried, 'Hester! I must go! My little girl is there; I hear her; she is coming! Hester, I must go!'

I held her tight with all my strength; with a set will, I held her. If I had died, my hands would have grasped her still, I was so resolved in my mind. Miss Furnivall stood listening, and paid no regard to my darling, who had got down to the ground, and whom I, upon my knees now, was holding with both my arms clasped round her neck; she still striving and crying to get free.

All at once the east door gave way with a thundering crash, as if torn open in a violent passsion, and there came into that broad and mysterious light, the figure of a tall old man, with grey hair and gleaming eyes. He drove before him, with many a relentless gesture of abhorrence, a stern and beautiful woman, with a little child clinging to her dress.

'O Hester! Hester!' cried Miss Rosamond. 'It's the lady! the lady below the holly-trees; and my little girl is with her. Hester! Hester! let me go to her; they are drawing me to them. I feel them—I feel them. I must go!'

Again she was almost convulsed by her efforts to get away; but I held her tighter and tighter, till I feared I should do her a hurt; but rather that than let her go towards those terrible phantoms. They passed along towards the great hall-door, where the winds howled and ravened for their prey; but before they reached that, the lady turned; and I could see that she defied the old man with a fierce and proud defiance; but then she quailed—and then she threw up her arms wildly and piteously to save her child—her little child—from a blow from his uplifted crutch.

And Miss Rosamond was torn as by a power stronger than mine, and writhed in my arms, and sobbed (for by this time the poor darling was growing faint).

'They want me to go with them on to the Fells—they are drawing me to them. Oh, my little girl! I would come, but cruel, wicked Hester holds me very tight.' But when she saw the uplifted crutch she swooned away, and I thanked God for it. Just at this moment—when the tall old man, his hair streaming as in the blast of a furnace, was going to strike the little shrinking child—Miss Furnivall, the old woman by my side, cried out, 'Oh, father! father! spare the little innocent child!' But just then I saw—we all saw—another phantom shape itself, and grow clear out of the blue and misty light that filled the hall; we had not seen her till now, for it was another lady who stood by the old man, with a look of relentless hate and triumphant scorn. That figure was very beautiful to look upon, with a soft white hat drawn down over the proud brows and a red and curling lip. It was dressed in an open robe of blue satin. I had seen that figure before. It was the likeness of Miss Furnivall in her youth; and the terrible phantoms moved on, regardless of old Miss Furnivall's wild entreaty—and the uplifted crutch fell on the right shoulder of the little child, and the younger sister looked on, stony and deadly serene. But at that moment the dim lights, and the fire that gave no heat, went out of themselves, and Miss Furnivall lay at our feet stricken down by the palsy—death-stricken.

Yes! she was carried to her bed that night never to rise again. She lay with her face to the wall muttering low but muttering alway: 'Alas! alas! what is done in youth can never be undone in age! What is done in youth can never be undone in age!'

An Account of Some Strange Disturbances in Aungier Street

J. S. LE FANU

It is not worth telling, this story of mine—at least, not worth writing. Told, indeed, as I have sometimes been called upon to tell it, to a circle of intelligent and eager faces, lighted up by a good after-dinner fire on a winter's evening, with a cold wind rising and wailing outside, and all snug and cosy within, it has gone off—though I say it, who should not—indifferent well. But it is a venture to do as you would have me. Pen, ink, and paper are cold vehicles for the marvellous, and a 'reader' decidedly a more critical animal than a 'listener'. If, however, you can induce your friends to read it after nightfall, and when the fireside talk has run for a while on thrilling tales of shapeless terror; in short, if you will secure me the *mollia tempora fandi*, I will go to my work, and say my say, with better heart. Well, then, these conditions presupposed, I shall waste no more words, but tell you simply how it all happened.

My cousin (Tom Ludlow) and I studied medicine together. I think he would have succeeded, had he stuck to the profession; but he preferred the Church, poor fellow, and died early, a sacrifice to contagion, contracted in the noble discharge of his duties. For my present purpose, I say enough of his character when I mention that he was of a sedate but frank and cheerful nature; very exact in his observance of truth, and not by any means like myself—of an excitable or nervous temperament.

My Uncle Ludlow—Tom's father—while we were attending lectures, purchased three or four old houses in Aungier Street, one of which was unoccupied. *He* resided in the country, and Tom proposed that we should take up our abode in the untenanted house, so long as it should continue unlet; a move which would accomplish the double end of settling us nearer alike to our lecture-rooms and to our amusements, and of relieving us from the weekly charge of rent for our lodgings.

Our furniture was very scant—our whole equipage remarkably modest and primitive; and, in short, our arrangements pretty nearly as simple as those of a bivouac. Our new plan was, therefore, executed almost as soon as conceived. The front drawing-room was our sitting-room.

I had the bedroom over it, and Tom the back bedroom on the same floor, which nothing could have induced me to occupy.

The house, to begin with, was a very old one. It had been, I believe, newly fronted about fifty years before; but with this exception, it had nothing modern about it. The agent who bought it and looked into the titles for my uncle, told me that it was sold, along with much other forfeited property, at Chichester House, I think, in 1702; and had belonged to Sir Thomas Hacket, who was Lord Mayor of Dublin in James II's time. How old it was *then*, I can't say; but, at all events, it had seen years and changes enough to have contracted all that mysterious and saddened air, at once exciting and depressing, which belongs to most old mansions.

There had been very little done in the way of modernizing details; and, perhaps, it was better so; for there was something queer and by-gone in the very walls and ceilings—in the shape of doors and windows—in the odd diagonal site of the chimney-pieces—in the beams and ponderous cornices—not to mention the singular solidity of all the woodwork, from the bannisters to the window-frames, which hopelessly defied disguise, and would have emphatically proclaimed their antiquity through any conceivable amount of modern finery and varnish.

An effort had, indeed, been made, to the extent of papering the drawing-rooms; but somehow, the paper looked raw and out of keeping; and the old woman, who kept a little dirt-pie of a shop in the lane, and whose daughter—a girl of two and fifty—was our solitary handmaid, coming in at sunrise, and chastely receding again as soon as she had made all ready for tea in our state apartment;—this woman, I say, remembered it, when old Judge Horrocks (who, having earned the reputation of a particularly 'hanging judge', ended by hanging himself, as the coroner's jury found, under an impulse of 'temporary insanity', with a child's skipping-rope, over the massive old bannisters) resided there, entertaining good company, with fine venison and rare old port. In those halcyon days, the drawing-rooms were hung with gilded leather, and, I dare say, cut a good figure, for they were really spacious rooms.

The bedrooms were wainscoted, but the front one was not gloomy; and in it the cosiness of antiquity quite overcame its sombre associations. But the back bedroom, with its two queerly-placed melancholy windows, staring vacantly at the foot of the bed, and with the shadowy recess to be found in most old houses in Dublin, like a large ghostly closet, which, from congeniality of temperament, had amalgamated with the bedchamber, and dissolved the partition. At night-time, this 'alcove'—as our 'maid' was wont to call it—had, in my eyes, a specially sinister and suggestive character. Tom's distant and solitary candle glimmered vainly into its darkness. *There* it was always overlooking him—always itself impenetrable. But this was only part of the effect. The whole room was, I can't tell how, repulsive to me. There was, I suppose, in its proportions and features, a latent discord—a certain mysterious and indescribable relation, which jarred indistinctly upon some secret sense of the fitting and the safe, and raised indefinable suspicions and apprehensions of the imagination. On the whole, as I began by saying, nothing could have induced me to pass a night alone in it.

I had never pretended to conceal from poor Tom my superstitious weakness; and he, on the other hand, most unaffectedly ridiculed my tremors. The sceptic was, however, destined to receive a lesson, as you shall hear.

We had not been very long in occupation of our respective dormitories, when I began to complain of uneasy nights and disturbed sleep. I was, I suppose, the more impatient under this annoyance, as I was usually a sound sleeper, and by no means prone to nightmares. It was now, however, my destiny, instead of enjoying my customary repose, every night to 'sup full of horrors'. After a preliminary course of disagreeable and frightful dreams, my troubles took a definite form, and the same vision, without an appreciable variation in a single detail, visited me at least (on an average) every second night in the week.

Now, this dream, nightmare, or infernal illusion—which you please—of which I was the miserable sport, was on this wise:

I saw, or thought I saw, with the most abominable distinctness, although at the time in profound darkness, every article of furniture and accidental arrangement of the chamber in which I lay. This, as you know, is incidental to ordinary nightmare. Well, while in this clairvoyant condition, which seemed but the lighting up of the theatre in which was to be exhibited the monotonous tableau of horror, which made my nights insupportable, my attention invariably became, I know not why, fixed upon the windows opposite the foot of my bed; and, uniformly with the same effect, a sense of dreadful anticipation always took slow but sure possession of me. I became somehow conscious of a sort of horrid but undefined preparation going forward in some unknown quarter, and by some unknown agency, for my torment; and, after an interval, which always seemed to me of the same length, a picture suddenly flew up to the window, where it remained fixed, as if by an electrical attraction, and my discipline of horror then commenced, to last perhaps for hours. The picture thus mysteriously glued to the window-panes, was the portrait of an old man, in a crimson flowered silk dressing-gown, the folds of which I could now describe, with a countenance embodying a strange mixture of intellect, sensuality, and power, but withal sinister and full of malignant omen. His nose was hooked, like the beak of a vulture; his eyes large, grey, and prominent, and lighted up with a more than mortal cruelty and coldness. These features were surmounted by a crimson velvet cap, the hair that peeped from under which was white with age, while the eyebrows retained their original blackness. Well I remember every line, hue, and shadow of that stony countenance, and well I may! The gaze of this hellish visage was fixed upon me, and mine returned it with the inexplicable fascination of nightmare, for what appeared to me to be hours of agony. At last—

The cock he crew, away then flew

the fiend who had enslaved me through the awful watches of the night; and, harassed and nervous, I rose to the duties of the day.

I had—I can't say exactly why, but it may have been from the exquisite anguish and profound impressions of unearthly horror, with which this strange phantasmagoria was associated—an insurmountable antipathy to describing the exact nature of my nightly troubles to my friend and comrade. Generally, however, I told him that I was haunted by abominable dreams; and, true to the imputed materialism of medicine, we put our heads together to dispel my horrors, not by exorcism, but by a tonic.

I will do this tonic justice, and frankly admit that the accursed portrait began to intermit its visits under its influence. What of that? Was this singular apparition—as full of character as of terror—therefore the creature of my fancy, or the invention of my poor stomach? Was it, in short, *subjective* (to borrow the technical slang of the day) and not the palpable aggression and intrusion of an external agent? That, good friend, as we will both admit, by no means follows. The evil spirit, who enthralled my senses in the shape of that portrait, may have been just as near me, just as energetic, just as malignant, though I saw him not. What means the whole moral code of revealed religion regarding the due keeping of our own bodies, soberness, temperance, etc.? here is an obvious connection between the material and the invisible; the healthy tone of the system, and its unimpaired energy, may, for aught we can tell, guard us against influences which would otherwise render life itself terrific. The mesmerist and the electro-biologist will fail upon an average with nine patients out of ten—so may the evil spirit. Special conditions of the corporeal system are indispensable to the production of certain spiritual phenomena. The operation succeeds sometimes—sometimes fails—that is all.

I found afterwards that my would-be sceptical companion had his troubles too. But of these I knew nothing yet. One night, for a wonder, I was sleeping soundly, when I was roused by a step on the lobby outside my room, followed by the loud clang of what turned out to be a large brass candlestick, flung with all his force by poor Tom Ludlow over the banisters, and rattling with a rebound down the second flight of stairs; and almost concurrently with this, Tom burst open my door, and bounced into my room backwards, in a state of extraordinary agitation.

I had jumped out of bed and clutched him by the arm before I had any distinct idea of my own whereabouts. There we were—in our shirts—standing before the open door—staring through the great old banister opposite, at the lobby window, through which the sickly light of a clouded moon was gleaming.

'What's the matter, Tom? What's the matter with you? What the devil's the matter with you, Tom?' I demanded shaking him with nervous impatience.

He took a long breath before he answered me, and then it was not very coherently.

'It's nothing, nothing at all—did I speak?—what did I say?—Where's the candle, Richard? It's dark; I—I had a candle!'

'Yes, dark enough,' I said; 'but what's the matter?—what *is* it?—why don't you speak, Tom?—have you lost your wits?—what is the matter?'

'The matter?—oh, it is all over. It must have been a dream—nothing at all but a dream—don't you think so? It could not be anything more than a dream.'

'Of *course*,' said I, feeling uncommonly nervous, 'it *was* a dream.'

'I thought,' he said, 'there was a man in my room, and—and I jumped out of bed; and—and—where's the candle?'

'In your room, most likely,' I said, 'shall I go and bring it?'

'No; stay here—don't go; it's no matter—don't, I tell you; it was all a dream. Bolt the door, Dick; I'll stay here with you—I feel nervous. So, Dick, like a good fellow, light your candle and open the window—I am in a *shocking state*.'

I did as he asked me, and robing himself like Granuaile in one of my blankets, he seated himself close beside my bed.

Everybody knows how contagious is fear of all sorts, but more especially that particular kind of fear under which poor Tom was at that moment labouring. I would not have heard, nor I believe would he have recapitulated, just at that moment, for half the world, the details of the hideous vision which had so unmanned him.

'Don't mind telling me anything about your nonsensical dream, Tom,' said I, affecting contempt, really in a panic; 'let us talk about something else; but it is quite plain that this dirty old house disagrees with us both, and hang me if I stay here any longer, to be pestered with indigestion and—and—bad nights, so we may as well look out for lodgings—don't you think so?—at once.'

Tom agreed, and, after an interval, said—

'I have been thinking, Richard, that it is a long time since I saw my father, and I have made up my mind to go down tomorrow and return in a day or two, and you can take rooms for us in the meantime.'

I fancied that this resolution, obviously the result of the vision which had so profoundly scared him, would probably vanish next morning with the damps and shadows of night. But I was mistaken. Off went Tom at peep of day to the country, having agreed that so soon as I had secured suitable lodgings, I was to recall him by letter from his visit to my Uncle Ludlow.

Now, anxious as I was to change my quarters, it so happened, owing to a series of petty procrastinations and accidents, that nearly a week elapsed before my bargain was made and my letter of recall on the wing to Tom; and, in the meantime, a trifling adventure or two had occurred to your humble servant, which, absurd as they now appear, diminished by distance, did certainly at the time serve to whet my appetite for change considerably.

A night or two after the departure of my comrade, I was sitting by my bedroom fire, the door locked, and the ingredients of a tumbler of hot whisky-punch upon the crazy spider-table; for, as the best mode of keeping the

> Black spirits and white,
> Blue spirits and grey,

with which I was environed, at bay, I had adopted the practice recommended by the wisdom of my ancestors, and 'kept my spirits up by pouring spirits down'. I had thrown aside my volume of Anatomy, and was treating myself by way of a tonic, preparatory to my punch and bed, to half-a-dozen pages of the *Spectator*, when I heard a step on the flight of stairs descending from the attics. It was two o'clock, and the streets were as silent as a churchyard—the sounds were, therefore, perfectly distinct. There was a slow, heavy tread, characterized by the emphasis and deliberation of age, descending by the narrow staircase from above; and, what made the sound more singular, it was plain that the feet which produced it were perfectly bare, measuring the descent with something between a pound and a flop, very ugly to hear.

I knew quite well that my attendant had gone away many hours before, and that nobody but myself had any business in the house. It was quite plain also that the person who was coming down stairs had no intention whatever of concealing his movements; but, on the contrary, appeared disposed to make even more noise, and proceed more deliberately, than was at all necessary. When the step reached the foot of the stairs outside my room, it seemed to stop; and I expected every moment to see my door open spontaneously, and give admission to the original of my detested portrait. I was, however, relieved in a few seconds by hearing the descent renewed, just in the same manner, upon the staircase leading down to the drawing-rooms, and thence, after another pause, down the next flight, and so on to the hall, whence I heard no more.

Now, by the time the sound had ceased, I was wound up, as they say, to a very unpleasant pitch of excitement. I listened, but there was not a stir. I screwed up my courage to a decisive experiment—opened my door, and in a stentorian voice bawled over the banisters, 'Who's there?' There was no answer but the ringing of my own voice through the empty old house—no renewal of the movement; nothing, in short, to give my unpleasant sensations a definite direction. There is, I think, something most disagreeably disenchanting in the sound of one's own voice under such circumstances, exerted in solitude, and in vain. It redoubled my sense of isolation, and my misgivings increased on perceiving that the door, which I certainly thought I had left open, was closed behind me; in a vague alarm, lest my retreat should be cut off, I got again into my room as quickly as I could, where I remained in a state of imaginary blockade, and very uncomfortable indeed, till morning.

Next night brought no return of my barefooted fellow-lodger; but the night following, being in my bed, and in the dark—somewhere, I suppose, about the same hour as before, I distinctly heard the old fellow again descending from the garrets.

This time I had had my punch, and the *morale* of the garrison was consequently excellent. I jumped out of bed, clutched the poker as I passed the expiring fire, and in a moment was upon the lobby. The sound had ceased by this time—the dark and chill were discouraging; and, guess my horror, when I saw, or thought I saw, a black monster, whether in the shape of a man or a bear I could not say, standing, with its back to the wall, on the lobby, facing me, with a pair of great greenish eyes shining dimly out. Now, I must be frank, and confess that the cupboard which displayed our plates and cups stood just there, though at the moment I did not recollect it. At the same time I must honestly say, that making every allowance for an excited imagination, I never could satisfy myself that I was made the dupe of my own fancy in this matter; for this apparition, after one or two shiftings of shape, as if in the act of incipient transformation, began, as it seemed on second thoughts, to advance upon me in its original form. From an instinct of terror rather than of courage, I hurled the poker, with all my force, at its head; and to the music of a horrid crash made my way into my room, and double-locked the door. Then, in a minute more, I heard the horrid bare feet walk down the stairs, till the sound ceased in the hall, as on the former occasion.

If the apparition of the night before was an ocular delusion of my fancy sporting with the dark outlines of our cupboard, and if its horrid eyes were nothing but a pair of inverted teacups, I had, at all events, the satisfaction of having launched the poker with admirable effect, and in true 'fancy' phrase, 'knocked its two daylights into one', as the commingled fragments of my tea-service testified. I did my best to gather comfort and courage from these evidences; but it would not do. And then what could I say of those horrid bare feet, and the regular tramp, tramp, tramp, which measured the distance of the entire staircase through the solitude of my haunted dwelling, and at an hour when no good influence was stirring? Confound it!—the whole affair was abominable. I was out of spirits, and dreaded the approach of night.

It came, ushered ominously in with a thunderstorm and dull torrents of depressing rain. Earlier than usual the streets grew silent; and by twelve o'clock nothing but the comfortless pattering of the rain was to be heard.

I made myself as snug as I could. I lighted *two* candles instead of one. I forswore bed, and held myself in readiness for a sally, candle in hand; for, *coute qui coute*, I was resolved to *see* the being, if visible at all, who troubled the nightly stillness of my mansion. I was fidgety and nervous and, tried in vain to interest myself with my books. I walked up and down my room, whistling in turn martial and hilarious music, and listening ever and anon for the dreaded noise. I sate down and stared at the square label on the solemn and reserved-looking black bottle, until 'FLANAGAN & CO.'S BEST OLD MALT WHISKY' grew into a sort of subdued accompaniment to all the fantastic and horrible speculations which chased one another through my brain.

Silence, meanwhile, grew more silent, and darkness darker. I listened in vain for the rumble of a vehicle, or the dull clamour of a distant row.

There was nothing but the sound of a rising wind, which had succeeded the thunderstorm that had travelled over the Dublin mountains quite out of hearing. In the middle of this great city I began to feel myself alone with nature, and Heaven knows what beside. My courage was ebbing. Punch, however, which makes beasts of so many, made a man of me again—just in time to hear with tolerable nerve and firmness the lumpy, flabby, naked feet deliberately descending the stairs again.

I took a candle, not without a tremor. As I crossed the floor I tried to extemporize a prayer, but stopped short to listen, and never finished it. The steps continued. I confess I hesitated for some seconds at the door before I took heart of grace and opened it. When I peeped out the lobby was perfectly empty—there was no monster standing on the staircase; and as the detested sound ceased, I was reassured enough to venture forward nearly to the banisters. Horror of horrors! within a stair or two beneath the spot where I stood the unearthly tread smote the floor. My eye caught something in motion; it was about the size of Goliath's foot—it was grey, heavy, and flapped with a dead weight from one step to another. As I am alive, it was the most monstrous grey rat I ever beheld or imagined.

Shakespeare says—'Some men there are cannot abide a gaping pig, and some that are mad if they behold a cat.' I went well-nigh out of my wits when I beheld this *rat*; for, laugh at me as you may, it fixed upon me, I thought, a perfectly human expression of malice; and, as it shuffled about and looked up into my face almost from between my feet, I saw, I could swear it—I felt it then, and know it now, the infernal gaze and the accursed countenance of my old friend in the portrait, transfused into the visage of the bloated vermin before me.

I bounced into my room again with a feeling of loathing and horror I cannot describe, and locked and bolted my door as if a lion had been at the other side. D——n him or *it*; curse the portrait and its original! I felt in my soul that the rat—yes, the *rat*, the RAT I had just seen, was that evil being in masquerade, and rambling through the house upon some infernal night lark.

Next morning I was early trudging through the miry streets; and, among other transactions, posted a peremptory note recalling Tom. On my return, however, I found a note from my absent 'chum', announcing his intended return next day. I was doubly rejoiced at this, because I had succeeded in getting rooms; and because the change of scene and return of my comrade were rendered specially pleasant by the last night's half ridiculous half horrible adventure.

I slept extemporaneously in my new quarters in Digges' Street that night, and next morning returned for breakfast to the haunted mansion, where I was certain Tom would call immediately on his arrival.

I was quite right—he came; and almost his first question referred to the primary object of our change of residence.

'Thank God,' he said with genuine fervour, on hearing that all was arranged. 'On *your* account I am delighted. As to myself, I assure you that no earthly consideration could have induced me ever again to pass a night in this disastrous old house.'

'Confound the house!' I ejaculated, with a genuine mixture of fear and detestation, 'we have not had a pleasant hour since we came to live here'; and so I went on, and related incidentally my adventure with the plethoric old rat.

'Well, if that were *all*,' said my cousin, affecting to make light of the matter, 'I don't think I should have minded it very much.'

'Ay, but its eye—its countenance, my dear Tom,' urged I; 'if you had seen *that*, you would have felt it might be *anything* but what it seemed.'

'I inclined to think the best conjurer in such a case would be an able-bodied cat,' he said, with a provoking chuckle.

'But let us hear your own adventure,' I said tartly.

At this challenge he looked uneasily round him. I had poked up a very unpleasant recollection.

'You shall hear it, Dick; I'll tell it to you,' he said. 'Begad, sir, I should feel quite queer, though, telling it *here*, though we are too strong a body for ghosts to meddle with just now.'

Though he spoke this like a joke, I think it was serious calculation. Our Hebe was in a corner of the room, packing our cracked delft tea and dinner-services in a basket. She soon suspended operations, and with mouth and eyes wide open became an absorbed listener. Tom's experiences were told nearly in these words:

'I saw it three times, Dick—three distinct times; and I am perfectly certain it meant me some infernal harm. I was, I say, in danger—in *extreme* danger; for, if nothing else had happened, my reason would most certainly have failed me, unless I had escaped so soon. Thank God. I *did* escape.

'The first night of this hateful disturbance, I was lying in the attitude of sleep, in that lumbering old bed. I hate to think of it. I was really wide awake, though I had put out my candle, and was lying as quietly as if I had been asleep; and although accidentally restless, my thoughts were running in a cheerful and agreeable channel.

'I think it must have been two o'clock at least when I thought I heard a sound in that—that odious dark recess at the far end of the bedroom. It was as if someone was drawing a piece of cord slowly along the floor, lifting it up, and dropping it softly down again in coils. I sat up once or twice in my bed, but could see nothing, so I concluded it must be mice in the wainscot. I felt no emotion graver than curiosity, and after a few minutes ceased to observe it.

'While lying in this state, strange to say; without at first a suspicion of anything supernatural, on a sudden I saw an old man, rather stout and square, in a sort of roan-red dressing-gown, and with a black cap on his head, moving stiffly and slowly in a diagonal direction, from the recess, across the floor of the bedroom, passing my bed at the foot, and entering the lumber-closet at the left. He had something under his arm; his head hung a little at one side; and, merciful God! when I saw his face.'

Tom stopped for a while, and then said—

'That awful countenance, which living or dying I never can forget, disclosed what he was. Without turning to the right or left, he passed beside me, and entered the closet by the bed's head.

'While this fearful and indescribable type of death and guilt was passing, I felt that I had no more power to speak or stir than if I had been myself a corpse. For hours after it had disappeared, I was too terrified and weak to move. As soon as daylight came, I took courage, and examined the room, and especially the course which the frightful intruder had seemed to take, but there was not a vestige to indicate anybody's having passed there; no sign of any disturbing agency visible among the lumber that strewed the floor of the closet.

'I now began to recover a little. I was fagged and exhausted, and at last, overpowered by a feverish sleep. I came down late; and finding you out of spirits, on account of your dreams about the portrait, whose *original* I am now certain disclosed himself to me, I did not care to talk about the infernal vision. In fact, I was trying to persuade myself that the whole thing was an illusion, and I did not like to revive in their intensity the hated impressions of the past night—or, to risk the constancy of my scepticism, by recounting the tale of my sufferings.

'It required some nerve, I can tell you, to go to my haunted chamber next night, and lie down quietly in the same bed,' continued Tom. 'I did so with a degree of trepidation, which, I am not ashamed to say, a very little matter would have sufficed to stimulate to downright panic. This night, however, passed off quietly enough, as also the next; and so too did two or three more. I grew more confident, and began to fancy that I believed in the theories of spectral illusions, with which I had at first vainly tried to impose upon my convictions.

'The apparition had been, indeed, altogether anomalous. It had crossed the room without any recognition of my presence: I had not disturbed *it*, and *it* had no mission to *me*. What, then, was the imaginable use of its crossing the room in a visible shape at all? Of course it might have *been* in the closet instead of *going* there, as easily as it introduced itself into the recess without entering the chamber in a shape discernible by the senses. Besides, how the deuce *had* I seen it? It was a dark night; I had no candle; there was no fire; and yet I saw it as distinctly, in colouring and outline, as ever I beheld human form! A cataleptic dream would explain it all; and I was determined that a dream it should be.

'One of the most remarkable phenomena connected with the practice of mendacity is the vast number of deliberate lies we tell ourselves, whom, of all persons, we can least expect to deceive. In all this, I need hardly tell you, Dick, I was simply lying to myself, and did not believe one word of the wretched humbug. Yet I went on, as men will do, like persevering charlatans and impostors, who tire people into credulity by the mere force of reiteration; so I hoped to win myself over at last to a comfortable scepticism about the ghost.

'He had not appeared a second time—that certainly was a comfort; and what, after all, did I care for him, and his queer old toggery and strange looks? Not a fig! I was nothing the worse for having seen him, and a good story the better. So I tumbled into bed, put out my candle, and, cheered by a loud drunken quarrel in the back lane, went fast asleep.

'From this deep slumber I awoke with a start. I knew I had had a horrible dream; but what it was I could not remember. My heart was thumping furiously; I felt bewildered and feverish; I sat up in the bed and looked about the room. A broad flood of moonlight came in through the curtainless window; everything was as I had last seen it; and though the domestic squabble in the back lane was, unhappily for me, allayed, I yet could hear a pleasant fellow singing, on his way home, the then popular comic ditty called, "Murphy Delany". Taking advantage of this diversion I lay down again, with my face towards the fireplace, and closing my eyes, did my best to think of nothing else but the song, which was every moment growing fainter in the distance:

'Twas Murphy Delany, so funny and frisky,
Stept into a shebeen shop to get his skin full;
He reeled out again pretty well lined with whiskey,
As fresh as a shamrock, as blind as a bull.

'The singer, whose condition I dare say resembled that of his hero, was soon too far off to regale my ears any more; and as his music died away, I myself sank into a doze, neither sound nor refreshing. Somehow the song had got into my head, and I went meandering on through the adventures of my respectable fellow-countryman, who, on emerging from the "shebeen shop", fell into a river, from which he was fished up to be "sat upon" by a coroner's jury, who having learned from a "horse-doctor" that he was "dead as a door-nail, so there was an end", returned their verdict accordingly, just as he returned to his senses, when an angry altercation and a pitched battle between the body and the coroner winds up the lay with due spirit and pleasantry.

'Through this ballad I continued with a weary monotony to plod, down to the very last line, and then *da capo*, and so on, in my uncomfortable half-sleep, for how long, I can't conjecture. I found myself at last, however, muttering, "*dead* as a door-nail, so there was an end"; and something like another voice within me, seemed to say, very faintly, but sharply, "dead! dead! *dead!* and may the Lord have mercy on your soul!" and instantaneously I was wide awake, and staring right before me from the pillow.

'Now—will you believe it, Dick?—I saw the same accursed figure standing full front, and gazing at me with its stony and fiendish countenance, not two yards from the bedside.'

Tom stopped here, and wiped the perspiration from his face. I felt very queer. The girl was as pale as Tom; and, assembled as we were in the very scene of these adventures, we were all, I dare say, equally grateful for the clear daylight and the resuming bustle out of doors.

'For about three seconds only I saw it plainly; then it grew indistinct; but, for a long time, there was something like a column of dark vapour where it had been standing, between me and the wall; and I felt sure that he was still there. After a good while, this appearance went too. I took my clothes downstairs to the hall, and dressed there, with the door half open; then went out into the street, and walked about the town till morning, when I came back, in a miserable state of nervousness and exhaustion. I was such a fool, Dick, as to be ashamed to tell you how I came to be so upset. I thought you would laugh at me; especially as I had always talked philosophy, and treated *your* ghosts with contempt. I concluded you would give me no quarter; and so kept my tale of horror to myself.

'Now, Dick, you will hardly believe me, when I assure you, that for many nights after this last experience, I did not go to my room at all. I used to sit up for a while in the drawing-room after you had gone up to your bed; and then steal down softly to the hall-door, let myself out, and sit in the "Robin Hood" tavern until the last guest went off; and then I got through the night like a sentry, pacing the streets till morning.

'For more than a week I never slept in bed. I sometimes had a snooze on a form in the "Robin Hood", and sometimes a nap in a chair during the day; but regular sleep I had absolutely none.

'I was quite resolved that we should get into another house; but I could not bring myself to tell you the reason, and I somehow put it off from day to day, although my life was, during every hour of this procrastination, rendered as miserable as that of a felon with the constables on his track. I was growing absolutely ill from this wretched mode of life.

'One afternoon I determined to enjoy an hour's sleep upon your bed. I hated mine; so that I had never, except in a stealthy visit every day to unmake it, lest Martha should discover the secret of my nightly absence, entered the ill-omened chamber.

'As ill-luck would have it, you had locked your bedroom, and taken away the key. I went into my own to unsettle the bedclothes, as usual, and give the bed the appearance of having been slept in. Now, a variety of circumstances concurred to bring about the dreadful scene through which I was that night to pass. In the first place, I was literally overpowered with fatigue, and longing for sleep; in the next place, the effect of this extreme exhaustion upon my nerves resembled that of a narcotic, and rendered me less susceptible than, perhaps I should in any other condition have been, of the exciting fears which had become habitual to me. Then again, a little bit of the window was open, a pleasant freshness pervaded the room, and, to crown all, the cheerful sun of day was making the room quite pleasant. What was to prevent my enjoying an hour's nap *here*? The whole air was resonant with the cheerful hum of life, and the broad matter-of-fact light of day filled every corner of the room.

'I yielded—stifling my qualms—to the almost overpowering temptation; and merely throwing off my coat, and loosening my cravat, I lay down, limiting myself to *half*-an-hour's doze in the unwonted enjoyment of a feather bed, a coverlet, and a bolster.

'It was horribly insidious; and the demon, no doubt, marked my infatuated preparations. Dolt that I was, I fancied, with mind and body worn out for want of sleep, and an arrear of a full week's rest to my credit, that such measure as *half*-an-hour's sleep, in such a situation, was possible. My sleep was death-like, long, and dreamless.

'Without a start or fearful sensation of any kind, I waked gently, but completely. It was, as you have good reason to remember, long past midnight—I believe, about two o'clock. When sleep has been deep and long enough to satisfy nature thoroughly, one often wakens in this way, suddenly, tranquilly, and completely.

'There was a figure seated in that lumbering, old sofa-chair, near the fireplace. Its back was rather towards me, but I could not be mistaken; it turned slowly round, and, merciful heavens! there was the stony face, with its infernal lineaments of malignity and despair, gloating on me. There was now no doubt as to its consciousness of my presence, and the hellish malice with which it was animated, for it arose, and drew close to the bedside. There was a rope about its neck, and the other end, coiled up, it held stiffly in its hand.

'My good angel nerved me for this horrible crisis. I remained for some seconds transfixed by the gaze of this tremendous phantom. He came close to the bed, and appeared on the point of mounting upon it. The next instant I was upon the floor at the far side, and in a moment more was, I don't know how, upon the lobby.

'But the spell was not yet broken; the valley of the shadow of death was not yet traversed. The abhorred phantom was before me there; it was standing near the banisters, stooping a little, and with one end of the rope round its own neck, was poising a noose at the other, as if to throw over mine; and while engaged in this baleful pantomime, it wore a smile so sensual, so unspeakably dreadful, that my senses were nearly overpowered. I saw and remember nothing more, until I found myself in your room.

'I had a wonderful escape, Dick—there is no disputing *that*—an escape for which, while I live, I shall bless the mercy of heaven. No one can conceive or imagine what it is for flesh and blood to stand in the presence of such a thing, but one who has had the terrific experience. Dick, Dick, a shadow has passed over me—a chill has crossed my blood and marrow, and I will never be the same again—never, Dick—never!'

Our handmaid, a mature girl of two-and-fifty, as I have said, stayed her hand, as Tom's story proceeded, and by little and little drew near to us, with open mouth, and her brows contracted over her little, beady black eyes, till stealing a glance over her shoulder now and then, she established herself close behind us. During the relation, she had made various earnest comments, in an undertone; but these and her ejaculations, for the sake of brevity and simplicity, I have omitted in my narration.

'It's often I heard tell of it,' she now said, 'but I never believed it rightly till now—though, indeed, why should not I? Does not my mother, down there in the lane, know quare stories, God bless us, beyant telling about it? But you ought not to have slept in the back bedroom. She was loath to let me be going in and out of that room even in the day time, let alone for any Christian to spend the night in it; for sure she says it was his own bedroom.'

'*Whose* own bedroom?' we asked, in a breath.

'Why, *his*—the ould Judge's—Judge Horrocks's, to be sure, God rest his sowl'; and she looked fearfully round.

'Amen!' I muttered. 'But did he die there?'

'Die there! No, not quite *there*,' she said. 'Shure, was not it over the bannisters he hung himself, the ould sinner, God be merciful to us all? and was not it in the alcove they found the handles of the skipping-rope cut off, and the knife where he was settling the cord, God bless us, to hang himself with? It was his housekeeper's daughter owned the rope, my mother often told me, and the child never throve after, and used to be starting up out of her sleep, and screeching in the night time, wid dhrames and frights that cum an her; and they said how it was the speerit of the ould Judge that was tormentin' her; and she used to be roaring and yelling out to hould back the big ould fellow with the crooked neck; and then she'd screech "Oh, the master! the master! he's stampin' at me, and beckoning to me! Mother, darling, don't let me go!" And so the poor crathure died at last, and the docthers said it was wather on the brain, for it was all they could say.'

'How long ago was all this?' I asked.

'Oh, then, how would I know?' she answered. 'But it must be a wondherful long time ago, for the housekeeper was an ould woman, with a pipe in her mouth, and not a tooth left, and better nor eighty years ould when my mother was first married; and they said she was a rale buxom, fine-dressed woman when the ould Judge come to his end; an', indeed, my mother's not far from eighty years ould herself this day; and what made it worse for the unnatural ould villain, God rest his soul, to frighten the little girl out of the world the way he did, was what was mostly thought and believed by every one. My mother says how the poor little crathure was his own child; for he was by all accounts an ould villain every way, an' the hangin'est judge that ever was known in Ireland's ground.'

'From what you said about the danger of sleeping in that bedroom,' said I, 'I suppose there were stories about the ghost having appeared there to others.'

'Well, there *was* things said—quare things, surely,' she answered, as it seemed, with some reluctance. 'And why would not there? Sure was it not up in that same room he slept for more than twenty years? and was it not in the *alcove* he got the rope ready that done his own business at last, the way he done many a betther man's in his lifetime?—and was not the body lying in the same bed after death, and put in the coffin there, too, and carried out to his grave from it in Pether's churchyard, after the coroner was done? But there was quare stories—my mother has them all—about how one Nicholas Spaight got into trouble on the head of it.'

'And what did they say of this Nicholas Spaight?' I asked.

'Oh, for that matther, it's soon told,' she answered.

And she certainly did relate a very strange story, which so piqued my curiosity, that I took occasion to visit the ancient lady, her mother, from whom I learned many very curious particulars. Indeed, I am tempted to tell the tale, but my fingers are weary, and I must defer it. But if you wish to hear it another time, I shall do my best.

When we had heard the strange tale I have *not* told you, we put one or two further questions to her about the alleged spectral visitations, to which the house had, ever since the death of the wicked old Judge, been subjected.

'No one ever had luck in it,' she told us. 'There was always cross accidents, sudden deaths, and short times in it. The first that tuck it was a family—I forget their name—but at any rate there was two young ladies and their papa. He was about sixty, and a stout healthy gentleman as you'd wish to see at that age. Well, he slept in that unlucky back bedroom; and, God between us an' harm! sure enough he was found dead one morning, half out of the bed, with his head as black as a sloe, and swelled like a puddin', hanging down near the floor. It was a fit, they said. He was as dead as a mackerel, and so *he* could not say what it was; but the ould people was all sure that it was nothing at all but the ould Judge, God bless us! that frightened him out of his senses and his life together.

'Some time after there was a rich old maiden lady took the house. I don't know which room *she* slept in, but she lived alone; and at any rate, one morning, the servants going down early to their work, found her sitting on the passage-stairs, shivering and talkin' to herself, quite mad; and never a word more could any of *them* or her friends get from her ever afterwards but, "Don't ask me to go, for I promised to wait for him." They never made out from her who it was she meant by *him*, but of course those that knew all about the ould house were at no loss for the meaning of all that happened to her.

'Then afterwards, when the house was let out in lodgings, there was Micky Byrne that took the same room, with his wife and three little children; and sure I heard Mrs Byrne myself telling how the children used to be lifted up in the bed at night, she could not see by what mains; and how they were starting and screeching every hour, just all as one as the housekeeper's little girl that died, till at last one night poor Micky had a dhrop in him, the way he used now and again; and what do you think in the middle of the night he thought he heard a noise on the stairs, and being in liquor, nothing less id do him but out he must go himself to see what was wrong. Well, after that, all she ever heard of him was himself sayin', "Oh, God!" and a tumble that shook the very house; and there, sure enough, he was lying on the lower stairs, under the lobby, with his neck smashed double undher him, where he was flung over the banisters.'

Then the handmaiden added—

'I'll go down to the lane, and send up Joe Gavvey to pack up the rest of the taythings, and bring all the things across to your new lodgings.'

And so we all sallied out together, each of us breathing more freely, I have no doubt, as we crossed that ill-omened threshold for the last time.

Now, I may add thus much, in compliance with the immemorial usage of the realm of fiction, which sees the hero not only through his adventures, but fairly out of the world. You must have perceived that what the flesh, blood, and bone hero of romance proper is to the regular compounder of fiction, this old house of brick, wood, and mortar is to the humble recorder of this true tale. I, therefore, relate, as in duty bound, the catastrophe which ultimately befell it, which was simply this—that about two years subsequently to my story it was taken by a quack doctor, who called himself Baron Duhlstoerf, and filled the parlour windows with bottles of indescribable horrors preserved in brandy, and the newspapers with the usual grandiloquent and mendacious advertisements. This gentleman among his virtues did not reckon sobriety, and one night, being overcome with much wine, he set fire to his bed curtains, partially burned himself, and totally consumed the house. It was afterwards rebuilt, and for a time an undertaker established himself in the premises.

I have now told you my own and Tom's adventures, together with some valuable collateral particulars; and having acquitted myself of my engagement, I wish you a very good night, and pleasant dreams.

The Miniature

J. Y. AKERMAN

Calling one day on a friend, who had amassed a large collection of autographs, and other manuscript curiosities, he showed me a small quarto volume, which had been bequeathed to him by a relative, a physician, who for many years had been in extensive practice in London.

'He attended the patients at a private asylum for insane persons of the better classes,' said my friend, 'and I have often heard him speak of the writer of that beautiful MS, a gentleman of good family, who had been an inmate of —— House upwards of thirty years,' at the time he was first called to attend him.

On looking over the volume, I found it filled with scraps of poetry, extracts from classic authors, and even from the Talmudic writers; but what interested me most was a narrative of several pages, which appeared so circumstantially related as to leave little doubt of its being partly, if not wholly, founded on fact. I begged permission to make a transcript, which was readily granted, and the result is before the reader.

'We laugh at what we call the folly of our ancestors, and their notions of destiny, and the malignant influences of the stars. For what will our children deride us? Perhaps for dreaming that friendship was a reality, and that constant love dwelt upon earth. I once believed that friendship was not a vain name, and thought, with the antique sage, that one mind sometimes dwelt in two bodies. I dreamt, and woke to find that I had been dreaming!

'George S —— was my chum at school, and my inseparable companion at college. We quitted it at the same time, he to proceed to London, where he was in expectation of obtaining a lucrative appointment in one of the English colonies, and I to return for a short period to the family mansion. When I reached —— Hall, I found several visitors, among whom was my cousin, Maria D——. She had grown a woman since I had last met her, and I now thought I had never seen a more perfect figure, or a more bewitching countenance. Then she sang like a siren, and was an elegant horsewoman. Will those who read this wonder that I fell in love with her, that I spent nearly the whole of the day in her company, and that I could think of nothing in the world besides.

'Something occurred to delay my friend George's departure from England, and, as he was idling about town, I invited him to —— Hall. Great as was my regard for him, I now, however, discovered that I could live less in his company. No marvel! I preferred the society of my lovely cousin, upon whose heart, I had the happiness to learn, my constant attentions had already made a sensible impression. I hesitated to make her an offer, though I had every reason to believe our attachment was mutual, partly, perhaps, from that excessive delicacy which constantly attends on true love, and partly because I wished to do so when my friend should have left us less exposed to intrusion. Would that the deep sea had swallowed him up, or that he had rotted under a tropical sun, ere he had come to —— Hall!

'One morning I arose earlier than usual, and was looking from my chamber window on the beautiful prospect which the house commanded. Wrapped in a delightful reverie, of which my lovely cousin was the principal subject, I paid but little attention to the sound of voices below. Suddenly, however, I awoke to consciousness: for the sweet tones of a woman in earnest conversation struck on my ear. Yes, it was hers—it was Maria's. What could have called her forth at so early an hour? As I looked earnestly towards the walk which ran through the plantation, I saw emerge from it my cousin and my friend! My heart rose to my lips, and choked my utterance, or I should have cried out at the sight. I withdrew from the window, and threw myself on the sofa, tormented with surmises a thousand times more painful even than realities.

'At the breakfast table I was moody and thoughtful, which my friend perceiving, attempted a joke; but I was in no humour to receive it, when Maria, in a compassionating tone, remarked that I looked unwell, and that I should take a walk or ride before breakfast, adding, that she and George S —— had walked for an hour and more in the plantation near the house. Though this announcement was certainly but ill calculated to afford perfect ease to my mind, it was yet made with such an artless air, that my more gloomy surmises vanished, and I rallied; but I wished my friend would take his departure. Right truly says the Italian proverb, "Love's guerdon is jealousy."

'After breakfast, George S —— proposed a stroll on foot to the ruins of the Cistercian Abbey, about a mile distant from the Hall, to which I at once assented. As we walked along the beautiful and shady lane which led to the ruin, George was as loquacious as ever, talked of everybody and everything, and of his confident expectation of realizing a fortune abroad. I was, however, in no humour for talking, and made few remarks in reply; but he appeared not to heed my taciturnity, and, when he arrived at the spot, broke forth into raptures at the sight of the noble ruin.

'And truly it was a scene the contemplation of which might have lulled the minds of most men! A thousand birds were caroling around us; the grass near the ruin was not long and rank, but short, close, studded with trefoil, and soft as a rich carpet. Luxuriant ivy climbed the shattered walls, bleached by the winds of centuries; and the lizards, basking in the sun, darted beneath the fallen fragments at the sound of our footsteps as we approached the spot.

'We both sat down on a large stone, and surveyed the noble oriel. I was passionately fond of Gothic architecture, and had often admired this window, but I thought I had never seen it look so beautiful before. My moody thoughts fled, and I was wrapped in the contemplation of the exquisite tracery, when I was suddenly roused by my friend, who, patting me familiarly on the back, exclaimed,

'"It *is* a beautiful ruin, Dick! How I wish thy sweet cousin, Maria, had accompanied us!"

'I was struck dumb by this declaration; but my look was sufficiently eloquent to be understood by him, and he did not fail to interpret it aright. He appeared confused, and I, regaining my self-possession, arose from my seat with the laconic remark, "Indeed!"

'George S —— attempted a laugh, but it failed; he was evidently as much disconcerted and disquieted as myself. How lynx-eyed is love! We mutually read each other's hearts at the same moment.

'"I am sorry for you, Dick," said he, after a short pause, affecting very awkwardly an air of indifference; "'pon my soul, I am; but I'm over head and ears in love with the girl, and should die at the bare thought of her encouraging another."

'I wished for the strength of Milo, that I might have dashed out his brains against the huge stone on which we ' ad been sitting. I felt my very blood seethe and simmer at the declar on, and with my clenched fist I struck him a violent and stunning blow, which, though it did not beat him to the ground, sent him staggering several paces backward.

'"Liar!" screamed I frantically, "take that! You dare not proceed with your folly."

'Recovering his feet, George S —— laid his hand on his sword, which he half unsheathed; but, as if conscious of there being no witness present, or wishing, perhaps, still further to convince me of the advantage he possessed, he did not draw.

'"Nay," said I, "out with your weapon; nothing less will do. I would rather lose my birthright than yield to thee one, without whom life would be valueless."

'He smiled bitterly, wiped his bruised and bloody face, and slowly drew from his bosom a small miniature, encircled with diamonds, which he held before my eyes. One glance was sufficient, it was a portrait of Maria! It was that face which, sleeping or waking, has haunted me these thirty years past.

'"Villain!" I cried, clutching at the portrait with my left hand, while I snatched with my right hand my sword from its sheath, "you have stolen it."

'With assumed coolness, which it was impossible he could feel, he smiled again, put back the miniature in his bosom, and drew his sword. The next moment our weapons crossed with an angry clash, and were flashing in the morning's sun.

'My adversary was a perfect master of his weapon, and he pressed upon me with a vigour which any attempt to retaliate would have rendered dangerous in one so much inferior to him in skill. Maddened as I was, I yet restrained myself, and stood on my guard, my eyes fixed on his, and watching every glance: my wish to destroy him was intense. The fiend nerved my arm, and, while he warmed with the conflict, I became more cool and vigilant. At length he appeared to grow weary, and then I pressed upon him with the fixed determination of taking his life; but he rallied instantly, and, in returning a thrust, which I intended for his heart, and which he parried scarcely in time, his foot slipped, and he fell on one knee, the point of my sword entering the left breast by accident. It was not a deep wound, and perhaps he felt it not; for he attempted to master my sword with his left hand, while he shortened his own weapon, and thrust fiercely at my throat, making at the same time a spring to regain his feet. But his fate was sealed: as he rose, I dashed aside the thrust intended for me, and sheathed my weapon in his left breast. I believe I must have pierced his heart; for he sank on his knees with a gasp, and the next moment fell heavily on his face, with his sword still clutched tightly in his hand.

'Wearied, and panting from the effects of the violent struggle, I threw myself on the large stone which had so recently served us for a seat, and looked on the body of my adversary. He was dead!—that fatal thrust had destroyed all rivalry, but at the price of murder, the murder of one who had been my friend from boyhood upwards! A thousand conflicting emotions racked me as I beheld the piteous sight. Hatred was extinguished, and remorse succeeded; yet I still thought of the audacity of him who had provoked such deadly resentment. Fear, too, fear of the consequences of this fatal encounter in a solitary spot, without witnesses, added to the intensity of my misery, and I groaned in anguish. What was to be done? Should I go and deliver myself up to justice, and declare the whole truth? Should I fly, and leave the body of my friend to tell the dismal tale?—or should I bury him secretly, and leave it to be supposed that he had been robbed and murdered? As each suggestion was canvassed and rejected, in my despair, I even thought of dying by my own hand.

'"Ah! miserable wretch!" I exclaimed, "what hast thou done?—to what dire necessity has a fair and false face driven thee? Yet I will look once more on those bewitching features which have brought me to this wretched pass!"

'I stooped, and turned the dead man on his back. His pallid face was writhen and distorted, his lips were bloody, and his eyes, which were wide open, seemed still to glare with hatred and defiance, as when he stood before me in the desperate struggle for life and death. I tore open his vest, and discovered the wound which had killed him. It had collapsed, and looked no bigger than the puncture of a bodkin: but one little round crimson spot was visible, the haemorrhage was internal. There lay the miniature which, a few minutes before, had been held up exultingly to my frantic gaze. I seized, and pressed it to my lips, forgetting in my transports how dearly I had purchased it.

'This delirium, however, soon subsided, and my next thoughts were of the dead body. I looked about me for some nook where I might deposit it. There was a chasm in the ground among the ruins a few yards off, where the vaulted roof of the crypt had fallen in. It was scarcely large enough to admit the corpse; but I raised it in my arms, bore it thither, and with some difficulty thrust it through the aperture. I heard it fall, as if to some distance, with a dull, heavy sound; and, casting in after it my adversary's hat and sword, I hurried from the spot like another Cain.

'At dinner, one glance from Maria, as I replied, in answer to her enquiry after George S——, that he was gone to make a call a few miles off—one glance, I say, thrilled through my very soul, and almost caused me to betray myself. All noticed my perturbed look, and, complaining of violent headache, I withdrew from the table ere the meal was ended, and betook myself to my chamber.

'How shall I paint the horror of that evening, of the night that succeeded it, and the mental darkness which fell upon my wretched self ere the morning dawned! Night came; I rang for lights, and attempted to read, but in vain; and, after pacing my chamber for some hours, overpowered by fatigue, I threw myself on the bed and slept, how long I know not. A succession of hideous dreams haunted my slumbers, still I was not awakened by them; the scenes shifted when arrived at their climax, and a new ordeal of horrors succeeded, yet, like him who suffers from nightmare, with a vague consciousness that all was not real, I wished to awake. Last of all, I dreamt that I was arraigned for the murder of my friend. The judge summed up the evidence, which, though purely circumstantial, was sufficient to condemn me; and, amidst the silence of the crowded court, broken only by the sobs of anxious and sympathizing friends and relatives, I received sentence of death, and was hurried back to my cell. Here, abandoned by all hope, I lay grovelling on my straw bed, and cursed the hour of my birth. A figure entered, and in gentle accents, which I thought I recognized, bade me arise, quit my prison-house, and follow. The figure was that of a woman closely veiled. She led the way, and passed the gaolers, who seemed buried in profound sleep. We left the town, crossed the common, and entered a wood, when I threw myself at the feet of my deliverer, and passionately besought her to unveil. She shook her head mournfully, bade me wait a while till she should return with a change of apparel, and departed.

'I cast myself down at the foot of an aged oak, drew from my bosom the portrait of Maria, and, rapt in the contemplation of those lovely features, I did not perceive the approach of a man, the ranger of the forest, who, recognizing my prison-dress, darted upon me, exclaiming, "Villain! you have escaped from gaol, and stolen that miniature from the Hall!"

'I sprang to my feet, thrust the fatal portrait into my bosom, and would have fled; but he seized, and closed with me. In the struggle which followed we both fell, I undermost. At that moment I awoke; I was in reality struggling with some one, but whom I could not tell; for my candles had burnt out, and the chamber was in total darkness! A powerful, bony hand grasped me tightly by the throat, while another was thrust into my bosom, as if in search of the miniature, which I had placed there previous to lying down.

'With a desperate effort I disengaged myself, and leaped from the bed; but I was again seized, and again my assailant attempted to reach my fatal prize. We struggled violently; at one time I seemed to be overpowering him, and for several moments there was a pause, during which I heard my own breathing, and felt my own heart throbbing violently; but he with whom I contended seemed to breathe not, nor to feel like a warm and living man. An indescribable tremor shook my frame; I attempted to cry out, but my throat was rigid, and incapable of articulation. I made another effort to disengage myself from the grasp of my assailant, and in doing so drew him, as I found by the curtains, near to the window. Again the hand was thrust into my bosom, and again I repelled it.

'Panting with the violence of the struggle, while a cold sweat burst out at every pore, I disengaged my right hand, and, determined to see whom I was contending with, I dashed aside the curtain. The dim light of the waning moon shone into the chamber; it fell upon the face of my antagonist, and one glance froze the blood in my veins. It was he!—it was George S——;—he whom I had murdered, glaring upon me with eyes which no mortal could look upon a second time! My brain whirled, a sound like the discharge of artillery shook the place, and I fell to the ground, blasted at the sight!'

Here follows a few incoherent sentences, which I have not deemed it necessary to transcribe. The reader will probably supply the sequel to this sad story.

The Last House in C—— Street.

DINAH MULOCK

I am not a believer in ghosts in general; I see no good in them. They come—that is, are reported to come—so irrelevantly, purposelessly—so ridiculously, in short—that one's common sense as regards this world, one's supernatural sense of the other, are alike revolted. Then nine out of ten 'capital ghost stories' are so easily accounted for; and in the tenth, when all natural explanation fails, one who has discovered the extraordinary difficulty there is in all society in getting hold of that very slippery article called a *fact*, is strongly inclined to shake a dubious head, ejaculating, 'Evidence! a question of evidence!'

But my unbelief springs from no dogged or contemptuous scepticism as to the possibility—however great the improbability—of that strange impression upon or communication to, spirit in matter, from spirit wholly immaterialized, which is vulgarly called 'a ghost'. There is no credulity more blind, no ignorance more childish, than that of the sage who tries to measure 'heaven and earth and the things under the earth', with the small two-foot-rule of his own brains. Dare we presume to argue concerning any mystery of the universe, 'It is inexplicable, and therefore impossible'?

Premising these opinions, though simply as opinions, I am about to relate what I must confess is to me a thorough ghost story; its external and circumstantial evidence being indisputable, while its psychological causes and results, though not easy of explanation, are still more difficult to be explained away. The ghost, like Hamlet's, was 'an honest ghost'. From her daughter—an old lady, who, bless her good and gentle memory! has since learned the secrets of all things—I learnt this veritable tale.

'My dear,' said Mrs MacArthur to me—it was in the early days of table-moving, when young folk ridiculed and elder folk were shocked at the notion of calling up one's departed ancestors into one's dinner-table, and learning the wonders of the angelic world by the bobbings of a hat or the twirlings of a plate;—'My dear,' continued the old lady, 'I do not like playing at ghosts.'

'Why not. Do you believe in them?'

'A little.'

'Did you ever see one?'

'Never. But once I heard—'

She looked serious, as if she hardly liked to speak about it, either from a sense of awe or from fear of ridicule. But no one could have laughed at any illusions of the gentle old lady, who never uttered a harsh or satirical word to a living soul; and this evident awe was rather remarkable in one who had a large stock of common sense, little wonder, and no ideality.

I was rather curious to hear MacArthur's ghost story.

'My dear, it was a long time ago, so long that you may fancy I forget and confuse the circumstances. But I do not. Sometimes I think one recollects more clearly things that happened in one's teens—I was eighteen that year—than a great many nearer events. And besides, I had other reasons for remembering vividly everything belonging to this time,—for I was in love, you must know.'

She looked at me with a mild, deprecating smile, as if hoping my youthfulness would not consider the thing so very impossible or ridiculous. No; I was all interest at once.

'In love with Mr MacArthur,' I said, scarcely as a question, being at that Arcadian time of life when one takes as a natural necessity, and believes as an undoubted truth, that everybody marries his or her first love.

'No, my dear; not with Mr MacArthur.'

I was so astonished, so completely dumb-foundered—for I had woven a sort of ideal round my good old friend—that I suffered Mrs MacArthur to knit in silence for full five minutes. My surprise was not lessened when she said, with a little smile—

'He was a young gentleman of good parts; and he was very fond of me. Proud, too, rather. For though you might not think it, my dear, I was actually a beauty in those days.'

I had very little doubt of it. The slight lithe figure, the tiny hands and feet,—if you had walked behind Mrs MacArthur you might have taken her for a young woman still. Certainly, people lived slower and easier in the last generation than in ours.

'Yes, I was the beauty of Bath. Mr Everest fell in love with me there. I was much gratified; for I had just been reading Miss Burney's *Cecilia*, and I thought him exactly like Mortimer Delvil. A very pretty tale, *Cecilia*; did you ever read it?'

'No.' And, to arrive at her tale, I leaped to the only conclusion which could reconcile the two facts of her having had a lover named Everest, and being now Mrs MacArthur. 'Was it *his* ghost you saw?'

'No, my dear, no; thank goodness, he is alive still. He calls here sometimes; he has been a good friend to our family. Ah!' with a slow shake of the head, half pleased, half pensive, 'you would hardly believe, my dear, what a very pretty fellow he was.'

One could scarcely smile at the odd phrase, pertaining to last-century novels and to the loves of our great-grandmothers. I listened patiently to the wandering reminiscences which still further delayed the ghost-story.

'But, Mrs MacArthur, was it in Bath that you saw or heard what I think you were going to tell me? The ghost, you know?'

'Don't call it *that*; it sounds as if you were laughing at it. And you must not, for it is really true; as true as that I sit here, an old lady of seventy-five; and that then I was a young gentlewoman of eighteen. Nay, my dear, I will tell you all about it.'

'We had been staying in London, my father and mother, Mr Everest, and I. He had persuaded them to take me; he wanted to show me a little of the world, though it was but a narrow world, my dear—for he was a law student, living poorly and working hard. He took lodgings for us near the Temple; in C—— street, the last house there, looking on to the river. He was very fond of the river; and often of evenings, when his work was too heavy to let him take us to Ranelagh or to the play, he used to walk with my father and mother and me, up and down the Temple Gardens. Were you ever in the Temple Gardens? It is a pretty place now—a quiet, grey nook in the midst of noise and bustle; the stars look wonderful through those great trees; but still it is not like what it was then, when I was a girl.'

Ah! no; impossible.

'It was in the Temple Gardens, my dear, that I remember we took our last walk—my mother, Mr Everest, and I—before she went home to Bath. She was very anxious and restless to go, being too delicate for London gaieties. Besides, she had a large family at home, of which I was the eldest; and we were anxiously expecting the youngest in a month or two. Nevertheless, my dear mother had gone about with me, taken me to all the shows and sights that I, a hearty and happy girl, longed to see, and entered into them with almost as great enjoyment as my own.

'But tonight she was pale, rather grave, and steadfastly bent on returning home.

'We did all we could to persuade her to the contrary, for on the next night but one was to have been the crowning treat of all our London pleasures: we were to see *Hamlet* at Drury Lane, with John Kemble and Sarah Siddons! Think of that, my dear. Ah! you have no such sights now. Even my grave father longed to go, and urged in his mild way that we should put off our departure. But my mother was determined.

'At last Mr Everest said—(I could show you the very spot where he stood, with the river—it was high water—lapping against the wall, and the evening sun shining on the Southwark houses opposite.) He said—it was very wrong, of course, my dear; but then he was in love, and might be excused,—

'"Madam," said he, "it is the first time I ever knew you think of yourself alone."

'"Myself, Edmond?"

'"Pardon me, but would it not be possible for you to return home, leaving behind, for two days only, Mr Thwaite and Mistress Dorothy?"

'"Leave them behind—leave them behind!" She mused over the words. "What say you, Dorothy?"

'I was silent. In very truth, I had never been parted from her in all my life. It had never crossed my mind to wish to part from her, or to enjoy any pleasure without her, till—till within the last three months. "Mother, don't suppose I——"

'But here I caught sight of Mr Everest, and stopped.

'"Pray continue. Mistress Dorothy."

'No, I could not. He looked so vexed, so hurt; and we had been so happy together. Also, we might not meet again for years, for the journey between London and Bath was then a serious one, even to lovers; and he worked very hard—had few pleasures in his life. It did indeed seem almost selfish of my mother.

'Though my lips said nothing, perhaps my sad eyes said only too much, and my mother felt it.

'She walked with us a few yards, slowly and thoughtfully. I could see her now, with her pale, tired face, under the cherry-coloured ribbons of her hood. She had been very handsome as a young woman, and was most sweet-looking still—my dear, good mother!

'"Dorothy, we will no more discuss this. I am very sorry, but I must go home. However, I will persuade your father to remain with you till the week's end. Are you satisfied?"

'"No," was the first filial impulse of my heart; but Mr Everest pressed my arm with such an entreating look, that almost against my will I answered "Yes."

'Mr Everest overwhelmed my mother with his delight and gratitude. She walked up and down for some time longer, leaning on his arm—she was very fond of him; then stood looking on the river, upwards and downwards.

'"I suppose this is my last walk in London. Thank you for all the care you have taken of me. And when I am gone home—mind, oh mind, Edmond, that you take special care of Dorothy."

'These words, and the tone in which they were spoken, fixed themselves on my mind—first, from gratitude, not unmingled with regret, as if I had not been so considerate to her as she to me; *afterwards*—But we often err, my dear, in dwelling too much on that word. We finite creatures have only to deal with "now"—nothing whatever to do with "afterwards". In this case, I have ceased to blame myself or others. Whatever was, being past, was right to be, and could not have been otherwise.

'My mother went home next morning, alone. We were to follow in a few days, though she would not allow us to fix any time. Her departure was so hurried that I remember nothing about it, save her answer to my father's urgent desire—almost command—that if anything was amiss she would immediately let him know.

' "Under all circumstances, wife," he reiterated, "this you promise?"

' "I promise."

'Though when she was gone he declared she need not have said it so earnestly, since we should be at home almost as soon as the slow Bath coach could take her and bring us a letter. And besides, there was nothing likely to happen. But be fidgeted a good deal, being unused to her absence in their happy wedded life. He was, like most men, glad to blame anybody but himself, and the whole day, and the next, was cross at intervals with both Edmond and me; but we bore it—and patiently.

' "It will be all right when we get him to the theatre. He has no real cause for anxiety about her. What a dear woman she is, and a precious—your mother, Dorothy!"

'I rejoiced to hear my lover speak thus, and thought there hardly ever was young gentlewoman so blessed as I.

'We went to the play. Ah, you know nothing of what a play is, now-a-days. You never saw John Kemble and Mrs Siddons. Though in dresses and shows it was far inferior to the *Hamlet* you took me to see last week, my dear—and though I perfectly well remember being on the point of laughing when in the most solemn scene, it became clearly evident that the Ghost had been drinking. Strangely enough, no after events connected therewith—nothing subsequent ever drove from my mind the vivid impression of this my first play. Strange, also, that the play should have been *Hamlet*. Do you think that Shakespeare believed in—in what people call "ghosts"?'

I could not say; but I thought Mrs MacArthur's ghost very long in coming.

'Don't, my dear—don't; do anything but laugh at it.'

She was visibly affected, and it was not without an effort that she proceeded in her story.

'I wish you to understand exactly my position that night—a young girl, her head full of the enchantment of the stage—her heart of something not less engrossing. Mr Everest had supped with us, leaving us both in the best of spirits; indeed my father had gone to bed, laughing heartily at the remembrance of the antics of Mr Grimaldi, which had almost obliterated the Queen and Hamlet from his memory, on which the ridiculous always took a far stronger hold than the awful or sublime.

'I was sitting—let me see—at the window, chatting with my maid Patty, who was brushing the powder out of my hair. The window was open half-way, and looking out on the Thames; and the summer night being very warm and starry, made it almost like sitting out of doors. There was none of the awe given by the solitude of a midnight closed room, when every sound is magnified, and every shadow seems alive.

'As I said, we had been chatting and laughing; for Patty and I were both very young, and she had a sweetheart, too. She, like every one of our household, was a warm admirer of Mr Everest. I had just been half scolding, half smiling at her praises of him, when St Paul's great clock came booming over the silent river.

' "Eleven," counted Patty. "Terrible late we be, Mistress Dorothy: not like Bath hours, I reckon."

' "Mother will have been in bed an hour ago," said I, with a little self-reproach at not having thought of her till now.

'The next minute my maid and I both started up with a simultaneous exclamation.

' "Did you hear that?"

' "Yes, a bat flying against the window."

' "But the lattices are open, Mistress Dorothy."

'So they were; and there was no bird or bat or living thing about—only the quiet summer night, the river, and the stars.

' "I be certain sure I heard it. And I think it was like—just a bit like—somebody tapping."

' "Nonsense, Patty!" But it *had* struck me thus—though I said it was a bat. It was exactly like the sound of fingers against a pane—very soft, gentle fingers, such as, in passing into her flower-garden, my mother used often to tap outside the school-room casement at home.

' "I wonder, did father hear anything. It—the bird, you know, Patty—might have flown at his window, too?"

' "Oh, Mistress Dorothy!" Patty would not be deceived. I gave her the brush to finish my hair, but her hand shook too much. I shut the window, and we both sat down facing it.

'At that minute, distinct, clear, and unmistakable, like a person giving a summons in passing by, we heard once more the tapping on the pane. But nothing was seen; not a single shadow came between us and the open air, the bright starlight.

'Startled I was, and awed, but I was not frightened. The sound gave me even an inexplicable delight. But I had hardly time to recognize my feelings, still less to analyse them, when a loud cry came from my father's room.

' "Dolly,—Dolly!"

'Now my mother and I had both one name, but he always gave her the old-fashioned pet name—I was invariably Dorothy. Still I did not pause to think, but ran to his locked door, and answered.

'It was a long time before he took any notice, though I heard him talking to himself, and moaning. He was subject to bad dreams, especially before his attacks of gout. So my first alarm lightened. I stood listening, knocking at intervals, until at last he replied.

' "What do'ee want, child?"

' "Is anything the matter, father?"

' "Nothing. Go to thy bed, Dorothy."

' "Did you not call? Do you want any one?"

' "Not thee. O Dolly, my poor Dolly,'—and he seemed to be almost sobbing, "Why did I let thee leave me!"

' "Father, you are not going to be ill? It is not the gout, is it?" (for that was the time when he wanted my mother most, and indeed, when he was wholly unmanageable by any one but her.)

' "Go away. Get to thy bed, girl; I don't want 'ee."

'I thought he was angry with me for having been in some sort the cause of our delay, and retired very miserable. Patty and I sat up a good while longer, discussing the dreary prospect of my father's having a fit of the gout here in London lodgings, with only us to nurse him, and my mother away. Our alarm was so great that we quite forgot the curious circumstance which had first attracted us, till Patty spoke up, from her bed on the floor.

' "I hope master beant going to be very ill, and that—you know—came for a warning. Do 'ee think it *was* a bird, Mistress Dorothy?"

' "Very likely. Now, Patty, let us go to sleep."

'But I did not, for all night I heard my father groaning at intervals. I was certain it was the gout, and wished from the bottom of my heart that we had gone home with mother.

'What was my surprise when, quite early, I heard him rise and go down, just as if nothing was ailing him! I found him sitting at the breakfast-table in his travelling coat, looking very haggard and miserable, but evidently bent on a journey.

'"Father, you are not going to Bath?"

'"Yes, I be."

'"Not till the evening coach starts," I cried, alarmed. "We can't, you know?"

'"I'll take a post-chaise, then. We must be off in an hour."

'An hour! The cruel pain of parting—(my dear, I believe I used to feel things keenly when I was young)—shot through me—through and through. A single hour, and I should have said goodbye to Edmond—one of those heart-breaking farewells when we seem to leave half of our poor young life behind us, forgetting that the only real parting is when there is no love left to part from. A few years, and I wondered how I could have crept away and wept in such intolerable agony at the mere bidding goodbye to Edmond—Edmond, who loved me.

'Every minute seemed a day till he came in, as usual, to breakfast. My red eyes and my father's corded trunk explained all.

'"Doctor Thwaite, you are not going?"

'"Yes, I be," repeated my father. He sat moodily leaning on the table—would not taste his breakfast.

'"Not till the night coach, surely? I was to take you and Mistress Dorothy to see Mr Benjamin West, the king's painter."

'"Let kings and painters alone lad; I be going home to my Dolly."

'Mr Everest used many arguments, gay and grave, upon which I hung with earnest conviction and hope. He made things so clear always; he was a man of much brighter parts than my father, and had great influence over him.

'"Dorothy," he whispered, "help me to persuade the Doctor. It is so little time I beg for, only a few hours; and before so long a parting." Ay, longer than he thought, or I.

'"Children," cried my father at last, "you are a couple of fools. Wait till you have been married twenty years. I must go to my Dolly. I know there is something amiss at home."

'I should have felt alarmed, but I saw Mr Everest smile; and besides, I was yet glowing under his fond look, as my father spoke of our being "married twenty years".

'"Father, you have surely no reason for thinking this? If you have, tell us."

'My father just lifted his head, and looked me woefully in the face.

'"Dorothy, last night, as sure as I see you now, I saw your mother."

'"Is that all?" cried Mr Everest, laughing; "why, my good sir, of course you did; you were dreaming."

'"I had not gone to sleep."

'"How did you see her?"

'"Coming into the room just as she used to do in the bedroom at home, with the candle in her hand and the baby asleep on her arm."

'"Did she speak?" asked Mr Everest, with another and rather satirical smile; "remember, you saw *Hamlet* last night. Indeed, sir—indeed, Dorothy—it was a mere dream. I do not believe in ghosts; it would be an insult to common sense, to human wisdom—nay, even to Divinity itself."

'Edmond spoke so earnestly, so justly, so affectionately, that perforce I agreed; and even my father became to feel rather ashamed of his own weakness. He, a physician, the head of a family, to yield to a mere superstitious fancy, springing probably from a hot supper and an over-excited brain! To the same cause Mr Everest attributed the other incident, which somewhat hesitatingly I told him.

'"Dear, it was a bird; nothing but a bird. One flew in at my window last spring; it had hurt itself, and I kept it, and nursed it, and petted it. It was such a pretty, gentle little thing, it put me in mind of Dorothy."

'"Did it?" said I.

'"And at last it got well and flew away."

'"Ah! that was not like Dorothy."

'Thus, my father being persuaded, it was not hard to persuade me. We settled to remain till evening. Edmond and I, with my maid Patty, went about together, chiefly in Mr West's Gallery, and in the quiet shade of our favourite Temple Gardens. And if for those four stolen hours, and the sweetness in them, I afterwards suffered untold remorse and bitterness, I have entirely forgiven myself, as I know my dear mother would have forgiven me, long ago.'

Mrs MacArthur stopped, wiped her eyes, and then continued—speaking more in the matter-of-fact way that old people speak than she had been lately doing.

'Well, my dear, where was I?'

'In the Temple Gardens.'

'Yes, yes. Well, we came home to dinner. My father always enjoyed his dinner, and his nap afterwards; he had nearly recovered himself now: only looked tired from loss of rest. Edmond and I sat in the window, watching the barges and wherries down the Thames; there were no steam-boats then, you know.

'Some one knocked at the door with a message for my father, but he slept so heavily he did not hear. Mr Everest went to see what it was; I stood at the window. I remember mechanically watching the red sail of a Margate hoy that was going down the river, and thinking with a sharp pang how dark the room seemed, in a moment, with Edmond not there.

'Re-entering, after a somewhat long absence, he never looked at me, but went straight to my father.

'"Sir, it is almost time for you to start" (oh! Edmond). "There is a coach at the door; and, pardon me, but I think you should travel quickly."

'My father sprang to his feet.

'"Dear sir, indeed there is no need for anxiety now; but I have received news. You have another little daughter, sir, and—"

'"Dolly, my Dolly!" Without another word my father rushed away without his hat, leaped into the post-chaise that was waiting, and drove off.

'"Edmond!" I gasped.

'"My poor litle girl—my own Dorothy!"

'By the tenderness of his embrace, not lover-like, but brother-like—by his tears, for I could feel them on my neck—I knew, as well as if he had told me, that I should never see my dear mother any more.'

'She had died in childbirth,' continued the old lady after a long pause—'died at night, at the very hour and minute when I had heard the tapping on the window-pane, and my father had thought he saw her coming into his room with a baby on her arm.'

'Was the baby dead, too?'

'They thought so then, but it afterwards revived.'

'What a strange story!'

'I do not ask you to believe in it. How and why and what it was I cannot tell; I only know that it assuredly was so.'

'And Mr Everest?' I enquired, after some hesitation.

The old lady shook her head. 'Ah, my dear, you will soon learn how very, very seldom one marries one's first love. After that day, I did not see Mr Everest for twenty years.'

'How wrong—how——'

'Don't blame him; it was not his fault. You see, after that time my father took a prejudice against him—not unnatural, perhaps; and she was not there to make things straight. Besides, my own conscience was very sore, and there were the six children at home, and the little baby had no mother: so at last I made up my mind. I should have loved him just the same if we had waited twenty years: but he could not see things so. Don't blame him—my dear—don't blame him. It was as well, perhaps, as things turned out.'

'Did he marry?'

'Yes, after a few years; and loved his wife dearly. When I was about one-and-thirty, I married Mr MacArthur. So neither of us was unhappy, you see—at least, not more so than most people; and we became sincere friends afterwards. Mr and Mrs Everest come to see me, almost every Sunday. Why you foolish child, you are not crying?'

Ay, I was—but scarcely at the ghost story.

To be Taken with a Grain of Salt

CHARLES DICKENS

I have always noticed a prevalent want of courage, even among persons of superior intelligence and culture, as to imparting their own psychological experiences when those have been of a strange sort. Almost all men are afraid that what they could relate in such wise would find no parallel or response in a listener's internal life, and might be suspected or laughed at. A truthful traveller who should have seen some extraordinary creature in the likeness of a sea-serpent, would have no fear of mentioning it; but the same traveller having had some singular presentiment, impulse, vagary of thought, vision (so-called), dream, or other remarkable mental impression, would hesitate considerably before he would own to it. To this reticence I attribute much of the obscurity in which such subjects are involved. We do not habitually communicate our experiences of these subjective things, as we do our experiences of objective creation. The consequence is, that the general stock of experience in this regard appears exceptional, and really is so, in respect of being miserably imperfect.

In what I am going to relate I have no intention of setting up, opposing, or supporting, any theory whatever. I know the history of the Bookseller of Berlin, I have studied the case of the wife of a late Astronomer Royal as related by Sir David Brewster, and I have followed the minutest details of a much more remarkable case of Spectral Illusion occurring within my private circle of friends. It may be necessary to state as to this last that the sufferer (a lady) was in no degree, however distant, related to me. A mistaken assumption on that head, might suggest an explanation of a part of my own case—but only a part—which would be wholly without foundation. It cannot be referred to my inheritance of any developed peculiarity, nor had I ever before any at all similar experience, nor have I ever had any at all similar experience since.

It does not signify how many years ago, or how few, a certain Murder was committed in England, which attracted great attention. We hear more than enough of Murderers as they rise in succession to their atrocious eminence, and I would bury the memory of this particular brute, if I could, as his body was buried, in Newgate Jail.

I purposely abstain from giving any direct clue to the criminal's individuality.

When the murder was first discovered, no suspicion fell—or I ought rather to say, for I cannot be too precise in my facts, it was nowhere publicly hinted that any suspicion fell—on the man who was afterwards brought to trial. As no reference was at that time made to him in the newspapers, it is obviously impossible that any description of him can at that time have been given in the newspapers. It is essential that this fact be remembered.

Unfolding at breakfast my morning paper, containing the account of that first discovery, I found it to be deeply interesting, and I read it with close attention. I read it twice, if not three times. The discovery had been made in a bedroom, and, when I laid down the paper, I was aware of a flash—rush—flow—I do not know what to call it—no word I can find is satisfactorily descriptive—in which I seemed to see that bedroom passing through my room, like a picture impossibly painted on a running river. Though almost instantaneous in its passing, it was perfectly clear; so clear that I distinctly, and with a sense of relief, observed the absence of the dead body from the bed.

It was in no romantic place that I had this curious sensation, but in chambers in Piccadilly, very near to the corner of Saint James's Street. It was entirely new to me. I was in my easy-chair at the moment, and the sensation was accompanied with a peculiar shiver which started the chair from its position. (But it is to be noted that the chair ran easily on castors.) I went to one of the windows (there are two in the room, and the room is on the second floor) to refresh my eyes with the moving objects down in Piccadilly. It was a bright autumn morning, and the street was sparkling and cheerful. The wind was high. As I looked out, it brought down from the Park a quantity of fallen leaves, which a gust took, and whirled into a spiral pillar. As the pillar fell and the leaves dispersed, I saw two men on the opposite side of the way, going from West to East. They were one behind the other. The foremost man often looked back over his shoulder. The second man followed him, at a distance of some thirty paces, with his right hand menacingly raised. First, the singularity and steadiness of this threatening gesture in so public a thoroughfare, attracted my attention; and next, the more remarkable circumstance that nobody heeded it. Both men threaded their way among the other passengers, with a smoothness hardly consistent even with the action of walking on a pavement, and no single creature that I could see, gave them place, touched them, or looked after them. In passing before my windows, they both stared up at me. I saw their two faces very distinctly, and I knew that I could recognize them anywhere. Not that I had consciously noticed anything very remarkable in either face, except that the man who went first had an unusually lowering appearance, and that the face of the man who followed him was of the colour of impure wax.

I am a bachelor, and my valet and his wife constitute my whole establishment. My occupation is in a certain Branch Bank, and I wish that my duties as head of a Department were as light as they are popularly supposed to be. They kept me in town that autumn, when I stood in need of a change. I was not ill, but I was not well. My reader is to make the most that can be reasonably made of my feeling jaded, having a depressing sense upon me of a monotonous life, and being 'slightly dyspeptic'. I am assured by my renowned doctor that my real state of health at that time justifies no stronger description, and I quote his own from his written answer to my request for it.

As the circumstances of the Murder, gradually unravelling, took stronger and stronger possession of the public mind, I kept them away from mine, by knowing as little about them as was possible in the midst of the universal excitement. But I knew that a verdict of Wilful Murder had been found against the suspected Murderer, and that he had been committed to Newgate for trial. I also knew that his trial had been postponed over one Sessions of the Central Criminal Court, on the ground of general prejudice and want of time for the preparation of the defence. I may further have known, but I believe I did not, when, or about when, the Sessions to which his trial stood postponed would come on.

My sitting-room, bedroom, and dressing-room, are all on one floor. With the last, there is no communication but through the bedroom. True, there is a door in it, once communicating with the staircase; but a part of the fitting of my bath has been—and had then been for some years—fixed across it. At the same period, and as a part of the same arrangement, the door had been nailed up and canvassed over.

I was standing in my bedroom late one night, giving some directions to my servant before he went to bed. My face was towards the only available door of communication with the dressing-room, and it was closed. My servant's back was towards that door. While I was speaking to him I saw it open, and a man look in, who very earnestly and mysteriously beckoned to me. That man was the man who had gone second of the two along Piccadilly, and whose face was of the colour of impure wax.

The figure, having beckoned, drew back and closed the door. With no longer pause than was made by my crossing the bedroom, I opened the dressing-room door, and looked in. I had a lighted candle already in my hand. I felt no inward expectation of seeing the figure in the dressing-room, and I did not see it there.

Conscious that my servant stood amazed, I turned round to him, and said: 'Derrick, could you believe that in my cool senses I fancied I saw a——' As I there laid my hand upon his breast, with a sudden start he trembled violently, and said, 'O Lord yes sir! A dead man beckoning!'

Now, I do not believe that this John Derrick, my trusty and attached servant for more than twenty years, had any impression whatever of having seen any such figure, until I touched him. The change in him was so startling when I touched him, that I fully believe he derived his impression in some occult manner from me at that instant.

I bade John Derrick bring some brandy, and I gave him a dram, and was glad to take one myself. Of what had proceeded that night's phenomenon, I told him not a single word. Reflecting on it, I was absolutely certain that I had never seen that face before, except on the one occasion in Piccadilly. Comparing its expression when beckoning at the door, with its expression when it had stared up at me as I stood at my window, I came to the conclusion that on the first occasion it had sought to fasten itself upon my memory, and that on the second occasion it had made sure of being immediately remembered.

I was not very comfortable that night, though I felt a certainty, difficult to explain, that the figure would not return. At daylight, I fell into a heavy sleep, from which I was awakened by John Derrick's coming to my bedside with a paper in his hand.

This paper, it appeared, had been the subject of an altercation at the door between its bearer and my servant. It was a summons to me to serve upon a Jury at the forthcoming Sessions of the Central Criminal Court at the Old Bailey. I had never before been summoned on such a Jury, as John Derrick well knew. He believed—I am not certain at this hour whether with reason or otherwise—that that class of Jurors were customarily chosen on a lower qualification than mine, and he had at first refused to accept the summons. The man who served it had taken the matter very coolly. He had said that my attendance or non-attendance was nothing to him; there the summons was; and I should deal with it at my own peril, and not at his.

For a day or two I was undecided whether to respond to this call, or take no notice of it. I was not conscious of the slightest mysterious bias, influence, or attraction, one way or other. Of that I am as strictly sure as of every other statement that I make here. Ultimately I decided, as a break in the monotony of my life, that I would go.

The appointed morning was a raw morning in the month of November. There was a dense brown fog in Piccadilly, and it became positively black and in the last degree oppressive East of Temple Bar. I found the passages and staircases of the Court House flaringly lighted with gas, and the Court itself similarly illuminated. I *think* that until I was conducted by officers into the Old Court and saw its crowded state, I did not know that the Murderer was to be tried that day. I *think* that until I was so helped into the Old Court with considerable difficulty, I did not know into which of the two Courts sitting, my summons would take me. But this must not be received as a positive assertion, for I am not completely satisfied in my mind on either point.

I took my seat in the place appropriated to Jurors in waiting, and I looked about the Court as well as I could through the cloud of fog and breath that was heavy in it. I noticed the black vapour hanging like a murky curtain outside the great windows, and I noticed the stifled sound of wheels on the straw or tan that was littered in the street; also, the hum of the people gathered there, which a shrill whistle, or a louder song or hail than the rest, occasionally pierced. Soon afterwards the Judges, two in number, entered and took their seats. The buzz in the Court was awfully hushed. The direction was given to put the Murderer to the bar. He appeared there. And in that same instant I recognized in him, the first of the two men who had gone down Piccadilly.

If my name had been called then, I doubt if I could have answered to it audibly. But it was called about sixth or eighth in the panel, and I was by that time able to say 'Here!' Now, observe. As I stepped into the box, the prisoner, who had been looking on attentively but with no sign of concern, became violently agitated, and beckoned to his attorney. The prisoner's wish to challenge me was so manifest, that it occasioned a pause, during which the attorney, with his hand upon the dock, whispered with his client, and shook his head. I afterwards had it from that gentleman, that the prisoner's first affrighted words to him were, '*At all hazards challenge that man!*' But, that as he would give no reason for it, and admitted that he had not even known my name until he heard it called and I appeared, it was not done.

Both on the ground already explained, that I wish to avoid reviving the unwholesome memory of that Murderer, and also because a detailed account of his long trial is by no means indispensable to my narrative, I shall confine myself closely to such incidents in the ten days and nights during which we, the Jury, were kept together, as directly bear on my own curious personal experience. It is in that, and not in the Murderer, that I seek to interest my reader. It is to that, and not to a page of the Newgate Calendar, that I beg attention.

I was chosen Foreman of the Jury. On the second morning of the trial, after evidence had been taken for two hours (I heard the church clocks strike), happening to cast my eyes over my brother-jurymen, I found an inexplicable difficulty in counting them. I counted them several times, yet always with the same difficulty. In short, I made them one too many.

I touched the brother-juryman whose place was next to me, and I whispered to him, 'Oblige me by counting us.' He looked surprised by the request, but turned his head and counted. 'Why,' says he, suddenly, 'We are Thirt——; but no, it's not possible. No. We are twelve.'

According to my counting that day, we were always right in detail, but in the gross we were always one too many. There was no appearance—no figure—to account for it; but I had now an inward foreshadowing of the figure that was surely coming.

The Jury were housed at the London Tavern. We all slept in one large room on separate tables, and we were constantly in the charge and under the eye of the officer sworn to hold us in safe-keeping. I see no reason for suppressing the real name of that officer. He was intelligent, highly polite, and obliging, and (I was glad to hear) much respected in the City. He had an agreeable presence, good eyes, enviable black whiskers, and a fine sonorous voice. His name was Mr Harker.

When we turned into our twelve beds at night, Mr Harker's bed was drawn across the door. On the night of the second day, not being disposed to lie down, and seeing Mr Harker sitting on his bed, I went and sat beside him, and offered him a pinch of snuff. As Mr Harker's hand touched mine in taking it from my box, a peculiar shiver crossed him, and he said: 'Who is this!'

Following Mr Harker's eyes and looking along the room, I saw again the figure I expected—the second of the two men who had gone down Piccadilly. I rose, and advanced a few steps; then stopped, and looked round at Mr Harker. He was quite unconcerned, laughed, and said in a pleasant way, 'I thought for a moment we had a thirteenth juryman, without a bed. But I see it is the moonlight.'

Making no revelation to Mr Harker, but inviting him to take a walk with me to the end of the room, I watched what the figure did. It stood for a few moments by the bedside of each of my eleven brother-jurymen, close to the pillow. It always went to the right-hand side of the bed, and always passed out crossing the foot of the next bed. It seemed from the action of the head, merely to look down pensively at each recumbent figure. It took no notice of me, or of my bed, which was that nearest to Mr Harker's. It seemed to go out where the moonlight came in, through a high window, as by an aërial flight of stairs.

Next morning at breakfast, it appeared that everybody present had dreamed of the murdered man last night, except myself and Mr Harker.

I now felt as convinced that the second man who had gone down Piccadilly was the murdered man (so to speak), as if it had been borne into my comprehension by his immediate testimony. But even this took place, and in a manner for which I was not at all prepared.

On the fifth day of the trial, when the case for the prosecution was drawing to a close, a miniature of the murdered man, missing from his bedroom upon the discovery of the deed, and afterwards found in a hiding-place where the Murderer had been seen digging, was put in evidence. Having been identified by the witness under examination, it was handed up to the Bench, and thence handed down to be inspected by the Jury. As an officer in a black gown was making his way with it across to me, the figure of the second man who had gone down Piccadilly, impetuously started from the crowd, caught the miniature from the officer, and gave it to me with its own hands, at the same time saying in a low and hollow tone—before I saw the miniature, which was in a locket—'*I was younger then, and my face was not then drained of blood.*' It also came between me and the brother-juryman to whom I would have given the miniature, and between him and the brother-juryman to whom he would have given it, and so passed it on through the whole of our number, and back into my possession. Not one of them, however, detected this.

At table, and generally when we were shut up together in Mr Harker's custody, we had from the first naturally discussed the day's proceedings a good deal. On that fifth day, the case for the prosecution being closed, and we having that side of the question in a completed shape before us, our discussion was more animated and serious. Among our number was a vestryman—the densest idiot I have ever seen at large—who met the plainest evidence with the most preposterous objections, and who was sided with by two flabby parochial parasites; all the three empanelled from a district so delivered over to Fever that they ought to have been upon their own trial, for five hundred Murders. When these mischievous blockheads were at their loudest, which was towards midnight while some of us were already preparing for bed, I again saw the murdered man. He stood grimly behind them, beckoning to me. On my going towards them and striking into the conversation, he immediately retired. This was the beginning of a separate series of appearances, confined to that long room in which *we* were confined. Whenever a knot of my brother jurymen laid their heads together, I saw the head of the murdered man among theirs. Whenever their comparison of notes was going against him, he would solemnly and irresistibly beckon to me.

It will be borne in mind that down to the production of the miniature on the fifth day of the trial, I had never seen the Appearance in Court. Three changes occurred, now that we entered on the case for the defence. Two of them I will mention together, first. The figure was now in Court continually, and it never there addressed itself to me, but always to the person who was speaking at the time. For instance. The throat of the murdered man had been cut straight across. In the opening speech for the defence, it was suggested that the deceased might have cut his own throat. At that very moment, the figure with its throat in the dreadful condition referred to (this it had concealed before) stood at the speaker's elbow, motioning across and across its windpipe, now with the right hand, now with the left, vigorously suggesting to the speaker himself, the impossibility of such a wound having been self-inflicted by either hand. For another instance. A witness to character, a woman, deposed to the prisoner's being the most amiable of mankind. The figure at that instant stood on the floor before her, looking her full in the face, and pointing out the prisoner's evil countenance with an extended arm and an outstretched finger.

The third change now to be added, impressed me strongly, as the most marked and striking of all. I do not theorize upon it; I accurately state it, and there leave it. Although the Appearance was not itself perceived by those whom it addressed, its coming close to such persons was invariably attended by some trepidation or disturbance on their part. It seemed to me as if it were prevented by laws to which I was not amenable, from fully revealing itself to others, and yet as if it could, invisibly, dumbly and darkly, overshadow their minds. When the leading counsel for the defence suggested that hypothesis of suicide and the figure stood at the learned gentleman's elbow, frightfully sawing at its severed throat, it is undeniable that the counsel faltered in his speech, lost for a few seconds the thread of his ingenious discourse, wiped his forehead with his handkerchief, and turned extremely pale. When the witness to character was confronted by the Appearance, her eyes most certainly did follow the direction of its pointed finger, and rest in great hesitation and trouble upon the prisoner's face. Two additional illustrations will suffice. On the eighth day of the trial, after the pause which was every day made early in the afternoon for a few minutes' rest and refreshment, I came back into Court with the rest of the Jury, some little time before the return of the

Judges. Standing up in the box and looking about me, I thought the figure was not there, until, chancing to raise my eyes to the gallery, I saw it bending forward and leaning over a very decent woman, as if to assure itself whether the Judges had resumed their seats or not. Immediately afterwards, that woman screamed, fainted, and was carried out. So with the venerable, sagacious, and patient Judge who conducted the trial. When the case was over, and he settled himself and his papers to sum up, the murdered man entering by the Judges' door, advanced to his Lordship's desk, and looked eagerly over his shoulder at the pages of his notes which he was turning. A change came over his Lordship's face; his hand stopped; the peculiar shiver that I knew so well, passed over him; he faltered, 'Excuse me gentlemen, for a few moments. I am somewhat oppressed by the vitiated air;' and did not recover until he had drunk a glass of water.

Through all the monotony of six of those interminable ten days—the same Judges and others on the bench, the same Murderer in the dock, the same lawyers at the table, the same tones of question and answer rising to the roof of the court, the same scratching of the Judge's pen, the same ushers going in and out, the same lights kindled at the same hour when there had been any natural light of day, the same foggy curtain outside the great windows when it was foggy, the same rain pattering and dripping when it was rainy, the same footmarks of turnkeys and prisoner day after day on the same sawdust, the same keys locking and unlocking the same heavy doors—through all the wearisome monotony which made me feel as if I had been Foreman of the Jury for a vast period of time, and Piccadilly had flourished coevally with Babylon, the murdered man never lost one trace of his distinctness in my eyes, nor was he at any moment less distinct than anybody else. I must not omit, as a matter of fact, that I never once saw the Appearance which I call by the name of the murdered man, look at the Murderer. Again and again I wondered, 'Why does he not?' But he never did.

Nor did he look at me, after the production of the miniature, until the last closing minutes of the trial arrived. We retired to consider, at seven minutes before ten at night. The idiotic vestry-man and his two parochial parasites gave us so much trouble, that we twice returned into Court, to beg to have certain extracts from the Judge's notes re-read. Nine of us had not the smallest doubt about those passages, neither, I believe, had any one in Court; the dunder-headed triumvirate however, having no idea but obstruction, disputed them for that very reason. At length we prevailed, and finally the Jury returned into Court at ten minutes past twelve.

The murdered man at that time stood directly opposite the Jury-box, on the other side of the Court. As I took my place, his eyes rested on me, with great attention; he seemed satisfied, and slowly shook a great grey veil, which he carried on his arm for the first time, over his head and whole form. As I gave in our verdict 'Guilty', the veil collapsed, all was gone, and his place was empty.

The Murderer being asked by the Judge, according to usage, whether he had anything to say before sentence of Death should be passed upon him, indistinctly muttered something which was described in the leading newspapers of the following day as 'a few rambling, incoherent, and half-audible words, in which he was understood to complain that he had not had a fair trial because the Foreman of the Jury was prepossessed against him'. The remarkable declaration that he really made, was this: '*My Lord, I knew I was a doomed man when the Foreman of my Jury came into the box. My Lord, I knew he would never let me off, because, before I was taken, he somehow got to my bedside in the night, woke me, and put a rope round my neck.*'

The Botathen Ghost

R. S. HAWKER

There was something very painful and peculiar in the position of the clergy in the west of England throughout the seventeenth century. The Church of those days was in a transitory state, and her ministers, like her formularies, embodied a strange mixture of the old belief with the new interpretation. Their wide severance also from the great metropolis of life and manners, the city of London (which in those times was civilized England, much as the Paris of our own day is France), divested the Cornish clergy in particular of all personal access to the master-minds of their age and body. Then, too, the barrier interposed by the rude rough roads of their country, and by their abode in wilds that were almost inaccessible, rendered the existence of a bishop rather a doctrine suggested to their belief than a fact revealed to the actual vision of each in his generation. Hence it came to pass that the Cornish clergyman, insulated within his own limited sphere, often without even the presence of a country squire (and unchecked by the influence of the Fourth Estate—for until the beginning of this nineteenth century, *Flindell's Weekly Miscellany*, distributed from house to house from the pannier of a mule, was the only light of the West), became developed about middle life into an original mind and man, sole and absolute within his parish boundary, eccentric when compared with his brethren in civilized regions, and yet, in German phrase, 'a whole and seldom man' in his dominion of souls. He was 'the parson', in canonical phrase—that is to say, The Person, the somebody of consequence among his own people. These men were not, however, smoothed down into a monotonous aspect of life and manners by this remote and secluded existence. They imbibed, each in his own peculiar circle, the hue of surrounding objects, and were tinged into a distinctive colouring and character by many a contrast of scenery and people. There was the 'light of other days', the curate by the sea-shore, who professed to check the turbulence of the 'smugglers' landing' by his presence on the sands, and who 'held the lantern' for the guidance of his flock when the nights were dark, as the only proper ecclesiastical part he could take in the proceedings. He was soothed and silenced by the gift of a keg of hollands or a chest of tea. There was the merry minister of the mines, whose cure was honeycombed by the underground men. He must needs have been artist and poet in his way, for he had to enliven his people three or four times a-year, by mastering the arrangements of a 'guary', or religious mystery, which was duly performed in the topmost hollow of a green barrow or hill, of which many survive, scooped out into vast amphitheatres and surrounded by benches of turf which held two thousand spectators. Such were the historic plays, 'The Creation' and 'Noe's Flood', which still exist in the original Celtic as well as the English text, and suggest what critics and antiquaries these Cornish curates, masters of such revels, must have been—for the native language of Cornwall did not lapse into silence until the end of the seventeenth century. Then, moreover, here and there would be one parson more learned than his kind in the mysteries of a deep and thrilling lore of peculiar fascination. He was a man so highly honoured at college for natural gifts and knowledge of learned books which nobody else could read, that when he 'took his second orders' the bishop gave him a mantle of scarlet silk to wear upon his shoulders in church, and his lordship had put such power into it that, when the parson had it rightly on, he could 'govern any ghost or evil spirit', and even 'stop an earthquake'.

Such a powerful minister, in combat with supernatural visitations, was one Parson Rudall, of Launceston, whose existence and exploits we gather from the local tradition of his time, from surviving letters and other memoranda, and indeed from his own 'diurnal' which fell by chance into the hands of the present writer. Indeed the legend of Parson Rudall and the Botathen Ghost will be recognized by many Cornish people as a local remembrance of their boyhood.

It appears, then, from the diary of this learned master of the grammar school—for such was his office as well as perpetual curate of the parish—'that a pestilential disease did break forth in our town in the beginning of the year AD 1665; yea, and it likewise invaded my school, insomuch that therewithal certain of the chief scholars sickened and died'. 'Among others who yielded to the malign influence was Master John Eliot, the eldest son and the worshipful heir of Edward Eliot, Esquire, of Trebursey, a stripling of sixteen years of age, but of uncommon parts and hopeful ingenuity. At his own especial motion and earnest desire I did consent to preach his funeral sermon.' It should be remembered here that, howsoever strange and singular it may sound to us that a mere lad should formally solicit such a performance at the hands of his master, it was in consonance with the habitual usage of those times. The old services for the dead had been abolished by law, and in the stead of sacrament and ceremony, month's mind and year's mind, the sole substitute which survived was the general desire 'to partake', as they called it, of a posthumous discourse, replete with lofty eulogy and flattering remembrance of the living and the dead. The diary proceeds:

'I fulfilled my undertaking, and preached over the coffin in the presence of a full assemblage of mourners and lachrymose friends. An ancient gentleman, who was then and there in the church, a Mr Bligh, of Botathen, was much affected with my discourse, and he was heard to repeat to himself certain parentheses therefrom, especially a phrase from Maro Virgilius, which I had applied to the deceased youth, "*Et puer ipse fuit cantari dignus.*"

'The cause wherefore this old gentleman was moved by my applications was this: He had a first-born and only son—a child who, but a very few months before, had been not unworthy the character I drew of young Master Eliot, but who, by some strange accident, had of late quite fallen away from his parent's hopes, and become moody, and sullen, and distraught. When the funeral obsequies were over, I had no sooner come out of church than I was accosted by this aged parent, and he besought me incontinently, with a singular energy, that I would resort with him forthwith to his abode at Botathen that very night; nor could I have delivered myself from his importunity, had not Mr Eliot urged his claim to enjoy my company at his own house. Hereupon I got loose, but not until I had pledged a fast assurance that I would pay him, faithfully, an early visit the next day.'

'The Place', as it was called, of Botathen, where old Mr Bligh resided, was a low-roofed gabled manor-house of the fifteenth century, walled and mullioned, and with clustered chimneys of dark-grey stone from the neighbouring quarries of Ventor-gan. The mansion was flanked by a pleasance or enclosure in one space, of garden and lawn, and it was surrounded by a solemn grove of stag-horned trees. It had the sombre aspect of age and of solitude, and looked the very scene of strange and supernatural events. A legend might well belong to every gloomy glade around, and there must surely be a haunted room somewhere within its walls. Hither, according to his appointment, on the morrow, Parson Rudall betook himself. Another clergyman, as it appeared, had been invited to meet him, who, very soon after his arrival, proposed a walk together in the pleasance, on the pretext of showing him, as a stranger, the walks and trees, until the dinner-bell should strike. There, with much prolixity, and with many a solemn pause, his brother minister proceeded to 'unfold the mystery'.

A singular infelicity, he declared, had befallen young Master Bligh, once the hopeful heir of his parents and of the lands of Botathen.

Whereas he had been from childhood a blithe and merry boy, 'the gladness', like Isaac of old, of his father's age, he had suddenly, and of late, become morose and silent—nay, even austere and stern—dwelling apart, always solemn, often in tears. The lad had at first repulsed all questions as to the origin of this great change, but of late he had yielded to the importune researches of his parents, and had disclosed the secret cause. It appeared that he resorted every day, by a pathway across the fields, to this very clergyman's house, who had charge of his education, and grounded him in the studies suitable to his age. In the course of his daily walk he had to pass a certain heath or down where the road wound along through tall blocks of granite with open spaces of grassy sward between. There in a certain spot, and always in one and the same place, the lad declared that he encountered, every day, a woman with a pale and troubled face, clothed in a long loose garment of frieze, with one hand always stretched forth, and the other pressed against her side. Her name, he said, was Dorothy Dinglet, for he had known her well from his childhood, and she often used to come to his parents' house; but that which troubled him was, that she had now been dead three years, and he himself had been with the neighbours at her burial; so that, as the youth alleged, with great simplicity, since he had seen her body laid in the grave, this that he saw every day must needs be her soul or ghost. 'Questioned again and again,' said the clergyman, 'he never contradicts himself; but he relates the same and the simple tale as a thing that cannot be gain-said. Indeed, the lad's observance is keen and calm for a boy of his age. The hair of the appearance, sayeth he, is not like anything alive, but it is so soft and light that it seemeth to melt away while you look; but her eyes are set, and never blink—no, not when the sun shineth full upon her face. She maketh no steps, but seemeth to swim along the top of the grass; and her hand, which is stretched out alway, seemeth to point at something far away, out of sight. It is her continual coming; for she never faileth to meet him, and to pass on, that hath quenched his spirits; and although he never seeth her by night, yet cannot he get his natural rest.'

'Thus far the clergyman; whereupon the dinner-clock did sound, and we went into the house. After dinner, when young Master Bligh had withdrawn with his tutor, under excuse of their books, the parents did forthwith beset me as to my thoughts about their son. Said I, warily, "The case is strange but by no means impossible. It is one that I will study, and fear not to handle, if the lad will be free with me, and fulfil all that I desire." The mother was overjoyed, but I perceived that old Mr Bligh turned pale, and was downcast with some thought which, however, he did not express. Then they bade that Master Bligh should be called to meet me in the pleasance forthwith. The boy came, and he rehearsed to me his tale with an open countenance, and, withal, a pretty modesty of speech. Verily he seemed *ingenui vultus puer ingenuique pudoris*. Then I signified to him my purpose. "Tomorrow," said I, "we will go together to the place; and if, as I doubt not, the woman shall appear, it will be for me to proceed according to knowledge, and by rules laid down in my books."'

The unaltered scenery of the legend still survives, and, like the field of the forty footsteps in another history, the place is still visited by those who take interest in the supernatural tales of old. The pathway leads along a moorland waste, where large masses of rock stand up here and there from the grassy turf, and clumps of heath and gorse weave their tapestry of golden and purple garniture on every side. Amidst all these, and winding along between the rocks, is a natural footway worn by the scant, rare tread of the village traveller. Just midway, a somewhat larger stretch than usual of green sod expands, which is skirted by the path, and which is still identified as the legendary haunt of the phantom, by the name of Parson Rudall's Ghost.

But we must draw the record of the first interview between the minister and Dorothy from his own words. 'We met', thus he writes, 'in the pleasance very early, and before any others in the house were awake; and together the lad and myself proceeded towards the field. The youth was quite composed, and carried his Bible under his arm, from whence he read to me verses, which he said he had lately picked out, to have always in his mind. These were Job vii. 14, "Thou scarest me with dreams, and terrifiest me through visions", and Deuteronomy xxviii. 67, "In the morning thou shalt say, Would to God it were evening, and in the evening thou shalt say, Would to God it were morning; for the fear of thine heart wherewith thou shalt fear, and for the sight of thine eyes which thou shalt see."

'I was much pleased with the lad's ingenuity in these pious applications, but for mine own part I was somewhat anxious and out of cheer. For aught I knew this might be a *dæmonium meridianum*, the most stubborn spirit to govern and guide that any man can meet, and the most perilous withal. We had hardly reached the accustomed spot, when we both saw her at once gliding towards us; punctually as the ancient writers describe the motion of their "lemures, which swoon along the ground, neither marking the sand nor bending the herbage". The aspect of the woman was exactly that which had been related by the lad. There was the pale and stony face, the strange and misty hair, the eyes firm and fixed, that gazed, yet not on us, but on something that they saw far, far away; one hand and arm stretched out, and the other grasping the girdle of her waist. She floated along the field like a sail upon a stream, and glided past the spot where we stood, pausingly. But so deep was the awe that overcame me, as I stood there in the light of day, face to face with a human soul separate from her bones and flesh, that my heart and purpose both failed me. I had resolved to speak to the spectre in the appointed form of words, but I did not. I stood like one amazed and speechless, until she had passed clean out of sight. One thing remarkable came to pass. A spaniel dog, the favourite of young Master Bligh, had followed us, and lo! when the woman drew nigh, the poor creature began to yell and bark piteously, and ran backward and away, like a thing dismayed and appalled. We returned to the house, and after I had said all that I could to pacify the lad, and to soothe the aged people, I took my leave for that time, with a promise that when I had fulfilled certain business elsewhere, which I then alleged, I would return and take orders to assuage these disturbances and their cause.

'*January* 7, 1665. At my own house, I find, by my books, what is expedient to be done; and then Apage, Sathanas!

'*January* 9, 1665. This day I took leave of my wife and family, under pretext of engagements elsewhere, and made my secret journey to our diocesan city, wherein the good and venerable bishop then abode.

'*January* 10. *Deo gratias*, in safe arrival in Exeter; craved and obtained immediate audience of his lordship; pleading it was for counsel and admonition on a weighty and pressing cause; called to the presence; made obeisance; then and by command stated my case—the Botathen perplexity—which I moved with strong and earnest instances and solemn asseverations of that which I had myself seen and heard. Demanded by his lordship, what was the succour that I had come to entreat at his hands. Replied, licence for my exorcism, that so I might, ministerially, allay this spiritual visitant, and thus render to the living and the dead release from this surprise. "But," said our bishop, "on what authority do you allege that I am entrusted with faculty so to do? Our Church, as is well known, hath abjured certain branches of her ancient power, on grounds of perversion and abuse." "Nay, my lord," I humbly answered, "under favour, the seventy-second of the canons ratified and enjoined on us, the clergy, anno Domini 1604, doth expressly provide, that 'no minister, *unless he hath* the licence of his diocesan bishop, shall essay to exorcise a spirit, evil or good.' Therefore it was," I did here mildly allege, "that I did not presume to enter on such a work without lawful privilege under your lordship's hand and seal." Hereupon did our wise and learned bishop, sitting in his chair, condescend upon the theme at some length with many gracious interpretations from ancient writers and from Holy Scriptures, and I did humbly rejoin and reply, till the upshot was that he did call in his secretary and command him to draw the aforesaid faculty, forthwith and without further delay, assigning him a form, insomuch that the matter was incontinently done; and after I had disbursed into the secretary's hands certain moneys for signitary purposes, as the manner of such officers hath always been, the bishop did himself affix his signature under the *sigillum* of his see, and deliver the document into my hands. When I knelt down to receive his benediction, he softly said, "Let it be secret, Mr R. Weak brethren! weak brethren!"'

This interview with the bishop, and the success with which he vanquished his lordship's scruples, would seem to have confirmed Parson Rudall very strongly in his own esteem, and to have invested him with that courage which he evidently lacked at his first encounter with the ghost.

The entries proceed: '*January* 11, 1665. Therewithal did I hasten home and prepare my instruments, and cast my figures for the onset of the next day. Took out my ring of brass, and put it on the index-finger of my right hand, with the *scutum Davidis* traced thereon.

'*January* 12, 1665. Rode into the gateway at Botathen, armed at all points, but not with Saul's armour, and ready. There is danger from the demons, but so there is in the surrounding air every day. At early morning then, and alone—for so the usage ordains—I betook me towards the field. It was void, and I had thereby due time to prepare. First I paced and measured out my circle on the grass. Then I did mark my pentacle in the very midst, and at the intersection of the five angles I did set up and fix my crutch of *raun* [rowan]. Lastly, I took my station south, at the true line of the meridian, and stood facing due north. I waited and watched for a long time. At last there was a kind of trouble in the air, a soft and rippling sound, and all at once the shape appeared, and came on towards me gradually. I opened my parchment-scroll, and read aloud the command. She paused, and seemed to waver and doubt; stood still; then I rehearsed the sentence again, sounding out every syllable like a chant. She drew near my ring, but halted at first outside, on the brink. I sounded again, and now at the third time I gave the signal in Syriac—the speech which is used, they say, where such ones dwell and converse in thoughts that glide.

'She was at last obedient, and swam into the midst of the circle, and there stood still, suddenly. I saw, moreover, that she drew back her pointing hand. All this while I do confess that my knees shook under me, and the drops of sweat ran down my flesh like rain. But now, although face to face with the spirit, my heart grew calm, and my mind was composed. I knew that the pentacle would govern her, and the ring must bind, until I gave the word. Then I called to mind the rule laid down of old, that no angel or fiend, no spirit, good or evil, will ever speak until they have been first spoken to. *N.B.* This is the great law of prayer. God Himself will not yield reply until man hath made vocal entreaty, once and again. So I went on to demand, as the books advise; and the phantom made answer, willingly. Questioned wherefore not at rest. Unquiet, because of a certain sin. Asked what, and by whom. Revealed it; but it is *sub sigillo*, and therefore *nefas dictu*; more anon. Enquired, what sign she could give that she was a true spirit and not a false fiend. Stated, before next Yule-tide a fearful pestilence would lay waste the land and myriads of souls would be loosened from their flesh, until, as she piteously said, "our valleys will be full". Asked again, why she so terrified the lad. Replied: "It is the law: we must seek a youth or a maiden of clean life, and under age, to receive messages and admonitions." We conversed with many more words, but it is not lawful for me to set them down. Pen and ink would degrade and defile the thoughts she uttered, and which my mind received that day. I broke the ring and she passed, but to return once more next day. At even-song, a long discourse with that ancient transgressor, Mr B. Great horror and remorse; entire atonement and penance; whatsoever I enjoin; full acknowledgement before pardon.

'*January* 13, 1665.—At sunrise I was again in the field. She came in at once, and, as it seemed, with freedom. Enquired if she knew my thoughts, and what I was going to relate? Answered, "Nay, we only know what we perceive and hear; we cannot see the heart." Then I rehearsed the penitent words of the man she had come up to denounce, and the satisfaction he would perform. Then said she, "Peace in our midst." I went through the proper forms of dismissal, and fulfilled all as it was set down and written in my memoranda; and then, with certain fixed rites, I did dismiss that troubled ghost, until she peacefully withdrew, gliding towards the west. Neither did she ever afterward appear, but was allayed until she shall come in her second flesh to the valley of Armageddon on the last day.'

These quaint and curious details from the 'diurnal' of a simple-hearted clergyman of the seventeenth century appear to betoken his personal persuasion of the truth of what he saw and said, although the statements are strongly tinged with what some may term the superstition, and others the excessive belief, of those times. It is a singular fact, however, that the canon which authorizes exorcism under episcopal licence is still a part of the ecclesiastical law of the Anglican Church, although it might have a singular effect on the nerves of certain of our bishops if their clergy were to resort to them for the faculty which Parson Rudall obtained. The general facts stated in his diary are to this day matters of belief in that neighbourhood; and it has been always accounted a strong proof of the veracity of the Parson and the Ghost, that the plague, fatal to so many thousands, did break out in London at the close of that very year. We may well excuse a triumphant entry, on a subsequent page of the 'diurnal', with the date of July 10, 1665: 'How sorely must the infidels and heretics of this generation be dismayed when they know that this Black Death, which is now swallowing its thousands in the streets of the great city, was foretold six months agone, under the exorcisms of a country minister, by a visible and suppliant ghost! And what pleasures and improvements do such deny themselves who scorn and avoid all opportunity of intercourse with souls separate, and the spirits, glad and sorrowful, which inhabit the unseen world!'

The Truth, the Whole Truth, and Nothing but the Truth

RHODA BROUGHTON

Mrs De Wynt to Mrs Montresor

18, Eccleston Square,
May 5th

My dearest Cecilia,

Talk of the friendships of Orestes and Pylades, of Julie and Claire, what are they to ours? Did Pylades ever go *ventre à terre*, half over London on a day more broiling than any but an *âme damnée* could even imagine, in order that Orestes might be comfortably housed for the season? Did Claire ever hold sweet converse with from fifty to one hundred house agents, in order that Julie might have three windows to her drawing-room and a pretty *portière*. You see I am determined not to be done out of my full meed of gratitude.

Well, my friend, I had no idea till yesterday how closely we were packed in this great smoky beehive, as tightly as herrings in a barrel. Don't be frightened, however. By dint of squeezing and crowding, we have managed to make room for two more herrings in our barrel, and those two are yourself and your other self, i.e. your husband. Let me begin at the beginning. After having looked over, I verily believe, every undesirable residence in West London; after having seen nothing intermediate between what was suited to the means of a duke, and what was suited to the needs of a chimney-sweep; after having felt bed-ticking, and explored kitchen ranges till my brain reeled under my accumulated experience, I arrived at about half-past five yesterday afternoon at 32, —— Street, May Fair.

'Failure No. 253, I don't doubt,' I said to myself, as I toiled up the steps with my soul athirst for afternoon tea, and feeling as ill-tempered as you please. So much for my spirit of prophecy. Fate, I have noticed, is often fond of contradicting us flat, and giving the lie to our little predictions. Once inside, I thought I had got into a small compartment of Heaven by mistake. Fresh as a daisy, clean as a cherry, bright as a Seraph's face, it is all these, and a hundred more, only that my limited stock of similes is exhausted. Two drawing-rooms as pretty as ever woman crammed with people she did not care two straws about; white curtains with rose-coloured ones underneath, festooned in the sweetest way; marvellously, *immorally* becoming, my dear, as I ascertained entirely for your benefit, in the mirrors, of which there are about a dozen and a half; Persian mats, easy chairs, and lounges suited to every possible physical conformation, from the Apollo Belvedere to Miss Biffin; and a thousand of the important little trivialities that make up the sum of a woman's life: ormolu garden gates, handleless cups, naked boys and décolleté shepherdesses; not to speak of a family of china pugs, with blue ribbons round their necks, which ought of themselves to have added fifty pounds a year to the rent. Apropos, I asked, in fear and trembling, what the rent might be—'three hundred pounds a year'. A feather would have knocked me down. I could hardly believe my ears, and made the woman repeat it several times, that there might be no mistake. To this hour it is a mystery to me.

With that suspiciousness, which is so characteristic of you, you will immediately begin to hint that there must be some terrible unaccountable smell, or some odious inexplicable noise haunting the reception rooms. Nothing of the kind, the woman assured me, and she did not look as if she were telling stories. You will next suggest—remembering the rose-coloured curtains—that its last occupant was a member of the demi-monde. Wrong again. Its last occupant was an elderly and unexceptionable Indian officer, without a liver, and with a most lawful wife. They did not stay long, it is true, but then, as the housekeeper told me, he was a deplorable old hypochondriac, who never could bear to stay a fortnight in any one place. So lay aside that scepticism, which is your besetting sin, and give unfeigned thanks to St Brigitta, or St Gengulpha, or St Catherine of Sienna, or whoever is your tutelar saint, for having provided you with a palace at the cost of a hovel, and for having sent you such an invaluable friend as

Your attached,
ELIZABETH DE WYNT

P.S. I am so sorry I shall not be in town to witness your first raptures, but dear Artie looks so pale and thin and tall after the whooping-cough, that I am sending him off at once to the sea, and as I cannot bear the child out of my sight, I am going into banishment likewise.

MRS MONTRESOR TO MRS DE WYNT

32, —— Street, May Fair,
May 14th

DEAREST BESSY,

Why did not dear little Artie defer his whooping-cough convalescence, etc., till August? It is very odd, to me, the perverse way in which children always fix upon the most inconvenient times and seasons for their diseases. Here we are installed in our Paradise, and have searched high and low, in every hole and corner, for the serpent, without succeeding in catching a glimpse of his spotted tail. Most things in this world are disappointing, but 32, —— Street, May Fair, is not. The mystery of the rent is still a mystery. I have been for my first ride in the Row this morning: my horse was a little fidgety; I am half afraid that my nerve is not what it was. I saw heaps of people I knew. Do you recollect Florence Watson? What a wealth of red hair she had last year! Well, that same wealth is black as the raven's wing this year! I wonder how people can make such walking impositions of themselves, don't you? Adela comes to us next week; I am so glad. It is dull driving by oneself of an afternoon; and I always think that one young woman alone in a brougham, or with only a dog beside her, does not look *good*. We sent round our cards a fortnight before we came up, and have been already deluged with callers. Considering that we have been two years exiled from civilized life, and that London memories are not generally of the longest, we shall do pretty well, I think. Ralph Gordon came to see me on Sunday; he is in the ——th Hussars now. He has grown up such a *dear* fellow, and *so* good-looking! Just my style, large and fair and whiskerless! Most men nowadays make themselves as like monkeys, or Scotch terriers, as they possibly can. I intend to be quite a *mother* to him. Dresses are gored to as *indecent* an extent as ever; short skirts are rampant. I am so sorry; I hate them. They make tall women look *lank*, and short ones insignificant. A knock! Peace is a word that might as well be expunged from one's London dictionary.

Yours affectionately,
CECILIA MONTRESOR

MRS DE WYNT TO MRS MONTRESOR

The Lord Warden, Dover,
May 18th

DEAREST CECILIA,

You will perceive that I am about to devote only one small sheet of note-paper to you. This is from no dearth of time, Heaven knows! time is a drug in the market here, but from a total dearth of ideas. Any ideas that I ever have, come to me from without, from external objects; I am not clever enough to generate any within myself. My life here is not an eminently suggestive one. It is spent in digging with a wooden spade, and eating prawns. Those are my employments, at least; my relaxation is going down to the Pier, to see the Calais boat come in. When one is miserable oneself, it is decidedly consolatory to see some one more miserable still; and wretched, and bored, and reluctant vegetable as I am, I am not *sea-sick*. I always feel my spirits rise after having seen that peevish, draggled procession of blue, green and yellow fellow-Christians file past me. There is a wind here *always*, in comparison of which the wind that behaved so violently to the corners of Job's house was a mere zephyr. There are heights to climb which require more daring perseverance than ever Wolfe displayed, with his paltry heights of Abraham. There are glaring white houses, glaring white roads, glaring white cliffs. If any one knew how unpatriotically I detest the chalk-cliffs of Albion! Having grumbled through my two little pages—I have actually been reduced to writing very large in order to fill even them—I will send off my dreary little billet. How I wish I could get into the envelope myself too, and whirl up with it to dear, beautiful, filthy London. Not more heavily could Madame de Staël have sighed for Paris from among the shades of Coppet.

Your disconsolate Bessy

Mrs Montresor to Mrs De Wynt

32, —— Street, May Fair,
May 27th

Oh, my dearest Bessy, how I wish we were out of this dreadful, dreadful house! Please don't think me very ungrateful for saying this, after your taking such pains to provide us with a Heaven upon earth, as you thought.

What has happened could, of course, have been neither foretold, nor guarded against, by any human being. About ten days ago, Benson (my maid) came to me with a very long face, and said, 'If you please, 'm, did you know that this house was *haunted*?' I was *so* startled: you know what a coward I am. I said, 'Good Heavens! No! is it?' 'well, 'm, I'm pretty nigh sure it is,' she said, and the expression of her countenance was about as lively as an undertaker's; and then she told me that cook had been that morning to order in groceries from a shop in the neighbourhood, and on her giving the man the direction where to send the things to, he had said, with a very peculiar smile, 'No.

32, —— Street, eh? h'm! I wonder how long *you*'ll stand it; last lot held out just a fortnight.' He looked so odd that she asked him what he meant, but he only said 'Oh! nothing; only that parties never *did* stay long at 32. He had known parties go in one day, and out the next, and during the last four years he had never known any remain over the month.' Feeling a good deal alarmed by this information, she naturally enquired the reason; but he declined to give it, saying that if she had not found it out for herself, she had much better leave it alone, as it would only frighten her out of her wits; and on her insisting and urging him, she could only extract from him, that the house had such a villainously bad name, that the owners were glad to let it for a mere song. You know how firmly I believe in apparitions, and what an unutterable fear I have of them; anything material, tangible, that I can lay hold of—anything of the same fibre, blood, and bone as myself, I could, I think, confront bravely enough; but the mere thought of being brought face to face with the 'bodiless dead', makes my brain unsteady. The moment Henry came in, I ran to him, and told him; but he pooh-poohed the whole story, laughed at me, and asked whether we should turn out of the prettiest house in London, at the very height of the season, because a grocer said it had a bad name. Most good things that had ever been in the world had had a bad name in their day; and, moreover, the man had probably a motive for taking away the house's character, some friend for whom he coveted the charming situation and the low rent. He derided my 'babyish fears', as he called them, to such an extent that I felt half ashamed, and yet not quite comfortable, either; and then came the usual rush of London engagements, during which one has no time to think of anything but how to speak, and act, and look for the moment then present. Adela was to arrive yesterday, and in the morning our weekly hamper of flowers, fruit, and vegetables arrived from home. I always dress the flower-vases myself, servants are so tasteless; and as I was arranging them, it occurred to me—you know Adela's passion for flowers—to carry up one particular cornucopia of roses and mignonette and set it on her toilet-table, as a pleasant surprise for her. As I came downstairs, I had seen the housemaid—a fresh round-faced country girl—go into the room, which was being prepared for Adela, with a pair of sheets that she had been airing over her arm. I went upstairs very slowly, as my cornucopia was full of water, and I was afraid of spilling some. I turned the handle of the bedroom-door and entered, keeping my eyes fixed on my flowers, to see how they bore the transit, and whether any of them had fallen out. Suddenly a sort of shiver passed over me; and feeling frightened—I did not know why—I looked up quickly. The girl was standing by the bed, leaning forward a little with her hands clenched in each other, rigid, every nerve tense; her eyes, wide open, starting out of her head, and a look of unutterable stony horror in them; her cheeks and mouth not pale, but livid as those of one that died awhile ago in mortal pain. As I looked at her, her lips moved a little, and an awful hoarse voice, not like hers in the least, said, 'Oh! my God, I have seen it!' and then she fell down suddenly, like a log, with a heavy noise. Hearing the noise, loudly audible all through the thin walls and floors of a London house, Benson came running in, and between us we managed to lift her on to the bed, and tried to bring her to herself by rubbing her feet and hands, and holding strong salts to her nostrils. And all the while we kept glancing over our shoulders, in a vague cold terror of seeing some awful, shapeless apparition. Two long hours she lay in a state of utter unconsciousness. Meanwhile Harry, who had been down to his club, returned. At the end of the two hours we succeeded in bringing her back to sensation and life, but only to make the awful discovery that she was raving mad. She became so violent that it required all the combined strength of Harry and Phillips (our butler) to hold her down in the bed. Of course, we sent off instantly for a doctor, who, on her growing a little calmer towards evening, removed her in a cab to his own house. He has just been here to tell me that she is now pretty quiet, not from any return to sanity, but from sheer exhaustion. We are, of course, utterly in the dark as to *what* she saw, and her ravings are far too disconnected and unintelligible to afford us the slightest clue. I feel so completely shattered and upset by this awful occurrence, that you will excuse me, dear, I'm sure, if I write incoherently. One thing, I need hardly tell you, and that is, that no earthly consideration would induce me to allow Adela to occupy that terrible room. I shudder and run by quickly as I pass the door.

Yours, in great agitation,

CECILIA

MRS DE WYNT TO MRS MONTRESOR

The Lord Warden, Dover,

May 28th

DEAREST CECILIA,

Yours just come; how very dreadful! But I am still unconvinced as to the house being in fault. You know I feel a sort of godmother to it, and responsible for its good behaviour. Don't you think that what the girl had might have been a fit? Why not? I myself have a cousin who is subject to seizures of the kind, and immediately on being attacked his whole body becomes rigid, his eyes glassy and staring, his complexion livid, exactly as in the case you describe. Or, if not a fit, are you sure that she has not been subject to fits of madness? *Please* be sure and ascertain whether there is not insanity in her family. It is so common now-a-days, and so much on the increase, that nothing is more likely. You know my utter disbelief in ghosts. I am convinced that most of them, if run to earth, would turn out about as genuine as the famed Cock Lane one. But even allowing the possibility, nay, the actual unquestioned existence of ghosts in the abstract, is it likely that there should be anything to be seen so horribly fear-inspiring, as to send a perfectly sane person *in one instant* raving mad, which you, after three weeks' residence in the house, have never caught a glimpse of? According to your hypothesis, your whole household ought, by this time, to be stark, staring mad. Let me implore you not to give way to a panic which may, possibly, probably prove utterly groundless. Oh, how wish I were with you, to make you listen to reason! Artie ought to be the best prop ever woman's old age was furnished with, to indemnify me, for all he and his whooping-cough have made me suffer. Write immediately, please, and tell me how the poor patient progresses. Oh, had I the wings of a dove! I shall be on wires till I hear again.

Yours,
BESSY

MRS MONTRESOR TO MRS DE WYNT

No. 5, Bolton Street, Piccadilly,
June 12th

DEAREST BESSY,

You will see that we have left that terrible, hateful, fatal house. How I wish we had escaped from it sooner! Oh, my dear Bessy, I shall never be the same woman again if I live to be a hundred. Let me try to be coherent, and to tell you connectedly what has happened. And first, as to the housemaid, she has been removed to a lunatic asylum, where she remains in much the same state. She has had several lucid intervals, and during them has been closely, pressingly questioned as to what it was she saw; but she has maintained an absolute, hopeless silence, and only shudders, moans, and hides her face in her hands when the subject is broached. Three days ago I went to see her, and on my return was sitting resting in the drawing-room, before going to dress for dinner, talking to Adela about my visit, when Ralph Gordon walked in. He has always been walking in the last ten days, and Adela has always flushed up and looked happy, poor little cat, whenever he made his appearance. He looked very handsome, dear fellow, just come in from the park in a coat that fitted like a second skin, lavender gloves, and a gardenia. He seemed in tremendous spirits, and was as sceptical as even you could be, as to the ghostly origin of Sarah's seizure. 'Let me come here tonight and sleep in that room; *do*, Mrs Montresor,' he said, looking very eager and excited, 'with the gas lit and a poker, I'll engage to exorcize every demon that shows his ugly nose; even if I should find

> Seven white ghostisses
> Sitting on seven white postisses.

'You don't mean really?' I asked, incredulously. 'Don't I? that's all,' he answered, emphatically. 'I should like nothing better. Well, is it a bargain?' Adela turned quite pale. 'Oh, don't,' she said, hurriedly, '*Please*, don't; why should you run such a risk? How do you know that you might not be sent mad too?' He laughed very heartily, and coloured a little with pleasure at seeing the interest she took in his safety. 'Never fear,' he said, 'it would take more than a whole squadron of departed ones, with the old gentleman at their head, to send me crazy.' He was so eager, so persistent, so thoroughly in earnest, that I yielded at last, though with a certain strong reluctance to his entreaties. Adela's blue eyes filled with tears, and she walked away hastily to the conservatory, and stood picking bits of heliotrope to hide them. Nevertheless, Ralph got his own way; it was so difficult to refuse him anything. We gave up all our engagements for the evening, and he did the same with his. At about ten o'clock he arrived, accompanied by a friend and brother officer, Captain Burton, who was anxious to see the result of the experiment. 'Let me go up at once,' he said, looking very happy and animated. 'I don't know when I have felt in such good tune; a new sensation is a luxury not to be had every day of one's life; turn the gas up as high as it will go; provide a good stout poker, and leave the issue to Providence and me.' We did as he bid. 'It's all ready now,' Henry said, coming downstairs after having obeyed his orders; 'the room is nearly as light as day. Well, good luck to you, old fellow!' 'Goodbye, Miss Bruce,' Ralph said, going over to Adela, and taking her hand with a look, half laughing, half sentimental—

> Fare thee well, and if for ever,
> Then for ever, fare thee well,

that is my last dying speech and confession. 'Now mind,' he went on, standing by the table, and addressing us all; if I ring once, *don't* come. I may be flurried, and lay hold of the bell without thinking; if I ring twice, *come*.' Then he went, jumping up the stairs three steps at a time, and humming a tune. As for us, we sat in different attitudes of expectation and listening about the drawing-room. At first we tried to talk a little, but it would not do; our whole souls seemed to have passed into our ears. The clock's ticking sounded as loud as a great church bell close to one's ear. Addy lay on the sofa, with her dear little white face hidden in the cushions. So we sat for exactly an hour; but it seemed like two years, and just as the clock began to strike eleven, a sharp ting, ting, ting rang clear and shrill through the house. 'Let us go,' said Addy, starting up, and running to the door. 'Let us go,' I cried too, following her. But Captain Burton stood in the way, and intercepted our progress. 'No,' he said, decisively, 'you must not go; remember Gordon told us distinctly, if he rang once *not* to come. I know the sort of fellow he is, and that nothing would annoy him more than having his directions disregarded.'

'Oh, nonsense!' Addy cried, passionately, 'he would never have rung if he had not seen something dreadful; do, *do* let us go!' she ended, clasping her hands. But she was overruled, and we all went back to our seats. Ten minutes more of suspense, next door to unendurable, I felt a lump in my throat, a gasping for breath—ten minutes on the clock, but a thousand centuries on our hearts. Then again, loud, sudden, violent the bell rang! We made a simultaneous rush to the door. I don't think we were one second flying upstairs. Addy was first. Almost simultaneously she and I burst into the room. There he was, standing in the middle of the floor, rigid, petrified, with that same look—that look that is burnt into my heart in letters of fire—of awful, unspeakable, stony fear on his brave young face. For one instant he stood thus; then stretching out his arms stiffly before him, he groaned in a terrible husky voice, 'Oh, my God, I have seen it!' and fell down *dead*. Yes, *dead*. Not in a swoon or in a fit, but *dead*. Vainly we tried to bring back the life to that strong young heart; it will never come back again till that day when the earth and the sea give up the dead that are therein. I cannot see the page for the tears that are blinding me; he was such a dear fellow! I can't write any more today.

Your broken-hearted CECILIA.

This is a true story.

The Romance of Certain Old Clothes

HENRY JAMES

I

Towards the middle of the eighteenth century there lived in the Province of Massachusetts a widowed gentlewoman, the mother of three children, by name Mrs Veronica Wingrave. She had lost her husband early in life, and had devoted herself to the care of her progeny. These young persons grew up in a manner to reward her tenderness and to gratify her highest hopes. The first-born was a son, whom she had called Bernard, after his father. The others were daughters—born at an interval of three years apart. Good looks were traditional in the family, and this youthful trio were not likely to allow the tradition to perish. The boy was of that fair and ruddy complexion and that athletic structure which in those days (as in these) were the sign of good English descent—a frank, affectionate young fellow, a deferential son, a patronizing brother, a steadfast friend. Clever, however, he was not; the wit of the family had been apportioned chiefly to his sisters. The late Mr Wingrave had been a great reader of Shakespeare, at a time when this pursuit implied more freedom of thought than at the present day, and in a community where it required much courage to patronize the drama even in the closet: and he had wished to call attention to his admiration of the great poet by calling his daughters out of his favourite plays. Upon the elder he had bestowed the romantic name of Rosalind, and the younger he had called Perdita, in memory of a little girl born between them, who had lived but a few weeks.

When Bernard Wingrave came to his sixteenth year his mother put a brave face upon it and prepared to execute her husband's last injunction. This had been a formal command that, at the proper age, his son should be sent out to England, to complete his education at the university of Oxford, where he himself had acquired his taste for elegant literature. It was Mrs Wingrave's belief that the lad's equal was not to be found in the two hemispheres, but she had the old traditions of literal obedience. She swallowed her sobs, and made up her boy's trunk and his simple provincial outfit, and sent him on his way across the seas. Bernard presented himself at his father's college, and spent five years in England, without great honour, indeed, but with a vast deal of pleasure and no discredit. On leaving the university he made the journey to France. In his twenty-fourth year he took ship for home, prepared to find poor little New England (New England was very small in those days) a very dull, unfashionable residence. But there had been changes at home, as well as in Mr Bernard's opinions. He found his mother's house quite habitable, and his sisters grown into two very charming young ladies, with all the accomplishments and graces of the young women of Britain, and a certain native-grown originality and wildness, which, if it was not an accomplishment, was certainly a grace the more. Bernard privately assured his mother that his sisters were fully a match for the most genteel young women in the old country; whereupon poor Mrs Wingrave, you may be sure, bade them hold up their heads. Such was Bernard's opinion, and such, in a tenfold higher degree, was the opinion of Mr Arthur Lloyd. This gentleman was a college-mate of Mr Bernard, a young man of reputable family, of a good person and a handsome inheritance; which latter appurtenance he proposed to invest in trade in the flourishing colony. He and Bernard were sworn friends; they had crossed the ocean together, and the young American had lost no time in presenting him at his mother's house, where he had made quite as good an impression as that which he had received and of which I have just given a hint.

The two sisters were at this time in all the freshness of their youthful bloom; each wearing, of course, this natural brilliancy in the manner that became her best. They were equally dissimilar in appearance and character. Rosalind, the elder—now in her twenty-second year—was tall and white, with calm grey eyes and auburn tresses; a very faint likeness to the Rosalind of Shakespeare's comedy, whom I imagine a brunette (if you will), but a slender, airy creature, full of the softest, quickest impulses. Miss Wingrave, with her slightly lymphatic fairness, her fine arms, her majestic height, her slow utterance, was not cut out for adventures. She would never have put on a man's jacket and hose; and, indeed, being a very plump beauty, she may have had reasons apart from her natural dignity. Perdita, too, might very well have exchanged the sweet melancholy of her name against something more in consonance with her aspect and disposition. She had the cheek of a gypsy and the eye of an eager child, as well as the smallest waist and lightest foot in all the country of the Puritans. When you spoke to her she never made you wait, as her handsome sister was wont to do (while she looked at you with a cold fine eye), but gave you your choice of a dozen answers before you had uttered half your thought.

The young girls were very glad to see their brother once more; but they found themselves quite able to spare part of their attention for their brother's friend. Among the young men their friends and neighbours, the *belle jeunesse* of the Colony, there were many excellent fellows, several devoted swains, and some two or three who enjoyed the reputation of universal charmers and conquerors. But the homebred arts and somewhat boisterous gallantry of these honest colonists were completely eclipsed by the good looks, the fine clothes, the punctilious courtesy, the perfect elegance, the immense information, of Mr Arthur Lloyd. He was in reality no paragon; he was a capable, honourable, civil youth, rich in pounds sterling, in his health and complacency and his little capital of uninvested affections. But he was a gentleman; he had a handsome person; he had studied and travelled; he spoke French, he played the flute, and he read verses aloud with very great taste. There were a dozen reasons why Miss Wingrave and her sister should have thought their other male acquaintance made but a poor figure before such a perfect man of the world. Mr Lloyd's anecdotes told our little New England maidens a great deal more of the ways and means of people of fashion in European capitals than he had any idea of doing. It was delightful to sit by and hear him and Bernard talk about the fine people and fine things they had seen. They would all gather round the fire after tea, in the little wainscoted parlour, and the two young men would remind each other, across the rug, of this, that and the other adventure. Rosalind and Perdita would often have given their ears to know exactly what adventure it was, and where it happened, and who was there, and what the ladies had on; but in those days a well-bred young woman was not expected to break into the conversation of her elders, or to ask too many questions; and the poor girls used therefore to sit fluttering behind the more languid—or more discreet—curiosity of their mother.

II

That they were both very fine girls Arthur Lloyd was not slow to discover; but it took him some time to make up his mind whether he liked the big sister or the little sister best. He had a strong presentiment—an emotion of a nature entirely too cheerful to be called a foreboding—that he was destined to stand up before the parson with one of them; yet he was unable to arrive at a preference, and for such a consummation a preference was certainly necessary, for Lloyd had too much young blood in his veins to make a choice by lot and be cheated of the satisfaction of falling in love. He resolved to take things as they came—to let his heart speak. Meanwhile he was on very pleasant footing. Mrs Wingrave showed a dignified indifference to his 'intentions', equally remote from a carelessness of her daughter's honour and from that sharp alacrity to make him come to the point, which, in his quality of young man of property, he had too often encountered in the worldly matrons of his native islands. As for Bernard, all that he asked was that his friend should treat his sisters as his own; and as for the poor girls themselves, however each may have secretly longed that their visitor should do or say something 'marked', they kept a very modest and contented demeanour.

Towards each other, however, they were somewhat more on the offensive. They were good friends enough, and accommodating bedfellows (they shared the same four-poster), betwixt whom it would take more than a day for the seeds of jealousy to sprout and bear fruit; but they felt that the seeds had been sown on the day that Mr Lloyd came into the house. Each made up her mind that, if she should be slighted, she would bear her grief in silence, and that no one should be any the wiser; for if they had a great deal of ambition, they had also a large share of pride. But each prayed in secret, nevertheless, that upon *her* the selection, the distinction, might fall. They had need of a vast deal of patience, of self-control, of dissimulation. In those days a young girl of decent breeding could make no advances whatever, and barely respond, indeed, to those that were made. She was expected to sit still in her chair, with her eyes on the carpet, watching the spot where the mystic handkerchief should fall. Poor Arthur Lloyd was obliged to carry on his wooing in the little wainscoted parlour, before the eyes of Mrs Wingrave, her son, and his prospective sister-in-law. But youth and love are so cunning that a hundred signs and tokens might travel to and fro, and not one of these three pairs of eyes detect them in their passage. The two maidens were almost always together, and had plenty of chances to betray themselves. That each knew she was being watched, made not a grain of difference in the little offices they mutually rendered, or in the various household tasks they performed in common. Neither flinched nor fluttered beneath the silent battery of her sister's eyes. The only apparent change in their habits was that they had less to say to each other. It was impossible to talk about Mr Lloyd, and it was ridiculous to talk about anything else. By tacit agreement they began to wear all their choice finery, and to devise such little implements of conquest, in the way of ribbons and top-knots and kerchiefs, as were sanctioned by indubitable modesty. They executed in the same inarticulate fashion a contract of fair play in this exciting game. 'Is it better so?' Rosalind would ask, tying a bunch of ribbons on her bosom, and turning about from her glass to her sister. Perdita would look up gravely from her work and examine the decoration. 'I think you had better give it another loop,' she would say, with great solemnity, looking hard at her sister with eyes that added, 'upon my honour!' So they were for ever stitching and turning their petticoats, and pressing out their muslins, and contriving washes and ointments and cosmetics, like the ladies in the household of the vicar of Wakefield. Some three or four months went by; it grew to be midwinter, and as yet Rosalind knew that if Perdita had nothing more to boast of than she, there was not much to be feared from her rivalry. But Perdita by this time—the charming Perdita—felt that her secret had grown to be tenfold more precious than her sister's.

One afternoon Miss Wingrave sat alone—that was a rare accident—before her toilet-glass, combing out her long hair. It was getting too dark to see; she lit the two candles in their sockets, on the frame of her mirror, and then went to the window to draw her curtains. It was a grey December evening; the landscape was bare and bleak, and the sky heavy with snow-clouds. At the end of the large garden into which her window looked was a wall with a little postern door, opening into a lane. The door stood ajar, as she could vaguely see in the gathering darkness, and moved slowly to and fro, as if someone were swaying it from the lane without. It was doubtless a servant-maid who had been having a tryst with her sweetheart. But as she was about to drop her curtain Rosalind saw her sister step into the garden and hurry along the path which led to the house. She dropped the curtain, all save a little crevice for her eyes. As Perdita came up the path she seemed to be examining something in her hand, holding it close to her eyes. When she reached the house she stopped a moment, looked intently at the object, and pressed it to her lips.

Poor Rosalind slowly came back to her chair and sat down before her glass where, if she had looked at it less abstractly, she would have seen her handsome features sadly disfigured by jealousy. A moment afterwards the door opened behind her and her sister came into the room, out of breath, her cheeks aglow with the chilly air.

Perdita started. 'Ah,' said she, 'I thought you were with our mother.' The ladies were to go to a tea-party, and on such occasions it was the habit of one of the girls to help their mother to dress. Instead of coming in, Perdita lingered at the door.

'Come in, come in,' said Rosalind. 'We have more than an hour yet. I should like you very much to give a few strokes to my hair.' She knew that her sister wished to retreat, and that she could see in the glass all her movements in the room. 'Nay, just help me with my hair,' she said, 'and I will go to mamma.'

Perdita came reluctantly, and took the brush. She saw her sister's eyes, in the glass, fastened hard upon her hands. She had not made three passes when Rosalind clapped her own right hand upon her sister's left, and started out of her chair. 'Whose ring is that?' she cried, passionately, drawing her towards the light.

On the young girl's third finger glistened a little gold ring, adorned with a very small sapphire. Perdita felt that she need no longer keep her secret, yet that she must put a bold face on her avowal. 'It's mine,' she said proudly.

'Who gave it to you?'cried the other.

Perdita hesitated a moment. 'Mr Lloyd.'

'Mr Lloyd is generous, all of a sudden.'

'Ah no,' cried Perdita, with spirit, 'not all of a sudden! He offered it to me a month ago.'

'And you needed a month's begging to take it?' said Rosalind, looking at the little trinket, which indeed was not especially elegant, although it was the best that the jeweller of the Province could furnish. 'I wouldn't have taken it in less than two.'

'It isn't the ring,' Perdita answered, 'it's what it means!'

'It means that you are not a modest girl!' cried Rosalind. 'Pray, does your mother know of your intrigue? does Bernard?'

'My mother has approved my "intrigue", as you call it. My Lloyd has asked for my hand, and mamma has given it. Would you have had him apply to you, dearest sister?'

Rosalind gave her companion a long look, full of passionate envy and sorrow. Then she dropped her lashes on her pale cheeks and turned away. Perdita felt that it had not been a pretty scene; but it was her sister's fault. However, the elder girl rapidly called back her pride, and turned herself about again. 'You have my very best wishes,' she said, with a low curtsey. 'I wish you every happiness, and a very long life.'

Perdita gave a bitter laugh. 'Don't speak in that tone!' she cried. 'I would rather you should curse me outright. Come, Rosy,' she added, 'he couldn't marry both of us.'

'I wish you very great joy,' Rosalind repeated, mechanically, sitting down to her glass again, 'and a very long life, and plenty of children.'

There was something in the sound of these words not at all to Perdita's taste. 'Will you give me a year to live at least?' she said. 'In a year I can have one little boy—or one little girl at least. If you will give me your brush again I will do your hair.'

'Thank you,' said Rosalind. 'You had better go to mamma. It isn't becoming that a young lady with a promised husband should wait on a girl with none.'

'Nay,' said Perdita good-humouredly, 'I have Arthur to wait upon me. You need my service more than I need yours.'

But her sister motioned her away, and she left the room. When she had gone poor Rosalind fell on her knees before her dressing-table, buried her head in her arms, and poured out a flood of tears and sobs. She felt very much the better for this effusion of sorrow. When her sister came back she insisted on helping her to dress—on her wearing her prettiest things. She forced upon her acceptance a bit of lace of her own, and declared that now that she was to be married she should do her best to appear worthy of her lover's choice. She discharged these offices in stern silence; but, such as they were, they had to do duty as an apology and an atonement; she never made any other.

Now that Lloyd was received by the family as an accepted suitor nothing remained but to fix the wedding-day. It was appointed for the following April, and in the interval preparations were diligently made for the marriage. Lloyd, on his side, was busy with his commercial arrangements, and with establishing a correspondence with the great mercantile house to which he had attached himself in England. He was therefore not so frequent a visitor at Mrs Wingrave's as during the months of his diffidence and irresolution, and poor Rosalind had less to suffer than she had feared from the sight of the mutual endearments of the young lovers. Touching his future sister-in-law Lloyd had a perfectly clear conscience. There had not been a particle of love-making between them, and he had not the slightest suspicion that he had dealt her a terrible blow. He was quite at his ease; life promised so well, both domestically and financially. The great revolt of the Colonies was not yet in the air, and that his connubial felicity should take a tragic turn it was absurd, it was blasphemous, to apprehend. Meanwhile, at Mrs Wingrave's, there was a greater rustling of silks, a more rapid clicking of scissors and flying of needles, than ever. The good lady had determined that her daughter should carry from home the genteelest outfit that her money could buy or that the country could furnish. All the sage women in the Province were convened, and their united taste was brought to bear on Perdita's wardrobe. Rosalind's situation, at this moment, was assuredly not to be envied. The poor girl had an inordinate love of dress, and the very best taste in the world, as her sister perfectly well knew. Rosalind was tall, she was stately and sweeping, she was made to carry stiff brocade and masses of heavy lace, such as belong to the toilet of a rich man's wife. But Rosalind sat aloof, with her beautiful arms folded and her head averted, while her mother and sister and the venerable women aforesaid worried and wondered over their materials, oppressed by the multitude of their resources. One day there came in a beautiful piece of white silk, brocaded with heavenly blue and silver sent by the bridegroom himself—it not being thought amiss in those days that the husband-elect should contribute to the bride's trousseau. Perdita could think of no form or fashion which would do sufficient honour to the splendour of the material.

'Blue's your colour, sister, more than mine,' she said, with appealing eyes. 'It is a pity it's not for you. You would know what to do with it.'

Rosalind got up from her place and looked at the great shining fabric, as it lay spread over the back of a chair. Then she took it up in her hands and felt it—lovingly, as Perdita could see—and turned about towards the mirror with it. She let it roll down to her feet, and flung the other end over her shoulder, gathering it in about her waist with her white arm, which was bare to the elbow. She threw back her head, and looked at her image, and a hanging tress of her auburn hair fell upon the gorgeous surface of the silk. It made a dazzling picture. The women standing about uttered a little 'Look, look!' of admiration. 'Yes, indeed,' said Rosalind, quietly, 'blue is my colour.' But Perdita could see that her fancy had been stirred, and that she would now fall to work and solve all their silken riddles. And indeed she behaved very well, as Perdita, knowing her insatiable love of millinery, was quite ready to declare. Innumerable yards of lustrous silk and satin, of muslin, velvet and lace, passed through her cunning hands, without a jealous word coming from her lips. Thanks to her industry, when the wedding-day came Perdita was prepared to espouse more of the vanities of life than any fluttering young bride who had yet received the sacramental blessing of a New England divine.

It had been arranged that the young couple should go out and spend the first days of their wedded life at the country-house of an English gentleman—a man of rank and a very kind friend to Arthur Lloyd. He was a bachelor; he declared he should be delighted to give up the place to the influence of Hymen. After the ceremony at church—it had been performed by an English clergyman—young Mrs Lloyd hastened back to her mother's house to change her nuptial robes for a riding-dress. Rosalind helped her to effect the change, in the little homely room in which they had spent their undivided younger years. Perdita then hurried off to bid farewell to her mother, leaving Rosalind to follow. Then parting was short; the horses were at the door, and Arthur was impatient to start. But Rosalind had not followed, and Perdita hastened back to her room, opening the door abruptly. Rosalind, as usual, was before the glass, but in a position which caused the other to stand still, amazed. She had dressed herself in Perdita's cast-off wedding veil and wreath, and on her neck she had hung the full string of pearls which the young girl had received from her husband as a wedding-gift. These things had been hastily laid aside, to await their possessor's disposal on her return from the country. Bedizened by this unnatural garb Rosalind stood before the mirror, plunging a long look into its depths and reading heaven knows what audacious visions. Perdita was horrified. It was a hideous image of their old rivalry come to life again. She made a step towards her sister, as if to pull off the veil and the flowers. But catching her eyes in the glass, she stopped.

'Farewell, sweetheart,' she said. 'You might at least have waited till I had got out of the house!' And she hurried away from the room.

Mr Lloyd had purchased in Boston a house which to the taste of those days appeared as elegant as it was commodious; and here he very soon established himself with his young wife. He was thus separated by a distance of twenty miles from the residence of his mother-in-law. Twenty miles, in that primitive era of roads and conveyances, were as serious a matter as a hundred at the present day, and Mrs Wingrave saw but little of her daughter during the first twelvemonth of her marriage. She suffered in no small degree from Perdita's absence; and her affliction was not diminished by the fact that Rosalind had fallen into terribly low spirits and was not to be roused or cheered but by change of air and company. The real cause of the young lady's dejection the reader will not be slow to suspect. Mrs Wingrave and her gossips, however, deemed her complaint a mere bodily ill, and doubted not that she would obtain relief from the remedy just mentioned. Her mother accordingly proposed, on her behalf, a visit to certain relatives on the paternal side, established in New York, who had long complained that they were able to see so little of their New England cousins. Rosalind was despatched to these good people, under a suitable escort, and remained with them for several months. In the interval her brother Bernard, who had begun the practice of the law, made up his mind to take a wife. Rosalind came home to the wedding, apparently cured of her heartache, with bright roses and lilies in her face and a proud smile on her lips. Arthur Lloyd came over from Boston to see his brother-in-law married, but without his wife, who was expecting very soon to present him with an heir. It was nearly a year since Rosalind had seen him. She was glad—she hardly knew why—that Perdita had stayed at home. Arthur looked happy, but he was more grave and important than before his marriage. She thought he looked 'interesting'—for although the word, in its modern sense, was not then invented, we may be sure that the idea was. The truth is, he was simply anxious about his wife and her coming ordeal. Nevertheless, he by no means failed to observe Rosalind's beauty and splendour, and to note how she effaced the poor little bride. The allowance that Perdita had enjoyed for her dress had now been transferred to her sister, who turned it to wonderful account. On the morning after the wedding he had a lady's saddle put on the horse of the servant who had come with him from town, and went out with the young girl for a ride. It was a keen, clear morning in January; the ground was bare and hard, and the horses in good condition—to say nothing of Rosalind, who was charming in her hat and plume, and her dark blue riding coat, trimmed with fur. They rode all the morning, lost their way and were obliged to stop for dinner at a farm-house. The early winter dusk had fallen when they got home. Mrs Wingrave met them with a long face. A messenger had arrived at noon from Mrs Lloyd; she was beginning to be ill, she desired her husband's immediate return. The young man, at the thought that he had lost several hours, and that by hard riding he might already have been with his wife, uttered a passionate oath. He barely consented to stop for a mouthful of supper, but mounted the messenger's horse and started off at a gallop.

He reached home at midnight. His wife had been delivered of a little girl. 'Ah, why weren't you with me?' she said, as he came to her bedside.

'I was out of the house when the man came. I was with Rosalind,' said Lloyd, innocently.

Mrs Lloyd made a little moan, and turned away. But she continued to do very well, and for a week her improvement was uninterrupted. Finally, however, through some indiscretion in the way of diet or exposure, it was checked, and the poor lady grew rapidly worse. Lloyd was in despair. It very soon became evident that she was breathing her last. Mrs Lloyd came to a sense of her approaching end, and declared that she was reconciled with death. On the third evening after the change took place she told her husband that she felt she should not get through the night. She dismissed her servants, and also requested her mother to withdraw—Mrs Wingrave having arrived on the preceding day. She had had her infant placed on the bed beside her, and she lay on her side, with the child against her breast, holding her husband's hands. The night-lamp was hidden behind the heavy curtains of the bed, but the room was illuminated with a red glow from the immense fire of logs on the hearth.

'It seems strange not to be warmed into life by such a fire as that,' the young woman said, feebly trying to smile. 'If I had but a little of it in my veins! But I have given all *my* fire to this little spark of mortality.' And she dropped her eyes on her child. Then raising them she looked at her husband with a long, penetrating gaze. The last feeling which lingered in her heart was one of suspicion. She had not recovered from the shock which Arthur had given her by telling her that in the hour of her agony he had been with Rosalind. She trusted her husband very nearly as well as she loved him; but now that she was called away forever she felt a cold horror of her sister. She felt in her soul that Rosalind had never ceased to be jealous of her good fortune; and a year of happy security had not effaced the young girl's image, dressed in her wedding-garments, and smiling with simulated triumph. Now that Arthur was to be alone, what might not Rosalind attempt? She was beautiful, she was engaging; what arts might she not use, what impression might she not make upon the young man's saddened heart? Mrs Lloyd looked at her husband in silence. It seemed hard, after all, to doubt of his constancy. His fine eyes were filled with tears; his face was convulsed with weeping; the clasp of his hands was warm and passionate. How noble he looked, how tender, how faithful and devoted! 'Nay,' thought Perdita, 'he's not for such a one as Rosalind. He'll never forget me. Nor does Rosalind truly care for him; she cares only for vanities and finery and jewels.' And she lowered her eyes on her white hands, which her husband's liberality had covered with rings, and on the lace ruffles which trimmed the edge of her night-dress. 'She covets my rings and my laces more than she covets my husband.'

At this moment the thought of her sister's rapacity seemed to cast a dark shadow between her and the helpless figure of her little girl. 'Arthur,' she said, 'you must take off my rings. I shall not be buried in them. One of these days my daughter shall wear them—my rings and my laces and silks. I had them all brought out and shown me today. It's a great wardrobe—there's not such another in the Province; I can say it without vanity, now that I have done with it. It will be a great inheritance for my daughter when she grows into a young woman. There are things there that a man never buys twice, and if they are lost you will never again see the like. So you will watch them well. Some dozen things I have left to Rosalind; I have named them to my mother. I have given her that blue and silver; it was meant for her; I wore it only once, I looked ill in it. But the rest are to be sacredly kept for this little innocent. It's such a providence that she should be my colour; she can wear my gowns; she has her mother's eyes. You know the same fashions come back every twenty years. She can wear my gowns as they are. They will lie there quietly waiting till she grows into them—wrapped in camphor and rose-leaves, and keeping their colours in the sweet-scented darkness. She shall have black hair, she shall wear my carnation satin. Do you promise me, Arthur?'

'Promise you what, dearest?'

'Promise me to keep your poor little wife's old gowns.'

'Are you afraid I shall sell them?'

'No, but that they may get scattered. My mother will have them properly wrapped up, and you shall lay them away under a double-lock. Do you know the great chest in the attic, with the iron bands? There is no end to what it will hold. You can put them all there. My mother and the housekeeper will do it, and give you the key. And you will keep the key in your secretary, and never give it to anyone but your child. Do you promise me?'

'Ah, yes, I promise you,' said Lloyd, puzzled at the intensity with which his wife appeared to cling to this idea.

'Will you swear?' repeated Perdita.

'Yes, I swear.'

'Well—I trust you—I trust you,' said the poor lady, looking into his eyes with eyes in which, if he had suspected her vague apprehensions, he might have read an appeal quite as much as an assurance.

Lloyd bore his bereavement rationally and manfully. A month after his wife's death, in the course of business, circumstances arose which offered him an opportunity of going to England. He took advantage of it, to change the current of his thoughts. He was absent nearly a year, during which his little girl was tenderly nursed and guarded by her grandmother. On his return he had his house again thrown open, and announced his intention of keeping the same state as during his wife's lifetime. It very soon came to be predicted that he would marry again, and there were at least a dozen young women of whom one may say that it was by no fault of theirs that, for six months after his return, the prediction did not come true. During this interval he still left his little daughter in Mrs Wingrave's hands, the latter assuring him that a change of residence at so tender an age would be full of danger for her health. Finally, however, he declared that his heart longed for his daughter's presence and that she must be brought up to town. He sent his coach and his housekeeper to fetch her home. Mrs Wingrave was in terror lest something should befall her on the road; and, in accordance with this feeling, Rosalind offered to accompany her. She could return the next day. So she went up to town with her little niece, and Mr Lloyd met her on the threshold of his house, overcome with her kindness and with paternal joy. Instead of returning the next day Rosalind stayed out the week; and when at last she reappeared, she had only come for her clothes. Arthur would not hear of her coming home, nor would the baby. That little person cried and choked if Rosalind left her; and at the sight of her grief Arthur lost his wits, and swore that she was going to die. In fine, nothing would suit them but that the aunt should remain until the little niece had grown used to strange faces.

It took two months to bring this consummation about; for it was not until this period had elapsed that Rosalind took leave of her brother-in-law. Mrs Wingrave had shaken her head over her daughter's absence; she had declared that it was not becoming, that it was the talk of the whole country. She had reconciled herself to it only because, during the girl's visit, the household enjoyed an unwonted term of peace. Bernard Wingrave had brought his wife home to live, between whom and her sister-in-law there was as little love as you please. Rosalind was perhaps no angel; but in the daily practice of life she was a sufficiently good-natured girl, and if she quarrelled with Mrs Bernard, it was not without provocation. Quarrel, however, she did, to the great annoyance not only of her antagonist, but of the two spectators of these constant altercations. Her stay in the household of her brother-in-law, therefore, would have been delightful, if only because it removed her from contact with the object of her antipathy at home. It was doubly—it was ten times—delightful, in that it kept her near the object of her early passion. Mrs Lloyd's sharp suspicions had fallen very far short of the truth. Rosalind's sentiment had been a passion at first, and a passion it remained—a passion of whose radiant heat, tempered to the delicate state of his feelings, Mr Lloyd very soon felt the influence. Lloyd, as I have hinted, was not a modern Petrarch; it was not in his nature to practise an ideal constancy. He had not been many days in the house with his sister-in-law before he began to assure himself that she was, in the language of that day, a devilish fine woman. Whether Rosalind really practised those insidious arts that her sister had been tempted to impute to her it is needless to enquire. It is enough to say that she found means to appear to the very best advantage. She used to seat herself every morning before the big fireplace in the dining-room, at work upon a piece of tapestry, with her little niece disporting herself on the carpet at her feet, or on the train of her dress, and playing with her woollen balls. Lloyd would have been a very stupid fellow if he had remained insensible to the rich suggestions of this charming picture. He was exceedingly fond of his little girl, and was never weary of taking her in his arms and tossing her up and down, and making her crow with delight. Very often, however, he would venture upon greater liberties than the young lady was yet prepared to allow, and then she would suddenly vociferate her displeasure. Rosalind, at this, would drop her tapestry, and put out her handsome hands with the serious smile of the young girl whose virgin fancy has revealed to her all a mother's healing arts. Lloyd would give up the child, their eyes would meet, their hands would touch, and Rosalind would extinguish the little girl's sobs upon the snowy folds of the kerchief that crossed her bosom. Her dignity was perfect, and nothing could be more discreet than the manner in which she accepted her brother-in-law's hospitality. It may almost be said, perhaps, that there was something harsh in her reserve. Lloyd had a provoking feeling that she was in the house and yet was unapproachable. Half-an-hour after supper, at the very outset of the long winter evenings, she would light her candle, make the young man a most respectful curtsey, and march off to bed. If these were arts, Rosalind was a great artist. But their effect was so gentle, so gradual, they were calculated to work upon the young widower's fancy with a *crescendo* so finely shaded, that, as the reader has seen, several weeks elapsed before Rosalind began to feel sure that her returns would cover her outlay. When this became morally certain she packed up her trunk and returned to her mother's house. For three days she waited: on the fourth Mr Lloyd made his appearance—a respectful but pressing suitor. Rosalind heard him to the end, with great humility, and accepted him with infinite modesty. It is hard to imagine that Mrs Lloyd would have forgiven her husband; but if anything might have disarmed her resentment it would have been the ceremonious continence of this interview. Rosalind imposed upon her lover but a short probation. They were married, as was becoming, with great privacy—almost with secrecy—in the hope perhaps, as was waggishly remarked at the time, that the late Mrs Lloyd wouldn't hear of it.

The marriage was to all appearance a happy one, and each party obtained what each had desired—Lloyd 'a devilish fine woman', and Rosalind—but Rosalind's desires, as the reader will have observed, had remained a good deal of a mystery. There were, indeed, two blots upon their felicity, but time would perhaps efface them. During the first three years of her marriage Mrs Lloyd failed to become a mother, and her husband on his side suffered heavy losses of money. This latter circumstance compelled a material retrenchment in his expenditure, and Rosalind was perforce less of a fine lady than her sister had been. She contrived, however, to carry it like a woman of considerable fashion. She had long since ascertained that her sister's copious wardrobe had been sequestrated for the benefit of her daughter, and that it lay languishing in thankless gloom in the dusty attic. It was a revolting thought that these exquisite fabrics should await the good pleasure of a little girl who sat in a high chair and ate bread-and-milk with a wooden spoon. Rosalind had the good taste, however, to say nothing about the matter until several months had expired. Then, at last, she timidly broached it to her husband. Was it not a pity that so much finery should be lost?—for lost it would be, what with colours fading, and moths eating it up, and the change of fashions. But Lloyd gave her so abrupt and peremptory a refusal, that she saw, for the present, her attempt was vain. Six months went by, however, and brought with them new needs and new visions. Rosalind's thoughts hovered lovingly about her sister's relics. She went up and looked at the chest in which they lay imprisoned. There was a sullen defiance in its three great padlocks and its iron bands which only quickened her cupidity. There was something exasperating in its incorruptible immobility. It was like a grim and grizzled old household servant, who locks his jaws over a family secret. And then there was a look of capacity in its vast extent, and a sound as of dense fullness, when Rosalind knocked its side with the toe of her little shoe, which caused her to flush with baffled longing. 'It's absurd,' she cried; 'it's improper, it's wicked'; and she forthwith resolved upon another attack upon her husband. On the following day, after dinner, when he had had his wine, she boldly began it. But he cut her short with great sternness.

'Once for all, Rosalind,' said he, 'it's out of the question. I shall be gravely displeased if you return to the matter.'

'Very good,' said Rosalind. 'I am glad to learn the esteem in which I am held. Gracious heaven,' she cried, 'I am a very happy woman! It's an agreeable thing to feel one's self sacrificed to a caprice!' And her eyes filled with tears of anger and disappointment.

Lloyd had a good-natured man's horror of a woman's sobs, and he attempted—I may say he condescended—to explain. 'It's not a caprice, dear, it's a promise,' he said—'an oath.'

'An oath? It's a pretty matter for oaths! and to whom, pray?'

'To Perdita,' said the young man, raising his eyes for an instant, and immediately dropping them.

'Perdita—ah, Perdita!' and Rosalind's tears broke forth. Her bosom heaved with stormy sobs—sobs which were the long-deferred sequel of the violent fit of weeping in which she had indulged herself on the night when she discovered her sister's betrothal. She had hoped, in her better moments, that she had done with her jealousy; but her temper, on that occasion, had taken an ineffaceable hold. 'And pray, what right had Perdita to dispose of my future?' she cried. 'What right had she to bind you to meanness and cruelty? Ah, I occupy a dignified place, and I make a very fine figure! I am welcome to what Perdita has left! And what has she left? I never knew till now how little! Nothing, nothing, nothing.'

This was very poor logic, but it was very good as a 'scene'. Lloyd put his arm around his wife's waist and tried to kiss her, but she shook him off with magnificent scorn. Poor fellow! he had coveted a 'devilish fine woman', and he had got one. Her scorn was intolerable. He walked away with his ears tingling—irresolute, distracted. Before him was his secretary, and in it the sacred key which with his own hand he had turned in the triple lock. He marched up and opened it, and took the key from a secret drawer, wrapped in a little packet which he had sealed with his own honest bit of glazonry. *Je garde*, said the motto—'I keep.' But he was ashamed to put it back. He flung it upon the table beside his wife.

'Put it back!' she cried. 'I want it not. I hate it!'

'I wash my hands of it,' cried her husband. 'God forgive me!'

Mrs Lloyd gave an indignant shrug of her shoulders, and swept out of the room, while the young man retreated by another door. Ten minutes later Mrs Lloyd returned, and found the room occupied by her little stepdaughter and the nursery-maid. The key was not on the table. She glanced at the child. Her little niece was perched on a chair, with the packet in her hands. She had broken the seal with her own small fingers. Mrs Lloyd hastily took possession of the key.

At the habitual supper-hour Arthur Lloyd came back from his counting-room. It was the month of June, and supper was served by daylight. The meal was placed on the table, but Mrs Lloyd failed to make her appearance. The servant whom his master sent to call her came back with the assurance that her room was empty, and that the women informed him that she had not been seen since dinner. They had, in truth, observed her to have been in tears, and, supposing her to be shut up in her chamber, had not disturbed her. Her husband called her name in various parts of the house, but without response. At last it occurred to him that he might find her by taking the way to the attic. The thought gave him a strange feeling of discomfort, and he bade his servants remain behind, wishing no witness in his quest. He reached the foot of the staircase leading to the topmost flat, and stood with his hands on the banisters, pronouncing his wife's name. His voice trembled. He called again louder and more firmly. The only sound which disturbed the absolute silence was a faint echo of his own tones, repeating his question under the great eaves. He nevertheless felt irresistibly moved to ascend the staircase. It opened upon a wide hall, lined with wooden closets, and terminating in a window which looked westward, and admitted the last rays of the sun. Before the window stood the great chest. Before the chest, on her knees, the young man saw with amazement and horror the figure of his wife. In an instant he crossed the interval between them, bereft of utterance. The lid of the chest stood open, exposing, amid their perfumed napkins, its treasure of stuffs and jewels. Rosalind had fallen backward from a kneeling posture, with one hand supporting her on the floor and the other pressed to her heart. On her limbs was the stiffness of death, and on her face, in the fading light of the sun, the terror of something more than death. Her lips were parted in entreaty, in dismay, in agony; and on her blanched brow and cheeks there glowed the marks of ten hideous wounds from two vengeful ghostly hands.

Pichon & Sons, of the Croix Rousse

ANONYMOUS

Giraudier, *pharmacien, première classe*, is the legend, recorded in huge, ill-proportioned letters, which directs the attention of the stranger to the most prosperous-looking shop in the grand *place* of La Croix Rousse, a well-known suburb of the beautiful city of Lyons, which has its share of the shabby gentility and poor pretence common to the suburban commerce of great towns.

Giraudier is not only *pharmacien* but *propriétaire*, though not by inheritance; his possession of one of the prettiest and most prolific of the small vineyards in the beautiful suburb, and a charming inconvenient house, with low ceilings, liliputian bedrooms, and a profusion of *persiennes, jalousies*, and *contrevents*, comes by purchase. This enviable little *terre* was sold by the Nation, when that terrible abstraction transacted the public business of France; and it was bought very cheaply by the strong-minded father of the Giraudier of the present, who was not disturbed by the evil reputation which the place had gained, at a time when the peasants of France, having been bullied into a renunciation of religion, eagerly cherished superstition. The Giraudier of the present cherishes the particular superstition in question affectionately; it reminds him of an uncommonly good bargain made in his favour, which is always a pleasant association of ideas, especially to a Frenchman still more especially to a Lyonnais; and it attracts strangers to his *pharmacie*, and leads to transactions in *Grand Chartreuse* and *Crême de Roses*, ensuing naturally on the narration of the history of Pichon and Sons. Giraudier is not of aristocratic principles and sympathies; on the contrary, he has decided republican leanings, and considers *Le Progrès* a masterpiece of journalistic literature; but, as he says simply and strongly, 'it is not because a man is a marquis that one is not to keep faith with him; a bad action is not good because it harms a good-for-nothing of a noble; the more when that good-for-nothing is no longer a noble, but *pour rire*'. At the easy price of acquiescence in these sentiments, the stranger hears one of the most authentic, best-remembered, most popular of the many traditions of the bad old times 'before General Buonaparte', as Giraudier, who has no sympathy with any later designation of *le grand homme*, calls the Emperor, whose statue one can perceive—a speck in the distance—from the threshold of the *pharmacie*.

The Marquis de Sénanges, in the days of the triumph of the great Revolution, was fortunate enough to be out of France, and wise enough to remain away from that country, though he persisted, long after the old *régime* was as dead as the Ptolemies, in believing it merely suspended, and the Revolution a lamentable accident of vulgar complexion, but happily temporary duration. The Marquis de Sénanges, who affected the *style régence*, and was the politest of infidels and the most refined of voluptuaries, got on indifferently in inappreciative foreign parts; but the members of his family—his brother and sisters, two of whom were guillotined, while the third escaped to Savoy and found refuge there in a convent of her order—got on exceedingly ill in France. If the *ci-devant* Marquis had had plenty of money to expend in such feeble imitations of his accustomed pleasures as were to be had out of Paris, he would not have been much affected by the fate of his relatives. But money became exceedingly scarce; the Marquis had actually beheld many of his peers reduced to the necessity of earning the despicable but indispensable article after many ludicrous fashions. And the duration of this absurd upsetting of law, order, privilege, and property began to assume unexpected and very unpleasant proportions.

The Château de Sénanges, with its surrounding lands, was confiscated to the Nation, during the third year of the 'emigration' of the Marquis de Sénanges; and the greater part of the estate was purchased by a thrifty, industrious, and rich *avocat*, named Prosper Alix, a widower with an only daughter. Prosper Alix enjoyed the esteem of the entire neighbourhood. First, he was rich; secondly, he was of a taciturn disposition, and of a neutral tint in politics. He had done well under the old *régime*, and he was doing well under the new—thank God, or the Supreme Being, or the First Cause, or the goddess Reason herself, for all—he would have invoked Dagon, Moloch, or Kali, quite as readily as the Saints and the Madonna, who had gone so utterly out of fashion of late. Nobody was afraid to speak out before Prosper Alix; he was not a spy; and though a cold-hearted man, except in the instance of his only daughter, he never harmed anybody.

Very likely it was because he was the last person in the vicinity whom anybody would have suspected of being applied to by the dispossessed family, that the son of the Marquis's brother, a young man of promise, of courage, of intellect, and of morals of decidedly a higher calibre than those actually and traditionally imputed to the family, sought the aid of the new possessor of the Château de Sénanges, which had changed its old title for that of the Maison Alix. The father of M. Paul de Sénanges had perished in the September massacres; his mother had been guillotined at Lyons; and he—who had been saved by the inter-position of a young comrade, whose father had, in the wonderful rotations of the wheel of Fate, acquired authority in the place where he had once esteemed the notice of the nephew of the Marquis a crowning honour for his son—had passed through the common vicissitudes of that dreadful time, which would take a volume for their recital in each individual instance.

Paul de Sénanges was a handsome young fellow, frank, high-spirited, and of a brisk and happy temperament; which, however, modified by the many misfortunes he had undergone, was not permanently changed. He had plenty of capacity for enjoyment in him still; and as his position was very isolated, and his mind had become enlightened on social and political matters to an extent in which the men of his family would have discovered utter degradation and the women diabolical possession, he would not have been very unhappy if, under the new condition of things, he could have lived in his native country and gained an honest livelihood. But he could not do that, he was too thoroughly 'suspect'; the antecedents of his family were too powerful against him: his only chance would have been to have gone into the popular camp as an extreme, violent partisan, to have out-Heroded the revolutionary Herods; and that Paul de Sénanges was too honest to do. So he was reduced to being thankful that he had escaped with his life, and to watching for an opportunity of leaving France and gaining some country where the reign of liberty, fraternity, and equality was not quite so oppressive.

The long-looked-for opportunity at length offered itself, and Paul de Sénanges was instructed by his uncle the Marquis that he must contrive to reach Marseilles, whence he should be transported to Spain—in which country the illustrious emigrant was then residing—by a certain named date. His uncle's communication arrived safely, and the plan proposed seemed a secure and eligible one. Only in two respects was it calculated to make Paul de Sénanges thoughtful. The first was, that his uncle should take any interest in the matter of his safety; the second, what could be the nature of a certain deposit which the Marquis's letter directed him to procure, if possible, from the Château de Sénanges. The fact of this injunction explained, in some measure, the first of the two difficulties. It was plain that whatever were the contents of this packet which he was to seek for, according to the indications marked on a ground-plan drawn by his uncle and enclosed in the letter, the Marquis wanted them, and could not procure them except by the agency of his nephew. That the Marquis should venture to direct Paul de Sénanges to put himself in communication with Prosper Alix, would have been surprising to any one acquainted only with the external and generally understood features of the character of the new proprietor of the Château de Sénanges. But a few people knew Prosper Alix thoroughly, and the Mārquis was one of the number; he was keen enough to know in theory that, in the case of a man with only one weakness, that is likely to be a very weak weakness indeed, and to apply the theory to the *avocat*. The beautiful, pious, and aristocratic mother of Paul de Sénanges—a lady to whose superiority the Marquis had rendered the distinguished testimony of his dislike, not hesitating to avow that she was 'much too good for *his* taste'—had been very fond of, and very kind to, the motherless daughter of Prosper Alix, and he held her memory in reverence which he accorded to nothing beside, human or divine, and taught his daughter the matchless worth of the friend she had lost. The Marquis knew this, and though he had little sympathy with the sentiment, he believed he might use it in the present instance to his own profit, with safety. The event proved that he was right. Private negotiations, with the manner of whose transaction we are not concerned, passed between the *avocat* and the *ci-devant* Marquis; and the young man, then leading a life in which skulking had a large share, in the vicinity of Dijon, was instructed to present himself at the Maison Alix, under the designation of Henri Glaire, and in the character of an artist in house-decoration. The circumstances of his life in childhood and boyhood had led to his being almost safe from recognition as a man at Lyons; and, indeed, all the people on the *ci-devant* visiting-list of the château had been pretty nearly killed off, in the noble and patriotic ardour of the revolutionary times.

The ancient Château de Sénanges was proudly placed near the summit of the 'Holy Hill', and had suffered terrible depredations when the church at Fourvières was sacked, and the shrine desecrated with that ingenious impiety which is characteristic of the French; but it still retained somewhat of its former heavy grandeur. The château was much too large for the needs, tastes, or ambition of its present owner, who was too wise, if even he had been of an ostentatious disposition, not to have sedulously resisted its promptings. The jealousy of the nation of brothers was easily excited, and departure from simplicity and frugality was apt to be commented upon by domiciliary visits, and the eager imposition of fanciful fines. That portion of the vast building occupied by Prosper Alix and the *citoyenne* Berthe, his daughter, presented an appearance of well-to-do comfort and modest ease, which contrasted with the grandiose proportions and the elaborate decorations of the wide corridors, huge flat staircases, and lofty panelled apartments. The *avocat* and his daughter lived quietly in the old place, hoping, after a general fashion, for better times, but not finding the present very bad; the father becoming day by day more pleasant with his bargain, the daughter growing fonder of the great house, and the noble *bocages*, of the scrappy little vineyards, struggling for existence on the sunny hill-side, and the place where the famous shrine had been. They had done it much damage; they had parted its riches among them; the once ever-open doors were shut, and the worn flags were untrodden; but nothing could degrade it, nothing could destroy what had been, in the mind of Berthe Alix, who was as devout as her father was unconcernedly unbelieving. Berthe was wonderfully well educated for a Frenchwoman of that period, and surprisingly handsome for a Frenchwoman of any. Not too tall to offend the taste of her compatriots, and not too short to be dignified and graceful, she had a symmetrical figure, and a small, well-poised head, whose profuse, shining, silken dark-brown hair she wore as nature intended, in a shower of curls, never touched by the hand of the coiffeur—curls which clustered over her brow, and fell far down on her shapely neck. Her features were fine; the eyes very dark, and the mouth very red; the complexion clear and rather pale, and the style of the face and its expression lofty. When Berthe Alix was a child, people were accustomed to say she was pretty and refined enough to belong to the aristocracy; nobody would have dared to say so now, prettiness and refinement, together with all the other virtues admitted to a place on the patriotic roll, having become national property.

Berthe loved her father dearly. She was deeply impressed with the sense of her supreme importance to him, and fully comprehended that he would be influenced by and through her when all other persuasion or argument would be unavailing. When Prosper Alix wished and intended to do anything rather mean or selfish, he did it without letting Berthe know; and when he wished to leave undone something which he knew his daughter would decide ought to be done, he carefully concealed from her the existence of the dilemma. Nevertheless, this system did not prevent the father and daughter being very good and even confidential friends. Prosper Alix loved his daughter immeasurably, and respected her more than he respected anyone in the world. With regard to her persevering religiousness, when such things were not only out of fashion and date, but illegal as well, he was very tolerant. Of course it was weak, and an absurdity; but every woman, even his beautiful, incomparable Berthe, was weak and absurd on some point or other; and, after all, he had come to the conclusion that the safest weakness with which a woman can be afflicted is that romantic and ridiculous *faiblesse* called piety. So these two lived a happy life together, Berthe's share of it being very secluded, and were wonderfully little troubled by the turbulence with which society was making its tumultuous way to the virtuous serenity of republican perfection.

The communication announcing the project of the *ci-devant* Marquis for the secure exportation of his nephew, and containing the skilful appeal before mentioned, grievously disturbed the tranquillity of Prosper, and was precisely one of those incidents which he would especially have liked to conceal from his daughter. But he could not do so; the appeal was too cleverly made; and utter indifference to it, utter neglect of the letter, which naturally suggested itself as the easiest means of getting rid of a difficulty, would have involved an act of direct and uncompromising dishonesty to which Prosper, though of sufficiently elastic conscience within the limit of professional gains, could not contemplate. The Château de Sénanges was indeed his own lawful property; his without prejudice to the former owners, dispossessed by no act of his. But the *ci-devant* Marquis—confiding in him to an extent which was quite astonishing, except on the *pis-aller* theory, which is so unflattering as to be seldom accepted—announced to him the existence of a certain packet, hidden in the château, acknowledging its value, and urging the need of its safe transmission. This was not his property. He heartily wished he had never learned its existence, but wishing that was clearly of no use; then he wished the nephew of the *ci-devant* might come soon, and take himself and the hidden wealth away with all possible speed. This latter was a more realizable desire, and Prosper settled his mind with it, communicated the interesting but decidedly dangerous secret to Berthe, received her warm sanction, and transmitted to the Marquis, by the appointed means, an assurance that his wishes should be punctually carried out. The absence of an interdiction of his visit before a certain date was to be the signal to M. Paul de Sénanges that he was to proceed to act upon his uncle's instructions; he waited the proper time, the reassuring silence was maintained unbroken, and he ultimately set forth on his journey, and accomplished it in safety.

Preparations had been made at the Maison Alix for the reception of M. Glaire, and his supposed occupation had been announced. The apartments were decorated in a heavy, gloomy style, and those of the *citoyenne* in particular (they had been occupied by a lady who had once been designated as *feue Madame la Marquise*, but who was referred to now as *la mère du ci-devant*) were much in need of renovation. The alcove, for instance, was all that was least gay and most far from simple. The *citoyenne* would have all that changed. On the morning of the day of the expected arrival, Berthe said to her father:

'It would seem as if the Marquis did not know the exact spot in which the packet is deposited. M. Paul's assumed character implies the necessity for a search.'

M. Henri Glaire arrived at the Maison Alix, was fraternally received, and made acquainted with the sphere of his operations. The young man had a good deal of both ability and taste in the line he had assumed, and the part was not difficult to play. Some days were judiciously allowed to pass before the real object of the masquerade was pursued, and during that time cordial relations established them selves between the *avocat* and his guest. The young man was handsome, elegant, engaging, with all the external advantages, and devoid of the vices, errors, and hopeless infatuated unscrupulousness, of his class; he had naturally quick intelligence, and some real knowledge and comprehension of life had been knocked into him by the hard-hitting blows of Fate. His face was like his mother's, Prosper Alix thought, and his mind and tastes were of the very pattern which, in theory, Berthe approved. Berthe, a very unconventional French girl—who though the new era of purity, love, virtue, and disinterestedness ought to do away with marriage by barter as one of its most notable reforms, and had been disenchanted by discovering that the abolition of marriage altogether suited the taste of the incorruptible Republic better—might like, might even love, this young man. She saw so few men, and had no fancy for patriots; she would certainly be obstinate about it if she did chance to love him. This would be a nice state of affairs. This would be a pleasant consequence of the confiding request of the *ci-devant*. Prosper wished with all his heart for the arrival of the concerted signal, which should tell Henri Glaire that he might fulfil the purpose of his sojourn at the Maison Alix, and set forth for Marseilles.

But the signal did not come, and the days—long, beautiful, sunny, soothing summer-days—went on. The painting of the panels of the *citoyenne*'s apartment, which she vacated for that purpose, progressed slowly; and M. Paul de Sénanges, guided by the ground-plan, and aided by Berthe, had discovered the spot in which the jewels of price, almost the last remnants of the princely wealth of the Sénanges, had been hidden by the *femme-de-chambre* who had perished with her mistress, having confided a general statement of the fact to a priest, for transmission to the Marquis. This spot had been ingeniously chosen. The sleeping-apartment of the late Marquis was extensive, lofty, and provided with an alcove of sufficiently large dimensions to have formed in itself a handsome room. This space, containing a splendid but gloomy bed, on an estrade, and hung with rich faded brocade, was divided from the general extent of the apartment by a low railing of black oak, elaborately carved, opening in the centre, and with a flat wide bar along the top, covered with crimson velvet. The curtains were contrived to hang from the ceiling, and, when let down inside the screen of railing, they matched the draperies which closed before the great stone balcony at the opposite end of the room. Since the *avocat*'s daughter had occupied this palatial chamber, the curtains of the alcove had never been drawn, and she had substituted for them a high folding screen of black-and-gold Japanese pattern, also a relic of the grand old times, which stood about six feet on the outside of the rails that shut in her bed. The floor was of shining oak, testifying to the conscientious and successful labours of successive generations of *frotteurs*; and on the spot where the railing of the alcove opened by a pretty quaint device sundering the intertwined arms of a pair of very chubby cherubs, a square space in the floor was also richly carved.

The seekers soon reached the end of their search. A little effort removed the square of carved oak, and underneath they found a casket, evidently of old workmanship, richly wrought in silver, much tarnished but quite intact. It was agreed that this precious deposit should be replaced, and the carved square laid down over it, until the signal for his departure should reach Paul. The little baggage which under any circumstances he could have ventured to allow himself in the dangerous journey he was to undertake, must be reduced, so as to admit of his carrying the casket without exciting suspicion.

The finding of the hidden treasure was not the first joint discovery made by the daughter of the *avocat* and the son of the *ci-devant*. The cogitations of Prosper Alix were very wise, very reasonable; but they were a little tardy. Before he had admitted the possibility of mischief, the mischief was done. Each had found out that the love of the other was indispensable to the happiness of life; and they had exchanged confidences, assurances, protestations, and promises, as freely, as fervently, and as hopefully, as if no such thing as a Republic, one and indivisible, with a keen scent and an unappeasable thirst for the blood of aristocrats, existed. They forgot all about 'Liberty, Fraternity, and Equality'—these egotistical, narrow-minded young people; they also forgot the characteristic alternative to those unparalleled blessings—'Death'. But Prosper Alix did not forget any of these things; and his consternation, his prevision of suffering for his beloved daughter, were terrible, when she told him, with a simple noble frankness which the *grandes dames* of the dead-and-gone time of great ladies had rarely had a chance of exhibiting, that she loved M. Paul de Sénanges, and intended to marry him when the better times should come. Perhaps she meant when that alternative of *death* should be struck off the sacred formula; of course she meant to marry him with the sanction of her father, which she made no doubt she should receive.

Prosper Alix was in pitiable perplexity. He could not bear to terrify his daughter by a full explanation of the danger she was incurring; he could not bear to delude her with false hope. If this young man could be got away at once safely, there was not much likelihood that he would ever be able to return to France. Would Berthe pine for him, or would she forget him, and make a rational, sensible, rich, republican marriage, which would not imperil either her reputation for pure patriotism or her father's? The latter would be the very best thing that could possibly happen, and therefore it was decidedly unwise to calculate upon it; but, after all, it was possible; and Prosper had not the courage, in such a strait, to resist the hopeful promptings of a possibility. How ardently he regretted that he had complied with the prayer of the *ci-devant*! When would the signal for M. Paul's departure come?

Prosper Alix had made many sacrifices, had exercised much self-control for his daughter's sake; but he had never sustained a more severe trial than this, never suffered more than he did now, under the strong necessity for hiding from her his absolute conviction of the impossibility of a happy result for this attachment, in that future to which the lovers looked so fearlessly. He could not even make his anxiety and apprehension known to Paul de Sénanges; for he did not believe the young man had sufficient strength of will to conceal anything so important from the keen and determined observation of Berthe.

The expected signal was not given, and the lovers were incautious. The seclusion of the Maison Alix had all the danger, as well as all the delight, of solitude, and Paul dropped his disguise too much and too often. The servants, few in number, were of the truest patriotic principles, and to some of them the denunciation of the *citoyen*, whom they condescended to serve because the sacred Revolution had not yet made them as rich as he, would have been a delightful duty, a sweet-smelling sacrifice to be laid on the altar of the country. They heard certain names and places mentioned; they perceived many things which led them to believe that Henri Glaire was not an industrial artist and pure patriot, worthy of respect, but a wretched *ci-devant*, resorting to the dignity of labour to make up for the righteous destruction of every other kind of dignity. One day a gardener, of less stoical virtue than his fellows, gave Prosper Alix a warning that the presence of a *ci-devant* upon his premises was suspected, and that he might be certain a domiciliary visit, attended with dangerous results to himself, would soon take place. Of course the *avocat* did not commit himself by any avowal to this lukewarm patriot; but he casually mentioned that Henri Glaire was about to take his leave. What was to be done? He must not leave the neighbourhood without receiving the instructions he was awaiting; but he must leave the house, and be supposed to have gone quite away. Without any delay or hesitation, Prosper explained the facts to Berthe and her lover, and insisted on the necessity for an instant parting. Then the courage and the readiness of the girl told. There was no crying, and very little trembling; she was strong and helpful.

'He must go to Pichon's, father,' she said, 'and remain there until the signal is given. Pichon is a master-mason, Paul,' she continued, turning to her lover, 'and his wife was my nurse. They are avaricious people; but they are fond of me in their way, and they will shelter you faithfully enough, when they know that my father will pay them handsomely. You must go at once, unseen by the servants; they are at supper. Fetch your valise, and bring it to my room. We will put the casket in it, and such of your things as you must take out to make room for it, we can hide under the plank. My father will go with you to Pichon's, and we will communicate with you there as soon as it is safe.'

Paul followed her to the large gloomy room where the treasure lay, and they took the casket from its hiding-place. It was heavy, though not large, and an awkward thing to pack away among linen in a small valise. They managed it, however, and, the brief preparation completed, the moment of parting arrived. Firmly and eloquently, though in haste, Berthe assured Paul of her changeless love and faith, and promised him to wait for him for any length of time in France, if better days should be slow of coming, or to join him in some foreign land, if they were never to come. Her father was present, full of compassion and misgiving. At length he said,

'Come, Paul, you must leave her; every moment is of importance.'

The young man and his betrothed were standing on the spot whence they had taken the casket; the carved rail with the heavy curtains might have been the outer sanctuary of an altar, and they bride and bridegroom before it, with earnest, loving faces, and clasped hands.

'Farewell, Paul,' said Berthe; 'promise me once more, in this the moment of our parting, that you will come to me again, if you are alive, when the danger is past.'

'Whether I am living or dead, Berthe,' said Paul de Sénanges, strongly moved by some sudden inexplicable instinct, 'I will come to you again.'

In a few more minutes, Prosper Alix and his guest, who carried, not without difficulty, the small but heavy leather valise, had disappeared in the distance, and Berthe was on her knees before the *priedieu* of the *ci-devant* Marquise, her face turned towards the 'Holy Hill' of Fourvières.

Pichon, *maître*, and his sons, *garçons-maçons*, were well-to-do people, rather morose, exceedingly avaricious, and of taciturn dispositions; but they were not ill spoken of by their neighbours. They had amassed a good deal of money in their time, and were just then engaged on a very lucrative job. This was the construction of several of the steep descents, by means of stairs, straight and winding, cut in the face of the *côteaux*, by which pedestrians are enabled to descend into the town. Pichon *père* was a *propriétaire* as well; his property was that which is now in the possession of Giraudier, *pharmacien, première classe*, and which was destined to attain a sinister celebrity during his proprietorship. One of the straightest and steepest of the stairways had been cut close to the *terre* which the mason owned, and a massive wall, destined to bound the high-road at the foot of the declivity, was in course of construction.

When Prosper Alix and Paul de Sénanges reached the abode of Pichon, the master-mason, with his sons and workmen, had just completed their day's work, and were preparing to eat the supper served by the wife and mother, a tall gaunt woman, who looked as if a more liberal scale of housekeeping would have done her good, but on whose features the stamp of that devouring and degrading avarice which is the commonest vice of the French peasantry, was set as plainly as on the hard faces of her husband and her sons. The *avocat* explained his business and introduced his companion briefly, and awaited the reply of Pichon *père* without any appearance of inquietude.

'You don't run any risk,' he said; 'at least, you don't run any risk which I cannot make it worth your while to incur. It is not the first time you have received a temporary guest on my recommendation. You know nothing about the citizen Glaire, except that he is recommended to you by me. I am responsible; you can, on occasion, make me so. The citizen may remain with you a short time; can hardly remain long. Say, citizen, is it agreed? I have no time to spare.'

It was agreed, and Prosper Alix departed, leaving M. Paul de Sénanges, convinced that the right, indeed the only, thing had been done, and yet much troubled and depressed.

Pichon *père* was a short, squat, powerfully built man, verging on sixty, whose thick dark grizzled hair, sturdy limbs, and hard hands, on which the muscles showed like cords, spoke of endurance and strength; he was, indeed, noted in the neighbourhood for those qualities. His sons resembled him slightly, and each other closely, as was natural, for they were twins. They were heavy, lumpish fellows, and they made but an ungracious return to the attempted civilities of the stranger, to whom the offer of their mother to show him his room was a decided relief. As he rose to follow the woman, Paul de Sénanges lifted his small valise with difficulty from the floor, on which he had placed it on entering the house, and carried it out of the room in both his arms. The brothers followed these movements with curiosity, and, when the door closed behind their mother and the stranger, their eyes met.

Twenty-four hours had passed away, and nothing new had occurred at the Maison Alix. The servants had not expressed any curiosity respecting the departure of the citizen Glaire, no domiciliary visit had taken place, and Berthe and her father were discussing the propriety of Prosper's venturing, on the pretext of an excursion in another direction, a visit to the isolated and quiet dwelling of the master-mason. No signal had yet arrived. It was agreed that after the lapse of another day, if their tranquillity remained undisturbed, Prosper Alix should visit Paul de Sénanges. Berthe, who was silent and preoccupied, retired to her own room early, and her father, who was uneasy and apprehensive, desperately anxious for the promised communication from the Marquis, was relieved by her absence.

The moon was high in the dark sky, and her beams were flung across the polished oak floor of Berthe's bedroom, through the great window with the stone balcony, when the girl, who had gone to sleep with her lover's name upon her lips in prayer, awoke with a sudden start, and sat up in her bed. An unbearable dread was upon her; and yet she was unable to utter a cry, she was unable to make another movement. Had she heard a voice? No, no one had spoken, nor did she fancy that she heard any sound. But within her, somewhere inside her heaving bosom, something said, 'Berthe!'

And she listened, and knew what it was. And it spoke, and said:

'I promised you that, living or dead, I would come to you again, And I have come to you; but not living.'

She was quite awake. Even in the agony of her fear she looked around, and tried to move her hands, to feel her dress and the bedclothes, and to fix her eyes on some familiar object, that she might satisfy herself, before this racing and beating, this whirling and yet icy chilliness of her blood should kill her outright, that she was really awake.

'I have come to you; but not living.'

What an awful thing that voice speaking within her was! She tried to raise her head and to look towards the place where the moonbeams marked bright lines upon the polished floor, which lost themselves at the foot of the Japanese screen. She forced herself to this effort, and lifted her eyes, wild and haggard with fear, and there, the moonbeams at his feet, the tall black screen behind him, she saw Paul de Sénanges. She saw him; she looked at him quite steadily; she rose, slowly, with a mechanical movement, and stood upright beside her bed, clasping her forehead with her hands, and gazing at him. He stood motionless, in the dress he had worn when he took leave of her, the light-coloured riding-coat of the period, with a short cape, and a large white cravat tucked into the double breast. The white muslin was flecked, and the front of the riding-coat was deeply stained, with blood. He looked at her, and she took a step forward—another—then, with a desperate effort, she dashed open the railing and flung herself on her knees before him, with her arms stretched out as if to clasp him. But he was no longer there; the moonbeams fell clear and cold upon the polished floor, and lost themselves where Berthe lay, at the foot of the screen, her head upon the ground, and every sign of life gone from her.

'Where is the citizen Glaire?' asked Prosper Alix of the *citoyenne* Pichon, entering the house of the master-mason abruptly, and with a stern and threatening countenance. 'I have a message for him; I must see him.'

'I know nothing about him,' replied the *citoyenne*, without turning in his direction, or relaxing her culinary labours. 'He went away from here the next morning, and I did not trouble myself to ask where; that is his affair.'

'He went away? Without letting me know! Be careful, *citoyenne*; this is a serious mattter.'

'So they tell me,' said the woman with a grin, which was not altogether free from pain and fear; 'for you! A serious thing to have a *suspect* in your house, and palm him off on honest people. However, he went away peaceably enough when he knew we had found him out, and that we had no desire to go to prison, or worse, on his account, or yours.'

She was strangely insolent, this woman, and the listener felt his helplessness; he had brought the young man there with such secrecy, he had so carefully provided for the success of concealment.

'Who carried his valise?', Prosper Alix asked her suddenly.

'How should I know?', she replied; but her hands lost their steadiness, and she upset a stew-pan; 'he carried it here, didn't he? and I suppose he carried it away again.'

Prosper Alix looked at her steadily—she shunned his gaze, but she showed no other sign of confusion; then horror and disgust of the woman came over him.

'I must see Pichon,' he said; 'where is he?'

'Where should he be but at the wall? he and the boys are working there, as always. The citizen can see them; but he will remember not to detain them; in a little quarter of an hour the soup will be ready.'

The citizen did see the master-mason and his sons, and after an interview of some duration he left the place in a state of violent agitation and complete discomfiture. The master-mason had addressed to him these words at parting:

'I assert that the man went away at his own free will; but if you do not keep very quiet, I shall deny that he came here at all—you cannot prove he did—and I will denounce you for harbouring a *suspect* and *ci-devant* under a false name. I know a de Sénanges when I see him as well as you, citizen Alix; and, wishing M. Paul a good journey, I hope you will consider about this matter, for truly, my friend, I think you will sneeze in the sack before I shall.'

'We must bear it, Berthe, my child,' said Prosper Alix to his daughter many weeks later, when the fever had left her, and she was able to talk with her father of the mysterious and frightful events which had occurred. 'We are utterly helpless. There is no proof, only the word of these wretches against mine, and certain destruction to me if I speak. We will go to Spain, and tell the Marquis all the truth, and never return, if you would rather not. But, for the rest, we must bear it.'

'Yes, my father,' said Berthe submissively, 'I know we must; but God need not, and I don't believe He will.'

The father and the daughter left France unmolested, and Berthe 'bore it' as well as she could. When better times came they returned, Prosper Alix an old man, and Berthe a stern, silent, handsome woman, with whom no one associated any notions of love or marriage. But long before their return the traditions of the Croix Rousse were enriched by circumstances which led to that before-mentioned capital bargain made by the father of the Giraudier of the present. These circumstances were the violent death of Pichon and his two sons, who were killed by the fall of a portion of the great boundary-wall on the very day of its completion, and the discovery, close to its foundation, at the extremity of Pichon's *terre*, of the corpse of a young man attired in a light-coloured riding-coat, who had been stabbed through the heart.

Berthe Alix lived alone in the Château de Sénanges, under its restored name, until she was a very old woman. She lived long enough to see the golden figure on the summit of the 'Holy Hill', long enough to forget the bad old times, but not long enough to forget or cease to mourn the lover who had kept his promise, and come back to her; the lover who rested in the earth which once covered the bones of the martyrs, and who kept a place for her by his side. She has filled that place for many years. You may see it, when you look down from the second gallery of the bell-tower at Fourvières, following the bend of the outstretched golden arm of Notre Dame.

The château was pulled down some years ago, and there is no trace of its former existence among the vines.

Good times, and bad times, and again good times have come for the Croix Rousse, for Lyons, and for France, since then; but the remembrance of the treachery of Pichon and Sons, and of the retribution which at once exposed and punished their crime, outlives all changes. And once, every year, on a certain summer night, three ghostly figures are seen, by any who have courage and patience to watch for them, gliding along by the foot of the boundary-wall, two of them carrying a dangling corpse, and the other, implements for mason's work and a small leather valise. Giraudier, *pharmacien*, has never seen these ghostly figures, but he describes them with much minuteness; and only the *esprits forts* of the Croix Rousse deny that the ghosts of Pichon and Sons are not yet laid.

Reality or Delusion?

MRS HENRY WOOD

This is a ghost story. Every word of it is true. And I don't mind confessing that for ages afterwards some of us did not care to pass the spot alone at night. Some people do not care to pass it yet.

It was autumn, and we were at Crabb Cot. Lena had been ailing; and in October Mrs Todhetley proposed to the Squire that they should remove with her there, to see if the change would do her good.

We Worcestershire people call North Crabb a village; but one might count the houses in it, little and great, and not find four-and-twenty. South Crabb, half a mile off, is ever so much larger; but the church and school are at North Crabb.

John Ferrar had been employed by Squire Todhetley as a sort of overlooker on the estate, or working bailiff. He had died the previous winter; leaving nothing behind him except some debts; for he was not provident; and his handsome son Daniel. Daniel Ferrar, who was rather superior as far as education went, disliked work: he would make a show of helping his father, but it came to little. Old Ferrar had not put him to any particular trade or occupation, and Daniel, who was as proud as Lucifer, would not turn to it himself. He liked to be a gentleman. All he did now was to work in his garden, and feed his fowls, ducks, rabbits, and pigeons, of which he kept a great quantity, selling them to the houses around and sending them to market.

But, as every one said, poultry would not maintain him. Mrs Lease, in the pretty cottage hard by Ferrar's, grew tired of saying it. This Mrs Lease and her daughter, Maria, must not be confounded with Lease the pointsman: they were in a better condition of life, and not related to him. Daniel Ferrar used to run in and out of their house at will when a boy, and he was now engaged to be married to Maria. She would have a little money, and the Leases were respected in North Crabb. People began to whisper a query as to how Ferrar got his corn for the poultry: he was not known to buy much: and he would have to go out of his house at Christmas, for its owner, Mr Coney, had given him notice. Mrs Lease, anxious about Maria's prospects, asked Daniel what he intended to do then, and he answered, 'Make his fortune: he should begin to do it as soon as he could turn himself round.' But the time was going on, and the turning round seemed to be as far off as ever.

After Midsummer, a niece of the schoolmistress's, Miss Timmens, had come to the school to stay: her name was Harriet Roe. The father, Humphrey Roe, was half-brother to Miss Timmens. He had married a Frenchwoman, and lived more in France than in England until his death. The girl had been christened Henriette; but North Crabb, not understanding much French, converted it into Harriet. She was a showy, free-mannered, good-looking girl, and made speedy acquaintance with Daniel Ferrar; or he with her. They improved upon it so rapidly that Maria Lease grew jealous, and North Crabb began to say he cared for Harriet more than for Maria. When Tod and I got home the latter end of October, to spend the Squire's birthday, things were in this state. James Hill, the bailiff who had been taken on by the Squire in John Ferrar's place (but a far inferior man to Ferrar; not much better, in fact, than a common workman), gave us an account of matters in general. Daniel Ferrar had been drinking lately, Hill added, and his head was not strong enough to stand it; and he was also beginning to look as if he had some care upon him.

'A nice lot, he, for them two women to be fighting for,' cried Hill, who was no friend to Ferrar. 'There'll be mischief between 'em if they don't draw in a bit. Maria Lease is next door to mad over it, I know; and t'other, finding herself the best liked, crows over her. It's something like the Bible story of Leah and Rachel, young gents, Dan Ferrar likes the one, and he's bound by promise to the t'other. As to the French jade,' concluded Hill, giving his head a toss, 'she'd make a show of liking any man that followed her, she would; a dozen of 'em on a string.'

It was all very well for surly Hill to call Daniel Ferrar a 'nice lot', but he was the best-looking fellow in church on Sunday morning—well-dressed too. But his colour seemed brighter; and his hands shook as they were raised, often, to push back his hair, that the sun shone upon through the south-window, turning it to gold. He scarcely looked up, not even at Harriet Roe, with her dark eyes roving everywhere, and her streaming pink ribbons. Maria Lease was pale, quiet, and nice, as usual; she had no beauty, but her face was sensible, and her deep grey eyes had a strange and curious earnestness. The new parson preached, a young man just appointed to the parish of Crabb. He went in for great observances of Saints' days, and told his congregation that he should expect to see them at church on the morrow, which would be the Feast of All Saints.

Daniel Ferrar walked home with Mrs Lease and Maria after service, and was invited to dinner. I ran across to shake hands with the old dame, who had once nursed me through an illness, and promised to look in and see her later. We were going back to school on the morrow. As I turned away, Harriet Roe passed, her pink ribbons and her cheap gay silk dress gleaming in the sunlight. She stared at me, and I stared back again. And now, the explanation of matters being over, the real story begins. But I have to tell some of it as it was told by others.

The tea-things waited on Mrs Lease's table in the afternoon; waited for Daniel Ferrar. He had left them shortly before to go and attend to his poultry. Nothing had been said about his coming back for tea: that he would do so had been looked upon as a matter of course. But he did not make his appearance, and the tea was taken without him. At half-past five the church-bell rang out for evening service, and Maria put her things on. Mrs Lease did not go out at night.

'You are starting early, Maria. You'll be in church before other people.'

'That won't matter, mother.'

A jealous suspicion lay on Maria—that the secret of Daniel Ferrar's absence was his having fallen in with Harriet Roe: perhaps had gone of his own accord to seek her. She walked slowly along. The gloom of dusk, and a deep dusk, had stolen over the evening, but the moon would be up later. As Maria passed the school-house, she halted to glance in at the little sitting-room window: the shutters were not closed yet, and the room was lighted by the blazing fire. Harriet was not there. She only saw Miss Timmens, the mistress, who was putting on her bonnet before a hand-glass propped upright on the mantelpiece. Without warning, Miss Timmens turned and threw open the window. It was only for the purpose of pulling-to the shutters, but Maria thought she must have been observed, and spoke.

'Good evening, Miss Timmens.'

'Who is it?' cried out Miss Timmens, in answer, peering into the dusk. 'Oh, it's you, Maria Lease! Have you seen anything of Harriet? She went off somewhere this afternoon, and never came in to tea.'

'I have not seen her.'

'She's gone to the Batleys', I'll be bound. She knows I don't like her to be with the Batley girls: they make her ten times flightier than she would otherwise be.'

Miss Timmens drew in her shutters with a jerk, without which they would not close, and Maria Lease turned away.

'Not at the Batleys', not at the Batleys', but with *him*,' she cried, in bitter rebellion, as she turned away from the church. From the church, not to it. Was Maria to blame for wishing to see whether she was right or not?—for walking about a little in the thought of meeting them? At any rate it is what she did. And had her reward; such as it was.

As she was passing the top of the withy walk, their voices reached her ear. People often walked there, and it was one of the ways to South Crabb. Maria drew back amidst the trees, and they came on: Harriet Roe and Daniel Ferrar, walking arm-in-arm.

'I think I had better take it off,' Harriet was saying. 'No need to invoke a storm upon my head. And that would come in a shower of hail from stiff old Aunt Timmens.'

The answer seemed one of quick accent, but Ferrar spoke low. Maria Lease had hard work to control herself: anger, passion, jealousy, all blazed up. With her arms stretched out to a friendly tree on either side—with her heart beating—with her pulses coursing on to fever-heat, she watched them across the bit of common to the road. Harriet went one way then; he another, in the direction of Mrs Lease's cottage. No doubt to fetch her—Maria—to church, with a plausible excuse of having been detained. Until now she had had no proof of his falseness; had never perfectly believed in it.

She took her arms from the trees and went forward, a sharp faint cry of despair breaking forth on the night air. Maria Lease was one of those silent-natured girls who can never speak of a wrong like this. She had to bury it within her; down, down, out of sight and show; and she went into church with her usual quiet step. Harriet Roe with Miss Timmens came next, quite demure, as if she had been singing some of the infant scholars to sleep at their own homes. Daniel Ferrar did not go to church at all: he stayed, as was found afterwards, with Mrs Lease.

Maria might as well have been at home as at church: better perhaps that she had been. Not a syllable of the service did she hear: her brain was a sea of confusion; the tumult within it rising higher and higher. She did not hear even the text, 'Peace, be still', or the sermon; both so singularly appropriate. The passions in men's minds, the preacher said, raged and foamed just like the angry waves of the sea in a storm, until Jesus came to still them.

I ran after Maria when church was over, and went in to pay the promised visit to old Mother Lease. Daniel Ferrar was sitting in the parlour. He got up and offered Maria a chair at the fire, but she turned her back and stood at the table under the window, taking off her gloves. An open Bible was before Mrs Lease: I wondered whether she had been reading aloud to Daniel.

'What was the text, child?' asked the old lady.

No answer.

'Do you hear, Maria! What was the text?'

Maria turned at that, as if suddenly awakened. Her face was white; her eyes had in them an uncertain terror.

'The text?' she stammered. 'I—I forget it, mother. It was from Genesis, I think.'

'Was it, Master Johnny?'

'It was from the fourth chapter of St Mark, "Peace, be still".'

Mrs Lease stared at me. 'Why, that is the very chapter I've been reading. Well now, that's curious. But there's never a better in the Bible, and never a better text was taken from it than those three words. I have been telling Daniel here, Master Johnny, that when once that peace, Christ's peace, is got into the heart, storms can't hurt us much. And you are going away again tomorrow, sir?' she added, after a pause. 'It's a short stay?'

I was not going away on the morrow. Tod and I, taking the Squire in a genial moment after dinner, had pressed to be let stay until Tuesday, Tod using the argument, and laughing while he did it, that it must be wrong to travel on All Saints' Day, when the parson had specially enjoined us to be at church. The Squire told us we were a couple of encroaching rascals, and if he did let us stay it should be upon condition that we did go to church. This I said to them.

'He may send you all the same, sir, when the morning comes,' remarked Daniel Ferrar.

'Knowing Mr Todhetley as you do Ferrar, you may remember that he never breaks his promises.'

Daniel laughed. 'He grumbles over them, though, Master Johnny.'

'Well, he may grumble tomorrow about our staying, say it is wasting time that ought to be spent in study, but he will not send us back until Tuesday.'

Until Tuesday! If I could have foreseen then what would have happened before Tuesday! If all of us could have foreseen! Seen the few hours between now and then depicted, as in a mirror, event by event! Would it have saved the calamity, the dreadful sin that could never be redeemed? Why, yes; surely it would. Daniel Ferrar turned and looked at Maria.

'Why don't you come to the fire?'

'I am very well here, thank you.'

She had sat down where she was, her bonnet touching the curtain. Mrs Lease, not noticing that anything was wrong, had begun talking about Lena, whose illness was turning to low fever, when the house door opened and Harriet Roe came in.

'What a lovely night it is!' she said, taking of her own accord the chair I had not cared to take, for I kept saying I must go. 'Maria, what went with you after church? I hunted for you everywhere.'

Maria gave no answer. She looked black and angry; and her bosom heaved as if a storm were brewing. Harriet Roe slightly laughed.

'Do you intend to take holiday tomorrow, Mrs Lease?'

'Me take holiday! what is there in tomorrow to take holiday for?' returned Mrs Lease.

'I shall,' continued Harriet, not answering the question: 'I have been used to it in France. All Saints' Day is a grand holiday there; we go to church in our best clothes, and pay visits afterwards. Following it, like a dark shadow, comes the gloomy Jour des Morts.'

'The what?' cried Mrs Lease, bending her ear.

'The day of the dead. All Souls' Day. But you English don't go to the cemeteries to pray.'

Mrs Lease put on her spectacles, which lay upon the open pages of the Bible, and stared at Harriet. Perhaps she thought they might help her to understand. The girl laughed.

'On All Souls' Day, whether it be wet or dry, the French cemeteries are full of kneeling women draped in black; all praying for the repose of their dead relatives, after the manner of the Roman Catholics.'

Daniel Ferrar, who had not spoken a word since she came in, but sat with his face to the fire, turned and looked at her. Upon which she tossed back her head and her pink ribbons, and smiled till all her teeth were seen. Good teeth they were. As to reverence in her tone, there was none.

'I have seen them kneeling when the slosh and wet have been ankle-deep. Did you ever see a ghost?' added she, with energy. 'The French believe that the spirits of the dead come abroad on the night of All Saints' Day. You'd scarcely get a French woman to go out of her house after dark. It is their chief superstition.'

'What *is* the superstition?' questioned Mrs Lease.

'Why, *that*', said Harriet. 'They believe that the dead are allowed to revisit the world after dark on the Eve of All Souls; that they hover in the air, waiting to appear to any of their living relatives, who may venture out, lest they should forget to pray on the morrow for the rest of their souls.'*

'Well, I never!' cried Mrs Lease, staring excessively. 'Did you ever hear the like of that, sir?' turning to me.

'Yes; I have heard of it.'

* A superstition obtaining amongst some of the lower orders in France.

Harriet Roe looked up at me; I was standing at the corner of the mantelpiece. She laughed a free laugh.

'I say, wouldn't it be fun to go out tomorrow night, and meet the ghosts? Only, perhaps they don't visit this country, as it is not under Rome.'

'Now just you behave yourself before your betters, Harriet Roe,' put in Mrs Lease, sharply. 'That gentleman is young Mr Ludlow of Crabb Cot.'

'And very happy I am to make young Mr Ludlow's acquaintance, returned easy Harriet, flinging back her mantle from her shoulders. 'How hot your parlour is, Mrs Lease.'

The hook of the cloak had caught in a thin chain of twisted gold that she wore round her neck, displaying it to view. She hurriedly folded her cloak together, as if wishing to conceal the chain. But Mrs Lease's spectacles had seen it.

'What's that you've got on, Harriet? A gold chain?'

A moment's pause, and then Harriet Roe flung back her mantle again, defiance upon her face, and touched the chain with her hand.

'That's what it is, Mrs Lease: a gold chain. And a very pretty one, too.'

'Was it your mother's?'

'It was never anybody's but mine. I had it made a present to me this afternoon; for a keepsake.'

Happening to look at Maria, I was startled at her face, it was so white and dark: white with emotion, dark with an angry despair that I for one did not comprehend. Harriet Roe, throwing at her a look of saucy triumph, went out with as little ceremony as she had come in, just calling back a general good night; and we heard her footsteps outside getting gradually fainter in the distance. Daniel Ferrar rose.

'I'll take my departure too, I think. You are very unsociable tonight, Maria.'

'Perhaps I am. Perhaps I have cause to be.'

She flung his hand back when he held it out; and in another moment, as if a thought struck her, ran after him into the passage to speak. I, standing near the door in the small room, caught the words.

'I must have an explanation with you, Daniel Ferrar. Now. Tonight. We cannot go on thus for a single hour longer.'

'Not tonight Maria; I have no time to spare. And I don't know what you mean.'

'You do know. Listen. I will not go to my rest, no, though it were for twenty nights to come, until we have had it out. I *vow* I will not. There.

You are playing with me. Others have long said so, and I know it now.'

He seemed to speak some quieting words to her, for the tone was low and soothing; and then went out, closing the door behind him. Maria came back and stood with her face and its ghastliness turned from us. And still the old mother noticed nothing.

'Why don't you take your things off, Maria?' she asked.

'Presently,' was the answer.

I said goodnight in my turn, and went away. Half-way home I met Tod with the two young Lexoms. The Lexoms made us go in and stay to supper, and it was ten o'clock before we left them.

'We shall catch it,' said Tod, setting off at a run. They never let us stay out late on a Sunday evening, on account of the reading.

But, as it happened, we escaped scot-free this time, for the house was in a commotion about Lena. She had been better in the afternoon, but at nine o'clock the fever returned worse than ever. Her little cheeks and lips were scarlet as she lay on the bed, her wide-open eyes were bright and glistening. The Squire had gone up to look at her, and was fuming and fretting in his usual fashion.

'The doctor has never sent the medicine,' said patient Mrs Todhetley, who must have been worn out with nursing. 'She ought to take it; I am sure she ought.'

'These boys are good to run over to Cole's for that,' cried the Squire. 'It won't hurt them; it's a fine night.'

Of course we were good for it. And we got our caps again; being charged to enjoin Mr Cole to come over the first thing in the morning.

'Do you care much about my going with you, Johnny?' Tod asked as we were turning out at the door. 'I am awfully tired.'

'Not a bit. I'd as soon go alone as not. You'll see me back in half-an-hour.'

I took the nearest way; flying across the fields at a canter, and startling the hares. Mr Cole lived near South Crabb, and I don't believe more than ten minutes had gone by when I knocked at his door. But to get back as quickly was another thing. The doctor was not at home. He had been called out to a patient at eight o'clock, and had not yet returned.

I went in to wait: the servant said he might be expected to come in from minute to minute. It was of no use to go away without the medicine; and I sat down in the surgery in front of the shelves, and fell asleep counting the white jars and physic bottles. The doctor's entrance awoke me.

'I am sorry you should have had to come over and to wait,' he said. 'When my other patient, with whom I was detained a considerable time, was done with, I went on to Crabb Cot with the child's medicine, which I had in my pocket.'

'They think her very ill tonight, sir.'

'I left her better, and going quietly to sleep. She will soon be well again, I hope.'

'Why! is that the time?' I exclaimed, happening to catch sight of the clock as I was crossing the hall. It was nearly twelve. Mr Cole laughed, saying time passed quickly when folk were asleep.

I went back slowly. The sleep, or the canter before it, had made me feel as tired as Tod had said he was. It was a night to be abroad in and to enjoy; calm, warm, light. The moon, high in the sky, illumined every blade of grass; sparkled on the water of the little rivulet; brought out the moss on the grey walls of the old church; played on its round-faced clock, then striking twelve.

Twelve o'clock at night at North Crabb answers to about three in the morning in London, for country people are mostly in bed and asleep at ten. Therefore, when loud and angry voices struck up in dispute, just as the last stroke of the hour was dying away on the midnight air, I stood still and doubted my ears.

I was getting near home then. The sounds came from the back of a building standing alone in a solitary place on the left-hand side of the road. It belonged to the Squire, and was called the yellow barn, its walls being covered with a yellow wash; but it was in fact used as a storehouse for corn. I was passing in front of it when the voices rose upon the air. Round the building I ran, and saw—Maria Lease: and something else that I could not at first comprehend. In the pursuit of her vow, not to go to rest until she had 'had it out' with Daniel Ferrar, Maria had been abroad searching for him. What ill fate brought her looking for him up near our barn?—perhaps because she had fruitlessly searched in every other spot.

At the back of this barn, up some steps, was an unused door. Unused partly because it was not required, the principal entrance being in front; partly because the key of it had been for a long time missing. Stealing out at this door, a bag of corn upon his shoulders, had come Daniel Ferrar in a smock-frock. Maria saw him, and stood back in the shade. She watched him lock the door and put the key in his pocket; she watched him give the heavy bag a jerk as he turned to come down the steps. Then she burst out. Her loud reproaches petrified him, and he stood there as one suddenly turned to stone. It was at that moment that I appeared.

I understood it all soon; it needed not Maria's words to enlighten me. Daniel Ferrar possessed the lost key and could come in and out at will in the midnight hours when the world was sleeping, and help himself to the corn. No wonder his poultry throve; no wonder there had been grumblings at Crabb Cot at the mysterious disappearance of the good grain.

Maria Lease was decidedly mad in those few first moments. Stealing is looked upon in an honest village as an awful thing; a disgrace, a crime; and there was the night's earlier misery besides. Daniel Ferrar was a thief! Daniel Ferrar was false to her! A storm of words and reproaches poured forth from her in confusion, none of it very distinct. 'Living upon theft! Convicted felon! Transportation for life! Squire Todhetley's corn! Fattening poultry on stolen goods! Buying gold chains with the profits for that bold, flaunting French girl, Harriet Roe! Taking his stealthy walks with her!'

My going up to them stopped the charge. There was a pause; and then Maria, in her mad passion, denounced him to me, as representative (so she put it) of the Squire—the breaker-in upon our premises! the robber of our stored corn!

Daniel Ferrar came down the steps; he had remained there still as a statue, immovable; and turned his white face to me. Never a word in defence said he: the blow had crushed him; he was a proud man (if any one can understand that), and to be discovered in this ill-doing was worse than death to him.

'Don't think of me more hardly than you can help, Master Johnny,' he said in a quiet tone. 'I have been almost tired of my life this long while.'

Putting down the bag of corn near the steps, he took the key from his pocket and handed it to me. The man's aspect had so changed; there was something so grievously subdued and sad about him altogether, that I felt as sorry for him as if he had not been guilty. Maria Lease went on in her fiery passion.

'You'll be more tired of it tomorrow when the police are taking you to Worcester gaol. Squire Todhetley will not spare you, though your father was his many-years bailiff. He could not, you know, if he wished; Master Ludlow has seen you in the act.'

'Let me have the key again for a minute, sir,' he said, as quietly as though he had not heard a word. And I gave it to him. I'm not sure but I should have given him my head had he asked for it.

He swung the bag on his shoulders, unlocked the granary door, and put the bag beside the other sacks. The bag was his own, as we found afterwards, but he left it there. Locking the door again, he gave me the key, and went away with a weary step.

'Goodbye, Master Johnny.'

I answered back goodnight civilly, though he had been stealing. When he was out of sight, Maria Lease, her passion full upon her still, dashed off towards her mother's cottage, a strange cry of despair breaking from her lips.

'Where have you been lingering, Johnny?' roared the Squire, who was sitting up for me. 'You have been throwing at the owls, sir, that's what you've been at; you have been scudding after the hares.'

I said I had waited for Mr Cole, and had come back slower than I went; but I said no more, and went up to my room at once. And the Squire went to his.

I know I am only a muff; people tell me so, often: but I can't help it; I did not make myself. I lay awake till nearly daylight, first wishing Daniel Ferrar could be screened, and then thinking it might perhaps be done. If he would only take the lesson to heart and go on straight for the future, what a capital thing it would be. We had liked old Ferrar; he had done me and Tod many a good turn: and, for the matter of that, we liked Daniel. So I never said a word when morning came of the past night's work.

'Is Daniel at home?' I asked, going to Ferrar's the first thing before breakfast. I meant to tell him that if he would keep right, I would keep counsel.

'He went out at dawn, sir,' answered the old woman who did for him, and sold his poultry at market. 'He'll be in presently; he have had no breakfast yet.'

'Then tell him when he comes, to wait in, and see me: tell him it's all right. Can you remember, Goody? "It is all right".'

'I'll remember, safe enough, Master Ludlow.'

Tod and I, being on our honour, went to church, and found about ten people in the pews. Harriet Roe was one, with her pink ribbons, the twisted gold chain showing outside a short-cut velvet jacket.

'No, sir; he has not been home yet; I can't think where he can have got to,' was the old Goody's reply when I went again to Ferrar's. And so I wrote a word in pencil, and told her to give it him when he came in, for I could not go dodging there every hour of the day.

After luncheon, strolling by the back of the barn: a certain reminiscence I suppose taking me there, for it was not a frequented spot: I saw Maria Lease coming along.

Well, it was a change! The passionate woman of the previous night had subsided into a poor, wild-looking, sorrow-stricken thing, ready to die of remorse. Excessive passion had wrought its usual consequences; a reaction: a reaction in favour of Daniel Ferrar. She came up to me, clasping her hands in agony—beseeching that, I would spare him; that

I would not tell of him; that I would give him a chance for the future: and her lips quivered and trembled, and there were dark circles round her hollow eyes.

I said that I had not told and did not intend to tell. Upon which she was going to fall down on her knees, but I rushed off.

'Do you know where he is?' I asked, when she came to her sober senses.

'Oh, I wish I did know! Master Johnny, he is just the man to go and do something desperate. He would never face shame; and I was a mad, hard-hearted, wicked girl to do what I did last night. He might run away to sea; he might go and enlist for a soldier.'

'I dare say he is at home by this time. I have left a word for him there, and promised to go in and see him tonight. It he will undertake not to be up to wrong things again, no one shall ever know of this from me.'

She went away easier, and I sauntered on towards South Crabb. Eager as Tod and I had been for the day's holiday, it did not seem to be turning out much of a boon. In going home again—there was nothing worth staying out for—I had come to the spot by the three-cornered grove where I saw Maria, when a galloping policeman overtook me. My heart stood still; for I thought he must have come after Daniel Ferrar.

'Can you tell me if I am near to Crabb Cot—Squire Todhetley's?' he asked, reining-in his horse.

'You will reach it in a minute or two. I live there. Squire Todhetley is not at home. What do you want with him?'

'It's only to give in an official paper, sir. I have to leave one personally upon all the county magistrates.'

He rode on. When I got in I saw the folded paper upon the hall-table; the man and horse had already gone onwards. It was worse indoors than out; less to be done. Tod had disappeared after church; the Squire was abroad; Mrs Todhetley sat upstairs with Lena: and I strolled out again. It was only three o'clock then.

An hour, or more, was got through somehow; meeting one, talking to another, throwing at the ducks and geese; anything. Mrs Lease had her head, smothered in a yellow shawl, stretched out over the palings as I passed her cottage.

'Don't catch cold, mother.'

'I am looking for Maria, sir. I can't think what has come to her today, Master Johnny,' she added, dropping her voice to a confidential tone. 'The girl seems demented: she has been going in and out ever since daylight like a dog in a fair.'

'If I meet her I will send her home.'

And in another minute I did meet her. For she was coming out of Daniel Ferrar's yard. I supposed he was at home again.

'No,' she said, looking more wild, worn, haggard than before; 'that's what I have been to ask. I am just out of my senses, sir. He has gone for certain. Gone!'

I did not think it. He would not be likely to go away without clothes.

'Well, I know he is, Master Johnny; something tells me. I've been all about everywhere. There's a great dread upon me, sir; I never felt anything like it.'

'Wait until night, Maria; I dare say he will go home then. Your mother is looking out for you; I said if I met you I'd send you in.'

Mechanically she turned towards the cottage, and I went on. Presently, as I was sitting on a gate watching the sunset, Harriet Roe passed towards the withy walk, and gave me a nod in her free but good-natured way.

'Are you going there to look out for the ghosts this evening?' I asked: and I wished not long afterwards I had not said it. 'It will soon be dark.'

'So it will,' she said, turning to the red sky in the west. 'But I have no time to give to the ghosts tonight.'

'Have you seen Ferrar today?' I cried, an idea occurring to me.

'No. And I can't think where he has got to; unless he is off to Worcester. He told me he should have to go there some day this week.'

She evidently knew nothing about him, and went on her way with another free-and-easy nod. I sat on the gate till the sun had gone down, and then thought it was time to be getting homewards.

Close against the yellow barn, the scene of last night's trouble, whom should I come upon but Maria Lease. She was standing still, and turned quickly at the sound of my footsteps. Her face was bright again, but had a puzzled look upon it.

'I have just seen him: he has not gone,' she said in a happy whisper. 'You were right, Master Johnny, and I was wrong.'

'Where did you see him?'

'Here; not a minute ago. I saw him twice. He is angry, very, and will not let me speak to him; both times he got away before I could reach him. He is close by somewhere.'

I looked round, naturally; but Ferrar was nowhere to be seen. There was nothing to conceal him except the barn, and that was locked up. The account she gave was this—and her face grew puzzled again as she related it.

Unable to rest indoors, she had wandered up here again, and saw Ferrar standing at the corner of the barn, looking very hard at her. She thought he was waiting for her to come up, but before she got close to him he had disappeared, and she did not see which way. She hastened past the front of the barn, ran round to the back, and there he was. He stood near the steps looking out for her; waiting for her, as it again seemed; and was gazing at her with the same fixed stare. But again she missed him before she could get quite up; and it was at that moment that I arrived on the scene.

I went all round the barn, but could see nothing of Ferrar. It was an extraordinary thing where he could have got to. Inside the barn he could not be: it was securely locked; and there was no appearance of him in the open country. It was, so to say, broad daylight yet, or at least not far short of it; the red light was still in the west. Beyond the field at the back of the barn, was a grove of trees in the form of a triangle; and this grove was flanked by Crabb Ravine, which ran right and left. Crabb Ravine had the reputation of being haunted; for a light was sometimes seen dodging about its deep descending banks at night that no one could account for. A lively spot altogether for those who liked gloom.

'Are you sure it was Ferrar, Maria?'

'Sure!' she returned in surprise. 'You don't think I could mistake him, Master Johnny, do you? He wore that ugly seal-skin winter-cap of his tied over his ears, and his thick grey coat. The coat was buttoned closely round him. I have not seen him wear either since last winter.'

That Ferrar must have gone into hiding somewhere seemed quite evident; and yet there was nothing but the ground to receive him. Maria said she lost sight of him the last time in a moment; both times in fact; and it was absolutely impossible that he could have made off to the triangle or elsewhere, as she must have seen him cross the open land. For that matter I must have seen him also.

On the whole, not two minutes had elapsed since I came up, though it seems to have been longer in telling it: when, before we could look further, voices were heard approaching from the direction of Crabb Cot; and Maria, not caring to be seen, went away quickly. I was still puzzling about Ferrar's hiding-place, when they reached me—the Squire, Tod, and two or three men. Tod came slowly up, his face dark and grave.

'I say, Johnny, what a shocking thing this is!'

'What is a shocking thing?'

You have not heard of it?—But I don't see how you could hear it.'

I had heard nothing. I did not know what there was to hear. Tod told me in a whisper.

'Daniel Ferrar's dead, lad.'

'*What?*'

'He has destroyed himself. Not more than half-an-hour ago. Hung himself in the grove.'

I turned sick, taking one thing with another, comparing this recollection with that; which I dare say you will think no one but a muff would do.

Ferrar was indeed dead. He had been hiding all day in the three-cornered grove: perhaps waiting for night to get away—perhaps only waiting for night to go home again. Who can tell? About half-past two, Luke Macintosh, a man who sometimes worked for us, sometimes for old Coney, happening to go through the grove, saw him there, and talked with him. The same man, passing back a little before sunset, found him hanging from a tree, dead. Macintosh ran with the news to Crabb Cot, and they were now flocking to the scene. When facts came to be examined there appeared only too much reason to think that the unfortunate appearance of the galloping policeman had terrified Ferrar into the act; perhaps—we all hoped it!—had scared his senses quite away. Look at it as we would, it was very dreadful.

But what of the appearance Maria Lease saw? At that time, Ferrar had been dead at least half-an-hour. Was it reality or delusion? That is (as the Squire put it), did her eyes see a real, spectral Daniel Ferrar; or were they deceived by some imagination of the brain? Opinions were divided. Nothing can shake her own steadfast belief in its reality; to her it remains an awful certainty, true and sure as heaven.

If I say that I believe in it too, I shall be called a muff and a double muff. But there is no stumbling-block difficult to be got over. Ferrar, when found, was wearing the seal-skin cap tied over the ears and the thick grey coat buttoned up round him, just as Maria Lease had described to me; and he had never worn them since the previous winter, or taken them out of the chest where they were kept. The old woman at his home did not know he had done it then. When told that he died in these things, she protested that they were in the chest, and ran up to look for them. But the things were gone.

Uncle Cornelius His Story

GEORGE MACDONALD

It was a dull evening in November. A drizzling mist had been falling all day about the old farm. Harry Heywood and his two sisters sat in the house-place, expecting a visit from their uncle, Cornelius Heywood. This uncle lived alone, occupying the first floor above a chemist's shop in the town, and had just enough of money over to buy books that nobody seemed ever to have heard of but himself; for he was a student in all those regions of speculation in which anything to be called knowledge is impossible.

'What a dreary night!' said Kate. 'I wish uncle would come and tell us a story.'

'A cheerful wish,' said Harry. 'Uncle Cornie is a lively companion—isn't he? He can't even blunder through a Joe Miller without tacking a moral to it, and then trying to persuade you that the joke of it depends on the moral.'

'Here he comes!' said Kate, as three distinct blows with the knob of his walking-stick announced the arrival of Uncle Cornelius. She ran to the door to open it.

The air had been very still all day, but as he entered he seemed to have brought the wind with him, for the first moan of it pressed against rather than shook the casement of the low-ceiled room.

Uncle Cornelius was very tall, and very thin, and very pale, with large grey eyes that looked greatly larger because he wore spectacles of the most delicate hair-steel, with the largest pebble-eyes that ever were seen. He gave them a kindly greeting, but too much in earnest even in shaking hands to smile over it. He sat down in the arm-chair by the chimney corner.

I have been particular in my description of him, in order that my reader may give due weight to his words. I am such a believer in words, that I believe everything depends on who says them. Uncle Cornelius Heywood's story told word for word by Uncle Timothy Warren, would not have been the same story at all. Not one of the listeners would have believed a syllable of it from the lips of round-bodied, red-faced, small-eyed, little Uncle Tim; whereas from Uncle Cornie—disbelieve one of his stories if you could!

One word more concerning him. His interest in everything conjectured or believed relative to the awful borderland of this world and the next, was only equalled by his disgust at the vulgar, unimaginative forms which curiosity about such subjects has assumed in the present day. With a yearning after the unseen like that of a child for the lifting of the curtain of a theatre, he declared that, rather than accept such a spirit-world as the would-be seers of the nineteenth century thought or pretended to reveal—the prophets of a pauperized, workhouse immortality, invented by a poverty-stricken soul, and a sense so greedy that it would gorge on carrion—he would rejoice to believe that a man had just as much of a soul as the cabbage of Iamblichus, namely, an aerial double of his body.

'I'm so glad you're come, uncle!' said Kate. 'Why wouldn't you come to dinner? We have been so gloomy!'

'Well, Katey, you know I don't admire eating. I never could bear to see a cow tearing up the grass with her long tongue.' As he spoke he looked very much like a cow. He had a way of opening his jaws while he kept his lips closely pressed together, that made his cheeks fall in, and his face look awfully long and dismal. 'I consider eating', he went on, 'such an animal exercise that it ought always to be performed in private. You never saw me dine, Kate.'

'Never, uncle; but I have seen you drink—nothing but water, I must confess.'

'Yes, that is another affair. According to one eye-witness, that is no more than the disembodied can do. I must confess, however, that, although well attested, the story is to me scarcely credible. Fancy a glass of Bavarian beer lifted into the air without a visible hand, turned upside down, and set empty on the table!—and no splash on the floor or anywhere else!'

A solitary gleam of humour shone through the great eyes of the spectacles as he spoke.

'Oh, uncle! how can you believe such nonsense!' said Janet.

'I did not say I believed it—did I? But why not? The story has at least a touch of imagination in it.'

'That is a strange reason for believing a thing, uncle,' said Harry.

'You might have a worse, Harry. I grant it is not sufficient; but it is better than that commonplace aspect which is the ground of most faith. I believe I did say that the story puzzled me.'

'But how can you give it any quarter at all, uncle?'

'It does me no harm. There it is—between the boards of an old German book. There let it remain.'

'Well, you will never persuade me to believe such things,' said Janet.'

'Wait till I ask you, Janet,' returned her uncle, gravely. 'I have not the slightest desire to convince you. How did we get into this unprofitable current of talk? We will change it at once. How are consols, Harry?'

'Oh, uncle!' said Kate, 'we were longing for a story, and just as I thought you were coming to one, off you go to consols!'

'I thought a ghost story at least was coming,' said Janet.

'You did your best to stop it, Janet,' said Harry.

Janet began an angry retort, but Cornelius interrupted her. 'You never heard me tell a ghost story, Janet.'

'You have just told one about a drinking ghost, uncle,' said Janet—in such a tone that Cornelius replied—

'Well, take that for your story, and let us talk of something else.'

Janet apparently saw that she had been rude, and said as sweetly as she might—'Ah! but you didn't make that one, uncle. You got it out of a German book.'

'Make it!—Make a ghost story!' repeated Cornelius. 'No; that I never did.'

'Such things are not to be trifled with, are they?' said Janet.

'I at least have no inclination to trifle with them.'

'But, really and truly, uncle,' persisted Janet, 'you don't believe in such things?'

'Why should I either believe or disbelieve in them? They are not essential to salvation, I presume.'

'You must do the one or the other, I suppose.'

'I beg your pardon. You suppose wrong. It would take twice the proof I have ever had to make me believe in them; and exactly your prejudice, and allow me to say ignorance, to make me disbelieve in them. Neither is within my reach. I postpone judgement. But you, young people, of course, are wiser, and know all about the question.'

'Oh, uncle! I'm so sorry!' said Kate. 'I'm sure I did not mean to vex you.'

'Not at all, not at all, my dear. It wasn't you.'

'Do you know', Kate went on, anxious to prevent anything unpleasant, for there was something very black perched on Janet's forehead, 'I have taken to reading about that kind of thing.'

'I beg you will give it up at once. You will bewilder your brains till you are ready to believe anything, if only it be absurd enough. Nay, you may come to find the element of vulgarity essential to belief. I should be sorry to the heart to believe concerning a horse or dog what they tell you nowadays about Shakespeare and Burns. What have you been reading, my girl?'

'Don't be alarmed, uncle. Only some Highland legends, which are too absurd either for my belief or for your theories.'

'I don't know that, Kate.'

'Why, what could you do with such shapeless creatures as haunt their fords and pools for instance? They are as featureless as the faces of the mountains.'

'And so much the more terrible.'

'But that does not make it easier to believe in them,' said Harry.

'I only said,' returned his uncle, 'that their shapelessness adds to their horror.'

'But you allowed—almost, at least, uncle,' said Kate, 'that you could find a place in your theories even for those shapeless creatures.'

Cornelius sat silent for a moment; then, having first doubled the length of his face, and restored it to its natural condition, said thoughtfully, 'I suspect, Katey, if you were to come upon an ichthyosaurus or a pterodactyl asleep in the shrubbery, you would hardly expect your report of it to be believed all at once either by Harry or Janet.'

'I suppose not, uncle. But I can't see what——'

'Of course such a thing could not happen here and now. But there was a time when and a place where such a thing may have happened. Indeed, in my time, a traveller or two have got pretty soundly disbelieved for reporting what they saw—the last of an expiring race, which had strayed over the natural verge of its history, coming to life in some neglected swamp, itself a remnant of the slime of Chaos.'

'I never heard you talk like that before, uncle,' said Harry. 'If you go on like that, you'll land me in a swamp, I'm afraid.'

'I wasn't talking to you at all, Harry. Kate challenged me to find a place for kelpies, and such like, in the theories she does me the honour of supposing I cultivate.'

'Then you think, uncle, that all these stories are only legends which, if you could follow them up, would lead you back to some one of the awful monsters that have since quite disappeared from the earth.'

'It is possible those stories may be such legends; but that was not what I intended to lead you to. I gave you that only as something like what I am going to say now. What if, mind, I only suggest it, what if the direful creatures, whose report lingers in these tales, should have an origin far older still? What if they were the remnants of a vanishing period of the earth's history long antecedent to the birth of mastodon and iguanodon; a stage, namely, when the world, as we call it, had not yet become quite visible, was not yet so far finished as to part from the invisible world that was its mother, and which, on its part, had not then become quite invisible—was only almost such; and when, as a credible consequence, strange shapes of those now invisible regions, Gorgons and Chimæras dire, might be expected to gloom out occasionally from the awful Fauna of an ever-generating world upon that one which was being born of it. Hence, the life-periods of a world being long and slow, some of these huge, unformed bulks of half-created matter might, somehow, like the megatherium of later times, a baby creation to them, roll at age-long intervals, clothed in a mighty terror of shapelessness into the half-recognition of human beings, whose consternation at the uncertain vision were barrier enough to prevent all further knowledge of its substance.'

'I begin to have some notion of your meaning, uncle,' said Kate.

'But then,' said Janet, 'all that must be over by this time. That world has been invisible now for many years.'

'Ever since you were born, I suppose, Janet. The changes of a world are not to be measured by the changes of its generations.'

'Oh, but, uncle, there can't be any such things. You know that as well as I do.'

'Yes, just as well, and no better.'

'There can't be any ghosts now. Nobody believes such things.'

'Oh, as to ghosts, that is quite another thing. I did not know you were talking with reference to them. It is no wonder if one can get nothing sensible out of you, Janet, when your discrimination is no greater than to lump everything marvellous, kelpies, ghosts, vampires, doubles, witches, fairies, nightmares, and I don't know what all, under the one head of ghosts; and we haven't been saying a word about them. If one were to disprove to you the existence of the afreets of Eastern tales, you would consider the whole argument concerning the reappearance of the departed upset. I congratulate you on your powers of analysis and induction, Miss Janet. But it matters very little whether we believe in ghosts, as you say, or not, provided we believe that we are ghosts—that within this body, which so many people are ready to consider their own very selves, their lies a ghostly embryo, at least, which has an inner side to it God only can see, which says I concerning itself, and which will soon have to know whether or not it can appear to those whom it has left behind, and thus solve the question of ghosts for itself, at least.'

'Then you do believe in ghosts, uncle?' said Janet, in a tone that certainly was not respectful.

'Surely I said nothing of the sort, Janet. The man most convinced that he had himself had such an interview as you hint at, would find—ought to find it impossible to convince any one else of it.'

'You are quite out of my depth, uncle,' said Harry. 'Surely any honest man ought to be believed?'

'Honesty is not all, by any means, that is necessary to being believed. It is impossible to convey a conviction of anything. All you can do is to convey a conviction that you are convinced. Of course, what satisfied you might satisfy another; but, till you can present him with the sources of your conviction, you cannot present him with the conviction—and perhaps not even then.'

'You can tell him all about it, can't you?'

'Is telling a man about a ghost, affording him the source of your conviction? Is it the same as a ghost appearing to him? Really, Harry!—You cannot even convey the impression a dream has made upon you.'

'But isn't that just because it is only a dream?'

'Not at all. The impression may be deeper and clearer on your mind than any fact of the next morning will make. You will forget the next day altogether, but the impression of the dream will remain through all the following whirl and storm of what you call facts. Now a conviction may be likened to a deep impression on the judgement or the reason, or both. No one can feel it but the person who is convinced. It cannot be conveyed.'

'I fancy that is just what those who believe in spirit-rapping would say.'

'There are the true and false of convictions, as of everything else. I mean that a man may take that for a conviction in his own mind which is not a conviction, but only resembles one. But those to whom you refer profess to appeal to facts. It is on the ground of those facts, and with the more earnestness the more reason they can give for receiving them as facts, that I refuse all their deductions with abhorrence. I mean that, if what they say is true, the thinker must reject with contempt the claim to anything like revelation therein.'

'Then you do not believe in ghosts, after all?' said Kate, in a tone of surprise.

'I did not say so, my dear. Will you be reasonable, or will you not?'

'Dear uncle, do tell us what you really think.'

'I have been telling you what I think ever since I came, Katey; and you won't take in a word I say.'

'I have been taking in every word, uncle, and trying hard to understand it as well. Did you ever see a ghost, uncle?'

Cornelius Heywood was silent. He shut his lips and opened his jaws till his cheeks almost met in the vacuum. A strange expression crossed the strange countenance, and the great eyes of his spectacles looked as if, at the very moment, they were seeing something no other spectacles could see. Then his jaws closed with a snap, his countenance brightened, a flash of humour came through the goggle eyes of pebble, and, at length, he actually smiled as he said—'Really, Katey, you must take me for a simpleton!'

'How, uncle?'

'To think, if I had ever seen a ghost, I would confess the fact before a set of creatures like you—all spinning your webs like so many spiders to catch and devour old Daddy Longlegs.'

By this time Harry had grown quite grave. 'Indeed, I am very sorry, uncle,' he said, 'if I have deserved such a rebuke.'

'No, no, my boy,' said Cornelius; 'I did not mean it more than half. If I had meant it, I would not have said it. If you really would like——' Here he paused.

'Indeed we should, uncle,' said Kate, earnestly. 'You should have heard what we were saying just before you came in.'

'All you were saying, Katey?'

'Yes,' answered Kate, thoughtfully. 'The worst we said was that you could not tell a story without——well, we did say tacking a moral to it.'

'Well, well! I mustn't push it. A man has no right to know what people say about him. It unfits him for occupying his real position amongst them. He, least of all, has anything to do with it. If his friends won't defend him, he can't defend himself. Besides, what people say is so often untrue!—I don't mean to others, but to themselves. Their hearts are more honest than their mouths. But Janet doesn't want a strange story, I am sure.'

Janet certainly was not one to have chosen for a listener to such a tale. Her eyes were so small that no satisfaction could possibly come of it. 'Oh! I don't mind, uncle,' she said, with half-affected indifference, as she searched in her box for silk to mend her gloves.

'You are not very encouraging, I must say,' returned her uncle, making another cow-face.

'I will go away, if you like,' said Janet, pretending to rise.

'No, never mind,' said her uncle hastily. 'If you don't want me to tell it, I want you to hear it; and, before I have done, that may have come to the same thing perhaps.'

'Then you really are going to tell us a ghost story!' said Kate, drawing her chair nearer to her uncle's; and then, finding this did not satisfy her sense of propinquity to the source of the expected pleasure, drawing a stool from the corner, and seating herself almost on the hearth-rug at his knee.

'I did not say so,' returned Cornelius, once more. 'I said I would tell you a strange story. You may call it a ghost story if you like; I do not pretend to determine what it is. I confess it will look like one, though.'

After so many delays, Uncle Cornelius now plunged almost hurriedly into his narration.

'In the year 1820,' he said, 'in the month of August, I fell in love.' Here the girls glanced at each other. The idea of Uncle Cornie in love, and in the very same century in which they were now listening to the confession, was too astonishing to pass without ocular remark; but, if he observed it, he took no notice of it; he did not even pause. 'In the month of September, I was refused. Consequently, in the month of October, I was ready to fall in love again. Take particular care of yourself, Harry, for a whole month, at least, after your first disappointment; for you will never be more likely to do a foolish thing. Please yourself after the second. If you are silly then, you may take what you get, for you will deserve it—except it be good fortune.'

'Did you do a foolish thing then, uncle?' asked Harry, demurely.

'I did, as you will see; for I fell in love again.'

'I don't see anything so very foolish in that.'

'I have repented it since, though. Don't interrupt me again, please. In the middle of October, then, in the year 1820, in the evening, I was walking across Russell Square, on my way home from the British Museum, where I had been reading all day. You see I have a full intention of being precise, Janet.'

'I'm sure I don't know why you make the remark to me, uncle,' said Janet, with an involuntary toss of her head. Her uncle only went on with his narrative.

'I begin at the very beginning of my story,' he said; 'for I want to be particular as to everything that can appear to have had anything to do with what came afterwards. I had been reading, I say, all the morning in the British Museum; and, as I walked, I took off my spectacles to ease my eyes. I need not tell you that I am short-sighted now, for that you know well enough. But I must tell you that I was short-sighted then, and helpless enough without my spectacles, although I was not quite so much so as I am now—for I find it all nonsense about short-sighted eyes improving with age. Well, I was walking along the south side of Russell Square, with my spectacles in my hand, and feeling a little bewildered in consequence—for it was quite the dusk of the evening, and short-sighted people require more light than others. I was feeling, in fact, almost blind. I had got more than half-way to the other side, when, from the crossing that cuts off the corner in the direction of Montagu Place, just as I was about to turn towards it, an old lady stepped upon the kerbstone of the pavement, looked at me for a moment, and passed—an occurrence not very remarkable, certainly. But the lady was remarkable and so was her dress. I am not good at observing, and I am still worse at describing dress, therefore I can only say that hers reminded me of an old picture—that is, I had never seen anything like it, except in old pictures. She had no bonnet, and looked as if she had walked straight out of an ancient drawing-room in her evening attire. Of her face I shall say nothing now. The next instant I met a man on the crossing, who stopped and addressed me. So short-sighted was I that, although I recognized his voice as one I ought to know, I could not identify him until I had put on my spectacles, which I did instinctively in the act of returning his greeting. At the same moment I glanced over my shoulder after the old lady. She was nowhere to be seen.

'"What are you looking at?" asked James Hetheridge.

'"I was looking after that old lady," I answered, "but I can't see her."

'"What old lady?" said Hetheridge, with just a touch of impatience.

'"You must have seen her," I returned. "You were not more than three yards behind her."

'"Where is she then?"

'"She must have gone down one of the areas, I think. But she looked a lady, though an old-fashioned one."

'"Have you been dining?" asked James, in a tone of doubtful enquiry.

'"No," I replied, not suspecting the insinuation; "I have only just come from the Museum."

'"Then I advise you to call on your medical man before you go home."

'"Medical man!" returned; "I have no medical man. What do you mean? I never was better in my life."

'"I mean that there was no old lady. It was an illusion, and that indicates something wrong. Besides, you did not know me when I spoke to you."

'"That is nothing," I returned. "I had just taken off my spectacles, and without them I shouldn't know my own father."

'"How was it you saw the old lady, then?"

'The affair was growing serious under my friend's cross-questioning. I did not at all like the idea of his supposing me subject to hallucinations. So I answered, with a laugh, "Ah! to be sure, that explains it. I am so blind without my spectacles, that I shouldn't know an old lady from a big dog."

'"There was no big dog," said Hetheridge, shaking his head, as the fact for the first time dawned upon me that, although I had seen the old lady clearly enough to make a sketch of her; even to the features of her care-worn, eager old face, I had not been able to recognize the well-known countenance of James Hetheridge.

'"That's what comes of reading till the optic nerve is weakened," he went on. "You will cause yourself serious injury if you do not pull up in time. I'll tell you what; I'm going home next week—will you go with me?"

'"You are very kind," I answered, not altogether rejecting the proposal, for I felt that a little change to the country would be pleasant, and I was quite my own master. For I had unfortunately means equal to my wants, and had no occasion to follow any profession—not a very desirable thing for a young man, I can tell you, Master Harry. I need not keep you over the commonplaces of pressing and yielding. It is enough to say that he pressed and that I yielded. The day was fixed for our departure together; but something or other, I forget what, occurred, to make him advance the date, and it was resolved that I should follow later in the month.

'It was a drizzly afternoon in the beginning of the last week of October when I left the town of Bradford in a post-chaise to drive to Lewton Grange, the property of my friend's father. I had hardly left the town, and the twilight had only begun to deepen, when, glancing from one of the windows of the chaise, I fancied I saw, between me and the hedge, the dim figure of a horse keeping pace with us. I thought, in the first interval of unreason, that it was a shadow from my own horse, but reminded myself the next moment that there could be no shadow where there was no light. When I looked again, I was at the first glance convinced that my eyes had deceived me. At the second, I believed once more that a shadowy something, with the movements of a horse in harness, was keeping pace with us. I turned away again with some discomfort, and not till we had reached an open moorland road, whence a little watery light was visible on the horizon, could I summon up courage enough to look out once more. Certainly then there was nothing to be seen, and I persuaded myself that it had been all a fancy, and lighted a cigar. With my feet on the cushions before me, I had soon lifted myself on the clouds of tobacco far above all the terrors of the night, and believed them banished for ever. But, my cigar coming to an end just as we turned into the avenue that led up to the Grange, I found myself once more glancing nervously out of the window. The moment the trees were about me, there was, if not a shadowy horse out there by the side of the chaise, yet certainly more than half that conviction in here in my consciousness. When I saw my friend, however, standing on the doorstep, dark against the glow of the hall fire, I forgot all about it; and I need not add that I did not make it a subject of conversation when I entered, for I was well aware that it was essential to a man's reputation that his senses should be accurate, though his heart might without prejudice swarm with shadows, and his judgement be a very stable of hobbies.

'I was kindly received. Mrs Hetheridge had been dead for some years, and Lætitia, the eldest of the family, was at the head of the household. She had two sisters, little more than girls. The father was a burly, yet gentlemanlike Yorkshire squire, who ate well, drank well, looked radiant, and hunted twice a week. In this pastime his son joined him when in the humour, which happened scarcely so often. I, who had never crossed a horse in my life, took his apology for not being able to mount me very coolly, assuring him that I would rather loiter about with a book than be in at the death of the best-hunted fox in Yorkshire.

'I very soon found myself at home with the Hetheridges; and very soon again I began to find myself not so much at home; for Miss Hetheridge—Lætitia as I soon ventured to call her—was fascinating. I have told you, Katey, that there was an empty place in my heart. Look to the door then, Katey. That was what made me so ready to fall in love with Lætitia. Her figure was graceful, and I think, even now, her face would have been beautiful but for a certain contraction of the skin over the nostrils, suggesting an invisible thumb and forefinger pinching them, which repelled me, although I did not then know what it indicated. I had not been with her one evening before the impression it made on me had vanished, and that so entirely that I could hardly recall the perception of the peculiarity which had occasioned it. Her observation was remarkably keen, and her judgement generally correct. She had great confidence in it herself; nor was she devoid of sympathy with some of the forms of human imagination, only they never seemed to possess for her any relation to practical life. That was to be ordered by the judgement alone. I do not mean she ever said so. I am only giving the conclusions I came to afterwards. It is not necessary that you should have any more thorough acquaintance with her mental character. One point in her moral nature, of special consequence to my narrative, will show itself by and by.

'I did all I could to make myself agreeable to her, and the more I succeeded the more delightful she became in my eyes. We walked in the garden and grounds together; we read, or rather I read and she listened;—read poetry, Katey—sometimes till we could not read any more for certain haziness and huskiness which look now, I am afraid, considerably more absurd than they really were, or even ought to look. In short, I considered myself thoroughly in love with her.'

'And wasn't she in love with you, uncle?'

'Don't interrupt me, child. I don't know. I hoped so then. I hope the contrary now. She liked me I am sure. That is not much to say. Liking is very pleasant and very cheap. Love is as rare as a star.'

'I thought the stars were anything but rare, uncle.'

'That's because you never went out to find one for yourself, Katey. They would prove a few miles apart then.'

'But it would be big enough when I did find it.'

'Right, my dear. That is the way with love. Lætitia was a good housekeeper. Everything was punctual as clockwork. I use the word advisedly. If her father, who was punctual to one date, the dinner-hour, made any remark to the contrary as he took up the carving-knife, Lætitia would instantly send one of her sisters to question the old clock in the hall, and report the time to half a minute. It was sure to be found that, if there was a mistake, the mistake was in the clock. But although it was certainly a virtue to have her household in such perfect order, it was not a virtue to be impatient with every infringement of its rules on the part of others. She was very severe, for instance, upon her two younger sisters if, the moment after the second bell had rung, they were not seated at the dinner-table, washed and aproned. Order was a very idol with her. Hence the house was too tidy for any sense of comfort. If you left an open book on the table, you would, on returning to the room a moment after, find it put aside. What the furniture of the drawing-room was like, I never saw; for not even on Christmas Day, which was the last day I spent there, was it uncovered. Everything in it was kept in bibs and pinafores. Even the carpet was covered with a cold and slippery sheet of brown holland. Mr Hetheridge never entered that room, and therein was wise. James remonstrated once. She answered him quite kindly, even playfully, but no change followed. What was worse, she made very wretched tea. Her father never took tea; neither did James. I was rather fond of it, but I soon gave it up. Everything her father partook of was first-rate. Everything else was somewhat poverty-stricken. My pleasure in Lætitia's society prevented me from making practical deductions from such trifles.'

'I shouldn't have thought you knew anything about eating, uncle,' said Janet.

'The less a man eats, the more he likes to have it good, Janet. In short, there can be no harm in saying it now, Lætitia was so far from being like the name of her baptism, and most names are so good that they are worth thinking about; no children are named after bad ideas,

Lætitia was so far unlike hers as to be stingy—an abominable fault. But, I repeat, the notion of such a fact was far from me then. And now for my story.

'The first of November was a very lovely day, quite one of the "halcyon days" of "St Martin's summer". I was sitting in a little arbour I had just discovered, with a book in my hand—not reading, however, but day-dreaming—when, lifting my eyes from the ground, I was startled to see, through a thin shrub in front of the arbour, what seemed the form of an old lady seated, apparently reading from a book on her knee. The sight instantly recalled the old lady of Russell Square. I started to my feet, and then, clear of the intervening bush, saw only a great stone such as abounded on the moors in the neighbourhood, with a lump of quartz set on the top of it. Some childish taste had put it there for an ornament. Smiling at my own folly, I sat down again, and reopened my book. After reading for a while, I glanced up again, and once more started to my feet, overcome by the fancy that there verily sat the old lady reading. You will say it indicated an excited condition of the brain. Possibly; but I was, as far as I can recall, quite collected and reasonable. I was almost vexed this second time, and sat down once more to my book. Still, every time I looked up, I was startled afresh. I doubt, however, if the trifle is worth mentioning, or has any significance even in relation to what followed.

'After dinner I strolled out by myself, leaving father and son over their claret. I did not drink wine; and from the lawn I could see the windows of the library, whither Lætitia commonly retired from the dinner-table. It was a very lovely soft night. There was no moon, but the stars looked wider awake than usual. Dew was falling, but the grass was not yet wet, and I wandered about on it for half an hour. The stillness was somehow strange. It had a wonderful feeling in it as if something were expected—as if the quietness were the mould in which some event or other was about to be cast.

'Even then I was a reader of certain sorts of recondite lore. Suddenly I remembered that this was the eve of All Souls. This was the night on which the dead came out of their graves to visit their old homes. "Poor dead!" I thought with myself; "have you any place to call a home now? If you have, surely you will not wander back here, where all that you called home has either vanished or given itself to others, to be their home now and yours no more! What an awful doom the old fancy has allotted you! To dwell in your graves all the year, and creep out, this one night, to enter at the midnight door, left open for welcome! A poor welcome truly!—just an open door, a clean-swept floor, and a fire to warm your rain-sodden limbs! The household asleep, and the house-place swarming with the ghosts of ancient times—the miser, the spendthrift, the profligate, the coquette—for the good ghosts sleep, and are troubled with no waking like yours! Not one man, sleepless like yourselves, to question you, and be answered after the fashion of the old nursery rhyme—

> "What makes your eyes so holed?"
> "I've lain so long among the mould."
> "What makes your feet so broad?"
> "I've walked more than ever I rode!"

'"Yet who can tell?" I went on to myself. "It may be your hell to return thus. It may be that only on this one night of all the year you can show yourselves to him who can see you, but that the place where you were wicked is the Hades to which you are doomed for ages." I thought and thought till I began to feel the air alive about me, and was enveloped in the vapours that dim the eyes of those who strain them for one peep through the dull mica windows that will not open on the world of ghosts. At length I cast my fancies away, and fled from them to the library, where the bodily presence of Lætitia made the world of ghosts appear shadowy indeed.

'"What a reality there is about a bodily presence!" I said to myself, as I took my chamber-candle in my hand. "But what is there more real in a body?" I said again, as I crossed the hall. "Surely nothing," I went on, as I ascended the broad staircase to my room. "The body must vanish. If there be a spirit, that will remain. A body can but vanish. A ghost can appear."

'I woke in the morning with a sense of such discomfort as made me spring out of bed at once. My foot lighted upon my spectacles. How they came to be on the floor I could not tell, for I never took them off when I went to bed. When I lifted them I found they were in two pieces; the bridge was broken. This was awkward. I was so utterly helpless without them! Indeed, before I could lay my hand on my hair-brush I had to peer through one eye of the parted pair. When I looked at my watch after I was dressed, I found I had risen an hour earlier than usual. I groped my way downstairs to spend the hour before breakfast in the library.

'No sooner was I seated with a book than I heard the voice of Lætitia scolding the butler, in no very gentle tones, for leaving the garden door open all night. The moment I heard this, the strange occurrences I am about to relate began to dawn upon my memory. The door had been open the night long between All Saints and All Souls. In the middle of that night I awoke suddenly. I knew it was not the morning by the sensations I had, for the night feels altogether different from the morning. It was quite dark. My heart was beating violently, and I either hardly could or hardly dared breathe. A nameless terror was upon me, and my sense of hearing was, apparently by the force of its expectation, unnaturally roused and keen. There it was—a slight noise in the room!—slight, but clear, and with an unknown significance about it! It was awful to think it would come again. I do believe it was only one of those creaks in the timbers which announce the torpid, age-long, sinking flow of every house back to the dust—a motion to which the flow of the glacier is as a torrent, but which is no less inevitable and sure. Day and night it ceases not; but only in the night, when house and heart are still, do we hear it. No wonder it should sound fearful! for are we not the immortal dwellers in ever-crumbling clay? The clay is so near us, and yet not of us, that its every movement starts a fresh dismay. For what will its final ruin disclose? When it falls from about us, where shall we find that we have existed all the time?

'My skin tingled with the bursting of the moisture from its pores. Something was in the room beside me. A confused, indescribable sense of utter loneliness, and yet awful presence, was upon me, mingled with a dreary, hopeless desolation, as of burnt-out love and aimless life. All at once I found myself sitting up. The terror that a cold hand might be laid upon me, or a cold breath blow on me, or a corpse-like face bend down through the darkness over me, had broken my bonds!—I would meet half-way whatever might be approaching. The moment that my will burst into action the terror began to ebb.

'The room in which I slept was a large one, perfectly dreary with tidiness. I did not know till afterwards that it was Lætitia's room, which she had given up to me rather than prepare another. The furniture, all but one article, was modern and commonplace. I could not help remarking to myself afterwards how utterly void the room was of the nameless charm of feminine occupancy. I had seen nothing to wake a suspicion of its being a lady's room. The article I have excepted was an ancient bureau, elaborate and ornate, which stood on one side of the large bow window. The very morning before, I had seen a bunch of keys hanging from the upper part of it, and had peeped in. Finding, however, that the pigeon-holes were full of papers, I closed it at once. I should have been glad to use it, but clearly it was not for me. At that bureau the figure of a woman was now seated in the posture of one writing. A strange dim light was around her, but whence it proceeded I never thought of enquiring. As if I, too, had stepped over the bourne, and was a ghost myself, all fear was now gone. I got out of bed, and softly crossed the room to where she was seated. "If she should be beautiful!" I thought—for I had often dreamed of a beautiful ghost that made love to me. The figure did not move. She was looking at a faded brown paper. "Some old love-letter," I thought, and stepped nearer. So cool was I now, that I actually peeped over her shoulder. With mingled surprise and dismay I found that the dim page over which she bent was that of an old account-book. Ancient household records, in rusty ink, held up to the glimpses of the waning moon, which shone through the parting in the curtains, their entries of shillings and pence!—Of pounds there was not one. No doubt pounds and farthings are much the same in the world of thought—the true spirit-world; but in the ghost-world this eagerness over shillings and pence must mean something awful! To think that coins which had since been worn smooth in other pockets and purses, which had gone back to the Mint, and been melted down, to come out again and yet again with the heads of new kings and queens—that dinners, eaten by men and women and children whose bodies had since been eaten by the worms—that polish for the floors inches of whose thickness had since been worn away—that the hundred nameless trifles of a life utterly vanished, should be perplexing, annoying, and worst of all, interesting the soul of a ghost who had been in Hades for centuries! The writing was very old-fashioned, and the words were contracted. I could read nothing but the moneys and one single entry—"Corinths, Vs."

'Currants for a Christmas pudding, most likely!—Ah, poor lady! the pudding and not the Christmas was her care; not the delight of the children over it, but the beggarly pence which it cost. And she cannot get it out of her head, although her brain was "powdered all as thin as flour" ages ago in the mortar of Death. "Alas, poor ghost!" It needs no treasured hoard left behind, no floor stained with the blood of the murdered child, no wickedly hidden parchment of landed rights! An old account-book is enough for the hell of the house-keeping gentlewoman!

'She never lifted her face, or seemed to know that I stood behind her. I left her, and went into the bow window, where I could see her face. I was right. It was the same old lady I had met in Russell Square, walking in front of James Hetheridge. Her withered lips went moving as if they would have uttered words had the breath been commissioned thither; her brow was contracted over her thin nose; and once and again her shining forefinger went up to her temple as if she were pondering some deep problem of humanity. How long I stood gazing at her I do not know, but at last I withdrew to my bed, and left her struggling to solve that which she could never solve thus. It was the symbolic problem of her own life, and she had failed to read it. I remember nothing more. She may be sitting there still, solving at the insolvable.

'I should have felt no inclination, with the broad sun of the squire's face, the keen eyes of James, and the beauty of Lætitia before me at the breakfast table, to say a word about what I had seen, even if I had not been afraid of the doubt concerning my sanity which the story would certainly awaken. What with the memories of the night and the want of my spectacles, I passed a very dreary day, dreading the return of the night, for, cool as I had been in her presence, I could not regard the possible reappearance of the ghost with equanimity. But when the night did come, I slept soundly till the morning.

'The next day, not being able to read with comfort, I went wandering about the place, and at length began to fit the outside and inside of the house together. It was a large and rambling edifice, parts of it very old, parts comparatively modern. I first found my own window, which looked out of the back. Below this window, on one side, there was a door. I wondered whither it led, but found it locked. At the moment James approached from the stables. "Where does this door lead?" I asked him. "I will get the key," he answered. "It is rather a queer old place. We used to like it when we were children." "There's a stair, you see," he said, as he threw the door open. "It leads up over the kitchen." I followed him up the stair. "There's a door into your room," he said, "but it's always locked now. And here's Grannie's room, as they call it, though why, I have not the least idea," he added, as he pushed open the door of an old-fashioned parlour, smelling very musty. A few old books lay on a side table. A china bowl stood beside them, with some shrivelled, scentless rose-leaves in the bottom of it. The cloth that covered the table was riddled by moths, and the spider-legged chairs were covered with dust.

'A conviction seized me that the old bureau must have belonged to this room, and I soon found the place where I judged it must have stood. But the same moment I caught sight of a portrait on the wall above the spot I had fixed upon. "By Jove!" I cried, involuntarily, "that's the very old lady I met in Russell Square!"

'"Nonsense!" said James. "Old-fashioned ladies are like babies—they all look the same. That's a very old portrait."

'"So I see," I answered. "It is like a Zucchero."

'"I don't know whose it is," he answered hurriedly, and I thought he looked a little queer.

'"Is she one of the family?" I asked.

'"They say so; but who or what she was, I don't know. You must ask Letty," he answered.

'"The more I look at it," I said, "the more I am convinced it is the same old lady."

'"Well," he returned with a laugh, "my old nurse used to say she was rather restless. But it's all nonsense."

'"That bureau in my room looks about the same date as this furniture." I remarked.

'"It used to stand just there," he answered, pointing to the space under the picture. "Well I remember with what awe we used to regard it; for they said the old lady kept her accounts at it still. We never dared touch the bundles of yellow papers in the pigeon-holes. I remember thinking Letty a very heroine once when she touched one of them with the tip of her forefinger. She had got yet more courageous by the time she had it moved into her own room."

'"Then that is your sister's room I am occupying?" I said.

'"Yes."

'"I am ashamed of keeping her out of it."

'"Oh! she'll do well enough."

'"If I were she though," I added, "I would send that bureau back to its own place."

'"What do you mean, Heywood? Do you believe every old wife's tale that ever was told?"

'"She may get a fright some day—that's all!" I replied.

'He smiled with such an evident mixture of pity and contempt that for the moment I almost disliked him; and feeling certain that Lætitia would receive any such hint in a somewhat similar manner, I did not feel inclined to offer her any advice with regard to the bureau.

'Little occurred during the rest of my visit worthy of remark. Somehow or other I did not make much progress with Lætitia. I believe I had begun to see into her character a little, and therefore did not get deeper in love as the days went on. I know I became less absorbed in her society, although I was still anxious to make myself agreeable to her—or perhaps, more properly, to give her a favourable impression of me. I do not know whether she perceived any difference in my behaviour, but I remember that I began again to remark the pinched look of her nose, and to be a little annoyed with her for always putting aside my book. At the same time, I daresay I was provoking, for I never was given to tidiness myself.

'At length Christmas Day arrived. After breakfast, the squire, James, and the two girls arranged to walk to church. Lætitia was not in the room at the moment. I excused myself on the ground of a headache, for I had had a bad night. When they left, I went up to my room, threw myself on the bed, and was soon fast asleep.

'How long I slept I do not know, but I woke again with that indescribable yet well-known sense of not being alone. The feeling was scarcely less terrible in the daylight than it had been in the darkness. With the same sudden effort as before, I sat up in the bed. There was the figure at the open bureau, in precisely the same position as on the former occasion. But I could not see it so distinctly. I rose as gently as I could, and approached it, after the first physical terror. I am not a coward. Just as I got near enough to see the account book open on the folding cover of the bureau, she started up, and, turning, revealed the face of Lætitia. She blushed crimson.

'"I beg your pardon, Mr Heywood," she said, in great confusion; "I thought you had gone to church with the rest."

'"I had lain down with a headache, and gone to sleep," I replied. "But, forgive me, Miss Hetheridge," I added, for my mind was full of the dreadful coincidence, "don't you think you would have been better at church than balancing your accounts on Christmas Day?"

'"The better day the better deed," she said, with a somewhat offended air, and turned to walk from the room.

'"Excuse me, Lætitia," I resumed, very seriously, "but I want to tell you something."

'She looked conscious. It never crossed me, that perhaps she fancied I was going to make a confession. Far other things were then in my mind. For I thought how awful it was, if she too, like the ancestral ghost, should have to do an age-long penance of haunting that bureau and those horrid figures, and I had suddenly resolved to tell her the whole story. She listened with varying complexion and face half turned aside. When I had ended, which I fear I did with something of a personal appeal, she lifted her head and looked me in the face, with just a slight curl on her thin lip, and answered me. "If I had wanted a sermon, Mr Heywood, I should have gone to church for it. As for the ghost, I am sorry for you." So saying she walked out of the room.

'The rest of the day I did not find very merry. I pleaded my headache as an excuse for going to bed early. How I hated the room now! Next morning, immediately after breakfast, I took my leave of Lewton Grange.'

'And lost a good wife, perhaps, for the sake of a ghost, uncle!' said Janet.

'If I lost a wife at all, it was a stingy one. I should have been ashamed of her all my life long.'

'Better than a spendthrift,' said Janet.

'How do you know that?' returned her uncle. 'All the difference I see is, that the extravagant ruins the rich, and the stingy robs the poor.'

'But perhaps she repented, uncle,' said Kate.

'I don't think she did, Katey. Look here.'

Uncle Cornelius drew from the breast pocket of his coat a black-edged letter.

'I have kept up my friendship with her brother,' he said. 'All he knows about the matter is, that either we had a quarrel, or she refused me—he is not sure which. I must say for Lætitia, that she was no tattler. Well, here's a letter I had from James this very morning. I will read it to you.

MY DEAR HEYWOOD—We have had a terrible shock this morning. Letty did not come down to breakfast, and Lizzie went to see if she was ill. We heard her scream, and, rushing up, there was poor Letty, sitting at the old bureau, quite dead. She had fallen forward on the desk, and her housekeeping-book was crumpled up under her. She had been so all night long, we suppose, for she was not undressed, and was quite cold. The doctors say it was disease of the heart.

'There!' said Uncle Cornie, folding up the letter.

'Do you think the ghost had anything to do with it, uncle?' asked Kate, almost under her breath.

'How should I know, my dear? Possibly.'

'It's very sad,' said Janet; 'but I don't see the good of it all. If the ghost had come to tell that she had hidden away money in some secret place in the old bureau, one would see why she had been permitted to come back. But what was the good of those accounts after they were over and done with? I don't believe in the ghost.'

'Ah, Janet, Janet! but those wretched accounts were not over and done with, you see. That is the misery of it.'

Uncle Cornelius rose without another word, bade them goodnight, and walked out into the wind.

The Shadow of a Shade

TOM HOOD

My sister Lettie has lived with me ever since I had a home of my own. She was my little housekeeper before I married. Now she is my wife's constant companion, and the 'darling auntie' of my children, who go to her for comfort, advice, and aid in all their little troubles and perplexities.

But, though she has a comfortable home, and loving hearts around her, she wears a grave, melancholy look on her face, which puzzles acquaintances and grieves friends.

A disappointment! Yes, the old story of a lost lover is the reason for Lettie's looks. She has had good offers often; but since she lost the first love of her heart she has never indulged in the happy dream of loving and being loved.

George Mason was a cousin of my wife's—a sailor by profession. He and Lettie met one another at our wedding, and fell in love at first sight. George's father had seen service before him on the great mysterious sea, and had been especially known as a good Arctic sailor, having shared in more than one expedition in search of the North Pole and the North-West Passage.

It was not a matter of surprise to me, therefore, when George volunteered to go out in the *Pioneer*, which was being fitted out for a cruise in search of Franklin and his missing expedition. There was a fascination about such an undertaking that I felt I could not have resisted had I been in his place. Of course, Lettie did not like the idea at all, but he silenced her by telling her that men who volunteered for Arctic search were never lost sight of, and that he should not make as much advance in his profession in a dozen years as he would in the year or so of this expedition. I cannot say that Lettie, even after this, was quite satisfied with the notion of his going, but, at all events, she did not argue against it any longer. But the grave look, which is now habitual with her, but was a rare thing in her young and happy days, passed over her face sometimes when she thought no one was looking.

My younger brother, Harry, was at this time an academy student. He was only a beginner then. Now he is pretty well known in the art world, and his pictures command fair prices. Like all beginners in art, he was full of fancies and theories. He would have been a pre-

Raphaelite, only pre-Raphaelism had not been invented then. His peculiar craze was for what he styled the Venetian School. Now, it chanced that George had a fine Italian-looking head, and Harry persuaded him to sit to him for his portrait. It was a fair likeness, but a very moderate work of art. The background was so very dark, and George's naval costume so very deep in colour, that the face came out too white and staring. It was a three-quarter picture; but only one hand showed in it, leaning on the hilt of a sword. As George said, he looked much more like the commander of a Venetian galley than a modern mate.

However, the picture pleased Lettie, who did not care much about art provided the resemblance was good. So the picture was duly framed—in a tremendously heavy frame, of Harry's ordering—and hung up in the dining-room.

And now the time for George's departure was growing nearer. The *Pioneer* was nearly ready to sail, and her crew only waited orders. The officers grew acquainted with each other before sailing, which was an advantage. George took up very warmly with the surgeon, Vincent Grieve, and, with my permission, brought him to dinner once or twice.

'Poor chap, he has no friends nearer than the Highlands, and it's precious lonely work.'

'Bring him by all means, George! You know that any friends of yours will be welcome here.'

So Vincent Grieve came. I am bound to say I was not favourably impressed by him, and almost wished I had not consented to his coming. He was a tall, pale, fair young man, with a hard Scotch face and a cold, grey eye. There was something in his expression, too, that was unpleasant—something cruel or crafty, or both.

I considered that it was very bad taste for him to pay such marked attention to Lettie, coming, as he did, as the friend of her fiancé. He kept by her constantly and anticipated George in all the little attentions which a lover delights to pay. I think George was a little put out about it, though he said nothing, attributing his friend's offence to lack of breeding.

Lettie did not like it at all. She knew that she was not to have George with her much longer, and she was anxious to have him to herself as much as possible. But as Grieve was her lover's friend she bore the infliction with the best possible patience.

The surgeon did not seem to perceive in the least that he was interfering where he had no business. He was quite self-possessed and happy, with one exception. The portrait of George seemed to annoy him. He had uttered a little impatient exclamation when he first saw it which drew my attention to him; and I noticed that he tried to avoid looking at it. At last, when dinner came, he was told to sit exactly facing the picture. He hesitated for an instant and then sat down, but almost immediately rose again.

'It's very childish and that sort of thing,' he stammered, 'but I cannot sit opposite that picture.'

'It is not high art,' I said, 'and may irritate a critical eye.'

'I know nothing about art,' he answered, 'but it is one of those unpleasant pictures whose eyes follow you about the room. I have an inherited horror of such pictures. My mother married against her father's will, and when I was born she was so ill she was hardly expected to live. When she was sufficiently recovered to speak without delirious rambling she implored them to remove a picture of my grandfather that hung in the room, and which she vowed made threatening faces at her. It's superstitious, but constitutional—I have a horror of such paintings!'

I believe George thought this was a ruse of his friend's to get a seat next to Lettie; but I felt sure it was not, for I had seen the alarmed expression of his face.

At night, when George and his friend were leaving, I took an opportunity to ask the former, half in a joke, if he should bring the surgeon to see us again. George made a very hearty assertion to the contrary, adding that he was pleasant enough company among men at an inn, or on board ship, but not where ladies were concerned.

But the mischief was done. Vincent Grieve took advantage of the introduction and did not wait to be invited again. He called the next day, and nearly every day after. He was a more frequent visitor than George now, for George was obliged to attend to his duties, and they kept him on board the *Pioneer* pretty constantly, whereas the surgeon, having seen to the supply of drugs, etc., was pretty well at liberty. Lettie avoided him as much as possible, but he generally brought, or professed to bring, some little message from George to her, so that he had an excuse for asking to see her.

On the occasion of his last visit—the day before the *Pioneer* sailed—Lettie came to me in great distress. The young cub had actually the audacity to tell her he loved her. He knew, he said, about her engagement to George, but that did not prevent another man from loving her too. A man could no more help falling in love than he could help taking a fever. Lettie stood upon her dignity and rebuked him severely; but he told her he could see no harm in telling her of his passion, though he knew it was a hopeless one.

'A thousand things may happen,' he said at last, 'to bring your engagement with George Mason to an end. Then perhaps you will not forget that another loves you!'

I was very angry, and was forthwith going to give him my opinion on his conduct, when Lettie told me he was gone, that she had bade him go and had forbidden him the house. She only told me in order to protect herself, for she did not intend to say anything to George, for fear it should lead to a duel or some other violence.

That was the last we saw of Vincent Grieve before the *Pioneer* sailed.

George came the same evening, and was with us till daybreak, when he had to tear himself away and join his ship.

After shaking hands with him at the door, in the cold, grey, drizzly dawn, I turned back into the dining-room, where poor Lettie was sobbing on the sofa.

I could not help starting when I looked at George's portrait, which hung above her. The strange light of daybreak could hardly account for the extraordinary pallor of the face. I went close to it and looked hard at it. I saw that it was covered with moisture, and imagined that that possibly made it look so pale. As for the moisture, I supposed poor Lettie had been kissing the beloved's portrait, and that the moisture was caused by her tears.

It was not till a long time after, when I was jestingly telling Harry how his picture had been caressed, that I learnt the error of my conjecture. Lettie assured me most solemnly that I was mistaken in supposing she had kissed it.

'It was the varnish blooming, I expect,' said Harry. And thus the subject was dismissed, for I said no more, though I knew well enough, in spite of my not being an artist, that the bloom of varnish was quite another sort of thing.

The *Pioneer* sailed. We received—or, rather, Lettie received—two letters from George, which he had taken the opportunity of sending by homeward-bound whalers. In the second he said it was hardly likely he should have an opportunity of sending another, as they were sailing into high latitudes—into the solitary sea, to which none but expedition ships ever penetrated. They were all in high spirits, he said, for they had encountered very little ice and hoped to find clear water further north than usual. Moreover, he added, Grieve had held a sinecure so far, for there had not been a single case of illness on board.

Then came a long silence, and a year crept away very slowly for poor Lettie. Once we heard of the expedition from the papers. They were reported as pushing on and progressing favourably by a wandering tribe of Esquimaux with whom the captain of a Russian vessel fell in. They had laid the ship up for the winter, and were taking the boats on sledges, and believed they had met with traces of the lost crews that seemed to show they were on the right track.

The winter passed again, and spring came. It was a balmy, bright spring such as we get occasionally, even in this changeable and uncertain climate of ours.

One evening we were sitting in the dining-room with the window open, for, although we had long given up fires, the room was so oppressively warm that we were glad of the breath of the cool evening breeze.

Lettie was working. Poor child, though she never murmured, she was evidently pining at George's long absence. Harry was leaning out of the window, studying the evening effect on the fruit blossom, which was wonderfully early and plentiful, the season was so mild. I was sitting at the table, near the lamp, reading the paper.

Suddenly there swept into the room a chill. It was not a gust of cold wind, for the curtain by the open window did not swerve in the least. But the deathly cold pervaded the room—came, and was gone in an instant. Lettie shuddered, as I did, with the intense icy feeling.

She looked up. 'How curiously cold it has got all in a minute,' she said.

'We are having a taste of poor George's Polar weather,' I said with a smile.

At the same moment I instinctively glanced towards his portrait. What I saw struck me dumb. A rush of blood, at fever heat, dispelled the numbing influence of the chill breath that had seemed to freeze me.

I have said the lamp was lighted; but it was only that I might read with comfort, for the violet twilight was still so full of sunset that the room was not dark. But as I looked at the picture I saw it had undergone a strange change. I saw it as plainly as possible. It was no delusion, coined for the eye by the brain.

I saw, in the place of George's head, a grinning skull! I stared at it hard; but it was no trick of fancy. I could see the hollow orbits, the gleaming teeth, the fleshless cheekbones—it was the head of death!

Without saying a word, I rose from my chair and walked straight up to the painting. As I drew nearer a sort of mist seemed to pass before it; and as I stood close to it, I saw only the face of George. The spectral skull had vanished.

'Poor George!' I said unconsciously.

Lettie looked up. The tone of my voice had alarmed her, the expression of my face did not reassure her.

'What do you mean? Have you heard anything? Oh, Robert, in mercy tell me!'

She got up and came over to me and, laying her hands on my arm, looked up into my face imploringly.

'No, my dear; how should I hear? Only I could not help thinking of the privation and discomfort he must have gone through. I was reminded of it by the cold——'

'Cold!' said Harry, who had left the window by this time. 'Cold! what on earth are you talking about? Cold, such an evening as this! You must have had a touch of ague, I should think.'

'Both Lettie and I felt it bitterly cold a minute or two ago. Did not you feel it?'

'Not a bit; and as I was three parts out of the window I ought to have felt it if anyone did.'

It was curious, but that strange chill had been felt only in the room. It was not the night wind, but some supernatural breath connected with the dread apparition I had seen. It was, indeed, the chill of polar winter—the icy shadow of the frozen North.

'What is the day of the month, Harry?' I asked.

'Today—the 23rd, I think,' he answered; then added, taking up the newspaper I had been reading: 'Yes, here you are. Tuesday, February the 23rd, if the *Daily News* tells truth, which I suppose it does. Newspapers can afford to tell the truth about dates, whatever they may do about art.' Harry had been rather roughly handled by the critic of a morning paper for one of his pictures a few days before, and he was a little angry with journalism generally.

Presently Lettie left the room, and I told Harry what I had felt and seen, and told him to take note of the date, for I feared that some mischance had befallen George.

'I'll put it down in my pocket-book, Bob. But you and Lettie must have had a touch of the cold shivers, and your stomach or fancy misled you—they're the same thing, you know. Besides, as regards the picture, there's nothing in that! There is a skull there, of course. As Tennyson says:

> Any face, however full,
> Padded round with flesh and fat,
> Is but modelled on a skull.

The skull's there—just as in every good figure-subject the nude is there under the costumes. You fancy that is a mere coat of paint. Nothing of the kind! Art lives, sir! That is just as much a real head as yours is with all the muscles and bones, just the same. That's what makes the difference between art and rubbish.'

This was a favourite theory of Harry's, who had not yet developed from the dreamer into the worker. As I did not care to argue with him, I allowed the subject to drop after we had written down the date in our pocket-books. Lettie sent down word presently that she did not feel well and had gone to bed. My wife came down presently and asked what had happened. She had been up with the children and had gone in to see what was the matter with Lettie.

'I think it was very imprudent to sit with the window open, dear. I know the evenings are warm, but the night air strikes cold at times—at any rate, Lettie seems to have caught a violent cold, for she is shivering very much. I am afraid she has got a chill from the open windows.'

I did not say anything to her then, except that both Lettie and I had felt a sudden coldness; for I did not care to enter into an explanation again, for I could see Harry was inclined to laugh at me for being so superstitious.

At night, however, in our own room, I told my wife what had occurred, and what my apprehensions were. She was so upset and alarmed that I almost repented having done so.

The next morning Lettie was better again, and as we did not either of us refer to the events of the preceding night the circumstance appeared to be forgotten by us all.

But from that day I was ever inwardly dreading the arrival of bad news. And at last it came, as I expected.

One morning, just as I was coming downstairs to breakfast, there came a knock at the door, and Harry made his appearance. It was a very early visit from him, for he generally used to spend his mornings at the studio, and drop in on his way home at night.

He was looking pale and agitated.

'Lettie's not down, is she, yet?' he asked; and then, before I could answer, added another question:

'What newspaper do you take?'

'The *Daily News*,' I answered. 'Why?'

'She's not down?'

'No.'

'Thank God! Look here!'

He took a paper from his pocket and gave it to me, pointing out a short paragraph at the bottom of one of the columns.

I knew what was coming the moment he spoke about Lettie.

The paragraph was headed, 'Fatal Accident to one of the Officers of the *Pioneer* Expedition Ship'. It stated that news had been received at the Admiralty stating that the expedition had failed to find the missing crews, but had come upon some traces of them. Want of stores and necessaries had compelled them to turn back without following those traces up; but the commander was anxious, as soon as the ship could be refitted, to go out and take up the trail where he left it. An unfortunate accident had deprived him of one of his most promising officers, Lieutenant Mason, who was precipitated from an iceberg and killed while out shooting with the surgeon. He was beloved by all, and his death had flung a gloom over the gallant little troop of explorers.

'It's not in the *News* today, thank goodness, Bob,' said Harry, who had been searching that paper while I was reading the one he brought—'but you must keep a sharp look-out for some days and not let Lettie see it when it appears, as it is certain to do sooner or later.'

Then we both of us looked at each other with tears in our eyes. 'Poor George!—poor Lettie!' we sighed softly.

'But she must be told at some time or other?' I said despairingly.

'I suppose so,' said Harry; 'but it would kill her to come on it suddenly like this. Where's your wife?'

She was with the children, but I sent up for her and told her the ill-tidings.

She had a hard struggle to conceal her emotion, for Lettie's sake. But the tears would flow in spite of her efforts.

'How shall I ever find courage to tell her?' she asked.

'Hush!' said Harry, suddenly grasping her arm and looking towards the door.

I turned. There stood Lettie, with her face pale as death, with her lips apart, and with a blind look about her eyes. She had come in without our hearing her. We never learnt how much of the story she had overheard; but it was enough to tell her the worst. We all sprang towards her; but she only waved us away, turned round, and went upstairs again without saying a word. My wife hastened up after her and found her on her knees by the bed, insensible.

The doctor was sent for, and restoratives were promptly administered. She came to herself again, but lay dangerously ill for some weeks from the shock.

It was about a month after she was well enough to come downstairs again that I saw in the paper an announcement of the arrival of the *Pioneer*. The news had no interest for any of us now, so I said nothing about it. The mere mention of the vessel's name would have caused the poor girl pain.

One afternoon shortly after this, as I was writing a letter, there came a loud knock at the front door. I looked up from my writing and listened; for the voice which enquired if I was in sounded strange, but yet not altogether unfamiliar. As I looked up, puzzling whose it could be, my eye rested accidentally upon poor George's portrait. Was I dreaming or awake?

I have told you that the one hand was resting on a sword. I could see now distinctly that the forefinger was raised, as if in warning. I looked at it hard, to assure myself it was no fancy, and then I perceived, standing out bright and distinct on the pale face, two large drops, as if of blood.

I walked up to it, expecting the appearance to vanish, as the skull had done. It did not vanish; but the uplifted finger resolved itself into a little white moth which had settled on the canvas. The red drops were fluid, and certainly not blood, though I was at a loss for the time to account for them.

The moth seemed to be in a torpid state, so I took it off the picture and placed it under an inverted wine-glass on the mantelpiece. All this took less time to do than to describe. As I turned from the mantelpiece the servant brought in a card, saying the gentleman was waiting in the hall to know if I would see him.

On the card was the name of 'Vincent Grieve, of the exploring vessel *Pioneer*'.

'Thank Heaven, Lettie is out,' thought I; and then added aloud to the servant, 'Show him in here; and Jane, if your mistress and Miss Lettie come in before the gentleman goes, tell them I have someone with me on business and do not wish to be disturbed.'

I went to the door to meet Grieve. As he crossed the threshold, and before he could have seen the portrait, he stopped, shuddered and turned white, even to his thin lips.

'Cover that picture before I come in,' he said hurriedly, in a low voice. 'You remember the effect it had upon me. Now, with the memory of poor Mason, it would be worse than ever.'

I could understand his feelings better now than at first; for I had come to look on the picture with some awe myself. So I took the cloth off a little round table that stood under the window and hung it over the portrait.

When I had done so Grieve came in. He was greatly altered. He was thinner and paler than ever; hollow-eyed and hollow-cheeked. He had acquired a strange stoop, too, and his eyes had lost the crafty look for a look of terror, like that of a hunted beast. I noticed that he kept glancing sideways every instant, as if unconsciously. It looked as if he heard someone behind him.

I had never liked the man; but now I felt an insurmountable

repugnance to him—so great a repugnance that, when I came to think of it, I felt pleased that the incident of covering the picture at his request had led to my not shaking hands with him.

I felt that I could not speak otherwise than coldly to him; indeed, I had to speak with painful plainness.

I told him that, of course, I was glad to see him back, but that I could not ask him to continue to visit us. I should be glad to hear the particulars of poor George's death, but that I could not let him see my sister, and hinted, as delicately as I could, at the impropriety of which he had been guilty when he last visited.

He took it all very quietly, only giving a long, weary sigh when I told him I must beg him not to repeat his visit. He looked so weak and ill that I was obliged to ask him to take a glass of wine—an offer which he seemed to accept with great pleasure.

I got out the sherry and biscuits and placed them on the table between us, and he took a glass and drank it off greedily.

It was not without some difficulty that I could get him to tell me of George's death. He related, with evident reluctance, how they had gone out to shoot a white bear which they had seen on an iceberg stranded along the shore. The top of the berg was ridged like the roof of a house, sloping down on one side to the edge of a tremendous overhanging precipice. They had scrambled along the ridge in order to get nearer the game, when George incautiously ventured on the sloping side.

'I called out to him', said Grieve, 'and begged him to come back, but too late. The surface was as smooth and slippery as glass. He tried to turn back, but slipped and fell. And then began a horrible scene. Slowly, slowly, but with ever-increasing motion, he began to slide down towards the edge. There was nothing to grasp at—no irregularity or projection on the smooth face of the ice. I tore off my coat, and hastily attaching it to the stock of my gun, pushed the latter towards him; but it did not reach far enough. Before I could lengthen it, by tying my cravat to it, he had slid yet further away, and more quickly. I shouted in agony; but there was no one within hearing. He, too, saw his fate was sealed; and he could only tell me to bring his last farewell to you, and—and to her!'—Here Grieve's voice broke—'and it was all over! He clung to the edge of the precipice instinctively for one second, and was gone!'

Just as Grieve uttered the last word, his jaw fell; his eyeballs seemed ready to start from his head; he sprang to his feet, pointed at something behind me, and then flinging up his arms, fell, with a scream, as if he had been shot. He was seized with an epileptic fit.

I instinctively looked behind me as I hurried to raise him from the floor. The cloth had fallen from the picture, where the face of George, made paler than ever by the gouts of red, looked sternly down.

I rang the bell. Luckily, Harry had come in; and, when the servant told him what was the matter, he came in and assisted me in restoring Grieve to consciousness. Of course, I covered the painting up again.

When he was quite himself again, Grieve told me he was subject to fits occasionally.

He seemed very anxious to learn if he had said or done anything extraordinary while he was in the fit, and appeared reassured when I said he had not. He apologized for the trouble he had given, and said as soon as he was strong enough he would take his leave. He was leaning on the mantelpiece as he said this. The little white moth caught his eye.

'So you have had someone else from the *Pioneer* here before me?' he said, nervously.

I answered in the negative, asking what made him think so.

'Why, this little white moth is never found in such southern latitudes. It is one of the last signs of life northward. Where did you get it?'

'I caught it here, in this room,' I answered.

'That is very strange. I never heard of such a thing before. We shall hear of showers of blood soon, I should not wonder.'

'What do you mean?' I asked.

'Oh, these little fellows emit little drops of a red-looking fluid at certain seasons, and sometimes so plentifully that the superstitious think it is a shower of blood. I have seen the snow quite stained in places. Take care of it, it is a rarity in the south.'

I noticed, after he left, which he did almost immediately, that there was a drop of red fluid on the marble under the wine-glass. The blood-stain on the picture was accounted for; but how came the moth here?

And there was another strange thing about the man, which I had scarcely been able to assure myself of in the room, where there were cross-lights, but about which there was no possible mistake, when I saw him walking away up the street.

'Harry, here—quick!' I called to my brother, who at once came to the window. 'You're an artist, tell me, is there anything strange about that man?'

'No; nothing that I can see,' said Harry, but then suddenly, in an altered tone, added, 'Yes, there is. By Jove, *he has a double shadow*!'

That was the explanation of his sidelong glances, of the habitual stoop. There was a something always at his side, which none could see, but which cast a shadow.

He turned, presently, and saw us at the window. Instantly, he crossed the road to the shady side of the street. I told Harry all that had passed, and we agreed that it would be as well not to say a word to Lettie.

Two days later, when I returned from a visit to Harry's studio, I found the whole house in confusion.

I learnt from Lettie that while my wife was upstairs, Grieve had called, had not waited for the servant to announce him, but had walked straight into the dining-room, where Lettie was sitting. She noticed that he avoided looking at the picture, and, to make sure of not seeing it, had seated himself on the sofa just beneath it. He had then, in spite of Lettie's angry remonstrances, renewed his offer of love, strengthening it finally by assuring her that poor George with his dying breath had implored him to seek her, and watch over her, and marry her.

'I was so indignant I hardly knew how to answer him,' said Lettie. 'When, suddenly, just as he uttered the last words, there came a twang like the breaking of a guitar—and—I hardly know how to describe it—but the portrait had fallen, and the corner of the heavy frame had struck him on the head, cutting it open, and rendering him insensible.'

They had carried him upstairs, by the direction of the doctor, for whom my wife at once sent on hearing what had occurred. He was laid on the couch in my dressing-room, where I went to see him. I intended to reproach him for coming to the house, despite my prohibition, but I found him delirious. The doctor said it was a queer case; for, though the blow was a severe one, it was hardly enough to account for the symptoms of brain-fever. When he learnt that Grieve had but just returned in the *Pioneer* from the North, he said it was possible that the privation and hardship had told on his constitution and sown the seeds of the malady.

We sent for a nurse, who was to sit up with him, by the doctor's directions.

The rest of my story is soon told. In the middle of the night I was roused by a loud scream. I slipped on my clothes, and rushed out to find the nurse, with Lettie in her arms, in a faint. We carried her into her room, and then the nurse explained the mystery to us.

It appears that about midnight Grieve sat up in bed, and began to talk. And he said such terrible things that the nurse became alarmed. Nor was she much reassured when she became aware that the light of her single candle flung what seemed to be two shadows of the sick man on the wall.

Terrified beyond measure, she had crept into Lettie's room, and confided her fears to her; and Lettie, who was a courageous and kindly girl, dressed herself, and said she would sit with her. She, too, saw the double shadow—but what she heard was far more terrible.

Grieve was sitting up in bed, gazing at the unseen figure to which the shadow belonged. In a voice that trembled with emotion, he begged the haunting spirit to leave him, and prayed its forgiveness.

'You know the crime was not premeditated. It was a sudden temptation of the devil that made me strike the blow, and fling you over the precipice. It was the devil tempting me with the recollection of her exquisite face—of the tender love that might have been mine, but for you. But she will not listen to me. See, she turns away from me, as if she knew I was your murderer, George Mason!'

It was Lettie who repeated in a horrified whisper this awful confession.

I could see it all now! As I was about to tell Lettie of the many strange things I had concealed from her, the nurse, who had gone to see her patient, came running back in alarm.

Vincent Grieve had disappeared. He had risen in his delirious terror, had opened the window, and leaped out. Two days later his body was found in the river.

A curtain hangs now before poor George's portrait, though it is no longer connected with any supernatural marvels; and never, since the night of Vincent Grieve's death, have we seen aught of that most mysterious haunting presence—the Shadow of a Shade.

At Chrighton Abbey

MARY ELIZABETH BRADDON

The Chrightons were very great people in that part of the country where my childhood and youth were spent. To speak of Squire Chrighton was to speak of a power in that remote western region of England. Chrighton Abbey had belonged to the family ever since the reign of Stephen, and there was a curious old wing and a cloistered quadrangle still remaining of the original edifice, and in excellent preservation. The rooms at this end of the house were low, and somewhat darksome and gloomy, it is true; but, though rarely used, they were perfectly habitable, and were of service on great occasions when the Abbey was crowded with guests.

The central portion of the Abbey had been rebuilt in the reign of Elizabeth, and was of noble and palatial proportions. The southern wing, and a long music-room with eight tall narrow windows added on to it, were as modern as the time of Anne. Altogether, the Abbey was a very splendid mansion, and one of the chief glories of our county.

All the land in Chrighton parish, and for a long way beyond its boundaries, belonged to the great Squire. The parish church was within the park walls, and the living in the Squire's gift—not a very valuable benefice, but a useful thing to bestow upon a younger son's younger son, once in a way, or sometimes on a tutor or dependent of the wealthy house.

I was a Chrighton, and my father, a distant cousin of the reigning Squire, had been rector of Chrighton parish. His death left me utterly unprovided for, and I was fain to go out into the bleak unknown world, and earn my living in a position of dependence—a dreadful thing for a Chrighton to be obliged to do.

Out of respect for the traditions and prejudices of my race, I made it my business to seek employment abroad, where the degradation of one solitary Chrighton was not so likely to inflict shame upon the ancient house to which I belonged. Happily for myself, I had been carefully educated, and had industriously cultivated the usual modern accomplishments in the calm retirement of the Vicarage. I was so fortunate as to obtain a situation at Vienna, in a German family of high rank; and here I remained seven years, laying aside year by year a considerable portion of my liberal salary. When my pupils had grown up, my kind mistress procured me a still more profitable position at St Petersburg, where I remained five more years, at the end of which time I yielded to a yearning that had been long growing upon me—an ardent desire to see my dear old country home once more.

I had no very near relations in England. My mother had died some years before my father; my only brother was far away, in the Indian Civil Service; sister I had none. But I was a Chrighton, and I loved the soil from which I had sprung. I was sure, moreover, of a warm welcome from friends who had loved and honoured my father and mother, and I was still further encouraged to treat myself to this holiday by the very cordial letters I had from time to time received from the Squire's wife, a noble warm-hearted woman, who fully approved the independent course I had taken, and who had ever shown herself my friend.

In all her letters for some time past Mrs Chrighton begged that, whenever I felt myself justified in coming home, I would pay a long visit to the Abbey.

'I wish you could come at Christmas,' she wrote, in the autumn of the year of which I am speaking. 'We shall be very gay, and I expect all kinds of pleasant people at the Abbey. Edward is to be married early in the spring—much to his father's satisfaction, for the match is a good and appropriate one. His *fiancée* is to be among our guests. She is a very beautiful girl; perhaps I should say handsome rather than beautiful. Julia Tremaine, one of the Tremaines of Old Court, near Hayswell—a very old family, as I daresay you remember. She has several brothers and sisters, and will have little, perhaps nothing, from her father; but she has a considerable fortune left her by an aunt, and is thought quite an heiress in the county—not, of course, that this latter fact had any influence with Edward. He fell in love with her at an assize ball in his usual impulsive fashion, and proposed to her in something less than a fortnight. It is, I hope and believe, a thorough love-match on both sides.'

After this followed a cordial repetition of the invitation to myself. I was to go straight to the Abbey when I went to England, and was to take up my abode there as long as ever I pleased.

This letter decided me. The wish to look on the dear scenes of my happy childhood had grown almost into a pain. I was free to take a holiday, without detriment to my prospects. So, early in December, regardless of the bleak dreary weather, I turned my face homewards, and made the long journey from St Petersburg to London, under the kind escort of Major Manson, a Queen's Messenger, who was a friend of my late employer, the Baron Fruydorff, and whose courtesy had been enlisted for me by that gentleman.

I was three-and-thirty years of age. Youth was quite gone; beauty I had never possessed; and I was content to think of myself as a confirmed old maid, a quiet spectator of life's great drama, disturbed by no feverish desire for an active part in the play. I had a disposition to which this kind of passive existence is easy. There was no wasting fire in my veins. Simple duties, rare and simple pleasures, filled up my sum of life. The dear ones who had given a special charm and brightness to my existence were gone. Nothing could recall *them*, and without them actual happiness seemed impossible to me. Everything had a subdued and neutral tint; life at its best was calm and colourless, like a grey sunless day in early autumn, serene but joyless.

The old Abbey was in its glory when I arrived there, at about nine o'clock on a clear starlit night. A light frost whitened the broad sweep of grass that stretched away from the long stone terrace in front of the house to a semicircle of grand old oaks and beeches. From the music-room at the end of the southern wing, to the heavily framed gothic windows of the old rooms on the north, there shone one blaze of light. The scene reminded me of some weird palace in a German legend; and I half expected to see the lights fade out all in a moment, and the long stone façade wrapped in sudden darkness.

The old butler, whom I remembered from my very infancy, and who did not seem to have grown a day older during my twelve years' exile, came out of the dining-room as the footman opened the hall-door for me, and gave me cordial welcome, nay insisted upon helping to bring in my portmanteau with his own hands, an act of unusual condescension, the full force of which was felt by his subordinates.

'It's a real treat to see your pleasant face once more, Miss Sarah,' said this faithful retainer, as he assisted me to take off my travelling-cloak, and took my dressing-bag from my hand. 'You look a trifle older than when you used to live at the Vicarage twelve year ago, but you're looking uncommon well for all that; and, Lord love your heart, miss, how pleased they all will be to see you! Missus told me with her own lips about your coming. You'd like to take off your bonnet before you go to the drawing-room, I daresay. The house is full of company. Call Mrs Marjorum, James, will you?'

The footman disappeared into the back regions, and presently reappeared with Mrs Marjorum, a portly dame, who, like Truefold the butler, had been a fixture at the Abbey in the time of the present Squire's father. From her I received the same cordial greeting, and by her I was led off up staircases and along corridors, till I wondered where I was being taken.

We arrived at last at a very comfortable room—a square tapes-tried chamber, with a low ceiling supported by a great oaken beam. The room looked cheery enough, with a bright fire roaring in the wide chimney; but it had a somewhat ancient aspect, which the supersti-tiously inclined might have associated with possible ghosts.

I was fortunately of a matter-of-fact disposition, utterly sceptical upon the ghost subject; and the old-fashioned appearance of the room took my fancy.

'We are in King Stephen's wing, are we not, Mrs Marjorum?' I asked; 'this room seems quite strange to me. I doubt if I have ever been in it before.'

'Very likely not, miss. Yes, this is the old wing. Your window looks out into the old stable-yard, where the kennel used to be in the time of our Squire's grandfather, when the Abbey was even a finer place than it is now, I've heard say. We are so full of company this winter, you see, miss, that we are obliged to make use of all these rooms. You'll have no need to feel lonesome. There's Captain and Mrs Cranwick in the next room to this, and the two Miss Newports in the blue room opposite.'

'My dear good Marjorum, I like my quarters excessively; and I quite enjoy the idea of sleeping in a room that was extant in the time of Stephen, when the Abbey really was an abbey. I daresay some grave old monk has worn these boards with his devout knees.'

The old woman stared dubiously, with the air of a person who had small sympathy with monkish times, and begged to be excused for leaving me, she had so much on her hands just now.

There was coffee to be sent in; and she doubted if the still-room maid would manage matters properly, if she, Mrs Marjorum, were not at hand to see that things were right.

'You've only to ring your bell, miss, and Susan will attend to you. She's used to help waiting on our young ladies sometimes, and she's very handy. Missus has given particular orders that she should be always at your service.'

'Mrs Chrighton is very kind; but I assure you, Marjorum, I don't require the help of a maid once in a month. I am accustomed to do everything for myself. There, run along, Mrs Marjorum, and see after your coffee; and I'll be down in the drawing-room in ten minutes. Are there many people there, by the bye?'

'A good many. There's Miss Tremaine, and her mamma and younger sister; of course you've heard all about the marriage—such a handsome young lady—rather too proud for my liking; but the Tremaines always were a proud family, and this one's an heiress. Mr Edward is so fond of her—thinks the ground is scarcely good enough for her to walk upon, I do believe; and somehow I can't help wishing he'd chosen someone else—someone who would have thought more of him, and who would not take all his attentions in such a cool off-hand way. But of course it isn't my business to say such things, and I wouldn't venture upon it to any one but you, Miss Sarah.'

She told me that I would find dinner ready for me in the breakfast-room, and then bustled off, leaving me to my toilet.

This ceremony I performed as rapidly as I could, admiring the perfect comfort of my chamber as I dressed. Every modern appliance had been added to the sombre and ponderous furniture of an age gone by, and the combination produced a very pleasant effect. Perfume-bottles of ruby-coloured Bohemian glass, china brush-trays and ring-stands brightened the massive oak dressing-table; a low luxurious chintz-covered easy-chair of the Victorian era stood before the hearth; a dear little writing-table of polished maple was placed conveniently near it; and in the background the tapestried walls loomed duskily, as they had done hundreds of years before my time.

I had no leisure for dreamy musings on the past, however, provocative though the chamber might be of such thoughts. I arranged my hair in its usual simple fashion, and put on a dark-grey silk dress, trimmed with some fine old black lace that had been given to me by the Baroness—an unobtrusive demi-toilette, adapted to any occasion. I tied a massive gold cross, an ornament that had belonged to my dear mother, round my neck with a scarlet ribbon; and my costume was complete. One glance at the looking-glass convinced me that there was nothing dowdy in my appearance; and then I hurried along the corridor and down the staircase to the hall, where Truefold received me and conducted me to the breakfast-room, in which an excellent dinner awaited me.

I did not waste much time over this repast, although I had eaten nothing all day; for I was anxious to make my way to the drawing-room. Just as I had finished, the door opened, and Mrs Chrighton sailed in, looking superb in a dark-green velvet dress richly trimmed with old point lace. She had been a beauty in her youth, and, as a matron, was still remarkably handsome. She had, above all, a charm of expression which to me was rarer and more delightful than her beauty of feature and complexion.

She put her arms round me, and kissed me affectionately.

'I have only this moment been told of your arrival, my dear Sarah,' she said; 'and I find you have been in the house half an hour. What must you have thought of me!'

'What can I think of you, except that you are all goodness, my dear Fanny? I did not expect you to leave your guests to receive me, and am really sorry that you have done so. I need no ceremony to convince me of your kindness.'

'But, my dear child, it is not a question of ceremony. I have been looking forward so anxiously to your coming, and I should not have liked to see you for the first time before all those people. Give me another kiss, that's a darling. Welcome to Chrighton. Remember, Sarah, this house is always to be your home, whenever you have need of one.'

'My dear kind cousin! And you are not ashamed of me, who have eaten the bread of strangers?'

'Ashamed of you! No, my love; I admire your industry and spirit. And now come to the drawing-room. The girls will be so pleased to see you.'

'And I to see them. They were quite little things when I went away, romping in the hay-fields in their short white frocks; and now, I suppose, they are handsome young women.'

'They are very nice-looking; not as handsome as their brother. Edward is really a magnificent young man. I do not think my maternal pride is guilty of any gross exaggeration when I say that.'

'And Miss Tremaine?' I said. 'I am very curious to see her.'

I fancied a faint shadow came over my cousin's face as I mentioned this name.

'Miss Tremaine—yes—you cannot fail to admire her,' she said, rather thoughtfully.

She drew my hand through her arm and led me to the drawing-room: a very large room, with a fireplace at each end, brilliantly lighted tonight, and containing about twenty people, scattered about in little groups, and all seeming to be talking and laughing merrily. Mrs Chrighton took me straight to one of the fireplaces, beside which two girls were sitting on a low sofa, while a young man of something more than six feet high stood near them, with his arm resting on the broad marble slab of the mantelpiece. A glance told me that this young man with the dark eyes and crisp waving brown hair was Edward Chrighton. His likeness to his mother was in itself enough to tell me who he was; but I remembered the boyish face and bright eyes which had so often looked up to mine in the days when the heir of the Abbey was one of the most juvenile scholars at Eton.

The lady seated nearest Edward Chrighton attracted my chief attention; for I felt sure that this lady was Miss Tremaine. She was tall and slim, and carried her head and neck with a stately air, which struck me more than anything in that first glance. Yes, she was handsome, undeniably handsome; and my cousin had been right when she said I could not fail to admire her; but to me the dazzlingly fair face with its perfect features, the marked aquiline nose, the short upper lip expressive of unmitigated pride, the full cold blue eyes, pencilled brows, and aureole of pale golden hair, were the very reverse of sympathetic. That Miss Tremaine must needs be universally admired, it was impossible to doubt; but I could not understand how any man could fall in love with such a woman.

She was dressed in white muslin, and her only ornament was a superb diamond locket, heart-shaped, tied round her long white throat with a broad black ribbon. Her hair, of which she seemed to have a great quantity, was arranged in a massive coronet of plaits, which surmounted the small head as proudly as an imperial crown.

To this young lady Mrs Chrighton introduced me.

'I have another cousin to present to you, Julia,' she said smiling—'Miss Sarah Chrighton, just arrived from St Petersburg.'

'From St Petersburg? What an awful journey! How do you do, Miss Chrighton? It was really very courageous of you to come so far. Did you travel alone?'

'No; I had a companion as far as London, and a very kind one. I came on to the Abbey by myself.'

The young lady had given me her hand with rather a languid air, I thought. I saw the cold blue eyes surveying me curiously from head to foot, and it seemed to me as if I could read the condemnatory summing-up—'A frump, and a poor relation'—in Miss Tremaine's face.

I had not much time to think about her just now; for Edward Chrighton suddenly seized both my hands, and gave me so hearty and loving a welcome, that he almost brought the tears 'up from my heart into my eyes'.

Two pretty girls in blue crape came running forward from different parts of the room, and gaily saluted me as 'Cousin Sarah'; and the three surrounded me in a little cluster, and assailed me with a string of questions—whether I remembered this, and whether I had forgotten that, the battle in the hayfield, the charity-school tea-party in the vicarage orchard, our picnics in Hawsley Combe, our botanical and entomological excursions on Chorwell-common, and all the simple pleasures of their childhood and my youth. While this catechism was going on, Miss Tremaine watched us with a disdainful expression, which she evidently did not care to hide.

'I should not have thought you capable of such Arcadian simplicity,

Mr Chrighton,' she said at last. 'Pray continue your recollections. These juvenile experiences are most interesting.'

'I don't expect you to be interested in them, Julia,' Edward answered, with a tone that sounded rather too bitter for a lover. 'I know what a contempt you have for trifling rustic pleasures. Were you ever a child yourself, I wonder, by the way? I don't believe you ever ran after a butterfly in your life.'

Her speech put an end to our talk of the past, somehow. I saw that Edward was vexed, and that all the pleasant memories of his boyhood had fled before that cold scornful face. A young lady in pink, who had been sitting next Julia Tremaine, vacated the sofa, and Edward slipped into her place, and devoted himself for the rest of the evening to his betrothed. I glanced at his bright expressive face now and then as he talked to her, and could not help wondering what charm he could discover in one who seemed to me so unworthy of him.

It was midnight when I went back to my room in the north wing, thoroughly happy in the cordial welcome that had been given me. I rose early next morning—for early rising had long been habitual to me—and, drawing back the damask-curtain that sheltered my window, looked out at the scene below.

I saw a stable-yard, a spacious quadrangle, surrounded by the closed doors of stables and dog-kennels: low massive buildings of grey stone, with the ivy creeping over them here and there, and with an ancient moss-grown look, that gave them a weird kind of interest in my eyes. This range of stabling must have been disused for a long time, I fancied. The stables now in use were a pile of handsome red-brick buildings at the other extremity of the house, to the rear of the music-room, and forming a striking feature in the back view of the Abbey.

I had often heard how the present Squire's grandfather had kept a pack of hounds, which had been sold immediately after his death; and I knew that my cousin, the present Mr Chrighton, had been more than once requested to follow his ancestor's good example; for there were no hounds now within twenty miles of the Abbey, though it was a fine country for fox-hunting.

George Chrighton, however—the reigning lord of the Abbey—was not a hunting man. He had, indeed, a secret horror of the sport; for more than one scion of the house had perished untimely in the hunting-field. The family had not been altogether a lucky one, in spite of its wealth and prosperity. It was not often that the goodly heritage had descended to the eldest son. Death in some form or other—on too many occasions a violent death—had come between the heir and his inheritance. And when I pondered on the dark pages in the story of the house, I used to wonder whether my cousin Fanny was ever troubled by morbid forebodings about her only and fondly loved son.

Was there a ghost at Chrighton—that spectral visitant without which the state and splendour of a grand old house seem scarcely complete? Yes, I had heard vague hints of some shadowy presence that had been seen on rare occasions within the precincts of the Abbey; but I had never been able to ascertain what shape it bore.

Those whom I questioned were prompt to assure me that they had seen nothing. They had heard stories of the past—foolish legends, most likely, not worth listening to. Once, when I had spoken of the subject to my cousin George, he told me angrily never again to let him hear any allusion to *that* folly from my lips.

That December passed merrily. The old house was full of really pleasant people, and the brief winter days were spent in one unbroken round of amusement and gaiety. To me the old familiar English country-house life was a perpetual delight—to feel myself amongst kindred an unceasing pleasure. I could not have believed myself capable of being so completely happy.

I saw a great deal of my cousin Edward, and I think he contrived to make Miss Tremaine understand that, to please him, she must be gracious to me. She certainly took some pains to make herself agreeable to me; and I discovered that, in spite of that proud disdainful temper, which she so rarely took the trouble to conceal, she was really anxious to gratify her lover.

Their courtship was not altogether a halcyon period. They had frequent quarrels, the details of which Edward's sisters Sophy and Agnes delighted to discuss with me. It was the struggle of two proud spirits for mastery; but my cousin Edward's pride was of the nobler kind—the lofty scorn of all things mean—a pride that does not ill-become a generous nature. To me he seemed all that was admirable, and I was never tired of hearing his mother praise him. I think my cousin Fanny knew this, and that she used to confide in me as fully as if I had been her sister.

'I daresay you can see I am not quite so fond as I should wish to be of Julia Tremaine,' she said to me one day; 'but I am very glad that my son is going to marry. My husband's has not been a fortunate family, you know, Sarah. The eldest sons have been wild and unlucky for generations past; and when Edward was a boy I used to have many a bitter hour, dreading what the future might bring forth. Thank God he has been, and is, all that I can wish. He has never given me an hour's anxiety by any act of his. Yet I am not the less glad of his marriage. The heirs of Chrighton who have come to an untimely end have all died unmarried. There was Hugh Chrighton, in the reign of George the Second, who was killed in a duel; John, who broke his back in the hunting-field thirty years later; Theodore, shot accidentally by a schoolfellow at Eton; Jasper, whose yacht went down in the Mediterranean forty years ago. An awful list, is it not, Sarah? I shall feel as if my son were safer somehow when he is married. It will seem as if he has escaped the ban that has fallen on so many of our house. He will have greater reason to be careful of his life when he is a married man.'

I agreed with Mrs Chrighton; but could not help wishing that Edward had chosen any other woman than the cold handsome Julia. I could not fancy his future life happy with such a mate.

Christmas came by and by—a real old English Christmas—frost and snow without, warmth and revelry within; skating on the great pond in the park, and sledging on the ice-bound high-roads, by day; private theatricals, charades, and amateur concerts, by night. I was surprised to find that Miss Tremaine refused to take any active part in these evening amusements. She preferred to sit among the elders as a spectator, and had the air and bearing of a princess for whose diversion all our entertainments had been planned. She seemed to think that she fulfilled her mission by sitting still and looking handsome. No desire to show-off appeared to enter her mind. Her intense pride left no room for vanity. Yet I knew that she could have distinguished herself as a musician if she had chosen to do so; for I had heard her sing and play in Mrs Chrighton's morning-room, when only Edward, his sisters, and myself were present; and I knew that both as a vocalist and a pianist she excelled all our guests.

The two girls and I had many a happy morning and afternoon, going from cottage to cottage in a pony-carriage laden with Mrs Chrighton's gifts to the poor of her parish. There was no public formal distribution of blanketing and coals, but the wants of all were amply provided for in a quiet friendly way. Agnes and Sophy, aided by an indefatigable maid, the Rector's daughter, and one or two other young ladies, had been at work for the last three months making smart warm frocks and useful under-garments for the children of the cottagers; so that on Christmas morning every child in the parish was arrayed in a complete set of new garments. Mrs Chrighton had an admirable faculty of knowing precisely what was most wanted in every household; and our pony-carriage used to convey a varied collection of goods, every parcel directed in the firm free hand of the châtelaine of the Abbey.

Edward used sometimes to drive us on these expeditions, and I found that he was eminently popular among the poor of Chrighton parish. He had such an airy pleasant way of talking to them, a manner which set them at their ease at once. He never forgot their names or relationships, or wants or ailments; had a packet of exactly the kind of tobacco each man liked best always ready in his coat-pockets; and was full of jokes, which may not have been particularly witty, but which used to make the small low-roofed chambers ring with hearty laughter.

Miss Tremaine coolly declined any share in these pleasant duties.

'I don't like poor people,' she said. 'I daresay it sounds very dreadful, but it's just as well to confess my iniquity at once. I never can get on with them, or they with me. I am not *simpatica*, I suppose. And then I cannot endure their stifling rooms. The close faint odour of their houses gives me a fever. And again, what is the use of visiting them? It is only an inducement to them to become hypocrites. Surely it is better to arrange on a sheet of paper what it is just and fair for them to have—blankets, and coals, and groceries, and money, and wine, and so on—and let them receive the things from some trustworthy servant. In that case, there need be no cringing on one side, and no endurance on the other.'

'But, you see, Julia, there are some kinds of people to whom that sort of thing is not a question of endurance,' Edward answered, his face flushing indignantly. 'People who like to share in the pleasure they give—who like to see the poor careworn faces lighted up with sudden joy—who like to make these sons of the soil feel that there is some friendly link between themselves and their masters—some point of union between the cottage and the great house. There is my mother, for instance: all these duties which you think so tiresome are to her an unfailing delight. There will be a change, I'm afraid, Julia, when you are mistress of the Abbey.'

'You have not made me that yet,' she answered; 'and there is plenty of time for you to change your mind, if you do not think me suited for the position. I do not pretend to be like your mother. It is better that I should not affect any feminine virtues which I do not possess.'

After this Edward insisted on driving our pony-carriage almost every day, leaving Miss Tremaine to find her own amusement; and I think this conversation was the beginning of an estrangement between them, which became more serious than any of their previous quarrels had been.

Miss Tremaine did not care for sledging, or skating, or billiard-playing. She had none of the 'fast' tendencies which have become so common lately. She used to sit in one particular bow-window of the drawing-room all the morning, working a screen in berlin-wool and beads, assisted and attended by her younger sister Laura, who was a kind of slave to her—a very colourless young lady in mind, capable of no such thing as an original opinion, and in person a pale replica of her sister.

Had there been less company in the house, the breach between Edward Chrighton and his betrothed must have become notorious; but with a house so full of people, all bent on enjoying themselves, I doubt if it was noticed. On all public occasions my cousin showed himself attentive and apparently devoted to Miss Tremaine. It was only I and his sisters who knew the real state of affairs.

I was surprised, after the young lady's total repudiation of all benevolent sentiments, when she beckoned me aside one morning, and slipped a little purse of gold—twenty sovereigns—into my hand.

'I shall be very much obliged if you will distribute that among your cottagers today, Miss Chrighton,' she said. 'Of course I should like to give them something; it's only the trouble of talking to them that I shrink from; and you are just the person for an almoner. Don't mention my little commission to any one, please.'

'Of course I may tell Edward,' I said; for I was anxious that he should know his betrothed was not as hard-hearted as she had appeared.

'To him least of all,' she answered eagerly. 'You know that our ideas vary on that point. He would think I gave the money to please him. Not a word, pray, Miss Chrighton.' I submitted, and distributed my sovereigns quietly, with the most careful exercise of my judgement.

So Christmas came and passed. It was the day after the great anniversary—a very quiet day for the guests and family at the Abbey, but a grand occasion for the servants, who were to have their annual ball in the evening—a ball to which all the humbler class of tenantry were invited. The frost had broken up suddenly, and it was a thorough wet day—a depressing kind of day for any one whose spirits are liable to be affected by the weather, as mine are. I felt out of spirits for the first time since my arrival at the Abbey.

No one else appeared to feel the same influence. The elder ladies sat in a wide semicircle round one of the fireplaces in the drawing-room; a group of merry girls and dashing young men chatted gaily before the other. From the billiard-room there came the frequent clash of balls, and cheery peals of stentorian laughter. I sat in one of the deep windows, half hidden by the curtains, reading a novel—one of a boxful that came from town every month.

If the picture within was bright and cheerful, the prospect was dreary enough without. The fairy forest of snow-wreathed trees, the white valleys and undulating banks of snow, had vanished, and the rain dripped slowly and sullenly upon a darksome expanse of sodden grass, and a dismal background of leafless timber. The merry sound of the sledge-bells no longer enlivened the air; all was silence and gloom.

Edward Chrighton was not amongst the billiard-players; he was pacing the drawing-room to and fro from end to end, with an air that was at once moody and restless.

'Thank heaven, the frost has broken up at last!' he exclaimed, stopping in front of the window where I sat.

He had spoken to himself, quite unaware of my close neighbourhood. Unpromising as his aspect was just then, I ventured to accost him.

'What bad taste, to prefer such weather as this to frost and snow!' I answered. 'The park looked enchanting yesterday—a real scene from fairyland. And only look at it today!'

'O yes, of course, from an artistic point of view, the snow was better. The place does look something like the great dismal swamp today; but I am thinking of hunting, and that confounded frost made a day's sport impossible. We are in for a spell of mild weather now, I think.'

'But you are not going to hunt, are you, Edward?'

'Indeed I am, my gentle cousin, in spite of that frightened look in your amiable countenance.'

'I thought there were no hounds hereabouts.'

'Nor are there; but there is as fine a pack as any in the country—the Daleborough hounds—five-and-twenty miles away.'

'And you are going five-and-twenty miles for the sake of a day's run?'

'I would travel forty, fifty, a hundred miles for that same diversion. But I am not going for a single day this time; I am going over to Sir Francis Wycherly's place—young Frank Wycherly and I were sworn chums at Christchurch—for three or four days. I am due today, but I scarcely cared to travel by cross-country roads in such rain as this. However, if the floodgates of the sky are loosened for a new deluge, I must go tomorrow.'

'What a headstrong young man!' I exclaimed. 'And what will Miss Tremaine say to this desertion?' I asked in a lower voice.

'Miss Tremaine can say whatever she pleases. She had it in her power to make me forget the pleasures of the chase, if she had chosen, though we had been in the heart of the shires, and the welkin ringing with the baying of hounds.'

'O, I begin to understand. This hunting engagement is not of long standing.'

'No; I began to find myself bored here a few days ago, and wrote to Frank to offer myself for two or three days at Wycherly. I received a most cordial answer by return, and am booked till the end of this week.'

'You have not forgotten the ball on the first?'

'O, no; to do that would be to vex my mother, and to offer a slight to our guests. I shall be here for the first, come what may.'

Come what may! so lightly spoken. The time came when I had bitter occasion to remember those words.

'I'm afraid you will vex your mother by going at all,' I said. 'You know what a horror both she and your father have of hunting.'

'A most un-country-gentleman-like aversion on my father's part. But he is a dear old book-worm, seldom happy out of his library. Yes, I admit they both have a dislike to hunting in the abstract; but they know I am a pretty good rider, and that it would need a bigger country than I shall find about Wycherly to floor me. You need not feel nervous, my dear Sarah; I am not going to give papa and mamma the smallest ground for uneasiness.'

'You will take your own horses, I suppose?'

'That goes without saying. No man who has cattle of his own cares to mount another man's horses. I shall take Pepperbox and the Druid.'

'Pepperbox has a queer temper, I have heard your sisters say.'

'My sisters expect a horse to be a kind of overgrown baa-lamb. Everything splendid in horseflesh and womankind is prone to that slight defect, an ugly temper. There is Miss Tremaine, for instance.'

'I shall take Miss Tremaine's part. I believe it is you who are in the wrong in the matter of this estrangement, Edward.'

'Do you? Well, wrong or right, my cousin, until the fair Julia comes to me with sweet looks and gentle words, we can never be what we have been.'

'You will return from your hunting expedition in a softer mood,' I answered; 'that is to say, if you persist in going. But I hope and believe you will change your mind.'

'Such a change is not within the limits of possibility, Sarah. I am fixed as Fate.'

He strolled away, humming some gay hunting-song as he went. I was alone with Mrs Chrighton later in the afternoon, and she spoke to me about this intended visit to Wycherly.

'Edward has set his heart upon it evidently,' she said regretfully, 'and his father and I have always made a point of avoiding anything that could seem like domestic tyranny. Our dear boy is such a good son, that it would be very hard if we came between him and his pleasures. You know what a morbid horror my husband has of the dangers of the hunting-field, and perhaps I am almost as weak-minded. But in spite of this we have never interfered with Edward's enjoyment of a sport which he is passionately fond of; and hitherto, thank God! he has escaped without a scratch. Yet I have had many a bitter hour, I can assure you, my dear, when my son has been away in Leicestershire hunting four days a week.'

'He rides well, I suppose.'

'Superbly. He has a great reputation among the sportsmen of our neighbourhood. I daresay when he is master of the Abbey he will start a pack of hounds, and revive the old days of his great-grandfather, Meredith Chrighton.'

'I fancy the hounds were kenneled in the stable-yard below my bedroom window in those days, were they not, Fanny?'

'Yes,' Mrs Chrighton answered gravely; and I wondered at the sudden shadow that fell upon her face.

I went up to my room earlier than usual that afternoon, and I had a clear hour to spare before it would be time to dress for the seven o'clock dinner. This leisure hour I intended to devote to letter-writing; but on arriving in my room I found myself in a very idle frame of mind, and instead of opening my desk, I seated myself in the low easy-chair before the fire, and fell into a reverie.

How long I had been sitting there I scarcely know; I had been half meditating, half dozing, mixing broken snatches of thought with brief glimpses of dreaming, when I was startled into wakefulness by a sound that was strange to me.

It was a huntsman's horn—a few low plaintive notes on a huntsman's horn—notes which had a strange far-away sound, that was more unearthly than anything my ears had ever heard. I thought of the music in *Der Freischutz*; but the weirdest snatch of melody Weber ever wrote had not so ghastly a sound as these few simple notes conveyed to my ear.

I stood transfixed, listening to that awful music. It had grown dusk, my fire was almost out, and the room in shadow. As I listened, a light flashed suddenly on the wall before me. The light was as unearthly as the sound—a light that never shone from earth or sky.

I ran to the window; for this ghastly shimmer flashed through the window upon the opposite wall. The great gates of the stable-yard were open, and men in scarlet coats were riding in, a pack of hounds crowding in before them, obedient to the huntsman's whip. The whole scene was dimly visible by the declining light of the winter evening and the weird gleams of a lantern carried by one of the men. It was this lantern which had shone upon the tapestried wall. I saw the stable-doors opened one after another; gentlemen and grooms alighting from their horses; the dogs driven into their kennel; the helpers hurrying to and fro; and that strange wan lantern-light glimmering here and there is the gathering dusk. But there was no sound of horse's hoof or of human voices—not one yelp or cry from the hounds. Since those faint far-away sounds of the horn had died out in the distance, the ghastly silence had been unbroken.

I stood at my window quite calmly, and watched while the group of men and animals in the yard below noiselessly dispersed. There was nothing supernatural in the manner of their disappearance. The figures did not vanish or melt into empty air. One by one I saw the horses led into their separate quarters; one by one the redcoats strolled out of the gates, and the grooms departed, some one way, some another. The scene, but for its noiselessness, was natural enough; and had I been a stranger in the house, I might have fancied that those figures were real—those stables in full occupation.

But I knew that stable-yard and all its range of building to have been disused for more than half a century. Could I believe that, without an hour's warning, the long-deserted quadrangle could be filled—the empty stalls tenanted?

Had some hunting-party from the neighbourhood sought shelter here, glad to escape the pitiless rain? That was not impossible, I thought. I was an utter unbeliever in all ghostly things—ready to credit any possibility rather than suppose that I had been looking upon shadows. And yet the noiselessness, the awful sound of that horn—the strange unearthly gleam of that lantern! Little superstitious as I might be, a cold sweat stood out upon my forehead, and I trembled in every limb.

For some minutes I stood by the window, statue-like, staring blankly into the empty quadrangle. Then I roused myself suddenly, and ran softly downstairs by a back staircase leading to the servants' quarters, determined to solve the mystery somehow or other. The way to Mrs Marjorum's room was familiar to me from old experience, and it was thither that I bent my steps, determined to ask the housekeeper the meaning of what I had seen. I had a lurking conviction that it would be well for me not to mention that scene to any member of the family till I had taken counsel with some one who knew the secrets of Chrighton Abbey.

I heard the sound of merry voices and laughter as I passed the kitchen and servants' hall. Men and maids were all busy in the pleasant labour of decorating their rooms for the evening's festival. They were putting the last touches to garlands of holly and laurel, ivy and fir, as I passed the open doors; and in both rooms I saw tables laid for a substantial tea. The housekeeper's room was in a retired nook at the end of a long passage—a charming old room, panelled with dark oak, and full of capacious cupboards, which in my childhood I had looked upon as storehouses of inexhaustible treasures in the way of preserves and other confectionery. It was a shady old room, with a wide old-fashioned fireplace, cool in summer, when the hearth was adorned with a great jar of roses and lavender; and warm in winter, when the logs burnt merrily all day long.

I opened the door softly, and went in. Mrs Marjorum was dozing in a high-backed arm-chair by the glowing hearth, dressed in her state gown of grey watered silk, and with a cap that was a perfect garden of roses. She opened her eyes as I approached her, and stared at me with a puzzled look for the first moment or so.

'Why, is that you, Miss Sarah?' she exclaimed; 'and looking as pale as a ghost, I can see, even by this firelight! Let me just light a candle, and then I'll get you some sal volatile. Sit down in my armchair, miss; why, I declare you're all of a tremble!'

She put me into her easy-chair before I could resist, and lighted the two candles which stood ready upon her table, while I was trying to speak. My lips were dry, and it seemed at first as if my voice was gone.

'Never mind the sal volatile, Marjorum,' I said at last. 'I am not ill; I've been startled, that's all; and I've come to ask you for an explanation of the business that frightened me.'

'What business, Miss Sarah?'

'You must have heard something of it yourself, surely. Didn't you hear a horn just now, a huntsman's horn?'

'A horn! Lord no, Miss Sarah. What ever could have put such a fancy into your head?'

I saw that Mrs Marjorum's ruddy cheeks had suddenly lost their colour, that she was now almost as pale as I could have been myself.

'It was no fancy,' I said; 'I heard the sound, and saw the people. A hunting-party has just taken shelter in the north quadrangle. Dogs and horses, and gentlemen and servants.'

'What were they like, Miss Sarah?' the housekeeper asked in a strange voice.

'I can hardly tell you that. I could see that they wore red coats; and I could scarcely see more than that. Yes, I did get a glimpse of one of the gentlemen by the light of the lantern. A tall man, with grey hair and whiskers, and a stoop in his shoulders. I noticed that he wore a short-waisted coat with a very high collar—a coat that looked a hundred years old.'

'The old Squire!' muttered Mrs Marjorum under her breath; and then turning to me, she said with a cheery resolute air, 'You've been dreaming, Miss Sarah, that's just what it is. You've dropped off in your chair before the fire, and had a dream, that's it.'

'No, Marjorum, it was no dream. The horn woke me, and I stood at my window and saw the dogs and huntsmen come in.'

'Do you know, Miss Sarah, that the gates of the north quadrangle have been locked and barred for the last forty years, and that no one ever goes in there except through the house?'

'The gates may have been opened this evening to give shelter to strangers,' I said.

'Not when the only keys that will open them hang yonder in my cupboard, miss,' said the housekeeper, pointing to a corner of the room.

'But I tell you, Marjorum, these people came into the quadrangle; the horses and dogs are in the stables and kennels at this moment. I'll go and ask Mr Chrighton, or my cousin Fanny, or Edward, all about it, since you won't tell me the truth.'

I said this with a purpose, and it answered. Mrs Marjorum caught me eagerly by the wrist.

'No, miss, don't do that; for pity's sake don't do that; don't breathe a word to missus or master.'

'But why not?'

'Because you've seen that which always brings misfortune and sorrow to this house, Miss Sarah. You've seen the dead.'

'What do you mean?' I gasped, awed in spite of myself.

'I daresay you've heard say that there's been *something* seen at times at the Abbey—many years apart, thank God; for it never came that trouble didn't come after it.'

'Yes,' I answered hurriedly; 'but I could never get any one to tell me what it was that haunted this place.'

'No, miss. Those that know have kept the secret. But you have seen it all tonight. There's no use in trying to hide it from you any longer. You have seen the old Squire, Meredith Chrighton, whose eldest son was killed by a fall in the hunting-field, brought home dead one December night, an hour after his father and the rest of the party had come safe home to the Abbey. The old gentleman had missed his son in the field, but had thought nothing of that, fancying that master John had had enough of the day's sport, and had turned his horse's head homewards. He was found by a labouring-man, poor lad, lying in a ditch with his back broken, and his horse beside him staked. The old Squire never held his head up after that day, and never rode to hounds again, though he was passionately fond of hunting. Dogs and horses were sold, and the north quadrangle has been empty from that day.'

'How long is it since this kind of thing has been seen?'

'A long time, miss. I was a slip of a girl when it last happened. It was in the winter-time—this very night—the night Squire Meredith's son was killed; and the house was full of company, just as it is now. There was a wild young Oxford gentleman sleeping in your room at that time, and he saw the hunting-party come into the quadrangle; and what did he do but throw his window wide open, and give them the view-hallo as loud as ever he could. He had only arrived the day before, and knew nothing about the neighbourhood; so at dinner he began to ask where were his friends the sportsmen, and to hope he should be allowed to have a run with the Abbey hounds next day. It was in the time of our master's father; and his lady at the head of the table turned as white as a sheet when she heard this talk. She had good reason, poor soul. Before the week was out her husband was lying dead. He was struck with a fit of apoplexy, and never spoke or knew any one afterwards.'

'An awful coincidence,' I said; 'but it may have been only a coincidence.'

'I've heard other stories, miss—heard them from those that wouldn't deceive—all proving the same thing: that the appearance of the old Squire and his pack is a warning of death to this house.'

'I cannot believe these things,' I exclaimed; 'I *cannot* believe them. Does Mr Edward know anything about this?'

'No, miss. His father and mother have been most careful that it should be kept from him.'

'I think he is too strong-minded to be much affected by the fact,' I said.

'And you'll not say anything about what you've seen to my master or my mistress, will you, Miss Sarah?' pleaded the faithful old servant. 'The knowledge of it would be sure to make them nervous and unhappy. And if evil is to come upon this house, it isn't in human power to prevent its coming.'

'God forbid that there is any evil at hand!' I answered. 'I am no believer in visions or omens. After all, I would sooner fancy that I was dreaming—dreaming with my eyes open as I stood at the window—than that I beheld the shadows of the dead.'

Mrs Marjorum sighed, and said nothing. I could see that she believed firmly in the phantom hunt.

I went back to my room to dress for dinner. However rationally I might try to think of what I had seen, its effect upon my mind and nerves was not the less powerful. I could think of nothing else; and a strange morbid dread of coming misery weighted me down like an actual burden.

There was a very cheerful party in the drawing-room when I went downstairs, and at dinner the talk and laughter were unceasing; but I could see that my cousin Fanny's face was a little graver than usual, and I had no doubt she was thinking of her son's intended visit to Wycherly.

At the thought of this a sudden terror flashed upon me. How if the shadows I had seen that evening were ominous of danger to him—to Edward, the heir and only son of the house? My heart grew cold as I thought of this, and yet in the next moment I despised myself for such weakness.

'It is natural enough for an old servant to believe in such things,' I said to myself; 'but for me—an educated woman of the world—preposterous folly.'

And yet from that moment I began to puzzle myself in the endeavour to devise some means by which Edward's journey might be prevented. Of my own influence I knew that I was powerless to hinder his departure by so much as an hour; but I fancied that Julia Tremaine could persuade him to any sacrifice of his inclination, if she could only humble her pride so far as to entreat it. I determined to appeal to her in the course of the evening.

We were very merry all that evening. The servants and their guests danced in the great hall, while we sat in the gallery above, and in little groups upon the staircase, watching their diversions. I think this arrangement afforded excellent opportunities for flirtation, and that the younger members of our party made good use of their chances—with one exception: Edward Chrighton and his affianced contrived to keep far away from each other all the evening.

While all was going on noisily in the hall below, I managed to get Miss Tremaine apart from the others in the embrasure of a painted window on the stairs, where there was a wide oaken seat. Seated here side by side, I described to her, under a promise of secrecy, the scene which I had witnessed that afternoon, and my conversation with Mrs Marjorum.

'But, good gracious me, Miss Chrighton!' the young lady exclaimed, lifting her pencilled eyebrows with unconcealed disdain, 'you don't mean to tell me that you believe in such nonsense—ghosts and omens, and old woman's folly like that!'

'I assure you, Miss Tremaine, it is most difficult for me to believe in the supernatural,' I answered earnestly; 'but that which I saw this evening was something more than human. The thought of it has made me very unhappy; and I cannot help connecting it somehow with my cousin Edward's visit to Wycherly. If I had the power to prevent his going, I would do it at any cost; but I have not. You alone have influence enough for that. For heaven's sake use it! do anything to hinder his hunting with the Daleborough hounds.'

'You would have me humiliate myself by asking him to forgo his pleasure, and that after his conduct to me during the last week?'

'I confess that he has done much to offend you. But you love him, Miss Tremaine, though you are too proud to let your love be seen: I am certain that you do love him. For pity's sake speak to him; do not let him hazard his life, when a few words from you may prevent the danger.'

'I don't believe he would give up this visit to please me,' she answered; 'and I shall certainly not put it in his power to humiliate me by a refusal. Besides, all this fear of yours is such utter nonsense. As if nobody had ever hunted before. My brothers hunt four times a week every winter, and not one of them has ever been the worse for it yet.'

I did not give up the attempt lightly. I pleaded with this proud obstinate girl for a long time, as long as I could induce her to listen to me; but it was all in vain. She stuck to her text—no one should persuade her to degrade herself by asking a favour of Edward Chrighton. He had chosen to hold himself aloof from her, and she would show him that she could live without him. When she left Chrighton Abbey, they would part as strangers.

So the night closed, and at breakfast next morning I heard that Edward had started for Wycherly soon after daybreak. His absence made, for me at least, a sad blank in our circle. For one other also, I think; for Miss Tremaine's fair proud face was very pale, though she tried to seem gayer than usual, and exerted herself in quite an unaccustomed manner in her endeavour to be agreeable to everyone.

The days passed slowly for me after my cousin's departure. There was a weight upon my mind, a vague anxiety, which I struggled in vain to shake off. The house, full as it was of pleasant people, seemed to me to have become dull and dreary now that Edward was gone. The place where he had sat appeared always vacant to my eyes, though another filled it, and there was no gap on either side of the long dinner-table. Lighthearted young men still made the billiard-room resonant with their laughter; merry girls flirted as gaily as ever, undisturbed in the smallest degree by the absence of the heir of the house. Yet for me all was changed. A morbid fancy had taken complete possession of me. I found myself continually brooding over the housekeeper's words;

those words which had told me that the shadows I had seen boded death and sorrow to the house of Chrighton.

My cousins, Sophy and Agnes, were no more concerned about their brother's welfare than were their guests. They were full of excitement about the New-Year's ball, which was to be a very grand affair. Every one of importance within fifty miles was to be present, every nook and corner of the Abbey would be filled with visitors coming from a great distance, while others were to be billeted upon the better class of tenantry round about. Altogether the organization of this affair was no small business; and Mrs Chrighton's mornings were broken by discussions with the housekeeper, messages from the cook, interviews with the head-gardener on the subject of floral decorations, and other details, which all alike demanded the attention of the châtelaine herself. With these duties, and with the claims of her numerous guests, my cousin Fanny's time was so fully occupied, that she had little leisure to indulge in anxious feelings about her son, whatever secret uneasiness may have been lurking in her maternal heart. As for the master of the Abbey, he spent so much of his time in the library, where, under the pretext of business with his bailiff, he read Greek, that it was not easy for any one to discover what he did feel. Once, and once only, I heard him speak of his son, in a tone that betrayed an intense eagerness for his return.

The girls were to have new dresses from a French milliner in Wigmore Street; and as the great event drew near, bulky packages of millinery were continually arriving, and feminine consultations and expositions of finery were being held all day long in bedrooms and dressing-rooms with closed doors. Thus, with a mind always troubled by the same dark shapeless foreboding, I was perpetually being called upon to give an opinion about pink tulle and lilies of the valley, or maize silk and apple-blossoms.

New-Year's morning came at last, after an interval of abnormal length, as it seemed to me. It was a bright clear day, an almost springlike sunshine lighting up the leafless landscape. The great dining-room was noisy with congratulations and good wishes as we assembled for breakfast on this first morning of a new year, after having seen the old one out cheerily the night before; but Edward had not yet returned, and I missed him sadly. Some touch of sympathy drew me to the side of Julia Tremaine on this particular morning. I had watched her very often during the last few days, and I had seen that her cheek grew paler every day. Today her eyes had the dull heavy look that betokens a sleepless night. Yes, I was sure that she was unhappy—that the proud relentless nature suffered bitterly.

'He must be home today,' I said to her in a low voice, as she sat in stately silence before an untasted breakfast.

'Who must?' she answered, turning towards me with a cold distant look.

'My cousin Edward. You know he promised to be back in time for the ball.'

'I know nothing of Mr Chrighton's intended movements,' she said in her haughtiest tone; 'but of course it is only natural that he should be here tonight. He would scarcely care to insult half the county by his absence, however little he may value those now staying in his father's house.'

'But you know that there is one here whom he does value better than any one else in the world, Miss Tremaine,' I answered, anxious to soothe this proud girl.

'I know nothing of the kind. But why do you speak so solemnly about his return? He will come, of course. There is no reason he should not come.'

She spoke in a rapid manner that was strange to her, and looked at me with a sharp enquiring glance, that touched me somehow, it was so unlike herself—it revealed to me so keen an anxiety.

'No, there is no reasonable cause for anything like uneasiness,' I said; 'but you remember what I told you the other night. That has preyed upon my mind, and it will be an unspeakable relief to me when I see my cousin safe at home.'

'I am sorry that you should indulge in such weakness, Miss Chrighton.'

That was all she said; but when I saw her in the drawing-room after breakfast, she had established herself in a window that commanded a view of the long winding drive leading to the front of the Abbey. From this point she could not fail to see anyone approaching the house. She sat there all day; everyone else was more or less busy with arrangements for the evening, or at any rate occupied with an appearance of business; but Julia Tremaine kept her place by the window, pleading a headache as an excuse for sitting still, with a book in her hand, all day, yet obstinately refusing to go to her room and lie down, when her mother entreated her to do so.

'You will be fit for nothing tonight, Julia,' Mrs Tremaine said, almost angrily; 'you have been looking ill for ever so long, and today you are as pale as a ghost.'

I knew that she was watching for *him*; and I pitied her with all my heart, as the day wore itself out, and he did not come.

We dined earlier than usual, played a game or two of billiards after dinner, made a tour of inspection through the bright rooms, lit with wax-candles only, and odorous with exotics; and then came a long interregnum devoted to the arts and mysteries of the toilet; while maids flitted to and fro laden with frilled muslin petticoats from the laundry, and a faint smell of singed hair pervaded the corridors. At ten o'clock the band were tuning their violins, and pretty girls and elegant-looking men were coming slowly down the broad oak staircase, as the roll of fast-coming wheels sounded louder without, and stentorian voices announced the best people in the county.

I have no need to dwell long upon the details of that evening's festival. It was very much like other balls—a brilliant success, a night of splendour and enchantment for those whose hearts were light and happy, and who could abandon themselves utterly to the pleasure of the moment; a far-away picture of fair faces and bright-hued dresses, a wearisome kaleidoscopic procession of form and colour for those whose minds were weighed down with the burden of a hidden care.

For me the music had no melody, the dazzling scene no charm. Hour after hour went by; supper was over, and the waltzers were enjoying those latest dances which always seem the most delightful, and yet Edward Chrighton had not appeared amongst us.

There had been innumerable enquiries about him, and Mrs Chrighton had apologized for his absence as best she might. Poor soul, I well knew that his non-return was now a source of poignant anxiety to her, although she greeted all her guests with the same gracious smile, and was able to talk gaily and well upon every subject. Once, when she was sitting alone for a few minutes, watching the dancers, I saw the smile fade from her face, and a look of anguish come over it. I ventured to approach her at this moment, and never shall I forget the look which she turned towards me.

'My son, Sarah!' she said in a low voice—'something has happened to my son!'

I did my best to comfort her; but my own heart was growing heavier and heavier, and my attempt was a very poor one.

Julia Tremaine had danced a little at the beginning of the evening, to keep up appearances, I believe, in order that no one might suppose that she was distressed by her lover's absence; but after the first two or three dances she pronounced herself tired, and withdrew to a seat amongst the matrons. She was looking very lovely in spite of her extreme pallor, dressed in white tulle, a perfect cloud of airy puffings, and with a wreath of ivy-leaves and diamonds crowning her pale golden hair.

The night waned, the dancers were revolving in the last waltz, when I happened to look towards the doorway at the end of the room. I was startled by seeing a man standing there, with his hat in his hand, not in evening costume; a man with a pale anxious-looking face, peering cautiously into the room. My first thought was of evil; but in the next moment the man had disappeared, and I saw no more of him.

I lingered by my cousin Fanny's side till the rooms were empty. Even Sophy and Aggy had gone off to their own apartments, their airy dresses sadly dilapidated by a night's vigorous dancing. There were only Mr and Mrs Chrighton and myself in the long suite of rooms, where the flowers were drooping and the wax-lights dying out one by one in the silver sconces against the walls.

'I think the evening went off very well,' Fanny said, looking rather anxiously at her husband, who was stretching himself and yawning with an air of intense relief.

'Yes, the affair went off well enough. But Edward has committed a terrible breach of manners by not being here. Upon my word, the young men of the present day think of nothing but their own pleasures. I suppose that something especially attractive was going on at Wycherly today, and he couldn't tear himself away.'

'It is so unlike him to break his word,' Mrs Chrighton answered. 'You are not alarmed, Frederick? You don't think that anything has happened—any accident?'

'What should happen? Ned is one of the best riders in the county. I don't think there's any fear of his coming to grief.'

'He might be ill.'

'Not he. He's a young Hercules. And if it were possible for him to be ill—which it is not—we should have had a message from Wycherly.'

The words were scarcely spoken when Truefold the old butler stood by his master's side, with a solemn anxious face.

'There is a—a person who wishes to see you, sir,' he said in a low voice, 'alone.'

Low as the words were, both Fanny and myself heard them.

'Someone from Wycherly?' she exclaimed. 'Let him come here.'

'But, madam, the person most particularly wished to see master alone. Shall I show him into the library, sir? The lights are not out there.'

'Then it *is* someone from Wycherly,' said my cousin, seizing my wrist with a hand that was icy cold. 'Didn't I tell you so, Sarah? Something has happened to my son. Let the person come here, Truefold, here; I insist upon it.'

The tone of command was quite strange in a wife who was always deferential to her husband, in a mistress who was ever gentle to her servants.

'Let it be so, Truefold,' said Mr Chrighton. 'Whatever ill news has come to us we will hear together.'

He put his arm round his wife's waist. Both were pale as marble, both stood in stony stillness waiting for the blow that was to fall upon them.

The stranger, the man I had seen in the doorway, came in. He was curate of Wycherly church, and chaplain to Sir Francis Wycherly; a grave middle-aged man. He told what he had to tell with all kindness, with all the usual forms of consolation which Christianity and an experience of sorrow could suggest. Vain words, wasted trouble. The blow must fall, and earthly consolation was unable to lighten it by a feather's weight.

There had been a steeplechase at Wycherly—an amateur affair with gentlemen riders—on that bright New-Year's-day, and Edward Chrighton had been persuaded to ride his favourite hunter Pepperbox. There would be plenty of time for him to return to Chrighton after the races. He had consented; and his horse was winning easily, when, at the last fence, a double one, with water beyond, Pepperbox baulked his leap, and went over head-foremost, flinging his rider over a hedge into a field close beside the course, where there was a heavy stone roller. Upon this stone roller Edward Chrighton had fallen, his head receiving the full force of the concussion. All was told. It was while the curate was relating the fatal catastrophe that I looked round suddenly, and saw Julia Tremaine standing a little way behind the speaker. She had heard all; she uttered no cry, she showed no signs of fainting, but stood calm and motionless, waiting for the end.

I know not how that night ended: there seemed an awful calm upon us all. A carriage was got ready, and Mr and Mrs Chrighton started for Wycherly to look upon their dead son. He had died while they were carrying him from the course to Sir Francis's house. I went with Julia Tremaine to her room, and sat with her while the winter morning dawned slowly upon us—a bitter dawning.

I have little more to tell. Life goes on, though hearts are broken. Upon Chrighton Abbey there came a dreary time of desolation. The master of the house lived in his library, shut from the outer world, buried almost as completely as a hermit in his cell. I have heard that Julia Tremaine was never known to smile after that day. She is still unmarried, and lives entirely at her father's country house; proud and reserved in her conduct to her equals, but a very angel of mercy and compassion amongst the poor of the neighbourhood. Yes; this haughty girl, who once declared herself unable to endure the hovels of the poor, is now a Sister of Charity in all but the robe. So does a great sorrow change the current of a woman's life.

I have seen my cousin Fanny many times since that awful New-Year's night; for I have always the same welcome at the Abbey. I have seen her calm and cheerful, doing her duty, smiling upon her daughter's children, the honoured mistress of a great household; but I know that the mainspring of life is broken, that for her there hath passed a glory from the earth, and that upon all the pleasures and joys of this world she looks with the solemn calm of one for whom all things are dark with the shadow of a great sorrow.

No Living Voice

THOMAS STREET MILLINGTON

'How do you account for it?'

'I don't account for it at all. I don't pretend to understand it.'

'You think, then, that it was really supernatural?'

'We know so little what Nature comprehends—what are its powers and limits—that we can scarcely speak of anything that happens as beyond it or above it.'

'And you are certain that this did happen?'

'Quite certain; of that I have no doubt whatever.'

These sentences passed between two gentlemen in the drawing-room of a country house, where a small family party was assembled after dinner; and in consequence of a lull in the conversation occurring at the moment they were distinctly heard by nearly everybody present. Curiosity was excited, and enquiries were eagerly pressed as to the nature or supernature of the event under discussion. 'A ghost story!' cried one; 'oh! delightful! we must and will hear it.' 'Oh! please, no,' said another; 'I should not sleep all night—and yet I am dying with curiosity.'

Others seemed inclined to treat the question rather from a rational or psychological point of view, and would have started a discussion upon ghosts in general, each giving his own experience; but these were brought back by the voice of the hostess, crying, 'Question, question!' and the first speakers were warmly urged to explain what particular event had formed the subject of their conversation.

'It was you, Mr Browne, who said you could not account for it; and you are such a very matter-of-fact person that we feel doubly anxious to hear what wonderful occurrence could have made you look so grave and earnest.'

'Thank you,' said Mr Browne. 'I am a matter-of-fact person, I confess; and I was speaking of a fact; though I must beg to be excused saying any more about it. It is an old story; but I never even think of it without a feeling of distress; and I should not like to stir up such keen and haunting memories merely for the sake of gratifying curiosity. I was relating to Mr Smith, in few words, an adventure which befell me in Italy many years ago, giving him the naked facts of the case, in refutation of a theory which he had been propounding.'

'Now we don't want theories, and we won't have naked facts; they are hardly proper at any time, and at this period of the year, with snow upon the ground,' they would be most unseasonable; but we must have that story fully and feelingly related to us, and we promise to give it a respectful hearing, implicit belief, and unbounded sympathy. So draw round the fire, all of you, and let Mr Browne begin.'

Poor Mr Browne turned pale and red, his lips quivered, his entreaties to be excused became quite plaintive; but his good nature and perhaps, also, the consciousness that he could really interest his hearers, led him to overcome his reluctance; and after exacting a solemn promise that there should be no jesting or levity in regard to what he had to tell, he cleared his throat twice or thrice, and in a hesitating nervous tone began as follows:

'It was in the spring of 18—. I had been at Rome during the Holy Week, and had taken a place in the diligence for Naples. There were two routes: one by way of Terracina and the other by the Via Latina, more inland. The diligence, which made the journey only twice a week, followed these routes alternately, so that each road was traversed only once in seven days. I chose the inland route, and after a long day's journey arrived at Ceprano, where we halted for the night. The next morning we started again very early, and it was scarcely yet daylight when we reached the Neapolitan frontier, at a short distance from the town. There our passports were examined, and to my great dismay I was informed that mine was not *en règle*. It was covered, indeed, with stamps and signatures, not one of which had been procured without some cost and trouble; but one '*visa*' yet was wanting, and that the all-important one, without which none could enter the kingdom of Naples. I was obliged therefore to alight, and to send my wretched passport back to Rome, my wretched self being doomed to remain under police surveillance at Ceprano, until the diligence should bring it back to me on that day week, at soonest.

'I took up my abode at the hotel where I had passed the previous night, and there I presently received a visit from the Capo di Polizia, who told me very civilly that I must present myself, every morning and evening at his *bureau*, but that I might have liberty to "circulate" in the neighbourhood during the day. I grew so weary of this dull place, that after I had explored the immediate vicinity of the town I began to extend my walks to a greater distance, and as I always reported myself to the police before night I met with no objection on their part.

'One day, however, when I had been as far as Alatri and was returning on foot, night overtook me. I had lost my way, and could not tell how far I might be from my destination. I was very tired and had a heavy knapsack on my shoulders, packed with stones and relics from the ruins of the old Pelasgic fortress which I had been exploring, besides a number of old coins and a lamp or two which I had purchased there. I could discern no signs of any human habitation, and the hills, covered with wood, seemed to shut me in on every side. I was beginning to think seriously of looking out for some sheltered spot under a thicket in which to pass the night, when the welcome sound of a footstep behind me fell upon my ears. Presently a man dressed in the usual long shaggy coat of a shepherd overtook me, and hearing of my difficulty offered to conduct me to a house at a short distance from the road, where I might obtain a lodging; before we reached the spot he told me that the house in question was an inn and that he was the landlord of it. He had not much custom, he said, so he employed himself in shepherding during the day; but he could make me comfortable, and give me a good supper also, better than I should expect, to look at him; but he had been in different circumstances once, and had lived in service in good families, and knew how things ought to be, and what a *signore* like myself was used to.

'The house to which he took me seemed like its owner to have seen better days. It was a large rambling place and much dilapidated, but it was tolerably comfortable within; and my landlord, after he had thrown off his sheepskin coat, prepared me a good and savoury meal, and sat down to look at and converse with me while I ate it. I did not much like the look of the fellow; but he seemed anxious to be sociable and told me a great deal about his former life when he was in service, expecting to receive similar confidences from me. I did not gratify him much, but one must talk of something, and he seemed to think it only proper to express an interest in his guests and to learn as much of their concerns as they would tell him.

'I went to bed early, intending to resume my journey as soon as it should be light. My landlord took up my knapsack, and carried it to my room, observing as he did so that it was a great weight for me to travel with. I answered jokingly that it contained great treasures, referring to my coins and relics; of course he did not understand me, and before I could explain he wished me a most happy little night, and left me.

'The room in which I found myself was situated at the end of a long passage; there were two rooms on the right side of this passage, and a window on the left, which looked out upon a yard or garden. Having taken a survey of the outside of the house while smoking my cigar after dinner, when the moon was up, I understood exactly the position of my chamber—the end room of a long narrow wing, projecting at right angles from the main building, with which it was connected only by the passage and the two side rooms already mentioned. Please to bear this description carefully in mind while I proceed.

'Before getting into bed, I drove into the floor close to the door a small gimlet which formed part of a complicated pocket-knife which I always carried with me, so that it would be impossible for any one to enter the room without my knowledge; there was a lock to the door, but the key would not turn in it; there was also a bolt, but it would not enter the hole intended for it, the door having sunk apparently from its proper level. I satisfied, myself, however, that the door was securely fastened by my gimlet, and soon fell asleep.

'How can I describe the strange and horrible sensation which oppressed me as I woke out of my first slumber? I had been sleeping soundly, and before I quite recovered consciousness I had instinctively risen from my pillow, and was crouching forward, my knees drawn up, my hands clasped before my face, and my whole frame quivering with horror. I saw nothing, felt nothing; but a sound was ringing in my ears which seemed to make my blood run cold. I could not have supposed it possible that any mere sound, whatever might be its nature, could have produced such a revulsion of feeling or inspired such intense horror as I then experienced. It was not a cry of terror that I heard—that would have roused me to action—nor the moaning of one in pain—that would have distressed me, and called forth sympathy rather than aversion. True, it was like the groaning of one in anguish and despair, but not like any mortal voice: it seemed too dreadful, too intense, for human utterance. The sound had begun while I was fast asleep—close to the head of my bed—close to my very pillow; it continued after I was wide awake—a long, loud, hollow, protracted groan, making the midnight air reverberate, and then dying gradually away until it ceased entirely. It was some minutes before I could at all recover from the terrible impression which seemed to stop my breath and paralyse my limbs. At length I began to look about me, for the night was not entirely dark, and I could discern the outlines of the room and the several pieces of furniture in it. I then got out of bed, and called aloud, "Who is there? What is the matter? Is anyone ill?" I repeated these enquiries in Italian and in French, but there was none that answered. Fortunately I had some matches in my pocket and was able to light my candle. I then examined every part of the room carefully, and especially the wall at the head of my bed, sounding it with my knuckles; it was firm and solid there, as in all other places. I unfastened my door, and explored the passage and the two adjoining rooms, which were unoccupied and almost destitute of furniture; they had evidently not been used for some time. Search as I would I could gain no clue to the mystery. Returning to my room I sat down upon the bed in great perplexity, and began to turn over in my mind whether it was possible I could have been deceived—whether the sounds which caused me such distress might be the offspring of some dream or nightmare; but to that conclusion I could not bring myself at all, much as I wished it, for the groaning had continued ringing in my ears long after I was wide awake and conscious. While I was thus reflecting, having neglected to close the door which was opposite to the side of my bed where I was sitting, I heard a soft footstep at a distance, and presently a light appeared at the further end of the passage. Then I saw the shadow of a man cast upon the opposite wall; it moved very slowly, and presently stopped. I saw the hand raised, as if making a sign to someone, and I knew from the fact of the shadow being thrown in advance that there must be a second person in the rear by whom the light was carried. After a short pause they seemed to retrace their steps, without my having had a glimpse of either of them, but only of the shadow which had come before and which followed them as they withdrew. It was then a little after one o'clock, and I concluded they were retiring late to rest, and anxious to avoid disturbing me, though I have since thought that it was the light from my room which caused their retreat. I felt half inclined to call to them, but I shrank, without knowing why, from making known what had disturbed me, and while I hesitated they were gone; so I fastened my door again, and resolved to sit up and watch a little longer by myself. But now my candle was beginning to burn low, and I found myself in this dilemma: either I must extinguish it at once, or I should be left without the means of procuring a light in case I should be again disturbed. I regretted that I had not called for another candle while there were people yet moving in the house, but I could not do so now without making explanations; so I grasped my box of matches, put out my light, and lay down, not without a shudder, in the bed.

'For an hour or more I lay awake thinking over what had occurred, and by that time I had almost persuaded myself that I had nothing but my own morbid imagination to thank for the alarm which I had suffered. "It is an outer wall," I said to myself; "they are all outer walls, and the house is built of stone; it is impossible that any sound could be heard through such a thickness. Besides, it seemed to be in my room, close to my ear. What an idiot I must be, to be excited and alarmed about nothing; I'll think no more about it." So I turned on my side, with a smile (rather a forced one) at my own foolishness, and composed myself to sleep.

'At that instant I heard, with more distinctness than I ever heard any other sound in my life, a gasp, a voiceless gasp, as if someone were in agony for breath, biting at the air, or trying with desperate efforts to cry out or speak. It was repeated a second and a third time; then there was a pause; then again that horrible gasping; and then a long-drawn breath, an audible drawing up of the air into the throat, such as one would make in heaving a deep sigh. Such sounds as these could not possibly have been heard unless they had been close to my ear; they seemed to come from the wall at my head, or to rise up out of my pillow. That fearful gasping, and that drawing in of the breath, in the darkness and silence of the night, seemed to make every nerve in my body thrill with dreadful expectation. Unconsciously I shrank away from it, crouching down as before, with my face upon my knees. It ceased, and immediately a moaning sound began, which lengthened out into an awful, protracted groan, waxing louder and louder, as if under an increasing agony, and then dying away slowly and gradually into silence; yet painfully and distinctly audible even to the last.

'As soon as I could rouse myself from the freezing horror which seemed to penetrate even to my joints and marrow, I crept away from the bed, and in the furthest corner of the room lighted with shaking hand my candle, looking anxiously about me as I did so, expecting some dreadful revelation as the light flashed up. Yet, if you will believe me, I did not feel alarmed or frightened; but rather oppressed, and penetrated with an unnatural, overpowering, sentiment of awe. I seemed to be in the presence of some great and horrible mystery, some bottomless depth of woe, or misery, or crime. I shrank from it with a sensation of intolerable loathing and suspense. It was a feeling akin to this which prevented me from calling to my landlord. I could not bring myself to speak to him of what had passed; not knowing how nearly he might be himself involved in the mystery. I was only anxious to escape as quietly as possible from the room and from the house. The candle was now beginning to flicker in its socket, but the stars were shining outside, and there was space and air to breathe there, which seemed to be wanting in my room; so I hastily opened my window, tied the bedclothes together for a rope, and lowered myself silently and safely to the ground.

'There was a light still burning in the lower part of the house; but I crept noiselessly along, feeling my way carefully among the trees, and in due time came upon a beaten track which led me to a road, the same which I had been travelling on the previous night. I walked on, scarcely knowing whither, anxious only to increase my distance from the accursed house, until the day began to break, when almost the first object I could see distinctly was a small body of men approaching me.

It was with no small pleasure that I recognized at their head my friend the Capo di Polizia. "Ah!" he cried, "unfortunate Inglese, what trouble you have given me! Where have you been? God be praised that I see you safe and sound! But how? What is the matter with you? You look like one possessed."

'I told him how I had lost my way, and where I had lodged.

'"And what happened to you there?" he cried, with a look of anxiety.

'"I was disturbed in the night. I could not sleep. I made my escape, and here I am. I cannot tell you more."

'"But you must tell me more, dear sir; forgive me; you must tell me everything. I must know all that passed in that house. We have had it under our surveillance for a long time, and when I heard in what direction you had gone yesterday, and had not returned, I feared you had got into some mischief there, and we were even now upon our way to look for you."

'I could not enter into particulars, but I told him I had heard strange sounds, and at his request I went back with him to the spot. He told me by the way that the house was known to be the resort of banditti; that the landlord harboured them, received their ill-gotten goods, and helped them to dispose of their booty.

'Arrived at the spot, he placed his men about the premises and instituted a strict search, the landlord and the man who was found in the house being compelled to accompany him. The room in which I had slept was carefully examined; the floor was of plaster or cement, so that no sound could have have passed through it; the walls were sound and solid, and there was nothing to be seen that could in any way account for the strange disturbance I had experienced. The room on the ground-floor underneath my bedroom was next inspected; it contained a quantity of straw, hay, firewood, and lumber. It was paved with brick, and on turning over the straw which was heaped together in a corner it was observed that the bricks were uneven, as if they had been recently disturbed.

'"Dig here," said the officer, "we shall find something hidden here, I imagine."

'The landlord was evidently much disturbed. "Stop," he cried. "I will tell you what lies there; come away out of doors, and you shall know all about it."

'"Dig, I say. We will find out for ourselves."

'"Let the dead rest," cried the landlord, with a trembling voice. "For the love of heaven come away, and hear what I shall tell you."

'"Go on with your work," said the sergeant to his men, who were now plying pickaxe and spade.

'"I can't stay here and see it," exclaimed the landlord once more. "Hear then! It is the body of my son, my only son—let him rest, if rest he can. He was wounded in a quarrel, and brought home here to die. I thought he would recover, but there was neither doctor nor priest at hand, and in spite of all that we could do for him he died. Let him alone now, or let a priest first be sent for; he died unconfessed, but it was not my fault; it may not be yet too late to make peace for him."

'"But why is he buried in this place?"

'"We did not wish to make a stir about it. Nobody knew of his death, and we laid him down quietly; one place I thought was as good as another when once the life was out of him. We are poor folk, and could not pay for ceremonies."

'The truth at length came out. Father and son were both members of a band of thieves; under this floor they concealed their plunder, and there too lay more than one mouldering corpse, victims who had occupied the room in which I slept, and had there met their death. The son was indeed buried in that spot; he had been mortally wounded in a skirmish with travellers, and had lived long enough to repent of his deeds and to beg for that priestly absolution which, according to his creed, was necessary to secure his pardon. In vain he had urged his father to bring the confessor to his bedside; in vain he had entreated him to break off from the murderous band with which he was allied and to live honestly in future; his prayers were disregarded, and his dying admonitions were of no avail. But for the strange mysterious warning which had roused me from my sleep and driven me out of the house that night another crime would have been added to the old man's tale of guilt. That gasping attempt to speak, and that awful groaning—whence did they proceed? It was *no living voice*. Beyond that I will express no opinion on the subject. I will only say it was the means of saving my life, and at the same time putting an end to the series of bloody deeds which had been committed in that house.

'I received my passport that evening by the diligence from Rome, and started the next morning on my way to Naples. As we were crossing the frontier a tall figure approached, wearing the long rough *cappotta* of the mendicant friars, with a hood over the face and holes for the eyes to look through. He carried a tin money-box in his hand, which he held out to the passengers, jingling a few coins in it, and crying in a monotonous voice, "*Anime in purgatorio! Anime in purgatorio!*" I do not believe in purgatory, nor in supplications for the dead; but I dropped a piece of silver into the box nevertheless, as I thought of that unhallowed grave in the forest, and my prayer went up to heaven in all sincerity—"*Requiescat in pace*!"'

Miss Jéromette and the Clergyman

WILKIE COLLINS

I

My brother, the clergyman, looked over my shoulder before I was aware of him, and discovered that the volume which completely absorbed my attention was a collection of famous Trials, published in a new edition and in a popular form.

He laid his finger on the Trial which I happened to be reading at the moment. I looked up at him; his face startled me. He had turned pale. His eyes were fixed on the open page of the book with an expression which puzzled and alarmed me.

'My dear fellow,' I said, 'what in the world is the matter with you?'

He answered in an odd absent manner, still keeping his finger on the open page.

'I had almost forgotten,' he said. 'And this reminds me.'

'Reminds you of what?' I asked. 'You don't mean to say you know anything about the Trial?'

'I know this,' he said. 'The prisoner was guilty.'

'Guilty?' I repeated. 'Why, the man was acquitted by the jury, with the full approval of the judge! What can you possibly mean?'

'There are circumstances connected with that Trial,' my brother answered, 'which were never communicated to the judge or the jury—which were never so much as hinted or whispered in court. *I* know them—of my own knowledge, by my own personal experience. They are very sad, very strange, very terrible. I have mentioned them to no mortal creature. I have done my best to forget them. You—quite innocently—have brought them back to my mind. They oppress, they distress me. I wish I had found you reading any book in your library, except *that* book!'

My curiosity was now strongly excited. I spoke out plainly.

'Surely,' I suggested, 'you might tell your brother what you are unwilling to mention to persons less nearly related to you. We have followed different professions, and have lived in different countries, since we were boys at school. But you know you can trust me.'

He considered a little with himself.

'Yes,' he said. 'I know I can trust you.' He waited a moment; and then he surprised me by a strange question.

'Do you believe,' he asked, 'that the spirits of the dead can return to earth, and show themselves to the living?'

I answered cautiously—adopting as my own the words of a great English writer, touching the subject of ghosts.

'You ask me a question,' I said, 'which, after five thousand years, is yet undecided. On that account alone, it is a question not to be trifled with.'

My reply seemed to satisfy him.

'Promise me,' he resumed, 'that you will keep what I tell you a secret as long as I live. After my death I care little what happens. Let the story of my strange experience be added to the published experience of those other men who have seen what I have seen, and who believe what I believe. The world will not be the worse, and may be the better, for knowing one day what I am now about to trust to your ear alone.'

My brother never again alluded to the narrative which he had confided to me, until the later time when I was sitting by his death-bed. He asked if I still remembered the story of Jéromette. 'Tell it to others,' he said, 'as I have told it to you.'

I repeat it, after his death—as nearly as I can in his own words.

II

On a fine summer evening, many years since, I left my chambers in the Temple, to meet a fellow-student, who had proposed to me a night's amusement in the public gardens at Cremorne.

You were then on your way to India; and I had taken my degree at Oxford. I had sadly disappointed my father by choosing the Law as my profession, in preference to the Church. At that time, to own the truth, I had no serious intention of following any special vocation. I simply wanted an excuse for enjoying the pleasures of a London life. The study of the Law supplied me with that excuse. And I chose the Law as my profession accordingly.

On reaching the place at which we had arranged to meet, I found that my friend had not kept his appointment. After waiting vainly for ten minutes, my patience gave way, and I went into the Gardens by myself.

I took two or three turns round the platform devoted to the dancers, without discovering my fellow-student, and without seeing any other person with whom I happened to be acquainted at that time.

For some reason which I cannot now remember, I was not in my usual good spirits that evening. The noisy music jarred on my nerves, the sight of the gaping crowd round the platform irritated me, the blandishments of the painted ladies of the profession of pleasure saddened and disgusted me. I opened my cigar-case, and turned aside into one of the quiet by-walks of the Gardens.

A man who is habitually careful in choosing his cigar has this advantage over a man who is habitually careless. He can always count on smoking the best cigar in his case, down to the last. I was still absorbed in choosing *my* cigar, when I heard these words behind me—spoken in a foreign accent and in a woman's voice:

'Leave me directly, sir! I wish to have nothing to say to you.'

I turned round and discovered a little lady very simply and tastefully dressed, who looked both angry and alarmed as she rapidly passed me on her way to the more frequented part of the Gardens. A man (evidently the worse for the wine he had drunk in the course of the evening) was following her, and was pressing his tipsy attentions on her with the coarsest insolence of speech and manner. She was young and pretty, and she cast one entreating look at me as she went by, which it was not in manhood—perhaps I ought to say, in young-manhood—to resist.

I instantly stepped forward to protect her, careless whether I involved myself in a discreditable quarrel with a blackguard or not. As a matter of course, the fellow resented my interference, and my temper gave way. Fortunately for me, just as I lifted my hand to knock him down, a policeman appeared who had noticed that he was drunk, and who settled the dispute officially by turning him out of the Gardens.

I led her away from the crowd that had collected. She was evidently frightened—I felt her hand trembling on my arm—but she had one great merit: she made no fuss about it.

'If I can sit down for a few minutes,' she said in her pretty foreign accent, 'I shall soon be myself again, and I shall not trespass any further on your kindness. I thank you very much, sir, for taking care of me.'

We sat down on a bench in a retired part of the Gardens, near a little fountain. A row of lighted lamps ran round the outer rim of the basin. I could see her plainly.

I have said that she was 'a little lady'. I could not have described her more correctly in three words.

Her figure was slight and small: she was a well-made miniature of a woman from head to foot. Her hair and her eyes were both dark. The hair curled naturally; the expression of the eyes was quiet, and rather sad; the complexion, as I then saw it, very pale; the little mouth perfectly charming. I was especially attracted, I remember, by the carriage of her head; it was strikingly graceful and spirited; it distinguished her, little as she was and quiet as she was, among the thousands of other women in the Gardens, as a creature apart. Even the one marked defect in her—a slight 'cast' in the left eye—seemed to add, in some strange way, to the quaint attractiveness of her face. I have already spoken of the tasteful simplicity of her dress. I ought now to add that it was not made of any costly material, and that she wore no jewels or ornaments of any sort. My little lady was not rich: even a man's eye could see that.

She was perfectly unembarrassed and unaffected. We fell as easily into talk as if we had been friends instead of strangers.

I asked how it was that she had no companion to take care of her. 'You are too young and too pretty,' I said in my blunt English way, 'to trust yourself alone in such a place as this.'

She took no notice of the compliment. She calmly put it away from her as if it had not reached her ears.

'I have no friend to take care of me,' she said simply. 'I was sad and sorry this evening, all by myself, and I thought I would go to the Gardens and hear the music, just to amuse me. It is not much to pay at the gate; only a shilling.'

'No friend to take care of you?' I repeated. 'Surely there must be one happy man who might have been here with you tonight?'

'What man do you mean?' she asked.

'The man,' I answered thoughtlessly, 'whom we call, in England, a sweetheart.'

I would have given worlds to have recalled those foolish words the moment they passed my lips. I felt that I had taken a vulgar liberty with her. Her face saddened; her eyes dropped to the ground. I begged her pardon.

'There is no need to beg my pardon,' she said. 'If you wish to know, sir—yes, I had once a sweetheart, as you call it in England. He has gone away and left me. No more of him, if you please. I am rested now. I will thank you again, and go home.'

She rose to leave me.

I was determined not to part with her in that way. I begged to be allowed to see her safely back to her own door. She hesitated. I took a man's unfair advantage of her, by appealing to her fears. I said, 'Suppose the blackguard who annoyed you should be waiting outside the gates?' That decided her. She took my arm. We went away together by the bank of the Thames, in the balmy summer night.

A walk of half an hour brought us to the house in which she lodged—a shabby little house in a by-street, inhabited evidently by very poor people.

She held out her hand at the door, and wished me goodnight. I was too much interested in her to consent to leave my little foreign lady without the hope of seeing her again. I asked permission to call on her the next day. We were standing under the light of the street-lamp. She studied my face with a grave and steady attention before she made any reply.

'Yes,' she said at last. 'I think I do know a gentleman when I see him. You may come, sir, if you please, and call upon me tomorrow.'

So we parted. So I entered—doubting nothing, foreboding nothing—on a scene in my life, which I now look back on with unfeigned repentance and regret.

III

I am speaking at this later time in the position of a clergyman, and in the character of a man of mature age. Remember that; and you will understand why I pass as rapidly as possible over the events of the next year of my life—why I say as little as I can of the errors and the delusions of my youth.

I called on her the next day. I repeated my visits during the days and weeks that followed, until the shabby little house in the by-street had become a second and (I say it with shame and self-reproach) a dearer home to me.

All of herself and her story which she thought fit to confide to me under these circumstances may be repeated to you in few words.

The name by which letters were addressed to her was 'Mademoiselle Jéromette'. Among the ignorant people of the house and the small tradesmen of the neighbourhood—who found her name not easy of pronunciation by the average English tongue—she was known by the friendly nickname of 'The French Miss'. When I knew her, she was resigned to her lonely life among strangers. Some years had elapsed since she had lost her parents, and had left France. Possessing a small, very small, income of her own, she added to it by colouring miniatures for the photographers. She had relatives still living in France; but she had long since ceased to correspond with them. 'Ask me nothing more about my family,' she used to say. 'I am as good as dead in my own country and among my own people.'

This was all—literally all—that she told me of herself. I have never discovered more of her sad story from that day to this.

She never mentioned her family name—never even told me what part of France she came from, or how long she had lived in England. That she was, by birth and breeding, a lady, I could entertain no doubt;

her manners, her accomplishments, her ways of thinking and speaking, all proved it. Looking below the surface, her character showed itself in aspects not common among young women in these days. In her quiet way, she was an incurable fatalist, and a firm believer in the ghostly reality of apparitions from the dead. Then again, in the matter of money, she had strange views of her own. Whenever my purse was in my hand, she held me resolutely at a distance from first to last. She refused to move into better apartments; the shabby little house was clean inside, and the poor people who lived in it were kind to her—and that was enough. The most expensive present that she ever permitted me to offer her was a little enamelled ring, the plainest and cheapest thing of the kind in the jeweller's shop. In all her relations with me she was sincerity itself. On all occasions, and under all circumstances, she spoke her mind (as the phrase is) with the same uncompromising plainness.

'I like you,' she said to me; 'I respect you; I shall always be faithful to you while you are faithful to me. But my love has gone from me. There is another man who has taken it away with him, I know not where.'

Who was the other man?

She refused to tell me. She kept his rank and his name strict secrets from me. I never discovered how he had met with her, or why he had left her, or whether the guilt was his of making her an exile from her country and her friends. She despised herself for still loving him; but the passion was too strong for her—she owned it and lamented it with the frankness which was so pre-eminently a part of her character. More than this, she plainly told me, in the early days of our acquaintance, that she believed he would return to her. It might be tomorrow, or it might be years hence. Even if he failed to repent of his own cruel conduct, the man would still miss her, as something lost out of his life; and, sooner or later, he would come back.

'And will you receive him if he does come back?' I asked.

'I shall receive him,' she replied, 'against my own better judgement—in spite of my own firm persuasion that the day of his return to me will bring with it the darkest days of my life.'

I tried to remonstrate with her.

'You have a will of your own,' I said. 'Exert it, if he attempts to return to you.'

'I have no will of my own,' she answered quietly, 'where *he* is concerned. It is my misfortune to love him.' Her eyes rested for a moment on mine, with the utter self-abandonment of despair. 'We have said enough about this,' she added abruptly. 'Let us say no more.'

From that time we never spoke again of the unknown man. During the year that followed our first meeting, she heard nothing of him directly or indirectly. He might be living, or he might be dead. There came no word of him, or from him. I was fond enough of her to be satisfied with this—he never disturbed us.

IV

The year passed—and the end came. Not the end as you may have anticipated it, or as I might have foreboded it.

You remember the time when your letters from home informed you of the fatal termination of our mother's illness? It is the time of which I am now speaking. A few hours only before she breathed her last, she called me to her bedside, and desired that we might be left together alone. Reminding me that her death was near, she spoke of my prospects in life; she noticed my want of interest in the studies which were then supposed to be engaging my attention, and she ended by entreating me to reconsider my refusal to enter the Church.

'Your father's heart is set upon it,' she said. 'Do what I ask of you, my dear, and you will help to comfort him when I am gone.'

Her strength failed her: she could say no more. Could I refuse the last request she would ever make to me? I knelt at the bedside, and took her wasted hand in mine, and solemnly promised her the respect which a son owes to his mother's last wishes.

Having bound myself by this sacred engagement, I had no choice but to accept the sacrifice which it imperatively exacted from me. The time had come when I must tear myself free from all unworthy associations. No matter what the effort cost me, I must separate myself at once and for ever from the unhappy woman who was not, who never could be, my wife.

At the close of a dull foggy day I set forth with a heavy heart to say the words which were to part us for ever.

Her lodging was not far from the banks of the Thames. As I drew near the place the darkness was gathering, and the broad surface of the river was hidden from me in a chill white mist. I stood for a while, with my eyes fixed on the vaporous shroud that brooded over the flowing water—I stood, and asked myself in despair the one dreary question: 'What am I to say to her?'

The mist chilled me to the bones. I turned from the river-bank, and made my way to her lodgings hard by. 'It must be done!' I said to myself, as I took out my key and opened the house door.

She was not at her work, as usual, when I entered her little sitting-room. She was standing by the fire, with her head down, and with an open letter in her hand.

The instant she turned to meet me, I saw in her face that something was wrong. Her ordinary manner was the manner of an unusually placid and self-restrained person. Her temperament had little of the liveliness which we associate in England with the French nature. She was not ready with her laugh; and, in all my previous experience, I had never yet known her to cry. Now, for the first time, I saw the quiet face disturbed; I saw tears in the pretty brown eyes. She ran to meet me, and laid her head on my breast, and burst into a passionate fit of weeping that shook her from head to foot.

Could she by any human possibility have heard of the coming change in my life? Was she aware, before I had opened my lips, of the hard necessity which had brought me to the house?

It was simply impossible; the thing could not be.

I waited until her first burst of emotion had worn itself out. Then I asked—with an uneasy conscience, with a sinking heart—what had happened to distress her.

She drew herself away from me, sighing heavily, and gave me the open letter which I had seen in her hand.

'Read that,' she said. 'And remember I told you what might happen when we first met.'

I read the letter.

It was signed in initials only; but the writer plainly revealed himself as the man who had deserted her. He had repented; he had returned to her. In proof of his penitence he was willing to do her the justice which he had hitherto refused—he was willing to marry her; on the condition that she would engage to keep the marriage a secret, so long as his parents lived. Submitting this proposal, he waited to know whether she would consent, on her side, to forgive and forget.

I gave her back the letter in silence. This unknown rival had done me the service of paving the way for our separation. In offering her the atonement of marriage, he had made it, on my part, a matter of duty to *her*, as well as to myself, to say the parting words. I felt this instantly. And yet, I hated him for helping me!

She took my hand, and led me to the sofa. We sat down, side by side. Her face was composed to a sad tranquillity. She was quiet; she was herself again.

'I have refused to see him,' she said, 'until I had first spoken to you. You have read his letter. What do you say?'

I could make but one answer. It was my duty to tell her what my own position was in the plainest terms. I did my duty—leaving her free to decide on the future for herself. Those sad words said, it was useless to prolong the wretchedness of our separation. I rose, and took her hand for the last time.

I see her again now, at that final moment, as plainly as if it had happened yesterday. She had been suffering from an affection of the throat; and she had a white silk handkerchief tied loosely round her neck. She wore a simple dress of purple merino, with a black-silk apron over it. Her face was deadly pale; her fingers felt icily cold as they closed round my hand.

'Promise me one thing,' I said, 'before I go. While I live, I am your friend—if I am nothing more. If you are ever in trouble, promise that you will let me know it.'

She started, and drew back from me as if I had struck her with a sudden terror.

'Strange!' she said, speaking to herself. 'He feels as I feel. *He* is afraid of what may happen to me, in my life to come.'

I attempted to reassure her. I tried to tell her what was indeed the truth—that I had only been thinking of the ordinary chances and changes of life, when I spoke.

She paid no heed to me; she came back and put her hands on my shoulders, and thoughtfully and sadly looked up in my face.

'My mind is not your mind in this matter,' she said. 'I once owned to you that I had my forebodings, when we first spoke of this man's return. I may tell you now, more than I told you then. I believe I shall die young, and die miserably. If I am right, have you interest enough still left in me to wish to hear of it?'

She paused, shuddering—and added these startling words:

'You *shall* hear of it.'

The tone of steady conviction in which she spoke alarmed and distressed me. My face showed her how deeply and how painfully I was affected.

'There, there!' she said, returning to her natural manner; 'don't take what I say too seriously. A poor girl who has led a lonely life like mine thinks strangely and talks strangely—sometimes. Yes; I give you my promise. If I am ever in trouble, I will let you know it. God bless you—you have been very kind to me—goodbye!'

A tear dropped on my face as she kissed me. The door closed between us. The dark street received me.

It was raining heavily. I looked up at her window, through the drifting shower. The curtains were parted: she was standing in the gap, dimly lit by the lamp on the table behind her, waiting for our last look at each other. Slowly lifting her hand, she waved her farewell at the window, with the unsought native grace which had charmed me on the night when we first met. The curtains fell again—she disappeared—nothing was before me, nothing was round me, but the darkness and the night.

V

In two years from that time, I had redeemed the promise given to my mother on her deathbed. I had entered the Church.

My father's interest made my first step in my new profession an easy one. After serving my preliminary apprenticeship as a curate, I was appointed, before I was thirty years of age, to a living in the West of England.

My new benefice offered me every advantage that I could possibly desire—with the one exception of a sufficient income. Although my wants were few, and although I was still an unmarried man, I found it desirable, on many accounts, to add to my resources. Following the example of other young clergymen in my position, I determined to receive pupils who might stand in need of preparation for a career at the Universities. My relatives exerted themselves; and my good fortune still befriended me. I obtained two pupils to start with. A third would complete the number which I was at present prepared to receive. In course of time, this third pupil made his appearance, under circumstances sufficiently remarkable to merit being mentioned in detail.

It was the summer vacation; and my two pupils had gone home. Thanks to a neighbouring clergyman, who kindly undertook to perform my duties for me, I too obtained a fortnight's holiday, which I spent at my father's house in London.

During my sojourn in the metropolis, I was offered an opportunity of preaching in a church, made famous by the eloquence of one of the popular pulpit-orators of our time. In accepting the proposal, I felt naturally anxious to do my best, before the unusually large and unusually intelligent congregation which would be assembled to hear me.

At the period of which I am now speaking, all England had been startled by the discovery of a terrible crime, perpetrated under circumstances of extreme provocation. I chose this crime as the main subject of my sermon. Admitting that the best among us were frail mortal creatures subject to evil promptings and provocations like the worst among us, my object was to show how a Christian man may find his certain refuge from temptation in the safeguards of his religion. I dwelt minutely on the hardship of the Christian's first struggle to resist the evil influence—on the help which his Christianity inexhaustibly held out to him in the worst relapses of the weaker and viler part of his nature—on the steady and certain gain which was the ultimate reward of his faith and his firmness—and on the blessed sense of peace and happiness which accompanied the final triumph. Preaching to this effect, with the fervent conviction which I really felt, I may say for myself, at least, that I did no discredit to the choice which had placed me in the pulpit. I held the attention of my congregation, from the first word to the last.

While I was resting in the vestry on the conclusion of the service, a note was brought to me written in pencil. A member of my congregation—a gentleman—wished to see me, on a matter of considerable importance to himself. He would call on me at any place, and at any hour, which I might choose to appoint. If I wished to be satisfied of his respectability, he would beg leave to refer me to his father, with whose name I might possibly be acquainted.

The name given in the reference was undoubtedly familiar to me, as the name of a man of some celebrity and influence in the world of London. I sent back my card, appointing an hour for the visit of my correspondent on the afternoon of the next day.

VI

The stranger made his appearance punctually. I guessed him to be some two or three years younger than myself. He was undeniably handsome; his manners were the manners of a gentleman—and yet, without knowing why, I felt a strong dislike to him the moment he entered the room.

After the first preliminary words of politeness had been exchanged between us, my visitor informed me as follows of the object which he had in view.

'I believe you live in the country, sir?' he began.

'I live in the West of England,' I answered.

'Do you make a long stay in London?'

'No. I go back to my rectory tomorrow.'

'May I ask if you take pupils?'

'Yes.'

'Have you any vacancy?'

'I have one vacancy.'

'Would you object to let me go back with you tomorrow, as your pupil?'

The abruptness of the proposal took me by surprise. I hesitated.

In the first place (as I have already said), I disliked him. In the second place, he was too old to be a fit companion for my other two pupils—both lads in their teens. In the third place, he had asked me to receive him at least three weeks before the vacation came to an end. I had my own pursuits and amusements in prospect during that interval, and saw no reason why I should inconvenience myself by setting them aside.

He noticed my hesitation, and did not conceal from me that I had disappointed him.

'I have it very much at heart,' he said, 'to repair without delay the time that I have lost. My age is against me, I know. The truth is—I have wasted my opportunities since I left school, and I am anxious, honestly anxious, to mend my ways, before it is too late. I wish to prepare myself for one of the Universities—I wish to show, if I can, that I am not quite unworthy to inherit my father's famous name. You are the man to help me, if I can only persuade you to do it. I was struck by your sermon yesterday; and, if I may venture to make the confession in your presence, I took a strong liking to you. Will you see my father, before you decide to say No? He will be able to explain whatever may seem strange in my present application; and he will be happy to see you this afternoon, if you can spare the time. As to the question of terms, I am quite sure it can be settled to your entire satisfaction.'

He was evidently in earnest—gravely, vehemently in earnest. I unwillingly consented to see his father.

Our interview was a long one. All my questions were answered fully and frankly.

The young man had led an idle and desultory life. He was weary of it, and ashamed of it. His disposition was a peculiar one. He stood sorely in need of a guide, a teacher, and a friend, in whom he was disposed to confide. If I disappointed the hopes which he had centred in me, he would be discouraged, and he would relapse into the aimless and indolent existence of which he was now ashamed. Any terms for which I might stipulate were at my disposal if I would consent to receive him, for three months to begin with, on trial.

Still hesitating, I consulted my father and my friends.

They were all of opinion (and justly of opinion so far) that the new connection would be an excellent one for me. They all reproached me for taking a purely capricious dislike to a well-born and well-bred young man, and for permitting it to influence me, at the outset of my career, against my own interests. Pressed by these considerations, I allowed myself to be persuaded to give the new pupil a fair trial. He accompanied me, the next day, on my way back to the rectory.

VII

Let me be careful to do justice to a man whom I personally disliked. My senior pupil began well: he produced a decidedly favourable impression on the persons attached to my little household.

The women, especially, admired his beautiful light hair, his crisply curling beard, his delicate complexion, his clear blue eyes, and his finely shaped hands and feet. Even the inveterate reserve in his manner, and the downcast, almost sullen, look which had prejudiced *me* against him, aroused a common feeling of romantic enthusiasm in my servants' hall. It was decided, on the high authority of the housekeeper herself, that 'the new gentleman' was in love—and, more interesting still, that he was the victim of an unhappy attachment which had driven him away from his friends and his home.

For myself, I tried hard, and tried vainly, to get over my first dislike to the senior pupil.

I could find no fault with him. All his habits were quiet and regular; and he devoted himself conscientiously to his reading. But, little by little, I became satisfied that his heart was not in his studies. More than this, I had my reasons for suspecting that he was concealing something from me, and that he felt painfully the reserve on his own part which he could not, or dared not, break through. There were moments when I almost doubted whether he had not chosen my remote country rectory, as a safe place of refuge from some person or persons of whom he stood in dread.

For example, his ordinary course of proceeding, in the matter of his correspondence, was, to say the least of it, strange.

He received no letters at my house. They waited for him at the village post-office. He invariably called for them himself, and invariably forbore to trust any of my servants with his own letters for the post. Again, when we were out walking together, I more than once caught him looking furtively over his shoulder, as if he suspected some person of following him, for some evil purpose. Being constitutionally a hater of mysteries, I determined, at an early stage of our intercourse, on making an effort to clear matters up. There might be just a chance of my winning the senior pupil's confidence, if I spoke to him while the last days of the summer vacation still left us alone together in the house.

'Excuse me for noticing it,' I said to him one morning, while we were engaged over our books—'I cannot help observing that you appear to have some trouble on your mind. Is it indiscreet, on my part, to ask if I can be of any use to you?'

He changed colour—looked up at me quickly—looked down again at his book—struggled hard with some secret fear or secret reluctance that was in him—and suddenly burst out with this extraordinary question:

'I suppose you were in earnest when you preached that sermon in London?'

'I am astonished that you should doubt it,' I replied.

He paused again; struggled with himself again; and startled me by a second outbreak, even stranger than the first.

'I am one of the people you preached at in your sermon,' he said. 'That's the true reason why I asked you to take me for your pupil. Don't turn me out! When you talked to your congregation of tortured and tempted people, you talked of Me.'

I was so astonished by the confession, that I lost my presence of mind. For the moment, I was unable to answer him.

'Don't turn me out!' he repeated. 'Help me against myself. I am telling you the truth. As God is my witness, I am telling you the truth!'

'Tell me the *whole* truth,' I said; 'and rely on my consoling and helping you—rely on my being your friend.'

In the fervour of the moment, I took his hand. It lay cold and still in mine: it mutely warned me that I had a sullen and a secret nature to deal with.

'There must be no concealment between us,' I resumed. 'You have entered my house, by your own confession, under false pretences. It is your duty to me, and your duty to yourself, to speak out.'

The man's inveterate reserve—cast off for the moment only—renewed its hold on him. He considered, carefully considered, his next words before he permitted them to pass his lips.

'A person is in the way of my prospects in life,' he began slowly, with his eyes cast down on his book. 'A person provokes me horribly. I feel dreadful temptations (like the man you spoke of in your sermon) when I am in the person's company. Teach me to resist temptation! I am afraid of myself, if I see the person again. You are the only man who can help me. Do it while you can.'

He stopped, and passed his handkerchief over his forehead.

'Will that do?' he asked—still with his eyes on his book.

'It will *not* do,' I answered. 'You are so far from really opening your heart to me, that you won't even let me know whether it is a man or a woman who stands in the way of your prospects in life. You use the word "person", over and over again—rather than say "he" or "she" when you speak of the provocation which is trying you. How can I help a man who has so little confidence in me as that?'

My reply evidently found him at the end of his resources. He tried, tried desperately, to say more than he had said yet. No! The words seemed to stick in his throat. Not one of them would pass his lips.

'Give me time,' he pleaded piteously. 'I can't bring myself to it, all at once. I mean well. Upon my soul, I mean well. But I am slow at this sort of thing. Wait till tomorrow.'

Tomorrow came—and again he put it off.

'One more day!' he said. 'You don't know how hard it is to speak plainly. I am half afraid; I am half ashamed. Give me one more day.'

I had hitherto only disliked him. Try as I might (and did) to make merciful allowance for his reserve, I began to despise him now.

VIII

The day of the deferred confession came, and brought an event with it, for which both he and I were alike unprepared. Would he really have confided in me but for that event? He must either have done it, or have abandoned the purpose which had led him into my house.

We met as usual at the breakfast-table. My housekeeper brought in my letters of the morning. To my surprise, instead of leaving the room again as usual, she walked round to the other side of the table, and laid a letter before my senior pupil—the first letter, since his residence with me, which had been delivered to him under my roof.

He started, and took up the letter. He looked at the address. A spasm of suppressed fury passed across his face; his breath came quickly; his hand trembled as it held the letter. So far, I said nothing. I waited to see whether he would open the envelope in my presence or not.

He was afraid to open it, in my presence. He got on his feet; he said, in tones so low that I could barely hear him: 'Please excuse me for a minute'—and left the room.

I waited for half an hour—for a quarter of an hour, after that—and then I sent to ask if he had forgotten his breakfast.

In a minute more, I heard his footstep in the hall. He opened the breakfast-room door, and stood on the threshold, with a small travelling-bag in his hand.

'I beg your pardon,' he said, still standing at the door. 'I must ask for leave of absence for a day or two. Business in London.'

'Can I be of any use?' I asked. 'I am afraid your letter has brought you bad news?'

'Yes,' he said shortly. 'Bad news. I have no time for breakfast.'

'Wait a few minutes,' I urged. 'Wait long enough to treat me like your friend—to tell me what your trouble is before you go.'

He made no reply. He stepped into the hall, and closed the door—then opened it again a little way, without showing himself.

'Business in London,' he repeated—as if he thought it highly important to inform me of the nature of his errand. The door closed for the second time. He was gone.

I went into my study, and carefully considered what had happened.

The result of my reflections is easily described. I determined on discontinuing my relations with my senior pupil. In writing to his father (which I did, with all due courtesy and respect, by that day's post), I mentioned as my reason for arriving at this decision: First, that I had found it impossible to win the confidence of his son. Secondly, that his son had that morning suddenly and mysteriously left my house for London, and that I must decline accepting any further responsibility towards him, as the necessary consequence.

I had put my letter in the post-bag, and was beginning to feel a little easier after having written it, when my housekeeper appeared in the study, with a very grave face, and with something hidden apparently in her closed hand.

'Would you please look, sir, at what we have found in the gentleman's bedroom, since he went away this morning?'

I knew the housekeeper to possess a woman's full share of that amiable weakness of the sex which goes by the name of 'Curiosity.' I had also, in various indirect ways, become aware that my senior pupil's strange departure had largely increased the disposition among the women of my household to regard him as the victim of an unhappy attachment. The time was ripe, as it seemed to me, for checking any further gossip about him, and any renewed attempts at prying into his affairs in his absence.

'Your only business in my pupil's bedroom,' I said to the housekeeper, 'is to see that it is kept clean, and that it is properly aired. There must be no interference, if you please, with his letters, or his papers, or with anything else that he has left behind him. Put back directly whatever you may have found in his room.'

The housekeeper had her full share of a woman's temper as well as of a woman's curiosity. She listened to me with a rising colour, and a just perceptible toss of the head.

'Must I put it back, sir, on the floor, between the bed and the wall?' she enquired, with an ironical assumption of the humblest deference to my wishes. '*That's* where the girl found it when she was sweeping the room. Anybody can see for themselves,' pursued the housekeeper indignantly, 'that the poor gentleman has gone away broken-hearted. And there, in my opinion, is the hussy who is the cause of it!'

With those words, she made me a low curtsy, and laid a small photographic portrait on the desk at which I was sitting.

I looked at the photograph.

In an instant, my heart was beating wildly—my head turned giddy—the housekeeper, the furniture, the walls of the room, all swayed and whirled round me.

The portrait that had been found in my senior pupil's bedroom was the portrait of Jéromette!

IX

I had sent the housekeeper out of my study. I was alone, with the photograph of the Frenchwoman on my desk.

There could surely be little doubt about the discovery that had burst upon me. The man who had stolen his way into my house, driven by the terror of a temptation that he dared not reveal, and the man who had been my unknown rival in the by-gone time, were one and the same!

Recovering self-possession enough to realize this plain truth, the inferences that followed forced their way into my mind as a matter of course. The unnamed person who was the obstacle to my pupil's prospects in life, the unnamed person in whose company he was assailed by temptations which made him tremble for himself, stood revealed to me now as being, in all human probability, no other than Jéromette. Had she bound him in the fetters of the marriage which he had himself proposed? Had she discovered his place of refuge in my house? And was the letter that had been delivered to him of her writing? Assuming these questions to be answered in the affirmative, what, in that case, was his 'business in London?' I remembered how he had spoken to me of his temptations, I recalled the expression that had crossed his face when he recognized the handwriting on the letter—and the conclusion that followed literally shook me to the soul. Ordering my horse to be saddled, I rode instantly to the railway-station.

The train by which he had travelled to London had reached the terminus nearly an hour since. The one useful course that I could take, by way of quieting the dreadful misgivings crowding one after another on my mind, was to telegraph to Jéromette at the address at which I had last seen her. I sent the subjoined message—prepaying the reply:

'If you are in any trouble, telegraph to me. I will be with you by the first train. Answer, in any case.'

There was nothing in the way of the immediate dispatch of my message. And yet the hours passed, and no answer was received. By the advice of the clerk, I sent a second telegram to the London office, requesting an explanation. The reply came back in these terms:

'Improvements in street. Houses pulled down. No trace of person named in telegram.'

I mounted my horse, and rode back slowly to the rectory.

'The day of his return to me will bring with it the darkest days of my life.' . . . 'I shall die young, and die miserably. Have you interest enough still left in me to wish to hear of it?' . . . 'You *shall* hear of it.' Those words were in my memory while I rode home in the cloudless moonlight night. They were so vividly present to me that I could hear again her pretty foreign accent, her quiet clear tones, as she spoke them. For the rest, the emotions of that memorable day had worn me out. The answer from the telegraph-office had struck me with a strange and stony despair. My mind was a blank. I had no thoughts. I had no tears.

I was about half-way on my road home, and I had just heard the clock of a village church strike ten, when I became conscious, little by little, of a chilly sensation slowly creeping through and through me to the bones. The warm balmy air of a summer night was abroad. It was the month of July. In the month of July, was it possible that any living creature (in good health) could feel cold? It was *not* possible—and yet, the chilly sensation still crept through and through me to the bones.

I looked up. I looked all round me.

My horse was walking along an open highroad. Neither trees nor waters were near me. On either side, the flat fields stretched away bright and broad in the moonlight.

I stopped my horse, and looked round me again.

Yes: I saw it. With my own eyes I saw it. A pillar of white mist—between five and six feet high, as well as I could judge—was moving beside me at the edge of the road, on my left hand. When I stopped, the white mist stopped. When I went on, the white mist went on. I pushed my horse to a trot—the pillar of mist was with me. I urged him to a gallop—the pillar of mist was with me. I stopped him again—the pillar of mist stood still.

The white colour of it was the white colour of the fog which I had seen over the river—on the night when I had gone to bid her farewell. And the chill which had then crept through me to the bones was the chill that was creeping through me now.

I went on again slowly. The white mist went on again slowly—with the clear bright night all round it.

I was awed rather than frightened. There was one moment, and one only, when the fear came to me that my reason might be shaken. I caught myself keeping time to the slow tramp of the horse's feet with the slow utterance of these words, repeated over and over again: 'Jéromette is dead. Jéromette is dead.' But my will was still my own: I was able to control myself, to impose silence on my own muttering lips. And I rode on quietly. And the pillar of mist went quietly with me.

My groom was waiting for my return at the rectory gate. I pointed to the mist, passing through the gate with me.

'Do you see anything there?' I said.

The man looked at me in astonishment.

I entered the rectory. The housekeeper met me in the hall. I pointed to the mist, entering with me.

'Do you see anything at my side?' I asked.

The housekeeper looked at me as the groom had looked at me.

'I am afraid you are not well, sir,' she said. 'Your colour is all gone—you are shivering. Let me get you a glass of wine.'

I went into my study, on the ground-floor, and took the chair at my desk. The photograph still lay where I had left it. The pillar of mist floated round the table, and stopped opposite to me, behind the photograph.

The housekeeper brought in the wine. I put the glass to my lips, and set it down again. The chill of the mist was in the wine. There was no taste, no reviving spirit in it. The presence of the housekeeper oppressed me. My dog had followed her into the room. The presence of the animal oppressed me. I said to the woman, 'Leave me by myself, and take the dog with you.'

They went out, and left me alone in the room.

I sat looking at the pillar of mist, hovering opposite to me.

It lengthened slowly, until it reached to the ceiling. As it lengthened, it grew bright and luminous. A time passed, and a shadowy appearance showed itself in the centre of the light. Little by little, the shadowy appearance took the outline of a human form. Soft brown eyes, tender and melancholy, looked at me through the unearthly light in the mist. The head and the rest of the face broke next slowly on my view. Then the figure gradually revealed itself, moment by moment, downward and downward to the feet. She stood before me as I had last seen her, in her purple-merino dress, with the black-silk apron, with the white handkerchief tied loosely round her neck. She stood before me, in the gentle beauty that I remembered so well; and looked at me as she had looked when she gave me her last kiss—when her tears had dropped on my cheek.

I fell on my knees at the table. I stretched out my hands to her imploringly. I said, 'Speak to me—O, once again speak to me, Jéromette.'

Her eyes rested on me with a divine compassion in them. She lifted her hand, and pointed to the photograph on my desk, with a gesture which bade me turn the card. I turned it. The name of the man who had left my house that morning was inscribed on it, in her own handwriting.

I looked up at her again, when I had read it. She lifted her hand once more, and pointed to the handkerchief round her neck. As I looked at it, the fair white silk changed horribly in colour—the fair white silk became darkened and drenched in blood.

A moment more—and the vision of her began to grow dim. By slow degrees, the figure, then the face, faded back into the shadowy appearance that I had first seen. The luminous inner light died out in the white mist. The mist itself dropped slowly downwards—floated a moment in airy circles on the floor—vanished. Nothing was before me but the familiar wall of the room, and the photograph lying face downwards on my desk.

X

The next day, the newspapers reported the discovery of a murder in London. A Frenchwoman was the victim. She had been killed by a wound in the throat. The crime had been discovered between ten and eleven o'clock on the previous night.

I leave you to draw your conclusion from what I have related. My own faith in the reality of the apparition is immovable. I say, and believe, that Jéromette kept her word with me. She died young, and died miserably. And I heard of it from herself.

Take up the Trial again, and look at the circumstances that were revealed during the investigation in court. His motive for murdering her is there.

You will see that she did indeed marry him privately; that they lived together contentedly, until the fatal day when she discovered that his fancy had been caught by another woman; that violent quarrels took place between them, from that time to the time when my sermon showed him his own deadly hatred towards her, reflected in the case of another man; that she discovered his place of retreat in my house, and threatened him by letter with the public assertion of her conjugal rights; lastly, that a man, variously described by different witnesses, was seen leaving the door of her lodgings on the night of the murder. The Law—advancing no further than this—may have discovered circumstances of suspicion, but no certainty. The Law, in default of direct evidence to convict the prisoner, may have rightly decided in letting him go free.

But *I* persist in believing that the man was guilty. *I* declare that he, and he alone, was the murderer of Jéromette. And now, you know why.

The Story of Clifford House

ANONYMOUS

This story I will tell to you now, as I have promised to do so, and yet I can hardly make you believe in the reluctance with which I even allow my thoughts to go back to the time which I spent in that house—my first town residence after I was married.

I had wished so much to go to town that spring—grown tired of my lovely country home, I suppose. Tired of wide lawns and quiet, glassy ponds and streams, bordered by luscious, blooming rhododendrons; of silent, mossy avenues, glorious with the flickering light that stole through pale green beech leaves; of rose gardens with grassy paths, jewel-sprinkled with shell-like petals of white, crimson, pink, and cream-like hues; of old-fashioned rooms with narrow, mullioned windows embowered in scarlet japonica and fragrant, starry jessamine.

I suppose I had grown tired of them all, and I begged George to see about getting a nice house in town for the season.

George 'saw about it'—viz.: he wrote one letter—from my dictation—to a house agent, and answered one advertisement, and yawned and grunted for a week afterwards about the 'bore of the thing'.

Of course I had to make him accompany me to town, and to the house agent's, and to the houses too. Let him smoke and yawn as much as he liked, I was determined to take a house, and take a nice one as well.

We had looked at—George said fourteen—but, in fact, seven, or eight houses I think, before we saw Clifford House.

I had found out a new house-agent's office, and this was the very first house we were shown—pressed upon our notice, too, by the enthusiastic encomiums of the said house-agent. It was certainly a very fine house, both as to exterior and interior appearances. Large, massively built, agreeably darkened in woodwork and masonry by Time's shading brush, in excellent repair, and the locality all that could be desired. Wide, lofty apartments, staircases, and landings; a handsome dining-room panelled in velvety dark-green 'flock' and gold; a handsome drawing-room panelled in pale cream-colour and gold; airy bed-chambers and dressing-rooms—one, in particular, attached to what seemed the principal bedroom, with a vast mirror occupying the whole side of the apartment which was opposite to the door leading into the bed-chamber.

'What a nice dressing-room!' I exclaimed, having a weakness, I confess, for large, handsome mirrors in the rooms I inhabit—George says impertinent things about my 'wishing to see as much of myself as I can'. I know I am not tall, in fact, rather what he should call *petite,* if he wished to be polite—but that is not my reason for liking a large mirror.

As I spoke the words I looked about mechanically for the house-agent's clerk who had been sent with us—a nervous-looking little man, with a pasty complexion, and orange-coloured hair meekly plastered down at each side of his face. He had been untiringly trotting up and down stairs, unlocking doors, answering questions, and keeping up a harmless soliloquy of chatter about the beauties and excellencies of the 'mansiond', as he called it, ever since he entered its doors, but now he was nowhere to be seen.

'What door have you open?' I said, speaking aloud to him, for suddenly a cold blast of air swept up the wide staircase and into the dressing-room, making me shudder.

'No door, ma'am—not one, indeed!' said the little clerk, hurrying to the dressing-room door, but not entering. His face looked whiter than before, and in his accents there was an almost terrified earnestness that puzzled me.

The shadows of the afternoon seemed to deepen. The aspect of the suites of rooms and long silent corridors, with their doors ajar, as if unseen inhabitants were stealthily crouching behind them, drearily impressed me with a sense of dull desolation; and it was with a sudden sensation of childish fear and loneliness that I rushed after my husband, and took his arm as he hastily descended the stairs.

'A spacious, handsome staircase, George?' I remarked.

'Yes; and a spacious, handsome rent, you may be sure,' George responded.

But, in this particular, he was exceedingly, and I agreeably, astonished.

The rent was but a hundred and fifty pounds a year; when, judging from the situation and appearance of the house, our lowest estimate had been double that sum.

'How cheap!' I whispered.

'A screw loose somewhere,' was George's oracular response.

He repeated his opinion to the clerk in a more business-like expression, to the effect that the rent seemed low, and that he trusted there was no—peculiar—eh?

'Drains, gas, water, all right, sir—right as—a—a trivet, sir,' said the clerk, looking over his shoulder oddly, as he spoke. 'Chimneys, ventilators, roof, tiles—everything in the perfectest repair and order, sir!'

'Hum!' said George, with a frown of thoroughly British dissatisfaction. 'Unpleasant neighbours, then?'

The little clerk coughed violently, and buried his nose and eyes in the depths of a red cotton handkerchief:

'Neighbours? Disagreeable, sir? Ah! dear me! Beg pardon, sir—a little cough. No, indeed, sir! Mrs Carmichael—very high lady—very rich, widow of young Mr William Carmichael, just opposite, sir—old Lady Broadleigh within two doors—Sir Thomas——'

'Oh, very well!' said George impatiently. 'Come, Helen.'

Nevertheless, I was rather surprised to see how many faces were clustered at the windows of our aristocratic neighbours' houses, and with what intently curious looks they watched our exit and departure, as if visitors, or would-be tenants for Clifford House, were some very wonderful people indeed.

However, wonderful or not, the house seemed all that we could desire; the lowness of the rent made it a decided bargain, the season was advancing, our low-ceiled, country rooms seemed contracted, old-fashioned, and shabby, after those lofty, handsome suites of apartments; and, in three weeks, huge furniture vans, and a clever upholsterer, had carpeted, curtained, and furnished our town mansion from garret to basement, and George and I, our two babies, a nurse, two maids, a cook, and a butler, were installed in Clifford House.

Dear George had been very generous—nay, almost extravagant—in his provisions for the comfort and pleasure of his wife and children; and my dressing-room and their nursery were fitted up so luxuriously and tastefully, that my feeling at the first inspection of them was that of self-gratulation on being such a fortunate woman, in having such a home, such babies, and such a husband.

I arrayed myself for dinner that evening quite gleefully; standing before my splendid mirror amid the blue drapery, cushions, and couches of my charming dressing-room. I put on George's favourite dress—a bronze-brown lustrous silk, with sparkling gold ornaments: he invariably kissed me when he saw it on, stroked my brown curls and brunette face, and called me 'Maid Marian'—and was still standing before the glass smiling at myself, like the happy, foolish little woman I was, when I perceived to my discomfiture that George was standing in the doorway watching my doings, and grinning very visibly under his moustache.

'Don't mind me, my dear, I beg! don't mind me in the least. But when you *have* done admiring Mrs George Russell, perhaps you will be kind enough to let me know'—then, suddenly changing his tone, he exclaimed, 'Have you the window open, Helen, this chilly evening?'

'No, George,' I replied, glancing at it to make sure of the fact.

'Change in the weather, then,' my husband said. 'Come, Helen, there is no use in making yourself any prettier!' He had just uttered the last words when I saw him spring aside suddenly, and look around.

'What is the matter?' I said—'George, dear, what is the matter?' For his face had grown quite white, and with his back against the wall, he was staring about him wildly.

'I don't know—Helen—something'—he ejaculated in a low tone; then recovering himself, with a laugh, he cried—'I struck myself against the door, I suppose! I declare one would think I was composed of old china, or wax, or sugar candy, it hurt and stunned me so! Come, dearest.'

He had not struck himself, for I had been watching him going out on the lobby, and I felt an uneasy conviction that he knew he had not done so, and only spoke as he did in order to deceive or satisfy me. Why? Why did I think so? As I live I cannot tell why I thought so *then*—I *know* now. We had the 'babies'—as George always called them—in with the dessert, after the time-honoured fashion of making olives as well as olive branches of them; and then, when the little ones had gone to bed, we sat side by side in the summer twilight, I lazily fanning myself, George bending over me like the lover-husband he was. Then came the lamps, and I played for him, and we sang duets and spent as happy an evening in our new home as a married pair could wish to spend. I cannot tell why I felt so disinclined to go upstairs that night, tired as I was, too—for we had had a long journey up from the country. However, as eleven struck, I routed George out of the easy chair where he had been indulging in a preliminary doze, and, ringing for my maid, went up to my dressing-room.

I like gas in my dressing-room, though not in my bedroom, and the globes at either side the great mirror were a blaze of light. As I entered I caught the reflection of a woman's figure in the depth of the glass, not my maid's. The glimpse I had was of a tall woman, strongly built, and broad-shouldered, a quantity of light hair hanging in a disordered manner on her neck, and the profile of a white, hard, masculine face, with the keen glittering eye turned watchfully towards the door.

This may seem an elaborately detailed description for the momentary glance I obtained, but it is well known with what lightning rapidity the organs of vision will, in moments of terror and amazement, convey impressions to the startled brain, impressions accurate and indelible.

I had taken but one step on entering, the next step the figure had vanished, and the mirror reflected but my own terrified face, and the homely, cheerful one of my maid Harriet, as she stooped over the dressing-table opening a jewel case.

I dropped down on the nearest chair, and, in answer to the girl's alarmed questions, replied that I did not feel very well. I was sick and shuddering from head to foot.

Suddenly it flashed across me that it was from a similar cause I had seen my husband's face grow ghastly, and that strange, terrified look come into his eyes,—he, who had been a soldier and unflinchingly had fought amidst the dead and dying on bloody Indian battlefields, almost boy as he was then! What was it? What had he seen?

Nonsense! was I going to believe I had seen a ghost? Nonsense, a thousand times over! I heard my husband's cheery voice as he ascended the stairs, and, quite angry with myself for giving way to such folly, I threw on my dressing gown, and, snatching up the brush from Harriet, I pulled my hair down and brushed it quite savagely, until my head ached well—for punishment.

If the bright morning light disperses sweet illusions formed overnight, as people say it does, it disperses gloomy ones as well. With the warmth and brightness of the unclouded summer's sun streaming in through softly coloured blinds, bringing out the velvety green of soft new carpets and lounges, the rainbow tints of glittering chandeliers, vases, and ornaments, the gilding on bright fresh wallpaper, and the spotless folds of snowy window drapery, it was impossible for an instant to connect anything dark or dismal with Clifford House. Why, my dressing-room even, where I had been so silly last evening, was like a woodland bower, with its deep purple-blue hangings and rose painted china flower-vases, filled with bouquets from our country home. Clustering fragrant honeysuckle half-opended moss roses, drooping emerald-green fern, and masses of delicious jessamine dropping its over-blown blossoms on the white toilet cover, lace-flounced and tied with blue ribbons, as Harriet delighted to have it.

'I think this such a charming room and such a charming house altogether, George!' I said; 'and you have been such a dear, thoughtful old darling!' For I had perceived that the dear fellow had had his own half-length portrait hung over my writing-table. Quite a pleasant surprise for me, for I thought he intended it to be hung in the dining-room, and I delighted in having the dear pleasant brown eyes looking down at me when I was busy writing or sewing.

'I am so glad you like everything, Nellie,' said he.

'Why, George, don't you?'

But George had walked off whistling, and presently I heard uproarious baby-laughter, and baby-chatter, and thumping, trotting of small fat feet, as George put the tidy nursery into dire confusion by his morning game of romps with his son and heir, and red-cheeked baby-daughter. And it did seem as if I must have been dreaming or delirious, when this day and many a succeeding one passed away swiftly and pleasantly, without the slightest recurring event to remind me of my strange alarm on the night of our arrival.

We had been in Clifford House about a fortnight, when one morning I received a visit from our opposite neighbour—the young widow, Mrs Carmichael. A very pretty, lady-like person she was, and as we had some common acquaintances we chattered away very freely and pleasantly for half-an-hour or so. As she rose to go she asked suddenly if we liked the house. I replied in the affirmative rather warmly.

She was opposite the light, and I saw an involuntary elevation of her eye-brows and compression of her lips that puzzled me. I fancied it was because I had spoken so enthusiastically. Yet her own manner was anything but languidly fashionable, being very cordial and decided.

'Yes; it is a very nice house, roomy and well-built,' said she, after a moment's pause; 'I am so glad you like it—we may be permanent neighbours.'

We went out to dinner at a friend's house in Seymour Street that evening, and when we returned about half-past eleven, in spite of a yawning remonstrance from George, I tripped off softly to have a peep at my darlings, before I went to bed.

The nursery was a large, pleasant room at the end of the long corridor leading from our own apartments, and, gently turning the handle and gathering my rustling silk dress around me, I opened the door and went in. There was the night-lamp burning clearly, shining softly on the tiny cribs with the sweet flushed infant faces, the long golden-brown lashes lying on the dimpled apple-bloom cheeks, the waxen hands and little rounded arms thrown above the tossed golden curls, and the heavenly calm of the little sleeping forms and pure, peaceful breathing.

I wonder would any mother, no matter how cold and careless, have neglected doing what I did, as I bent over my treasures, and prayed

God that His angels might keep watch over each cherub head on its little, soft, white pillow?

I had looked at and kissed them, and turned to go, when I glanced towards the nurse's bed.

'Are you not well, Mary? What is the matter?' I said in an anxious whisper.

She was a very respectable and trustworthy servant, as well as being a kind, gentle creature with the little ones, and consequently highly valued by me, but her health was never very good, and she was subject to severe attacks of nervous headache and sleeplessness. She was sitting up in bed, her hands grasping the bedclothes, her face and lips ashy white, and her eyes staring wildly, as if they would start from their sockets.

'Mary! Good Heavens! what is the matter?' I gasped.

'Ma'am! Oh, ma'am—oh, mistress, I am dying!' And with a stifled cry the poor girl fell back on the pillow, her eyes still retaining their frenzied stare. It was but the work of a few moments to ring bells and summon the household, to dispatch the man-servant for a doctor, and to have the sleeping children taken into my own bedchamber, while Harriet and I administered restoratives, and chafed the half-senseless girl's damp, cold hands.

I could imagine no cause for her sudden illness, and the other servants were very voluble in exclamations and laments. But when the physician—a pale, kindly, grave-looking man arrived—after a moment's examination, he demanded if she had been frightened? I replied in the negative, and was proceeding to describe to him the state in which I had found her, when I heard the housemaid and Harriet whispering energetically together.

'She has!'

'Hush!'

'I know she has!'

'What is it? Speak out at once my good girl!' said the doctor sternly to the housemaid; 'you know something of this.'

Both servants looked apprehensively at me and at George.

'Speak up at once, Margaret; the girl's life may depend on it! Tell the truth, my girl, and don't be afraid,' said her master kindly, but firmly.

'I don't know nothing, sir—indeed, no, ma'am,' said Margaret confusedly; 'but—I think, ma'am—she's seen the ghost, sir!'

'The *what!*' cried George angrily.

'She have, sir!' persisted Margaret eagerly, now that her confession was made. 'We're all afraid, sir; but she's been worser nor the rest of us. And she says to me only this morning, "Margaret", she says, "if I see it, I'll die!"'

'What ghost, you fool?' cried George more angrily. 'A pretty set you are!—great, grown men and women, afraid of some bogie story you have heard when you were gossipping with the servants on the terrace, I suppose!'

'No, indeed, sir,' said Margaret; 'I wasn't gossippin', sir; but the parlour-maid over the way, sir—Mrs Carmichael's parlour-maid, ma'am—she told me that there was somethin'——

'I thought so!' interrupted George. 'You ought to be ashamed of yourselves not to have an ounce of brains among you.'

'But, sir!' Margaret burst out again, unheeding her master's rather uncomplimentary phrenological verdict, 'we didn't mind, sir, though we was a bit frightened, until we seen it, sir! The butler see it, and he ran, and cook ran.'

'And you ran after them?' said George, with an indignant laugh.

'I did, sir, for I saw it too—a big woman with fair hair all over her shoulders,' said Margaret, in an awestruck whisper to Harriet, who nodded her head.

The doctor looked up, gravely and without a smile. The servants clustered together near the door, and muttered in undertones. George looked at me with a forced smile, which died away in an instant:

'You are not so foolish as to credit any of this nonsense, Helen?' he said.

The servants all turned eagerly to hear their mistress's opinion. I am afraid it was written in my pallid face. Was it true? Was it what I had seen? Could there be any reality in this, that here, in our pleasant, happy home, here, beneath the roof with our helpless little ones, was a dreadful, unblessed presence—a shadowy horror; that that *thing* with the watchful, cruel eyes had not been a mere vision of imagination, the mere offspring of an active brain, and the unstrung nerves of an overtired frame?

'Oh! they imagined something from the stories they heard, I dare say,' I faltered.

The butler shook his head solemnly:

'I could swear to it, ma'am.'

'And so could I, ma'am!' chorused the cook and housemaid.

'Hush!' said the doctor, as the nurse, roused, at length, from her stupor, lay quietly, with closed eyes, from which the tears streamed down her face. 'Some one must sit up with her now,' said the doctor, looking around.

'I will, sir, if my mistress allows me,' said Harriet.

'Certainly, Harriet,' said I at once.

He communicated his instructions to her and took his leave, promising to call in the morning.

'Did you ever hear anything like this folly, doctor?' said George, as he shook hands with him at the head of the stairs.

'Oh! yes, sir, I often hear such stories,' said the doctor quietly, as he bade us both goodnight.

'George! what has frightened the girl? What has she seen?' I whispered, clasping my husband's arm.

'Nellie, go to bed, and don't be a goose,' was George's reply.

'George—I saw that thing—that woman, in my dressing-room,' I said, trembling, 'and oh! think if the children were to see it and be frightened like poor Mary!'

'Well, Helen,' said my husband sharply, 'if you are going to listen to ignorant servants' superstitions and run out of your house, just as we are comfortably settled in it, on account of a foolish sickly woman fainting from hearing a ghost story—I say—it is a pity you ever came into it.'

He spoke very decidedly and sternly, and yet I felt in my inmost heart that he uttered what he wished me to believe, not what he believed himself.

I said no more, but went to my bedroom—not into the dreaded dressing-room—and lay awake listening and fevered with nervous anxiety until the morning dawned.

The nurse was better and able to speak next day, though extremely weak and unnerved yet. The doctor forbade much questioning, and all that could be got from her at intervals was that something had come up the staircase and ran through the corridor, that she heard struggling and scuffling outside, and then the nursery door opened and she saw a woman's face peering in, the eyes gleaming wickedly at her, and it had the yellow hair that 'belonged to the ghost'.

'The woman has had a bad fit of nightmare—that is all, Helen,' said George, rattling his paper unconcernedly, when I repeated to him the story I had just heard from poor Mary's trembling lips.

It might be so; but why were they all agreed as to what they had seen? Why did they all speak of the tangled fair hair, and the wicked gleaming eyes? Was our house haunted? Was this the mysterious cause of the exceedingly moderate rent and the house-agent's profuse civility?

The nurse did not recover strength, and being worse than useless in her present weak, hysterical condition, I sent her down to her country home for change of air, and hired another temporarily in her place.

The newcomer was a stout, small, cheerful woman of about forty. I liked her face the moment I saw her; for, besides its smiling, honest expression, there was a good deal of decided character in the large firm features. 'You appear to be a sensible person,' I said, when giving her her first instructions in the nursery, 'and I think I can rely on you. You know my nurse is leaving because of illness, and that illness was caused by her being frightened by—a ghost-story.' I paused; but the woman remained unmoved, listening to me in respectful silence.

'The servants downstairs have got some nonsense of the kind into their heads,' I went on; 'they will try to frighten you, too, and tell you they have seen——' I could not go on. For my life I could not calmly give her the description of that shadowy image of fear.

'They cannot frighten me, ma'am,' said my new nurse quietly. 'I am not afraid of spirits.'

I thought she spoke in jest, and smiled.

'I am not indeed, ma'am,' she repeated. 'I have lived where there were such things seen, but they never harmed me.'

'You don't mean to say you believe such nonsense?' said I, hypocritically trying to speak carelessly.

'Oh yes, ma'am, I do! I could not disbelieve it,' said the nurse, opening her eyes with earnestness, 'I know the story of this house, ma'am.'

'What story?' I cried.

The woman coloured and looked confused.

'I beg your pardon, ma'am—I mean what people say is seen here.'

'What do they say? Do not frighten me,' I said, and my voice quivered in spite of me; 'I have heard nothing but what the servants said.'

The nurse looked deeply concerned.

'I am very stupid, ma'am; I beg your pardon for repeating such stories to you—I daresay it is only idle people's gossip.'

She went about her duties, and I went—not into my dressing-room—but down into the drawing-room, where I sat by the window looking out until my husband returned.

Two or three weeks more passed away without any more alarms. The summer had deepened into its longest days and hottest sunshine; the gay season had reached and passed its meridian of wealth, beauty, luxury, extravagance, success, misery, hopes, and disappointments. I had enjoyed it very much at first; but I soon wearied of it as my bodily strength weakened in the ordeal of constant excitement, late hours, hot rooms, heavy perfumed atmosphere, ices, and diaphanous ball-dresses.

'Poor Maid Marian,' George said, 'she is pining for her green wild woods.' However, by following the doctor's advice—the same whom he had summoned the night of the nurse's illness, and whom we both liked very much—and living more quietly, I was able to enjoy quiet entertainments and my favourite operas very fairly, although my red brunette cheeks had faded dismally.

'An invitation for us, Helen, I know, and that is Willesden's writing.'

It was a sultry morning at the close of June. I felt tired and languid, and it was with a bad grace I tore open the envelope lying beside the breakfast tray.

'Yes, "Colonel and Mrs Willesden request the pleasure"——Why, George, it is for this evening!'

'Written the day before yesterday, though—delayed somehow,' said George, reading over my shoulder. 'Well, Helen, what do you say? It is only for a quiet, friendly dinner, and I like Willesden very much.'

'No, dear,' I replied wearily. 'You can go and make apologies for me. I am tired of dinner-parties, and, besides, George is not well.'

'My dear, the young urchin is far better than yourself,' replied George, dissecting a sardine with amazing relish; 'but just as you like, Nellie. There's "Mudie's last" on the sofa-table, and perhaps it is as well you should stay quiet this evening, and amuse yourself reading it.'

But 'Mudie's last' failed to possess either interest or the power of amusing me in the long, quiet evening hours, after I had fidgeted about George whilst he was dressing, until he spoiled two white ties, and played with my darlings, and heard them lisp their prayers, and sang them asleep; after the clock had struck eight, and through the open windows the echoes of footsteps in the hot, dusty street grew fewer and fewer. No, 'Mudie's last' was a failure, as far as I was concerned; and, after a faint attempt at practising an intricate *Morceau de Salon*, I lay down on my pet chintz-covered couch, near the window, to look at the sky and the stars—when they came.

The house was as still as the grave, save for the far-off sound of some of the servants' voices; for I had given leave to Harriet and the housemaid for an evening out, escorted and protected by Charles—gravest and most stupid of butlers, between whom and my maid there existed tender relations, which were to be consummated by 'the goodwill of a public' from master, and a silk wedding-dress from mistress, some happy future day.

Accordingly they had donned all their finery, and set off in high glee; at least, I had heard much giggling and rustling of ribbons, and Charles's dignified cockney accents, as he opened the area gate wide for the young ladies' crinolines, and then dead silence again. Cook and the nurse were ensconced in one of the garret windows comparing notes and chatting busily, and all the lower part of the house was left to darkness and to me.

Dead silence—and the 'ting, ting' of the little French clock on the mantelpiece marked the half-hour after eight. Dear me, how dark it was growing! this brooding storm I supposed, which had been making me feel so languid and restless. I wish it would come down and cool the air—not tonight, though. Dear me, how lonely it is! I wish George were home. Those women are talking very loudly—I wonder nurse would—here I got drowsy, and my eyes ached looking for the stars that had not come.

In a few minutes I roused again, my maternal anxiety changing into indignation as I heard the women's voices growing louder and shriller, and some doors opened and shut violently.

What can nurse be thinking of? They will wake the children most certainly, and Georgie was so long in falling asleep—quite feverish, my own boy! I shall really reprove her very plainly. I never needed to do so before. What could she be thinking of?

Dead silence again. Well, this was lonely; I was inclined to ring for lights, and turn on all the burners in the chandelier by way of company. Then I remembered there were some wax matches in one of the drawers of a writing-tray just at hand, and thought I would light the gas myself instead of bringing the servants down—yes—but—I wanted company. It was so dark and dreary, and—and—I was afraid.

Afraid to stir—afraid to get off the couch on which I was lying—afraid to look at the door! a numbing, chilling tide of icy fear ebbing through every vein—afraid to draw a breath—afraid to move hand or foot, in a nightmare of supernatural terror. At last, by a violent effort, I sprang at the bell-handle, and pulled it frantically, and as soon as I had done so, with a sudden revulsion of feeling, I felt thoroughly ashamed of my childish cowardice, although I could not have helped it, and it had overcome me as suddenly as unexpectedly. How George would have laughed at me!

There were those servants talking again, tramping about and banging the doors as before. Really, this was unbearable; cook must be in one of her fits of temper, and certainly had forgotten herself strangely.

And, as the quarrelsome tones grew louder and louder—evidently in bitter recrimination, although I could not catch a word—my own anger rose proportionately, and, forgetting loneliness and darkness in my indignant anxiety lest my children should be waked by this most unseemly behaviour of the servants, I ran hastily out of the room and up the wide staircase.

The dim light from the clouded evening sky, still further subdued by the gold and purple-stained glass of the conservatory door, streamed faintly down the steps from the first landing, and by it, just as I had ascended half way, I discovered the short, thick-set figure of the nurse rushing down—of course, in answer to my ring, I supposed.

Involuntarily I stepped aside to avoid coming in violent contact with her as she fled past. No, it was not the nurse; and the woman following her in headlong haste, sweeping by me so that the current of air from their floating dresses struck icily cold on my brow where the clammy dew of perspiration had started in great drops, was—was——Merciful Heavens! What was that tall figure, with the coarse, disordered, yellow hair, the white face, and glittering, steel-blue eyes, that glinted fiendishly on me for one dreadful instant, and then vanished? Vanished as the pursued and pursuing figures had vanished in the shadows of the wide, lofty hall, without sound of voice or footstep?

I would have cried out—would have shrieked, if every nerve had not been paralyzed. I could not doubt the evidence of my senses—if I could have done so the cold, unearthy horror which sickened my very soul would have borne its undeniable testimony that I had beheld the impersonation of the hidden curse that rested on this dwelling.

I stood there rigid and immovable, as if that blighting Medusa-glance had indeed changed me into stone.

It may have been but a very few minutes—it seemed to me a cycle of painful ages, when the light of a brightly burning lamp shone before me, and I heard the cheerful sound of the new nurse's voice in my ears:

'Come along, cook. Bless your heart, my dear! you needn't be nervous; there's no occasion. Mrs Russell, ma'am, aren't you well, ma'am?'

'No,' I said faintly, staggering to the woman's outstretched hands. 'Not down there—upstairs to the children.'

She turned as I bade her, and supported me up the stairs and into the nursery, the cook following close at my skirts, muttering fervent prayers and ejaculations.

The sight of the peacefully sleeping little ones did far more to restore me than all the essences and chafing and unlacing which the two women busily administered.

I had got suddenly ill when coming upstairs was the explanation I gave, which the cook, I plainly perceived, most thoroughly doubted, at least without the cause she suspected being assigned, which, even in the midst of my terror-stricken condition, I refrained from giving. I did not speak to the nurse either of what had happened, but I felt that she knew as well as if she had been by my side all the time. But when George returned I told him.

Distressed and alarmed on my account though he was, yet he did not, as before, refuse credence to my story. 'We must leave the house, George. I should die here very soon,' I said.

'Yes, Helen; of course we must leave if you have anything to distress or terrify you in this manner, though it does seem absurd to be driven out of one's house and home by a thing of this kind. Someone's practical joke, or a trick prompted by malice against the owner of the property in order to lessen its value. I have heard of such things often.'

'George, it is nothing of the kind,' I said earnestly; 'you know it is not.'

'No, I don't,' said George shortly and grimly, as he opened his case of revolvers, 'and I wish I did.'

The night passed away quietly, to our ears at least; but next morning when George had concluded the usual morning prayers, instead of the usual move of the servants, they remained clustered at the door, Charles with an exceedingly elongated visage standing slightly in advance of the group as spokesman.

'Please, sir and ma'am, we can't tell what to do.'

'Why, go and do your work,' retorted George, with a nervous tug at his moustache and an uneasy glance at me.

Charles shook his head slowly. 'It can't be done, sir—can't be done, ma'am. Why, no living Christian, not to speak of humble, but respectable servants,' said Charles with a flourish, quite unconscious of the nice distinction he had made, 'could stand it any longer.'

'What is the matter, pray?' said my husband.

'Ghosts, sir—spirits, sir—unclean spirits,' said Charles, in an awe-struck whisper which was re-echoed in the cook's 'Lor' 'a' mercy!' as she dodged back from the doorway with the housemaid holding fast to one of her ample sleeves, and the lady's maid holding fast to the other. The new nurse, quietly dandling the baby in her arms, was alone unmoved.

'What stories have you been listening to now?' said their master, with a slight laugh and a frown.

'No stories, sir; but what we've seen with our eyes and understanded with our ears, and—and—comprehended with our hearts,' said Charles, with an unsuccessful attempt at quoting Scripture. 'What was it as walked the floors last night between one and two, sir? What was it as talked and shrieked and run and raced? What was it as frightened the mistress on the stairs last evening?' And the whole *posse* of them turned to me, triumphantly awaiting my testimony.

I was feeling very ill, and looking so, I daresay, having struggled downstairs in order to prevent the servants having any additional confirmation of their surmises.

'That is no affair of yours,' said George gravely; 'your mistress is in delicate health, and was feeling unwell all day.'

'Will you allow me to speak, please, sir?' said the nurse, and, as her master nodded assent, she turned to the frightened group with a pleasant smile.

'You have no cause to be afraid, cook, or Mr Charles, or any of you,' said she, addressing the most important functionary first—'not in the least. I am only a servant like the rest, and here a shorter time than any one; but I think you are very foolish to unsettle yourselves in a good situation and frighten yourselves. You needn't think they'll harm you. Fear God and do your duty, and you needn't mind wandering, poor, lonely souls——'

'Lor' 'a' mercy! 'ow you do talk, Mrs Hamley!' said the cook indignantly.

'I've seen them more times than one—many and many a time, Mrs Cook; and they never harmed a hair of my head,' said the nurse, 'nor they'll never harm yours.'

'Well, then,' said the cook, packing into the hall, followed by her satellites, 'not to be made Queen Victorier of, nor Hemperor of Rooshia neither, would I stay to be frightened out of my seven senses, and made into a lunatic creature like poor Mary was!'

'Please to make better omelettes for luncheon, cook, than you did yesterday,' said George calmly, though he looked pale and angry enough, 'and leave me to deal with the ghosts—I'll settle accounts with them!'

The nurse turned quickly and looked earnestly at him: 'I would not say that, sir—God forbid,' said she in an undertone, and the next moment was singing softly and blithely as she carried the children away to their morning bath.

George and I looked at each other in silence.

'I wish we had never come into this house, dear,' I said.

'I wish from my heart that we never had, Helen,' he responded; 'but we must manage to stay the season out, at all events. It would be too absurd to run away like frightened hares, not to speak of the expense and trouble we have gone to.'

'We can get it taken off our hands without loss, perhaps,' I suggested. 'See the house-agent, George.'

'I have seen him,' he replied.

'Well?'

'Oh! all politeness and amiability, of course. Deeply regretted that we should have any occasion to find fault. No other tenants ever did. Happy to do anything in the way of clearing up this little mystery, *etcetera*. Of course he was laughing at me in his sleeve.'

Again, as after our previous alarms, days passed on and lengthened into weeks in undisturbed quietude. George had a good many business matters to arrange; the children looked as rosy and healthy as in their country home, from their constant walking and playing in the airy, pleasant parks. My own health was not very good; and Dr Winchester was kindest and wisest of grave, gentlemanly doctors; so, all things considered, we stayed in London until August—very willingly, too—and only spoke of an excursion of a few weeks to the Isle of Man as a probability in September. Only on my husband's account, I wished for any change. Something seemed to affect his health strangely, although he never complained of anything beyond the usual lassitude and want of tone which a gay London season might be expected to bequeath him. He was sleepless, frequently depressed, nervous, and irritable; and still he vehemently declared he was quite well, and seemed almost annoyed when I urged him to put his business aside for the present and leave town.

He had been induced to enter into a large mining speculation, and had, besides, some heavy money matters to arrange, connected with his sister's marriage settlements, which he expected would be required about Christmas. So, all things considered, he had some cause for looking as haggard as he did.

'It will be as well for him to leave London, Mrs Russell, as soon as he can,' said Dr Winchester at the close of one of his pleasant 'run-in' visits. 'His nerves are shaky. We men get nervous nearly as often as the ladies, though we don't confess to the fact quite so openly. A little unstrung, you know—nothing more. A few weeks in sea or mountain air will quite brace him up again.'

And as I dressed for dinner that evening, I determined that if wifely entreaties, arguments, and authority, should not fail for the first time in our wedded life, George should have the sea or mountain air without another week's delay; and, of course, I determined, likewise, to back up entreaties, arguments, and authority with the prettiest dress I could put on. I cannot tell why wives, and young wives too, will neglect their personal appearance when 'only one's husband' is present. It is unpolitic, unbecoming, and unloving; and men and husbands don't like neglect—direct or implied, be sure of that, ladies—young, middle-aged, or old.

'Your brown silk, ma'am?—it is rather cold this evening for that cream-coloured grenadine,' said Harriet, rustling at my wardrobe.

'No, Harriet, I won't have that brown, I am tired of it,' I replied. If I had said I was afraid of it, I should have kept closer to the truth. It so happened that it was this dress which I had worn on the three occasions when I had been terrified by the strange occurrences in this house; and I had acquired a superstitious aversion for this particular robe. So Harriet arrayed me in a particularly charming demi-toilette of pale yellow silk grenadine and white lace; and I felt myself to be a most amiable and affectionate little wife, as I went downstairs to await George's return for dinner.

I never sat in my pretty dressing-room alone. Truth to tell, I disliked the apartment secretly and intensely, and only for fear of troubling and displeasing George I would have shut it up from the first evening I spent in it.

He was late for dinner, and I was quite shocked to see how thin and ill he looked by the gas-light; and, as soon as it was concluded, and that by the aid of excellent coffee and a vast amount of petting, I had coaxed him into his usual smiles and good-humour, I began my petition—that he would leave town for his own sake.

He listened to me in silence, and then said, 'Very well, Helen, we will go as soon as we can get the house disposed of; I suppose you will not come back here again?'

'Oh! no, I think not,' I replied, 'we will spend the winter in Hertfordshire, in our dear old house, George.'

'Very well,' he said wearily, 'though you must know, Helen, I am not going on account of this thing. I would hardly quit my house, indeed, because of ghostly or bodily sights or sounds.'

He had started up from the couch on which he was lying, flushed and excited as he always was when the subject was mentioned, his eyes gleaming as brightly as the flashing scabbard which hung on the wall before him.

'Certainly not, dearest,' I said soothingly.

'I wish I could solve the mystery,' he pursued, more excitedly; 'I would make somebody suffer for it! One's peace destroyed, and people terrified, and servants driven away, as if one was living in the dark ages, with some cursed necromancer next door!'

'Oh! well, it is some time ago now, and the servants have got over their fright. Pray, don't distress yourself about it, dear George.'

'Ah, well—you don't—never mind,' he muttered; 'but I mean to have tangible evidence before ever I leave this house—I have sworn it!'

He was not easily roused, and I felt both surprise and alarm to see him so now, and for so inadequate a cause. I had almost fancied he had forgotten the matter, as we, by tacit consent, never alluded to it.

'Don't you allow yourself to be alarmed, Helen, that is all I care about,' he went on, pacing the floor. 'I have been half mad with anxiety on your account, for fear those idiotic servants should manage to startle you to death some dark evening—cowards, every one of them; but I mean to have someone to stay here and sit up——'

He paused suddenly, and listened, then stepped noiselessly to the door, and opening it, listened again intently.

'George,' I whispered.

He took no heed of me; but rapidly unlocking a cabinet drawer, he drew out a six-barrelled revolver, loaded and capped, and with his finger on the trigger stole softly to the door and into the hall, whither I followed him.

Everything was silent, and the hall and stairs lamps were burning clear and high. I could hear the throbbing of my own heart as I stood there watching. Suddenly we both heard heavy rapid footsteps, seemingly overhead; and then confused noises, as of struggling, and quarrelling, and sobbing, mingled in a swelling clamour which sounded now near, deafeningly near, and then far, far away; now overhead, now beside us, now beneath, undistinguishable, indescribable, and unearthly.

Then the rushing footsteps came nearer and nearer. And, clenching his teeth, while his face grew rigid and white in desperate resolve, George sprang up the staircase with a bound like a tiger.

It had all passed in less than half the time I have taken to relate it, and while I yet stood breathless and with straining eyes, George had nearly reached the last step when I saw him stagger backwards, the revolver raised in his hand.

There was a struggle, a rushing, swooping sound, two shots fired in rapid succession, a floating cloud of white smoke, through which I saw the streaming yellow hair and steel-blue eyes flash downwards, and then a shriek rang out—the dreadful cry of a man in mortal terror—a crashing fall, beneath which the house trembled to its foundations, and I saw my husband's body stretched before the conservatory door, whither he had toppled backwards—whether dead or dying I knew not.

I remember dimly hearing my own voice in agonized screams, and the terror-stricken servants hurrying from the kitchens below. I remember the kind face of my new nurse as she bravely rushed down and dispatched someone for the doctor, and made others help her to carry the senseless figure, with blood slowly dripping from the parted lips and staining the snowy linen shirt-front in great gouts and splashes, up to the chamber, where they laid him on his bed, and I, a wretched frenzied woman, knelt beside him with the sole, ceaseless prayer that brain or lips could form—'God help me!'

I remember the physician's arrival, and the grave face and low clear voice of Dr Winchester, as he made his enquiries; and then another physician summoned, and the low frightened voices, and peering frightened faces, and the lighted candles guttering away in currents of air form opening and shutting doors, and the long hours of night, and the cold grey dawning, and the heart-rending suspense, and speechless, tearless, wordless agony, and the sun rose, gloriously cloudless, smiling in radiance, as if there was not the shadow of death over the weary world beneath his rays, and I heard the verdict—

'There was scarcely a hope.'

But God was merciful to me and to him, and my darling did not die.

With a fevered brain and a shattered limb he lay there for weeks—lay there with the dark portals half open to receive him; lay there, when I could no longer watch beside him, but lay prostrate and suffering in another apartment, tended by kind relatives and friends; but at length, when the mellow sunshine, and the crisp clear air of the soft shadowy October days stole into the sick room, George was able to be dressed and sit up for an hour or two amongst the pillows of his easy-chair by the window.

And there he was, longing to be gone away from London.

'Helen, darling, weak or strong I must go,' he said in his trembling uncertain voice, and with a restless longing in his faded eyes, 'I shall never get better in this house.'

And so a few days afterwards, accompanied by the doctor and two nurses, we went down in a pleasant swift railroad journey to our dear, beautiful, peaceful home in Hertfordshire.

George never spoke of that night of horror but once, when Dr Winchester told us the story connected with Clifford House.

Thirty years before, the man who was both proprietor and tenant of Clifford House died, leaving his two daughters all he possessed.

He had been a bad man, led a bad wild life, and died in a fit brought on by drunkenness; and these two daughters, grown to womanhood, inherited with his ill-gotten gold his evil nature.

They were only half-sisters, and were believed to have been illegitimate also. The elder, a tall, masculine, strongly built woman, with masses of coarse fair hair, and bright, glittering blue eyes; and the younger, a plump, dark-haired rather pretty girl, but as treacherous, vain, and bold, as her elder sister was fierce, passionate, and cruel. They lived in this house, with only their servants, for several years after their father's death, a life of quarrelling and bickering, jealousy and heart-burnings, on various accounts. The elder strove to tyrannize over the younger, who repaid it by deceit and crafty selfishness. At length a lover came, whom the elder sister favoured; whom she loved as fiercely and rashly as such wild untamed natures do; and by falsehood and deep-laid treachery the younger sister won the man's fickle fancy from the great, harsh-featured, haughty, passionate elder one.

The elder woman soon perceived it, and there were dreadful scenes between the two sisters, when the younger taunted the elder, and the elder cursed the younger; and at length one night—when there had been a fiercer encounter of words than usual, and the dark-haired girl maddened her sister by insults, and the sudden information that she intended leaving the house in the morning, to stay with a relative until her marriage, which was to take place in one week from that time—the wronged woman, demon-possessed from that moment, waited in her dressing-room until her sister entered, and then she sprang on her, and, screaming and struggling, they both wrestled until they reached the staircase, where the younger sister, escaping for an instant, rushed wildly down, followed by her murderess, who overpowered her in spite of her frantic struggles, and with her strong, cruel, bony hands deliberately strangled her, until she lay a disfigured palpitating corpse at her feet.

The officers of justice arrested the murderess a few hours afterwards, but she died by poison self-administered on the second day of her imprisonment.

Clifford House had been shut up and silent for many a year afterwards, and when, at length, an enterprising landlord put it in habitable order, and found tenants for it again, he only found them to lose them.

Year after year passed away, its evil fame darkening with its massive masonry, for none could be found to sanctify with the sacred name and pleasures of home that dwelling blighted by an abiding curse.

'I never told you, Helen,' George said, 'although I told Dr Winchester, that from the first evening I led a haunted life in that dreadful house, and the more I struggled to disbelieve the evidence of my senses, and to keep the knowledge from you, the more unbearable it became, until I felt myself going mad. I knew I was haunted, but until that last night I had never witnessed what I dreaded day and night to see. And then, Helen, when I fired, and I saw the devilish murderess face, with its demon eyes blazing on me, and the tall unearthly figure hurrying down to meet me, dragging the other struggling, writhing figure, with her long sinewy fingers seemingly pressed around the convulsed face, then I knew it was all over with me. If there had been a flaming furnace beside me I think I should have leaped into it to escape that awful sight.'

That is years ago now. We have spent many a pleasant month in the great metropolis since, but love our country home best of all. But we never speak of that terrible time when we learned the story of Clifford House.

Was it an Illusion?

A Parson's Story

AMELIA B. EDWARDS

The facts which I am about to relate happened to myself some sixteen or eighteen years ago, at which time I served Her Majesty as an Inspector of Schools. Now, the Provincial Inspector is perpetually on the move; and I was still young enough to enjoy a life of constant travelling. There are, indeed, many less agreeable ways in which an unbeneficed parson may contrive to scorn delights and live laborious days. In remote places where strangers are scarce, his annual visit is an important event; and though at the close of a long day's work he would sometimes prefer the quiet of a country inn, he generally finds himself the destined guest of the rector or the squire. It rests with himself to turn these opportunities to account. If he makes himself pleasant, he forms agreeable friendships and sees English home-life under one of its most attractive aspects; and sometimes, even in these days of universal common-placeness, he may have the luck to meet with an adventure.

My first appointment was to a West of England district largely peopled with my personal friends and connections. It was, therefore, much to my annoyance that I found myself, after a couple of years of very pleasant work, transferred to what a policeman would call 'a new beat,' up in the North. Unfortunately for me, my new beat—a rambling, thinly populated area of something under 1,800 square miles—was three times as large as the old one, and more than proportionately unmanageable. Intersected at right angles by two ranges of barren hills and cut off to a large extent from the main lines of railway, it united about every inconvenience that a district could possess. The villages lay wide apart, often separated by long tracts of moorland; and in place of the well-warmed railway compartment and the frequent manor-house, I now spent half my time in hired vehicles and lonely country inns.

I had been in possession of this district for some three months or so, and winter was near at hand, when I paid my first visit of inspection to

Pit End, an outlying hamlet in the most northerly corner of my county, just twenty-two miles from the nearest station. Having slept overnight at a place called Drumley, and inspected Drumley schools in the morning, I started for Pit End, with fourteen miles of railway and twenty-two of hilly cross-roads between myself and my journey's end. I made, of course, all the enquiries I could think of before leaving; but neither the Drumley schoolmaster nor the landlord of the Drumley 'Feathers' knew much more of Pit End than its name. My predecessor, it seemed, had been in the habit of taking Pit End 'from the other side', the roads, though longer, being less hilly that way. That the place boasted some kind of inn was certain; but it was an inn unknown to fame, and to mine host of the 'Feathers'. Be it good or bad, however, I should have to put up at it.

Upon this scant information I started. My fourteen miles of railway journey soon ended at a place called Bramsford Road, whence an omnibus conveyed passengers to a dull little town called Bramsford Market. Here I found a horse and 'trap' to carry me on to my destination; the horse being a raw-boned grey with a profile like a camel, and the trap a ricketty high gig which had probably done commercial travelling in the days of its youth. From Bramsford Market the way lay over a succession of long hills, rising to a barren, high-level plateau. It was a dull, raw afternoon of mid-November, growing duller and more raw as the day waned and the east wind blew keener.

'How much further now, driver?' I asked, as we alighted at the foot of a longer and a stiffer hill than any we had yet passed over.

He turned a straw in his mouth, and grunted something about 'fower or foive mile by the rooad'.

And then I learned that by turning off at a point which he described as 't'owld tollus', and taking a certain footpath across the fields, this distance might be considerably shortened. I decided, therefore, to walk the rest of the way; and, setting off at a good pace, I soon left driver and trap behind. At the top of the hill I lost sight of them, and coming presently to a little road-side ruin which I at once recognized as the old toll-house, I found the footpath without difficulty. It led me across a barren slope divided by stone fences, with here and there a group of shattered sheds, a tall chimney, and a blackened cinder-mound, marking the site of a deserted mine. A light fog, meanwhile, was creeping up from the east, and the dusk was gathering fast.

Now, to lose one's way in such a place and at such an hour would be disagreeable enough, and the footpath—a trodden track already half obliterated—would be indistinguishable in the course of another ten minutes. Looking anxiously ahead, therefore, in the hope of seeing some sign of habitation, I hastened on, scaling one stone stile after another, till I all at once found myself skirting a line of park-palings. Following these, with bare boughs branching out overhead and dead leaves rustling underfoot, I came presently to a point where the path divided; here continuing to skirt the enclosure, and striking off yonder across a space of open meadow.

Which should I take?

By following the fence, I should be sure to arrive at a lodge where I could enquire my way to Pit End; but then the park might be of any extent, and I might have a long distance to go before I came to the nearest lodge. Again, the meadow-path, instead of leading to Pit End, might take me in a totally opposite direction. But there was no time to be lost in hesitation; so I chose the meadow, the further end of which was lost to sight in a fleecy bank of fog.

Up to this moment I had not met a living soul of whom to ask my way; it was, therefore, with no little sense of relief that I saw a man emerging from the fog and coming along the path. As we neared each other—I advancing rapidly; he slowly—I observed that he dragged the left foot, limping as he walked. It was, however, so dark and so misty, that not till we were within half a dozen yards of each other could I see that he wore a dark suit and an Anglican felt hat, and looked something like a dissenting minister. As soon as we were within speaking distance, I addressed him.

'Can you tell me', I said, 'if I am right for Pit End, and how far I have to go?'

He came on, looking straight before him; taking no notice of my question; apparently not hearing it.

'I beg your pardon,' I said, raising my voice; 'but will this path take me to Pit End, and if so'——

He had passed on without pausing; without looking at me; I could almost have believed, without seeing me!

I stopped, with the words on my lips; then turned to look after—perhaps, to follow—him.

But instead of following, I stood bewildered.

What had become of him? And what lad was that going up the path by which I had just come—that tall lad, half-running, half-walking, with a fishing-rod over his shoulder? I could have taken my oath that I had neither met nor passed him. Where then had he come from? And where was the man to whom I had spoken not three seconds ago, and who, at his limping pace, could not have made more than a couple of yards in the time?

My stupefaction was such that I stood quite still, looking after the lad with the fishing-rod till he disappeared in the gloom under the park-palings.

Was I dreaming?

Darkness, meanwhile, had closed in apace, and, dreaming or not dreaming, I must push on, or find myself benighted. So I hurried forward, turning my back on the last gleam of daylight, and plunging deeper into the fog at every step. I was, however, close upon my journey's end. The path ended at a turnstile; the turnstile opened upon a steep lane; and at the bottom of the lane, down which I stumbled among stones and ruts, I came in sight of the welcome glare of a blacksmith's forge.

Here, then, was Pit End. I found my trap standing at the door of the village inn; the raw-boned grey stabled for the night; the landlord watching for my arrival.

The 'Greyhound' was a hostelry of modest pretensions, and I shared its little parlour with a couple of small farmers and a young man who informed me that he 'travelled in' Thorley's Food for Cattle. Here I dined, wrote my letters, chatted awhile with the landlord, and picked up such scraps of local news as fell in my way.

There was, it seemed, no resident parson at Pit End; the incumbent being a pluralist with three small livings, the duties of which, by the help of a rotatory curate, he discharged in a somewhat easy fashion. Pit End, as the smallest and furthest off, came in for but one service each Sunday, and was almost wholly relegated to the curate. The squire was a more confirmed absentee than even the vicar. He lived chiefly in Paris, spending abroad the wealth of his Pit End coal-fields. He happened to be at home just now, the landlord said, after five years' absence; but he would be off again next week, and another five years might probably elapse before they should again see him at Blackwater Chase.

Blackwater Chase!—the name was not new to me; yet I could not remember where I had heard it. When, however, mine host went on to say that, despite his absenteeism, Mr Wolstenholme was 'a pleasant gentleman and a good landlord', and that, after all, Blackwater Chase was 'a lonesome sort of world-end place for a young man to bury himself in', then I at once remembered Phil Wolstenholme of Balliol, who, in his grand way, had once upon a time given me a general invitation to the shooting at Blackwater Chase. That was twelve years ago, when I was reading hard at Wadham, and Wolstenholme—the idol of a clique to which I did not belong—was boating, betting, writing poetry, and giving wine parties at Balliol.

Yes; I remembered all about him—his handsome face, his luxurious rooms, his boyish prodigality, his utter indolence, and the blind faith of his worshippers, who believed that he had only 'to pull himself together' in order to carry off every honour which the University had to bestow. He did take the Newdigate; but it was his first and last achievement, and he left college with the reputation of having narrowly escaped a plucking. How vividly it all came back upon my memory—the old college life, the college friendships, the pleasant time that could never come again! It was but twelve years ago; yet it seemed like half a century. And now, after these twelve years, here were Wolstenholme and I as near neighbours as in our Oxford days! I wondered if he was much changed, and whether, if changed, it were for the better or the worse. Had his generous impulses developed into sterling virtues, or had his follies hardened into vices? Should I let him know where I was, and so judge for myself? Nothing would be easier than to pencil a line upon a card tomorrow morning, and send it up to the big house. Yet, merely to satisfy a purposeless curiosity, was it worthwhile to reopen the acquaintanceship?

Thus musing, I sat late over the fire, and by the time I went to bed, I had well nigh forgotten my adventure with the man who vanished so mysteriously and the boy who seemed to come from nowhere.

Next morning, finding I had abundant time at my disposal, I did pencil that line upon my card—a mere line, saying that I believed we had known each other at Oxford, and that I should be inspecting the National Schools from nine till about eleven. And then, having dispatched it by one of my landlord's sons, I went off to my work. The day was brilliantly fine. The wind had shifted round to the north, the sun shone clear and cold, and the smoke-grimed hamlet, and the gaunt buildings clustered at the mouths of the coalpits round about, looked as bright as they could look at any time of the year. The village was built up a long hill-side; the church and schools being at the top, and the 'Greyhound' at the bottom. Looking vainly for the lane by which I had come the night before, I climbed the one rambling street, followed a path that skirted the churchyard, and found myself at the schools. These, with the teachers' dwellings, formed three sides of a quadrangle; the fourth side consisting of an iron railing and a gate. An inscribed tablet over the main entrance-door recorded how 'These school-houses were re-built by Philip Wolstenhome, Esquire: AD 18—.'

'Mr Wolstenholme, sir, is the Lord of the Manor,' said a soft, obsequious voice.

I turned, and found the speaker at my elbow, a square-built, sallow man, all in black, with a bundle of copy-books under his arm.

'You are the—the schoolmaster?' I said; unable to remember his name, and puzzled by a vague recollection of his face.

'Just so, sir. I conclude I have the honour of addressing Mr Frazer?'

It was a singular face, very pallid and anxious-looking. The eyes, too, had a watchful, almost a startled, look in them, which struck me as peculiarly unpleasant.

'Yes,' I replied, still wondering where and when I had seen him. 'My name is Frazer. Yours, I believe, is—is—,' and I put my hand into my pocket for my examination papers.

'Skelton—Ebenezer Skelton. Will you please to take the boys first, sir?'

The words were commonplace enough, but the man's manner was studiously, disagreeably deferential; his very name being given, as it were, under protest, as if too insignificant to be mentioned.

I said I would begin with the boys; and so moved on. Then, for we had stood still till now, I saw that the schoolmaster was lame. In that moment I remembered him. He was the man I met in the fog.

'I met you yesterday afternoon, Mr Skelton,' I said, as we went into the school-room.

'Yesterday afternoon, sir?' he repeated.

'You did not seem to observe me,' I said, carelessly. 'I spoke to you, in fact; but you did not reply to me.'

'But—indeed, I beg your pardon, sir—it must have been someone else,' said the schoolmaster. 'I did not go out yesterday afternoon.'

How could this be anything but a falsehood? I might have been mistaken as to the man's face; though it was such a singular face, and I had seen it quite plainly. But how could I be mistaken as to his lameness? Besides, that curious trailing of the right foot, as if the ankle was broken, was not an ordinary lameness.

I suppose I looked incredulous, for he added, hastily:

'Even if I had not been preparing the boys for inspection, sir, I should not have gone out yesterday afternoon. It was too damp and foggy. I am obliged to be careful—I have a very delicate chest.'

My dislike to the man increased with every word he uttered. I did not ask myself with what motive he went on heaping lie upon lie; it was enough that, to serve his own ends, whatever those ends might be, he did lie with unparallelled audacity.

'We will proceed to the examination, Mr Skelton,' I said, contemptuously.

He turned, if possible, a shade paler than before, bent his head silently, and called up the scholars in their order.

I soon found that, whatever his shortcomings as to veracity, Mr Ebenezer Skelton was a capital schoolmaster. His boys were uncommonly well taught, and as regarded attendance, good conduct, and the like, left nothing to be desired. When, therefore, at the end of the examination, he said he hoped I would recommend the Pit End Boys' School for the Government grant, I at once assented. And now I thought I had done with Mr Skelton for, at all events, the space of one year. Not so, however. When I came out from the Girls' School, I found him waiting at the door.

Profusely apologizing, he begged leave to occupy five minutes of my valuable time. He wished, under correction, to suggest a little improvement. The boys, he said, were allowed to play in the quadrangle, which was too small, and in various ways inconvenient; but round at the back there was a piece of waste land, half an acre of which, if enclosed, would admirably answer the purpose. So saying, he led the way to the back of the building, and I followed him.

'To whom does this ground belong?' I asked.

'To Mr Wolstenholme, sir.'

'Then why not apply to Mr Wolstenholme? He gave the schools, and I dare say he would be equally willing to give the ground.'

'I beg your pardon, sir. Mr Wolstenholme has not been over here since his return, and it is quite possible that he may leave Pit End without honouring us with a visit. I could not take the liberty of writing to him, sir.'

'Neither could I in my report suggest that the Government should offer to purchase a portion of Mr Wolstenholme's land for a playground to schools of Mr Wolstenholme's own building.' I replied. 'Under other circumstances' . . .

I stopped and looked round.

The schoolmaster repeated my last words.

'You were saying, sir—under other circumstances?'——

I looked round again.

'It seemed to me that there was someone here,' I said; 'some third person, not a moment ago.'

'I beg your pardon, sir—a third person?'

'I saw his shadow on the ground, between yours and mine.'

The schools faced due north, and we were standing immediately behind the buildings, with our backs to the sun. The place was bare, and open, and high; and our shadows, sharply defined, lay stretched before our feet.

'A—a shadow?' he faltered. 'Impossible.'

There was not a bush or a tree within half a mile. There was not a cloud in the sky. There was nothing, absolutely nothing, that could have cast a shadow.

I admitted that it was impossible, and that I must have fancied it; and so went back to the matter of the playground.

'Should you see Mr Wolstenholme,' I said, 'you are at liberty to say that I thought it a desirable improvement.'

'I am much obliged to you, sir. Thank you—thank you very much,' he said, cringing at every word. 'But—but I had hoped that you might perhaps use your influence'——

'Look there!' I interrupted. 'Is *that* fancy?'

We were now close under the blank wall of the boys' schoolroom. On this wall, lying to the full sunlight, our shadows—mine and the schoolmaster's—were projected. And there, too—no longer between his and mine, but a little way apart, as if the intruder were standing back—there, as sharply defined as if cast by lime-light on a prepared background, I again distinctly saw, though but for a moment, that third shadow. As I spoke, as I looked round, it was gone!

'Did you not see it?' I asked.

He shook his head.

'I—I saw nothing,' he said, faintly. 'What was it?'

His lips were white. He seemed scarcely able to stand.

'But you *must* have seen it!' I exclaimed. 'It fell just there—where that bit of ivy grows. There must be some boy hiding—it was a boy's shadow, I am confident.'

'A boy's shadow!' he echoed, looking round in a wild, frightened way. 'There is no place—for a boy—to hide.'

'Place or no place,' I said, angrily, 'if I catch him, he shall feel the weight of my cane!'

I searched backwards and forwards in every direction, the schoolmaster, with his scared face, limping at my heels; but, rough and irregular as the ground was, there was not a hole in it big enough to shelter a rabbit.

'But what was it?' I said, impatiently.

'An—an illusion. Begging your pardon, sir—an illusion.'

He looked so like a beaten hound, so frightened, so fawning, that I felt I could with lively satisfaction have transferred the threatened caning to his own shoulders.

'But you saw it?' I said again.

'No, sir. Upon my honour, no, sir. I saw nothing—nothing whatever.'

His looks belied his words. I felt positive that he had not only seen the shadow, but that he knew more about it than he chose to tell. I was by this time really angry. To be made the object of a boyish trick, and to be hoodwinked by the connivance of the schoolmaster, was too much. It was an insult to myself and my office.

I scarcely knew what I said; something short and stern at all events. Then, having said it, I turned my back upon Mr Skelton and the schools, and walked rapidly back to the village.

As I neared the bottom of the hill, a dog-cart drawn by a high-stepping chestnut dashed up to the door of the 'Greyhound', and the next moment I was shaking hands with Wolstenholme, of Balliol. Wolstenholme, of Balliol, as handsome as ever, dressed with the same careless dandyism, looking not a day older than when I last saw him at Oxford! He gripped me by both hands, vowed that I was his guest for the next three days, and insisted on carrying me off at once to Blackwater Chase. In vain I urged that I had two schools to inspect tomorrow ten miles the other side of Drumley; that I had a horse and trap waiting; and that my room was ordered at the 'Feathers'. Wolstenholme laughed away my objections.

'My dear fellow,' he said, 'you will simply send your horse and trap back with a message to the "Feathers", and a couple of telegrams to be dispatched to the two schools from Drumley station. Unforeseen circumstances compel you to defer those inspections till next week!'

And with this, in his masterful way, he shouted to the landlord to send my portmanteau up to the manor-house, pushed me up before him into the dog-cart, gave the chestnut his head, and rattled me off to Blackwater Chase.

It was a gloomy old barrack of a place, standing high in the midst of a sombre deer-park some six or seven miles in circumference. An avenue of oaks, now leafless, led up to the house; and a mournful heron-haunted tarn in the loneliest part of the park gave to the estate its name of Blackwater Chase. The place, in fact, was more like a border fastness than an English north-country mansion. Wolstenholme took me through the picture gallery and reception rooms after luncheon, and then for a canter round the park; and in the evening we dined at the upper end of a great oak hall hung with antlers, and armour, and antiquated weapons of warfare and sport.

'Now, tomorrow,' said my host, as we sat over our claret in front of a blazing log-fire; 'tomorrow, if we have decent weather, you shall have a day's shooting on the moors; and on Friday, if you will but be persuaded to stay a day longer, I will drive you over to Broomhead and give you a run with the Duke's hounds. Not hunt? My dear fellow, what nonsense! All our parsons hunt in this part of the world. By the way, have you ever been down a coal pit? No? Then a new experience awaits you. I'll take you down Carshalton shaft, and show you the home of the gnomes and trolls.'

'Is Carshalton one of your own mines?' I asked.

'All these pits are mine,' he replied. 'I am king of Hades, and rule the under world as well as the upper. There is coal everywhere underlying these moors. The whole place is honeycombed with shafts and galleries. One of our richest seams runs under this house, and there are upwards of forty men at work in it a quarter of a mile below our feet here every day. Another leads right away under the park, heaven only knows how far! My father began working it five-and-twenty years ago, and we have gone on working it ever since; yet it shows no sign of failing.'

'You must be as rich as a prince with a fairy godmother!'

He shrugged his shoulders.

'Well,' he said, lightly, 'I am rich enough to commit what follies I please; and that is saying a good deal. But then, to be always squandering money—always rambling about the world—always gratifying the impulse of the moment—is that happiness? I have been trying the experiment for the last ten years; and with what result? Would you like to see?'

He snatched up a lamp and led the way through a long suite of unfurnished rooms, the floors of which were piled high with packing cases of all sizes and shapes, labelled with the names of various foreign ports and the addresses of foreign agents innumerable. What did they contain? Precious marbles from Italy and Greece and Asia Minor; priceless paintings by old and modern masters; antiquities from the Nile, the Tigris, and the Euphrates; enamels from Persia, porcelain from China, bronzes from Japan, strange sculptures from Peru; arms, mosaics, ivories, wood-carvings, skins, tapestries, old Italian cabinets, painted bride-chests, Etruscan terracottas; treasures of all countries, of all ages, never even unpacked since they crossed that threshold which the master's foot had crossed but twice during the ten years it had taken to buy them! Should he ever open them, ever arrange them, ever enjoy them? Perhaps—if he became weary of wandering—if he married—if he built a gallery to receive them. If not——well, he might found and endow a museum; or leave the things to the nation. What did it matter? Collecting was like fox-hunting; the pleasure was in the pursuit, and ended with it!

We sat up late that first night, I can hardly say conversing, for Wolstenholme did the talking, while I, willing to be amused, led him on to tell me something of his wanderings by land and sea. So the time passed in stories of adventure, of perilous peaks ascended, of deserts traversed, of unknown ruins explored, of 'hairbreadth 'scapes' from icebergs and earthquakes and storms; and when at last he flung the end of his cigar into the fire and discovered that it was time to go to bed, the clock on the mantel-shelf pointed far on among the small hours of the morning.

Next day, according to the programme made out for my entertainment, we did some seven hours' partridge-shooting on the moors; and the day next following I was to go down Carshalton shaft before breakfast, and after breakfast ride over to a place some fifteen miles distant called Picts' Camp, there to see a stone circle and the ruins of a prehistoric fort.

Unused to field sports, I slept heavily after those seven hours with the guns, and was slow to wake when Wolstenholme's valet came next morning to my bedside with the waterproof suit in which I was to effect my descent into Hades.

'Mr Wolstenholme says, sir, that you had better not take your bath till you come back,' said this gentlemanly vassal, disposing the ungainly garments across the back of a chair as artistically as if he were laying out my best evening suit. 'And you will be pleased to dress warmly underneath the waterproofs, for it is very chilly in the mine.'

I surveyed the garments with reluctance. The morning was frosty, and the prospect of being lowered into the bowels of the earth, cold, fasting, and unwashed, was anything but attractive. Should I send word that I would rather not go? I hesitated; but while I was hesitating, the gentlemanly valet vanished, and my opportunity was lost. Grumbling and shivering, I got up, donned the cold and shiny suit, and went downstairs.

A murmur of voices met my ear as I drew near the breakfast-room. Going in, I found some ten or a dozen stalwart colliers grouped near the door, and Wolstenholme, looking somewhat serious, standing with his back to the fire.

'Look here, Frazer,' he said, with a short laugh, 'here's a pleasant piece of news. A fissure has opened in the bed of Blackwater tarn; the lake has disappeared in the night; and the mine is flooded! No Carshalton shaft for you today!'

'Seven foot o' wayter in Jukes's seam, an' eight in th' owd north and south galleries,' growled a huge red-headed fellow, who seemed to be the spokesman.

'An' it's the Lord's own marcy a' happened o' noight-time, or we'd be dead men all,' added another.

'That's true, my man,' said Wolstenholme, answering the last speaker. 'It might have drowned you like rats in a trap; so we may thank our stars it's no worse. And now, to work with the pumps! Lucky for us that we know what to do, and how to do it.'

So saying, he dismissed the men with a good-humoured nod, and an order for unlimited ale.

I listened in blank amazement. The tarn vanished! I could not believe it. Wolstenholme assured me, however, that it was by no means a solitary phenomenon. Rivers had been known to disappear before now, in mining districts; and sometimes, instead of merely cracking, the ground would cave in, burying not merely houses, but whole hamlets in one common ruin. The foundations of such houses were, however, generally known to be insecure long enough before the crash came; and these accidents were not therefore often followed by loss of life.

'And now,' he said, lightly, 'you may doff your fancy costume; for I shall have time this morning for nothing but business. It is not every day that one loses a lake, and has to pump it up again!'

Breakfast over, we went round to the mouth of the pit, and saw the men fixing the pumps. Later on, when the work was fairly in train, we started off across the park to view the scene of the catastrophe. Our way lay far from the house across a wooded upland, beyond which we followed a broad glade leading to the tarn. Just as we entered this glade—Wolstenholme rattling on and turning the whole affair into jest—a tall, slender lad, with a fishing-rod across his shoulder, came out from one of the side paths to the right, crossed the open at a long slant, and disappeared among the tree-trunks on the opposite side. I recognized him instantly. It was the boy whom I saw the other day, just after meeting the schoolmaster in the meadow.

'If that boy thinks he is going to fish in your tarn,' I said, 'he will find out his mistake.'

'What boy?' asked Wolstenholme, looking back.

'That boy who crossed over yonder, a minute ago.'

'Yonder!—in front of us?'

'Certainly. You must have seen him?'

'Not I.'

'You did not see him?—a tall, thin boy, in a grey suit, with a fishing-rod over his shoulder. He disappeared behind those Scotch firs.'

Wolstenholme looked at me with surprise.

'You are dreaming!' he said. 'No living thing—not even a rabbit—has crossed our path since we entered the park gates.'

'I am not in the habit of dreaming with my eyes open,' I replied, quickly.

He laughed, and put his arm through mine.

'Eyes or no eyes,' he said, 'you are under an illusion this time!'

An illusion—the very word made use of by the schoolmaster! What did it mean? Could I, in truth, no longer rely upon the testimony of my senses? A thousand half-formed apprehensions flashed across me in a moment. I remembered the illusions of Nicolini, the bookseller, and other similar cases of visual hallucination, and I asked myself if I had suddenly become afflicted in like manner.

'By Jove! this *is* a queer sight!' exclaimed Wolstenholme.

And then I found that we had emerged from the glade, and were looking down upon the bed of what yesterday was Blackwater Tarn.

It was indeed a queer sight—an oblong, irregular basin of blackest slime, with here and there a sullen pool, and round the margin an irregular fringe of bulrushes. At some little distance along the bank—less than a quarter of a mile from where we were standing—a gaping crowd had gathered. All Pit End, except the men at the pumps, seemed to have turned out to stare at the bed of the vanished tarn.

Hats were pulled off and curtsies dropped at Wolstenholme's approach. He, meanwhile, came up smiling, with a pleasant word for everyone.

'Well,' he said, 'are you looking for the lake, my friends? You'll have to go down Carshalton shaft to find it! It's an ugly sight you've come to see, anyhow!'

''Tes an ugly soight, squoire,' replied a stalwart blacksmith in a leathern apron; 'but thar's summat uglier, mebbe, than the mud, ow'r yonder.'

'Something uglier than the mud?' Wolstenholme repeated.

'Wull yo be pleased to stan' this way, squoire, an' look strite across at yon little tump o' bulrashes—doan't yo see nothin'?'

'I see a log of rotten timber sticking half in and half out of the mud,' said Wolstenholme; 'and something—a long reed, apparently . . . by Jove! I believe it's a fishing rod!'

'It *is* a fishin' rod, squoire,' said the blacksmith with rough earnestness; 'an' if yon rotten timber bayn't an unburied corpse, mun I never stroike hammer on anvil agin!'

There was a buzz of acquiescence from the bystanders. 'Twas an unburied corpse, sure enough. Nobody doubted it.

Wolstenholme made a funnel with his hands, and looked through it long and steadfastly.

'It must come out, whatever it is,' he said presently. 'Five feet of mud, do you say? Then here's a sovereign apiece for the first two fellows who wade through it and bring that object to land!'

The blacksmith and another pulled off their shoes and stockings, turned up their trousers, and went in at once.

They were over their ankles at the first plunge, and, sounding their way with sticks, went deeper at every tread. As they sank, our excitement rose. Presently they were visible from only the waist upwards. We could see their chests heaving, and the muscular efforts by which each step was gained. They were yet full twenty yards from the goal when the mud mounted to their armpits . . . a few feet more, and only their heads would remain above the surface!

An uneasy movement ran through the crowd.

'Call 'em back, vor God's sake!' cried a woman's voice.

But at this moment—having reached a point where the ground gradually sloped upwards—they began to rise above the mud as rapidly as they had sunk into it. And now, black with clotted slime, they emerge waist-high . . . now they are within three or four yards of the spot . . . and now . . . now they are there!

They part the reeds—they stoop low above the shapeless object on which all eyes are turned—they half-lift it from its bed of mud—they hesitate—lay it down again—decide, apparently, to leave it there; and turn their faces shorewards. Having come a few paces, the blacksmith remembers the fishing-rod; turns back; disengages the tangled line with some difficulty, and brings it over his shoulder.

They had not much to tell—standing, all mud from head to heel, on dry land again—but that little was conclusive. It was, in truth, an unburied corpse; part of the trunk only above the surface. They tried to lift it; but it had been so long under water, and was in so advanced a stage of decomposition, that to bring it to shore without a shutter was impossible. Being cross-questioned, they thought, from the slenderness of the form, that it must be the body of a boy.

'Thar's the poor chap's rod, anyhow,' said the blacksmith, laying it gently down upon the turf.

I have thus far related events as I witnessed them. Here, however, my responsibility ceases. I give the rest of my story at second-hand, briefly, as I received it some weeks later, in the following letter from Philip Wolstenholme:

'Blackwater Chase, Dec. 20th, 18—.

Dear Frazer, My promised letter has been a long time on the road, but I did not see the use of writing till I had something definite to tell you. I think, however, we have now found out all that we are ever likely to know about the tragedy in the tarn; and it seems that—— but, no; I will begin at the beginning.

That is to say, with the day you left the Chase, which was the day following the discovery of the body.

You were but just gone when a police inspector arrived from Drumley (you will remember that I had immediately sent a man over to the sitting magistrate); but neither the inspector nor anyone else could do anything till the remains were brought to shore, and it took us the best part of a week to accomplish this difficult operation. We had to sink no end of big stones in order to make a rough and ready causeway across the mud. This done, the body was brought over decently upon a shutter. It proved to be the corpse of a boy of perhaps fourteen or fifteen years of age. There was a fracture three inches long at the back of the skull, evidently fatal. This might, of course, have been an accidental injury; but when the body came to be raised from where it lay, it was found to be pinned down by a pitchfork, the handle of which had been afterwards whittled off, so as not to show above the water, a discovery tantamount to evidence of murder. The features of the victim were decomposed beyond recognition; but enough of the hair remained to show that it had been short and sandy. As for the clothing, it was a mere mass of rotten shreds; but on being subjected to some chemical process, proved to have once been a suit of lightish grey cloth.

A crowd of witnesses came forward at this stage of the inquiry—for I am now giving you the main facts as they came out at the coroner's inquest—to prove that about a year or thirteen months ago, Skelton the schoolmaster had staying with him a lad whom he called his nephew, and to whom it was supposed that he was not particularly kind. This lad was described as tall, thin, and sandy-haired. He habitually wore a suit corresponding in colour and texture to the shreds of clothing discovered on the body in the tarn; and he was much addicted to angling about the pools and streams, wherever he might have the chance of a nibble.

And now one thing led quickly on to another. Our Pit End shoemaker identified the boy's boots as being a pair of his own making and selling. Other witnesses testified to angry scenes between the uncle and nephew. Finally, Skelton gave himself up to justice, confessed the deed, and was duly committed to Drumley gaol for wilful murder.

And the motive? Well, the motive is the strangest part of my story. The wretched lad was, after all, not Skelton's nephew, but Skelton's own illegitimate son. The mother was dead, and the boy lived with his maternal grandmother in a remote part of Cumberland. The old woman was poor, and the schoolmaster made her an annual allowance for his son's keep and clothing. He had not seen the boy for some years, when he sent for him to come over on a visit to Pit End. Perhaps he was weary of the tax upon his purse. Perhaps, as he himself puts it in his confession, he was disappointed to find the boy, if not actually half-witted, stupid, wilful, and ill brought-up. He at all events took a dislike to the poor brute, which dislike by and by developed into positive hatred. Some amount of provocation there would seem to have been. The boy was as backward as a child of five years old. That Skelton put him into the Boys' School, and could do nothing with him; that he defied discipline, had a passion for fishing, and was continually wandering about the country with his rod and line, are facts borne out by the independent testimony of various witnesses. Having hidden his fishing-tackle, he was in the habit of slipping away at school-hours, and showed himself the more cunning and obstinate the more he was punished.

At last there came a day when Skelton tracked him to the place where his rod was concealed, and thence across the meadows into the park, and as far as the tarn. His (Skelton's) account of what followed is wandering and confused. He owns to having beaten the miserable lad about the head and arms with a heavy stick that he had brought with him for the purpose; but denies that he intended to murder him. When his son fell insensible and ceased to breathe, he for the first time realized the force of the blows he had dealt. He admits that his first impulse was one, not of remorse for the deed, but of fear for his own safety. He dragged the body in among the bulrushes by the water's edge, and there concealed it as well as he could. At night, when the neighbours were in bed and asleep, he stole out by starlight, taking with him a pitchfork, a coil of rope, a couple of old iron-bars, and a knife. Thus laden, he struck out across the moor, and entered the park by a stile and footpath on the Stoneleigh side; so making a circuit of between three and four miles. A rotten old punt used at that time to be kept on the tarn. He loosed this punt from its moorings, brought it round, hauled in the body, and paddled his ghastly burden out into the middle of the lake as far as a certain clump of reeds which he had noted as a likely spot for his purpose. Here he weighted and sunk the corpse, and pinned it down by the neck with his pitchfork. He then cut away the handle of the fork; hid the fishing-rod among the reeds; and believed, as murderers always believe, that discovery was impossible. As regarded the Pit End folk, he simply gave out that his nephew had gone back to Cumberland; and no one doubted it. Now, however, he says that accident has only anticipated him; and that he was on the point of voluntarily confessing his crime. His dreadful secret had of late become intolerable. He was haunted by an invisible Presence. That Presence sat with him at table, followed him in his walks stood behind him in the school-room, and watched by his bedside. He never saw it; but he felt that it was always there. Sometimes he raves of a shadow on the wall of his cell. The gaol authorities are of opinion that he is of unsound mind.

I have now told you all that there is at present to tell. The trial will not take place till the spring assizes. In the meanwhile I am off tomorrow to Paris, and thence, in about ten days, on to Nice, where letters will find me at the Hotel des Empereurs.

Always, dear Frazer,
Yours, &c., &c.,
P. W.

P.S.—Since writing the above, I have received a telegram from Drumley to say that Skelton has committed suicide. No particulars given. So ends this strange eventful history.

By the way, that was a curious illusion of yours the other day when we were crossing the park; and I have thought of it many times. Was it an illusion?—that is the question.'

Ay, indeed! that *is* the question; and it is a question which I have never yet been able to answer. Certain things I undoubtedly saw—with my mind's eye, perhaps—and as I saw them, I have described them; withholding nothing, adding nothing, explaining nothing. Let those solve the mystery who can. For myself, I but echo Wolstenholme's question: Was it an illusion?

The Open Door

CHARLOTTE RIDDELL

Some people do not believe in ghosts. For that matter, some people do not believe in anything. There are persons who even affect incredulity concerning that open door at Ladlow Hall. They say it did not stand wide open—that they could have shut it; that the whole affair was a delusion; that they are sure it must have been a conspiracy; that they are doubtful whether there is such a place as Ladlow on the face of the earth; that the first time they are in Meadowshire they will look it up.

That is the manner in which this story, hitherto unpublished, has been greeted by my acquaintances. How it will be received by strangers is quite another matter. I am going to tell what happened to me exactly as it happened, and readers can credit or scoff at the tale as it pleases them. It is not necessary for me to find faith and comprehension in addition to a ghost story, for the world at large. If such were the case, I should lay down my pen.

Perhaps, before going further, I ought to premise there was a time when I did not believe in ghosts either. If you had asked me one summer's morning years ago when you met me on London Bridge if I held such appearances to be probable or possible, you would have received an emphatic 'No' for answer.

But, at this rate, the story of the Open Door will never be told; so we will, with your permission, plunge into it immediately.

'Sandy!'

'What do you want?'

'Should you like to earn a sovereign?'

'Of course I should.'

A somewhat curt dialogue, but we were given to curtness in the office of Messrs Frimpton, Frampton and Fryer, auctioneers and estate agents, St Benet's Hill, City.

(My name is not Sandy or anything like it, but the other clerks so styled me because of a real or fancied likeness to some character, an ill-looking Scotchman, they had seen at the theatre. From this it may be inferred I was not handsome. Far from it. The only ugly specimen in my family, I knew I was very plain; and it chanced to be no secret to me either that I felt grievously discontented with my lot.

I did not like the occupation of clerk in an auctioneer's office, and I did not like my employers.

We are all of us inconsistent, I suppose, for it was a shock to me to find they entertained a most cordial antipathy to me.)

'Because,' went on Parton, a fellow, my senior by many years—a fellow who delighted in chaffing me, 'I can tell you how to lay hands on one.'

'How?' I asked, sulkily enough, for I felt he was having what he called his fun.

'You know that place we let to Carrison, the tea-dealer?'

Carrison was a merchant in the China trade, possessed of fleets of vessels and towns of warehouses; but I did not correct Parton's expression, I simply nodded.

'He took it on a long lease, and he can't live in it; and our governor said this morning he wouldn't mind giving anybody who could find out what the deuce is the matter, a couple of sovereigns and his travelling expenses.'

'Where is the place?' I asked, without turning my head; for the convenience of listening I had put my elbows on the desk and propped up my face with both hands.

'Away down in Meadowshire, in the heart of the grazing country.'

'And what *is* the matter?' I further enquired.

'A door that won't keep shut.'

'What?'

'A door that will keep open, if you prefer that way of putting it,' said Parton.

'You are jesting.'

'If I am, Carrison is not, or Fryer either. Carrison came here in a nice passion, and Fryer was in a fine rage; I could see he was, though he kept his temper outwardly. They have had an active correspondence it appears, and Carrison went away to talk to his lawyer. Won't make much by that move, I fancy.'

'But tell me,' I entreated, 'why the door won't keep shut?'

'They say the place is haunted.'

'What nonsense!' I exclaimed.

'Then you are just the person to take the ghost in hand. I thought so while old Fryer was speaking.'

'If the door won't keep shut,' I remarked, pursuing my own train of thought, 'why can't they let it stay open?'

'I have not the slightest idea. I only know there are two sovereigns to be made, and that I give you a present of the information.'

And having thus spoken, Parton took down his hat and went out, either upon his own business or that of his employers.

There was one thing I can truly say about our office, we were never serious in it. I fancy that is the case in most offices nowadays; at all events, it was the case in ours. We were always chaffing each other, playing practical jokes, telling stupid stories, scamping our work, looking at the clock, counting the weeks to next St Lubbock's Day, counting the hours to Saturday.

For all that we were all very earnest in our desire to have our salaries raised, and unanimous in the opinion no fellows ever before received such wretched pay. I had twenty pounds a year, which I was aware did not half provide for what I ate at home. My mother and sisters left me in no doubt on the point, and when new clothes were wanted I always hated to mention the fact to my poor worried father.

We had been better off once, I believe, though I never remember the time. My father owned a small property in the country, but owing to the failure of some bank, I never could understand what bank, it had to be mortgaged; then the interest was not paid, and the mortgagees foreclosed, and we had nothing left save the half-pay of a major, and about a hundred a year which my mother brought to the common fund.

We might have managed on our income, I think, if we had not been so painfully genteel; but we were always trying to do something quite beyond our means, and consequently debts accumulated, and creditors ruled us with rods of iron.

Before the final smash came, one of my sisters married the younger son of a distinguished family, and even if they had been disposed to live comfortably and sensibly she would have kept her sisters up to the mark. My only brother, too, was an officer, and of course the family thought it necessary he should see we preserved appearances.

It was all a great trial to my father, I think, who had to bear the brunt of the dunning and harass, and eternal shortness of money; and it would have driven me crazy if I had not found a happy refuge when matters were going wrong at home at my aunt's. She was my father's sister, and had married so 'dreadfully below her' that my mother refused to acknowledge the relationship at all.

For these reasons and others, Parton's careless words about the two sovereigns stayed in my memory.

I wanted money badly—I may say I never had sixpence in the world of my own—and I thought if I could earn two sovereigns I might buy some trifles I needed for myself, and present my father with a new umbrella. Fancy is a dangerous little jade to flirt with, as I soon discovered.

She led me on and on. First I thought of the two sovereigns; then I recalled the amount of the rent Mr Carrison agreed to pay for Ladlow Hall; then I decided he would gladly give more than two sovereigns if he could only have the ghost turned out of possession. I fancied I might get ten pounds—twenty pounds. I considered the matter all day, and I dreamed of it all night, and when I dressed myself next morning I was determined to speak to Mr Fryer on the subject.

I did so—I told that gentleman Parton had mentioned the matter to me, and that if Mr Fryer had no objection, I should like to try whether I could not solve the mystery. I told him I had been accustomed to lonely houses, and that I should not feel at all nervous; that I did not believe in ghosts, and as for burglars, I was not afraid of them.

'I don't mind your trying,' he said at last. 'Of course you understand it is no cure, no pay. Stay in the house for a week; if at the end of that time you can keep the door shut, locked, bolted, or nailed up, telegraph for me, and I will go down—if not, come back. If you like to take a companion there is no objection.'

I thanked him, but said I would rather not have a companion.

'There is only one thing, sir, I should like,' I ventured.

'And that——?' he interrupted.

'Is a little more money. If I lay the ghost, or find out the ghost, I think I ought to have more than two sovereigns.'

'How much more do you think you ought to have?' he asked.

His tone quite threw me off my guard, it was so civil and conciliatory, and I answered boldly:

'Well, if Mr Carrison cannot now live in the place perhaps he wouldn't mind giving me a ten-pound note.'

Mr Fryer turned, and opened one of the books lying on his desk. He did not look at or refer to it in any way—I saw that.

'You have been with us how long, Edlyd?' he said.

'Eleven months tomorrow,' I replied.

'And our arrangement was, I think, quarterly payments, and one month's notice on either side?'

'Yes, sir.' I heard my voice tremble, though I could not have said what frightened me.

'Then you will please to take your notice now. Come in before you leave this evening, and I'll pay you three months' salary, and then we shall be quits.'

'I don't think I quite understand,' I was beginning, when he broke in:

'But I understand, and that's enough. I have had enough of you and your airs, and your indifference, and your insolence here. I never had a clerk I disliked as I do you. Coming and dictating terms, forsooth! No, you shan't go to Ladlow. Many a poor chap'—(he said 'devil')—'would have been glad to earn half a guinea, let alone two sovereigns; and perhaps you may be before you are much older.'

'Do you mean that you won't keep me here any longer, sir?' I asked in despair. 'I had no intention of offending you. I——'

'Now you need not say another word,' he interrupted, 'for I won't bandy words with you. Since you have been in this place you have never known your position, and you don't seem able to realize it. When I was foolish enough to take you, I did it on the strength of your connections, but your connections have done nothing for me. I have never had a penny out of any one of your friends—if you have any. You'll not do any good in business for yourself or anybody else, and the sooner you go to Australia'—(here he was very emphatic)—'and get off these premises, the better I shall be pleased.'

I did not answer him—I could not. He had worked himself to a white heat by this time, and evidently intended I should leave his premises then and there. He counted five pounds out of his cash-box, and, writing a receipt, pushed it and the money across the table, and bade me sign and be off at once.

My hand trembled so I could scarcely hold the pen, but I had presence of mind enough left to return one pound ten in gold, and three shillings and fourpence I had, quite by the merest good fortune, in my waistcoat pocket.

'I can't take wages for work I haven't done,' I said, as well as sorrow and passion would let me. 'Good-morning,' and I left his office and passed out among the clerks.

I took from my desk the few articles belonging to me, left the papers it contained in order, and then, locking it, asked Parton if he would be so good as to give the key to Mr Fryer.

'What's up?' he asked 'Are you going?'

I said, 'Yes, I am going'.

'Got the sack?'

'That is exactly what has happened.'

'Well, I'm——!' exclaimed Mr Parton.

I did not stop to hear any further commentary on the matter, but bidding my fellow-clerks goodbye, shook the dust of Frimpton's Estate and Agency Office from off my feet.

I did not like to go home and say I was discharged, so I walked about aimlessly, and at length found myself in Regent Street. There I met my father, looking more worried than usual.

'Do you think, Phil,' he said (my name is Theophilus), 'you could get two or three pounds from your employers?'

Maintaining a discreet silence regarding what had passed, I answered:

'No doubt I could.'

'I shall be glad if you will then, my boy,' he went on, 'for we are badly in want of it.'

I did not ask him what was the special trouble. Where would have been the use? There was always something—gas, or water, or poor-rates, or the butcher, or the baker, or the bootmaker. Well, it did not much matter, for we were well accustomed to the life; but, I thought, 'if ever I marry, we will keep within our means'. And then there rose up before me a vision of Patty, my cousin—the blithest, prettiest, most useful, most sensible girl that ever made sunshine in poor man's house.

My father and I had parted by this time, and I was still walking aimlessly on, when all at once an idea occurred to me. Mr Fryer had not treated me well or fairly. I would hoist him on his own petard. I would go to headquarters, and try to make terms with Mr Carrison direct.

No sooner thought than done. I hailed a passing omnibus, and was ere long in the heart of the city. Like other great men, Mr Carrison was difficult of access—indeed, so difficult of access, that the clerk to whom I applied for an audience told me plainly I could not see him at all. I might send in my message if I liked, he was good enough to add, and no doubt it would be attended to. I said I should not send in a message, and was then asked what I would do. My answer was simple. I meant to wait till I did see him. I was told they could not have people waiting about the office in this way.

I said I supposed I might stay in the street. 'Carrison didn't own that,' I suggested.

The clerk advised me not to try that game, or I might get locked up.

I said I would take my chance of it.

After that we went on arguing the question at some length, and we were in the middle of a heated argument, in which several of Carrison's 'young gentlemen', as they called themselves, were good enough to join, when we were all suddenly silenced by a grave-looking individual, who authoritatively enquired:

'What is all this noise about?'

Before anyone could answer I spoke up:

'I want to see Mr Carrison, and they won't let me.'

'What do you want with Mr Carrison?'

'I will tell that to himself only.'

'Very well, say on—I am Mr Carrison.'

For a moment I felt abashed and almost ashamed of my persistency; next instant, however, what Mr Fryer would have called my 'native audacity' came to the rescue, and I said, drawing a step or two nearer to him, and taking off my hat:

'I wanted to speak to you about Ladlow Hall, if you please, sir.'

In an instant the fashion of his face changed, a look of irritation succeeded to that of immobility; an angry contraction of the eyebrows disfigured the expression of his countenance.

'Ladlow Hall!' he repeated; 'and what have you got to say about Ladlow Hall?'

'That is what I wanted to tell you, sir,' I answered, and a dead hush seemed to fall on the office as I spoke.

The silence seemed to attract his attention, for he looked sternly at the clerks, who were not using a pen or moving a finger.

'Come this way, then,' he said abruptly; and next minute I was in his private office.

'Now, what is it?' he asked, flinging himself into a chair, and addressing me, who stood hat in hand beside the great table in the middle of the room.

I began—I will say he was a patient listener—at the very beginning, and told my story straight through. I concealed nothing. I enlarged on nothing. A discharged clerk I stood before him, and in the capacity of a discharged clerk I said what I had to say. He heard me to the end, then he sat silent, thinking.

At last he spoke.

'You have heard a great deal of conversation about Ladlow, I suppose?' he remarked.

'No sir; I have heard nothing except what I have told you.'

'And why do you desire to strive to solve such a mystery?'

'If there is any money to be made, I should like to make it, sir.'

'How old are you?'

'Two-and-twenty last January.'

'And how much salary had you at Frimpton's?'

'Twenty pounds a year.'

'Humph! More than you are worth, I should say.'

'Mr Fryer seemed to imagine so, sir, at any rate,' I agreed, sorrowfully.

'But what do you think?' he asked, smiling in spite of himself.

'I think I did quite as much work as the other clerks,' I answered.

'That is not saying much, perhaps,' he observed. I was of his opinion, but I held my peace.

'You will never make much of a clerk, I am afraid,' Mr Carrison proceeded, fitting his disparaging remarks upon me as he might on a lay figure. 'You don't like desk work?'

'Not much, sir.'

'I should judge the best thing you could do would be to emigrate,' he went on, eyeing me critically.

'Mr Fryer said I had better go to Australia or——' I stopped, remembering the alternative that gentleman had presented.

'Or where?' asked Mr Carrison.

'The——, sir,' I explained, softly and apologetically.

He laughed—he lay back in his chair and laughed—and I laughed myself, though ruefully.

After all, twenty pounds was twenty pounds, though I had not thought much of the salary till I lost it.

We went on talking for a long time after that; he asked me all about my father and my early life, and how we lived, and where we lived, and the people we knew; and, in fact, put more questions than I can well remember.

'It seems a crazy thing to do,' he said at last; 'and yet I feel disposed to trust you. The house is standing perfectly empty. I can't live in it, and I can't get rid of it; all my own furniture I have removed, and there is nothing in the place except a few old-fashioned articles belonging to Lord Ladlow. The place is a loss to me. It is of no use trying to let it, and thus, in fact, matters are at a deadlock. You won't be able to find out anything, I know, because, of course, others have tried to solve the mystery ere now; still, if you like to try you may. I will make this bargain with you. If you like to go down, I will pay your reasonable expenses for a fortnight; and if you do any good for me, I will give you a ten-pound note for yourself. Of course I must be satisfied that what you have told me is true and that you are what you represent. Do you know anybody in the city who would speak for you?'

I could think of no one but my uncle. I hinted to Mr Carrison he was not grand enough or rich enough, perhaps, but I knew nobody else to whom I could refer him.

'What!' he said, 'Robert Dorland, of Cullum Street. He does business with us. If he will go bail for your good behaviour I shan't want any further guarantee. Come along.' And to my intense amazement, he rose, put on his hat, walked me across the outer office and along the pavements till we came to Cullum Street.

'Do you know this youth, Mr Dorland?' he said, standing in front of my uncle's desk, and laying a hand on my shoulder.

'Of course I do, Mr Carrison,' answered my uncle, a little apprehensively; for, as he told me afterwards, he could not imagine what mischief I had been up to. 'He is my nephew.'

'And what is your opinion of him—do you think he is a young fellow I may safely trust?'

My uncle smiled, and answered, 'That depends on what you wish to trust him with.'

'A long column of addition, for instance.'

'It would be safer to give that task to somebody else.'

'Oh, uncle!' I remonstrated; for I had really striven to conquer my natural antipathy to figures—worked hard, and every bit of it against the collar.

My uncle got off his stool, and said, standing with his back to the empty fire-grate:

'Tell me what you wish the boy to do, Mr Carrison, and I will tell you whether he will suit your purpose or not. I know him, I believe, better than he knows himself.'

In an easy, affable way, for so rich a man, Mr Carrison took possession of the vacant stool, and nursing his right leg over his left knee, answered:

'He wants to go and shut the open door at Ladlow for me. Do you think he can do that?'

My uncle looked steadily back at the speaker, and said, 'I thought, Mr Carrison, it was quite settled no one could shut it?'

Mr Carrison shifted a little uneasily on his seat, and replied: '*I* did not set your nephew the task he fancies he would like to undertake.'

'Have nothing to do with it, Phil,' advised my uncle, shortly.

'You don't believe in ghosts, do you, Mr Dorland?' asked Mr Carrison, with a slight sneer.

'Don't you, Mr Carrison?' retorted my uncle.

There was a pause—an uncomfortable pause—during the course of which I felt the ten pounds, which, in imagination, I had really spent, trembling in the scale. I was not afraid. For ten pounds, or half the money, I would have faced all the inhabitants of spirit land. I longed to tell them so; but something in the way those two men looked at each other stayed my tongue.

'If you ask me the question here in the heart of the city, Mr Dorland,' said Mr Carrison, at length, slowly and carefully, 'I answer "No"; but if you were to put it to me on a dark night at Ladlow, I should beg time to consider. I do not believe in supernatural phenomena myself, and yet—the door at Ladlow is as much beyond my comprehension as the ebbing and flowing of the sea.'

'And you can't live at Ladlow?' remarked my uncle.

'I can't live at Ladlow, and what is more, I can't get anyone else to live at Ladlow.'

'And you want to get rid of your lease?'

'I want so much to get rid of my lease that I told Fryer I would give him a handsome sum if he could induce anyone to solve the mystery. Is there any other information you desire, Mr Dorland? Because if there is, you have only to ask and have. I feel I am not here in a prosaic office in the city of London, but in the Palace of Truth.'

My uncle took no notice of the implied compliment. When wine is good it needs no bush. If a man is habitually honest in his speech and in his thoughts, he desires no recognition of the fact.

'I don't think so,' he answered; 'it is for the boy to say what he will do. If he be advised by me he will stick to his ordinary work in his employers' office, and leave ghost-hunting and spirit-laying alone.'

Mr Carrison shot a rapid glance in my direction, a glance which, implying a secret understanding, might have influenced my uncle could I have stooped to deceive my uncle.

'I can't stick to my work there any longer,' I said. 'I got my marching orders today.'

'What *had* you been doing, Phil?' asked my uncle.

'I wanted ten pounds to go and lay the ghost!' I answered, so dejectedly, that both Mr Carrison and my uncle broke out laughing.

'Ten pounds!' cried my uncle, almost between laughing and crying. 'Why, Phil boy, I had rather, poor man though I am, have given thee ten pounds than that thou should'st go ghost-hunting or ghost-laying.'

When he was very much in earnest my uncle went back to thee and thou of his native dialect. I liked the vulgarism, as my mother called it, and I knew my aunt loved to hear him use the caressing words to her. He had risen, not quite from the ranks it is true, but if ever a gentleman came ready born into the world it was Robert Dorland, upon whom at our home everyone seemed to look down.

'What will you do, Edlyd?' asked Mr Carrison; 'you hear what your uncle says, "Give up the enterprise", and what I say; I do not want either to bribe or force your inclinations.'

'I will go, sir,' I answered quite steadily. 'I am not afraid, and I should like to show you——' I stopped. I had been going to say, 'I should like to show you I am not such a fool as you all take me for', but I felt such an address would be too familiar, and refrained.

Mr Carrison looked at me curiously. I think he supplied the end of the sentence for himself, but he only answered:

'I should like you to show me that door fast shut; at any rate, if you can stay in the place alone for a fortnight, you shall have your money.'

'I don't like it, Phil,' said my uncle: 'I don't like this freak at all.'

'I am sorry for that, uncle,' I answered, 'for I mean to go.'

'When?' asked Mr Carrison.

'Tomorrow morning,' I replied.

'Give him five pounds, Dorland, please, and I will send you my cheque. You will account to me for that sum, you understand,' added Mr Carrison, turning to where I stood.

'A sovereign will be quite enough,' I said.

'You will take five pounds, and account to me for it,' repeated Mr Carrison, firmly; 'also, you will write to me every day, to my private address, and if at any monent you feel the thing too much for you, throw it up. Good afternoon,' and without more formal leavetaking he departed.

'It is of no use talking to you, Phil, I suppose?' said my uncle.

'I don't think it is,' I replied; 'you won't say anything to them at home, will you?'

'I am not very likely to meet any of them, am I?' he answered, without a shade of bitterness—merely stating a fact.

'I suppose I shall not see you again before I start,' I said, 'so I will bid you goodbye now.'

'Goodbye, my lad; I wish I could see you a bit wiser and steadier.'

I did not answer him; my heart was very full, and my eyes too. I had tried, but office-work was not in me, and I felt it was just as vain to ask me to sit on a stool and pore over writing and figures as to think a person born destitute of musical ability could compose an opera.

Of course I went straight to Patty; though we were not then married, though sometimes it seemed to me as if we never should be married, she was my better half then as she is my better half now.

She did not throw cold water on the project; she did not discourage me. What she said, with her dear face aglow with excitement, was, 'I only wish, Phil, I was going with you.' Heaven knows, so did I.

Next morning I was up before the milkman. I had told my people overnight I should be going out of town on business. Patty and I settled the whole plan in detail. I was to breakfast and dress there, for I meant to go down to Ladlow in my volunteer garments. That was a subject upon which my poor father and I never could agree; he called volunteering child's play, and other things equally hard to bear; whilst my brother, a very carpet warrior to my mind, was never weary of ridiculing the force, and chaffing me for imagining I was 'a soldier'.

Patty and I had talked matters over, and settled, as I have said, that I should dress at her father's.

A young fellow I knew had won a revolver at a raffle, and willingly lent it to me. With that and my rifle I felt I could conquer an army.

It was a lovely afternoon when I found myself walking through leafy lanes in the heart of Meadowshire. With every vein of my heart I loved the country, and the country was looking its best just then: grass ripe for the mower, grain forming in the ear, rippling streams, dreamy rivers, old orchards, quaint cottages.

'Oh that I had never to go back to London,' I thought, for I am one of the few people left on earth who love the country and hate cities. I walked on, I walked a long way, and being uncertain as to my road, asked a gentleman who was slowly riding a powerful roan horse under arching trees—a gentleman accompanied by a young lady mounted on a stiff white pony—my way to Ladlow Hall.

'That is Ladlow Hall,' he answered, pointing with his whip over the fence to my left hand. I thanked him and was going on, when he said:

'No one is living there now.'

'I am aware of that,' I answered.

He did not say anything more, only courteously bade me good-day, and rode off. The young lady inclined her head in acknowledgement of my uplifted cap, and smiled kindly. Altogether I felt pleased, little things always did please me. It was a good beginning—half-way to a good ending!

When I got to the Lodge I showed Mr Carrison's letter to the woman, and received the key.

'You are not going to stop up at the Hall alone, are you, sir?' she asked.

'Yes, I am,' I answered, uncompromisingly, so uncompromisingly that she said no more.

The avenue led straight to the house; it was uphill all the way, and bordered by rows of the most magnificent limes I ever beheld. A light iron fence divided the avenue from the park, and between the trunks of the trees I could see the deer browsing and cattle grazing. Ever and anon there came likewise to my ear the sound of a sheep-bell.

It was a long avenue, but at length I stood in front of the Hall—a square, solid-looking, old-fashioned house, three stories high, with no basement; a flight of steps up to the principal entrance; four windows to the right of the door, four windows to the left; the whole building flanked and backed with trees; all the blinds pulled down, a dead silence brooding over the place: the sun westering behind the great trees studding the park. I took all this in as I approached, and afterwards as I stood for a moment under the ample porch; then, remembering the business which had brought me so far, I fitted the great key in the lock, turned the handle, and entered Ladlow Hall.

For a minute—stepping out of the bright sunlight—the place looked to me so dark that I could scarcely distinguish the objects by which I was surrounded; but my eyes soon grew accustomed to the comparative darkness, and I found I was in an immense hall, lighted from the roof; a magnificent old oak staircase conducted to the upper rooms.

The floor was of black and white marble. There were two fireplaces, fitted with dogs for burning wood; around the walls hung pictures, antlers, and horns, and in odd niches and corners stood groups of statues, and the figures of men in complete suits of armour.

To look at the place outside, no one would have expected to find such a hall. I stood lost in amazement and admiration, and then I began to glance more particularly around.

Mr Carrison had not given me any instructions by which to identify the ghostly chamber—which I concluded would most probably be found on the first floor.

I knew nothing of the story connected with it—if there were a story. On that point I had left London as badly provided with mental as with actual luggage—worse provided, indeed, for a hamper, packed by Patty, and a small bag were coming over from the station; but regarding the mystery I was perfectly unencumbered. I had not the faintest idea in which apartment it resided. Well, I should discover that, no doubt, for myself ere long.

I looked around me—doors—doors—doors. I had never before seen so many doors together all at once. Two of them stood open—one wide, the other slightly ajar.

'I'll just shut them as a beginning,' I thought, 'before I go upstairs.'

The doors were of oak, heavy, well-fitting, furnished with good locks and sound handles. After I had closed I tried them. Yes, they were quite secure. I ascended the great staircase feeling curiously like an intruder, paced the corridors, entered the many bedchambers—some quite bare of furniture, others containing articles of an ancient fashion, and no doubt of considerable value—chairs, antique dressing-tables, curious wardrobes, and such like. For the most part the doors were closed, and I shut those that stood open before making my way into the attics.

I was greatly delighted with the attics. The windows lighting them did not, as a rule, overlook the front of the Hall, but commanded wide views over wood, and valley, and meadow. Leaning out of one, I could see, that to the right of the Hall the ground, thickly planted, shelved down to a stream, which came out into the daylight a little distance beyond the plantation, and meandered through the deer park. At the back of the Hall the windows looked out on nothing save a dense wood and a portion of the stable-yard, whilst on the side nearest the point from whence I had come there were spreading gardens surrounded by thick yew hedges, and kitchen-gardens protected by high walls; and further on a farmyard, where I could perceive cows and oxen, and, further still, luxuriant meadows, and fields glad with waving corn.

'What a beautiful place!' I said. 'Carrison must have been a duffer to leave it.' And then I thought what a great ramshackle house it was for anyone to be in all alone.

Getting heated with my long walk, I suppose, made me feel chilly, for I shivered as I drew my head in from the last dormer window, and prepared to go downstairs again.

In the attics, as in the other parts of the house I had as yet explored, I closed the doors, when there were keys locking them; when there were not, trying them, and in all cases, leaving them securely fastened.

When I reached the ground floor the evening was drawing on apace, and I felt that if I wanted to explore the whole house before dusk I must hurry my proceedings.

'I'll take the kitchens next,' I decided, and so made my way to a wilderness of domestic offices lying to the rear of the great hall. Stone passages, great kitchens, an immense servants'-hall, larders, pantries, coal-cellars, beer-cellars, laundries, brewhouses, housekeeper's room—it was not of any use lingering over these details. The mystery that troubled Mr Carrison could scarcely lodge amongst cinders and empty bottles, and there did not seem much else left in this part of the building.

I would go through the living-rooms, and then decide as to the apartments I should occupy myself.

The evening shadows were drawing on apace, so I hurried back into the hall, feeling it was a weird position to be there all alone with those ghostly hollow figures of men in armour, and the statues on which the moon's beams must fall so coldly. I would just look through the lower apartments and then kindle a fire. I had seen quantities of wood in a cupboard close at hand, and felt that beside a blazing hearth, and after a good cup of tea, I should not feel the solitary sensation which was oppressing me.

The sun had sunk below the horizon by this time, for to reach Ladlow I had been obliged to travel by cross lines of railway, and wait besides for such trains as condescended to carry third-class passengers; but there was still light enough in the hall to see all objects distinctly. With my own eyes I saw that one of the doors I had shut with my own hands was standing wide!

I turned to the door on the other side of the hall. It was as I had left it—closed. *This, then, was the room—this with the open door.* For a second I stood appalled; I think I was fairly frightened.

That did not last long, however. There lay the work I had desired to undertake, the foe I had offered to fight; so without more ado I shut the door and tried it.

'Now I will walk to the end of the hall and see what happens,' I considered. I did so. I walked to the foot of the grand staircase and back again, and looked.

The door stood wide open.

I went into the room, after just a spasm of irresolution—went in and pulled up the blinds: a good-sized room, twenty by twenty (I knew, because I paced it afterwards), lighted by two long windows.

The floor, of polished oak, was partially covered with a Turkey carpet. There were two recesses beside the fireplace, one fitted up as a bookcase, the other with an old and elaborately carved cabinet. I was astonished also to find a bedstead in an apartment so little retired from the traffic of the house; and there were also some chairs of an obsolete make, covered, so far as I could make out, with faded tapestry. Beside the bedstead, which stood against the wall opposite to the door, I perceived another door. It was fast locked, the only locked door I had as yet met with in the interior of the house. It was a dreary, gloomy room: the dark panelled walls; the black, shining floor; the windows high from the ground; the antique furniture; the dull four-poster bedstead, with dingy velvet curtains; the gaping chimney; the silk counterpane that looked like a pall.

'Any crime might have been committed in such a room,' I thought pettishly; and then I looked at the door critically.

Someone had been at the trouble of fitting bolts upon it, for when I passed out I not merely shut the door securely, but bolted it as well.

'I will go and get some wood, and then look at it again,' I soliloquized. When I came back it stood wide open once more.

'Stay open, then!' I cried in a fury. 'I won't trouble myself any more with you tonight!'

Almost as I spoke the words, there came a ring at the front door. Echoing through the desolate house, the peal in the then state of my nerves startled me beyond expression.

It was only the man who had agreed to bring over my traps. I bade him lay them down in the hall, and, while looking out some small silver, asked where the nearest post-office was to be found. Not far from the park gates, he said; if I wanted any letter sent, he would drop it in the box for me; the mail-cart picked up the bag at ten o'clock.

I had nothing ready to post then, and told him so. Perhaps the money I gave was more than he expected, or perhaps the dreariness of my position impressed him as it had impressed me, for he paused with his hand on the lock, and asked:

'Are you going to stop here all alone, master?'

'All alone,' I answered, with such cheerfulness as was possible under the circumstances.

'That's the room, you know,' he said, nodding in the direction of the open door, and dropping his voice to a whisper.

'Yes, I know,' I replied.

'What, you've been trying to shut it already, have you? Well, you are a game one!' And with this complimentary if not very respectful comment he hastened out of the house. Evidently he had no intention of proffering his services towards the solution of the mystery.

I cast one glance at the door—it stood wide open. Through the windows I had left bare to the night, moonlight was beginning to stream cold and silvery. Before I did aught else I felt I must write to Mr Carrison and Patty, so straightway I hurried to one of the great tables in the hall, and lighting a candle my thoughtful little girl had provided, with many other things, sat down and dashed off the two epistles.

Then down the long avenue, with its mysterious lights and shades, with the moonbeams glinting here and there, playing at hide-and-seek round the boles of the trees and through the tracery of quivering leaf and stem, I walked as fast as if I were doing a match against time.

It was delicious, the scent of the summer odours, the smell of the earth; if it had not been for the door I should have felt too happy. As it was——

'Look here, Phil,' I said, all of a sudden; 'life's not child's play, as uncle truly remarks. That door is just the trouble you have now to face, and you must face it! But for that door you would never have been here. I hope you are not going to turn coward the very first night. Courage!—that is your enemy—conquer it.'

'I will try,' my other self answered back. 'I can but try. I can but fail.'

The post-office was at Ladlow Hollow, a little hamlet through which the stream I had remarked dawdling on its way across the park flowed swiftly, spanned by an ancient bridge.

As I stood by the door of the little shop, asking some questions of the postmistress, the same gentleman I had met in the afternoon mounted on his roan horse, passed on foot. He wished me goodnight as he went by, and nodded familiarly to my companion, who curtseyed her acknowledgements.

'His lordship ages fast,' she remarked, following the retreating figure with her eyes.

'His lordship,' I repeated. 'Of whom are you speaking?'

'Of Lord Ladlow,' she said.

'Oh! I have never seen him,' I answered, puzzled.

'Why, *that* was Lord Ladlow!' she exclaimed.

You may be sure I had something to think about as I walked back to the Hall—something beside the moonlight and the sweet night-scents, and the rustle of beast and bird and leaf, that make silence seem more eloquent than noise away down in the heart of the country.

Lord Ladlow! my word, I thought he was hundreds, thousands of miles away; and here I find him—he walking in the opposite direction from his own home—I an inmate of his desolate abode. Hi!—what was that? I heard a noise in a shrubbery close at hand, and in an instant I was in the thick of the underwood. Something shot out and darted into the cover of the further plantation. I followed, but I could catch never a glimpse of it. I did not know the lie of the ground sufficiently to course with success, and I had at length to give up the hunt—heated, baffled, and annoyed.

When I got into the house the moon's beams were streaming down upon the hall; I could see every statue, every square of marble, every piece of armour. For all the world it seemed to me like something in a dream; but I was tired and sleepy, and decided I would not trouble about fire or food, or the open door, till the next morning: I would go to sleep.

With this intention I picked up some of my traps and carried them to a room on the first floor I had selected as small and habitable. I went down for the rest, and this time chanced to lay my hand on my rifle. *It was wet.* I touched the floor—it was wet likewise.

I never felt anything like the thrill of delight which shot through me. I had to deal with flesh and blood, and I would deal with it, heaven helping me.

The next morning broke clear and bright. I was up with the lark—had washed, dressed, breakfasted, explored the house before the postman came with my letters.

One from Mr Carrison, one from Patty, and one from my uncle: I gave the man half a crown, I was so delighted, and said I was afraid my being at the Hall would cause him some additional trouble.

'No, sir,' he answered, profuse in his expressions of gratitude; 'I pass here every morning on my way to her ladyship's.'

'Who is her ladyship?' I asked.

'The Dowager Lady Ladlow,' he answered—'the old lord's widow.'

'And where is her place?' I persisted.

'If you keep on through the shrubbery and across the waterfall, you come to the house about a quarter of a mile further up the stream.'

He departed, after telling me there was only one post a day; and I hurried back to the room in which I had breakfasted, carrying my letters with me.

I opened Mr Carrison's first. The gist of it was, 'Spare no expense; if you run short of money telegraph for it.'

I opened my uncle's next. He implored me to return; he had always thought me hair-brained, but he felt a deep interest in and affection for me, and thought he could get me a good berth if I would only try to settle down and promise to stick to my work. The last was from Patty. O Patty, God bless you! Such women, I fancy, the men who fight best in battle, who stick last to a sinking ship, who are firm in life's struggles, who are brave to resist temptation, must have known and loved. I can't tell you more about the letter, except that it gave me strength to go on to the end.

I spent the forenoon considering that door. I looked at it from within and from without. I eyed it critically. I tried whether there was any reason why it should fly open, and I found that so long as I remained on the threshold it remained closed; if I walked even so far away as the opposite side of the hall, it swung wide.

Do what I would, it burst from latch and bolt. I could not lock it because there was no key. Well, before two o'clock I confess I was baffled.

At two there came a visitor—none other than Lord Ladlow himself. Sorely I wanted to take his horse round to the stables, but he would not hear of it.

'Walk beside me across the park, if you will be so kind,' he said; 'I want to speak to you.'

We went together across the park, and before we parted I felt I could have gone through fire and water for this simple-spoken nobleman.

'You must not stay here ignorant of the rumours which are afloat,' he said. 'Of course, when I let the place to Mr Carrison I knew nothing of the open door.'

'Did you not, sir?—my lord, I mean,' I stammered.

He smiled. 'Do not trouble yourself about my title, which, indeed, carries a very empty state with it, but talk to me as you might to a friend. I had no idea there was any ghost story connected with the Hall, or I should have kept the place empty.'

I did not exactly know what to answer, so I remained silent.

'How did you chance to be sent here?' he asked, after a pause.

I told him. When the first shock was over, a lord did not seem very different from anybody else. If an emperor had taken a morning canter across the park, I might, supposing him equally affable, have spoken as familiarly to him as to Lord Ladlow. My mother always said I entirely lacked the bump of veneration!

Beginning at the beginning, I repeated the whole story, from Parton's remark about the sovereign to Mr Carrison's conversation with my uncle. When I had left London behind in the narrative, however, and arrived at the Hall, I became somewhat more reticent. After all, it was *his* Hall people could not live in—*his* door that would not keep shut; and it seemed to me these were facts he might dislike being forced upon his attention.

But he would have it. What had *I* seen? What did *I* think of the matter? Very honestly I told him I did not know what to say. The door certainly would not remain shut, and there seemed no human agency to account for its persistent opening; but then, on the other hand, ghosts generally did not tamper with firearms, and my rifle, though not loaded, had been tampered with—I was sure of that.

My companion listened attentively. 'You are not frightened, are you?' he enquired at length.

'Not now,' I answered. 'The door did give me a start last evening, but I am not afraid of that since I find someone else is afraid of a bullet.'

He did not answer for a minute; then he said:

'The theory people have set up about the open door is this: As in that room my uncle was murdered, they say the door will never remain shut till the murderer is discovered.'

'Murdered!' I did not like the word at all; it made me feel chill and uncomfortable.

'Yes—he was murdered sitting in his chair, and the assassin has never been discovered. At first many persons inclined to the belief that I killed him; indeed, many are of that opinion still.'

'But you did not, sir—there is not a word of truth in that story, is there?'

He laid his hand on my shoulder as he said:

'No, my lad; not a word. I loved the old man tenderly. Even when he disinherited me for the sake of his young wife, I was sorry, but not angry; and when he sent for me and assured me he had resolved to repair that wrong, I tried to induce him to leave the lady a handsome sum in addition to her jointure. "If you do not, people may think she has not been the source of happiness you expected," I added.

"Thank you, Hal," he said. "You are a good fellow; we will talk further about this tomorrow." And then he bade me goodnight.

'Before morning broke—it was in the summer two years ago—the household was aroused by a fearful scream. It was his death-cry. He had been stabbed from behind in the neck. He was seated in his chair writing—writing a letter to me. But for that I might have found it harder to clear myself than was in the case; for his solicitors came forward and said he had signed a will leaving all his personalty to me—he was very rich—unconditionally, only three days previously. That, of course, supplied the motive, as my lady's lawyer put it. She was very vindictive, spared no expense in trying to prove my guilt, and said openly she would never rest till she saw justice done, if it cost her the whole of her fortune. The letter lying before the dead man, over which blood had spurted, she declared must have been placed on his table by me; but the coroner saw there was an animus in this, for the few opening lines stated my uncle's desire to confide in me his reasons for changing his will—reasons, he said, that involved his honour, as they had destroyed his peace. "In the statement you will find sealed up with my will in——" At that point he was dealt his death-blow. The papers were never found, and the will was never proved. My lady put in the former will, leaving her everything. Ill as I could afford to go to law, I was obliged to dispute the matter, and the lawyers are at it still, and very likely will continue at it for years. When I lost my good name, I lost my good health, and had to go abroad; and while I was away Mr Carrison took the Hall. Till I returned, I never heard a word about the open door. My solicitor said Mr Carrison was behaving badly; but I think now I must see them or him, and consider what can be done in the affair. As for yourself, it is of vital importance to me that this mystery should be cleared up, and if you are really not timid, stay on. I am too poor to make rash promises, but you won't find me ungrateful.'

'Oh, my lord!' I cried—the address slipped quite easily and naturally off my tongue—'I don't want any more money or anything, if I can only show Patty's father I am good for something——'

'Who is Patty?' he asked.

He read the answer in my face, for he said no more.

'Should you like to have a good dog for company?' he enquired after a pause.

I hesitated; then I said:

'No, thank you. I would rather watch and hunt for myself.'

And as I spoke, the remembrance of that 'something' in the shrubbery recurred to me, and I told him I thought there had been someone about the place the previous evening.

'Poachers,' he suggested; but I shook my head.

'A girl or a woman I imagine. However, I think a dog might hamper me.'

He went away, and I returned to the house. I never left it all day. I did not go into the garden, or the stable-yard, or the shrubbery, or anywhere; I devoted myself solely and exclusively to that door.

If I shut it once, I shut it a hundred times, and always with the same result. Do what I would, it swung wide. Never, however, when I was looking at it. So long as I could endure to remain, it stayed shut—the instant I turned my back, it stood open.

About four o'clock I had another visitor; no other than Lord Ladlow's daughter—the Honourable Beatrice, riding her funny little white pony.

She was a beautiful girl of fifteen or thereabouts, and she had the sweetest smile you ever saw.

'Papa sent me with this,' she said; 'he would not trust any other messenger,' and she put a piece of paper in my hand.

'*Keep your food under lock and key; buy what you require yourself. Get your water from the pump in the stable-yard.* I am going from home; but if you want anything, go or send to my daughter.'

'Any answer?' she asked, patting her pony's neck.

'Tell his lordship, if you please, I will "keep my powder dry"!' I replied.

'You have made papa look so happy,' she said, still patting that fortunate pony.

'If it is in my power, I will make him look happier still, Miss——' and I hesitated, not knowing how to address her.

'Call me Beatrice,' she said, with an enchanting grace; then added, slily, 'Papa promises me I shall be introduced to Patty ere long,' and before I could recover from my astonishment, she had tightened the bit and was turning across the park.

'One moment, please,' I cried. 'You can do something for me.'

'What is it?' and she came back, trotting over the great sweep in front of the house.

'Lend me your pony for a minute.'

She was off before I could even offer to help her alight—off, and gathering up her habit dexterously with one hand, led the docile old sheep forward with the other.

I took the bridle—when I was with horses I felt amongst my own kind—stroked the pony, pulled his ears, and let him thrust his nose into my hand.

Miss Beatrice is a countess now, and a happy wife and mother; but I sometimes see her, and the other night she took me carefully into a conservatory and asked:

'Do you remember Toddy, Mr Edlyd?'

'Remember him!' I exclaimed; 'I can never forget him!'

'He is dead!' she told me, and there were tears in her beautiful eyes as she spoke the words. 'Mr Edlyd, *I loved Toddy*!'

Well, I took Toddy up to the house, and under the third window to the right hand. He was a docile creature, and let me stand on the saddle while I looked into the only room in Ladlow Hall I had been unable to enter.

It was perfectly bare of furniture, there was not a thing in it—not a chair or table, not a picture on the walls, or ornament on the chimney-piece.

'That is where my grand-uncle's valet slept,' said Miss Beatrice. 'It was he who first ran in to help him the night he was murdered.'

'Where is the valet?' I asked.

'Dead,' she answered. 'The shock killed him. He loved his master more than he loved himself.'

I had seen all I wished, so I jumped off the saddle, which I had carefully dusted with a branch plucked from a lilac tree; between jest and earnest pressed the hem of Miss Beatrice's habit to my lips as I arranged its folds; saw her wave her hand as she went at a hand-gallop across the park; and then turned back once again into the lonely house, with the determination to solve the mystery attached to it or die in the attempt.

Why, I cannot explain, but before I went to bed that night I drove a gimlet I found in the stables hard into the floor, and said to the door:

'Now *I* am keeping you open.'

When I went down in the morning the door was close shut, and the handle of the gimlet, broken off short, lying in the hall.

I put my hand to wipe my forehead; it was dripping with perspiration. I did not know what to make of the place at all! I went out into the open air for a few minutes; when I returned the door again stood wide.

If I were to pursue in detail the days and nights that followed, I should weary my readers. I can only say they changed my life. The solitude, the solemnity, the mystery, produced an effect I do not profess to understand, but that I cannot regret.

I have hesitated about writing of the end, but it must come, so let me hasten to it.

Though feeling convinced that no human agency did or could keep the door open, I was certain that some living person had means of access to the house which I could not discover. This was made apparent in trifles which might well have escaped unnoticed had several, or even two people occupied the mansion, but that in my solitary position it was impossible to overlook. A chair would be misplaced, for instance; a path would be visible over a dusty floor; my papers I found were moved; my clothes touched—letters I carried about with me, and kept under my pillow at night; still, the fact remained that when I went to the post-office, and while I was asleep, someone did wander over the house. On Lord Ladlow's return I meant to ask him for some further particulars of his uncle's death, and I was about to write to Mr Carrison and beg permission to have the door where the valet had slept broken open, when one morning, very early indeed, I spied a hairpin lying close beside it.

What an idiot I had been! If I wanted to solve the mystery of the open door, of course I must keep watch in the room itself. The door would not stay wide unless there was a reason for it, and most certainly a hairpin could not have got into the house without assistance.

I made up my mind what I should do—that I would go to the post early, and take up my position about the hour I had hitherto started for Ladlow Hollow. I felt on the eve of a discovery, and longed for the day to pass, that the night might come.

It was a lovely morning; the weather had been exquisite during the whole week, and I flung the hall-door wide to let in the sunshine and the breeze. As I did so, I saw there was a basket on the top step—a basket filled with rare and beautiful fruit and flowers.

Mr Carrison had let off the gardens attached to Ladlow Hall for the season—he thought he might as well save something out of the fire, he said, so my fare had not been varied with delicacies of that kind. I was very fond of fruit in those days, and seeing a card addressed to me, I instantly selected a tempting peach, and ate it a little greedily perhaps.

I might say I had barely swallowed the last morsel, when Lord Ladlow's caution recurred to me. The fruit had a curious flavour—there was a strange taste hanging about my palate. For a moment, sky, trees and park swam before my eyes; then I made up my mind what to do.

I smelt the fruit—it had all the same faint odour; then I put some in my pocket—took the basket and locked it away—walked round to the farmyard—asked for the loan of a horse that was generally driven in a light cart, and in less than half an hour was asking in Ladlow to be directed to a doctor.

Rather cross at being disturbed so early, he was at first inclined to pooh-pooh my idea; but I made him cut open a pear and satisfy himself the fruit had been tampered with.

'It is fortunate you stopped at the first peach,' he remarked, after giving me a draught, and some medicine to take back, and advising me to keep in the open air as much as possible. 'I should like to retain this fruit and see you again tomorrow.'

We did not think then on how many morrows we should see each other!

Riding across to Ladlow, the postman had given me three letters, but I did not read them till I was seated under a great tree in the park, with a basin of milk and a piece of bread beside me.

Hitherto, there had been nothing exciting in my correspondence. Patty's epistles were always delightful, but they could not be regarded as sensational; and about Mr Carrison's there was a monotony I had begun to find tedious. On this occasion, however, no fault could be found on that score. The contents of his letter greatly surprised me. He said Lord Ladlow had released him from his bargain—that I could, therefore, leave the Hall at once. He enclosed me ten pounds, and said he would consider how he could best advance my interests; and that I had better call upon him at his private house when I returned to London.

'I do not think I shall leave Ladlow yet awhile,' I considered, as I replaced his letter in its envelope. 'Before I go I should like to make it hot for whoever sent me that fruit; so unless Lord Ladlow turns me out I'll stay a little longer.'

Lord Ladlow did not wish me to leave. The third letter was from him.

'I shall return home tomorrow night,' he wrote, 'and see you on Wednesday. I have arranged satisfactorily with Mr Carrison, and as the Hall is my own again, I mean to try to solve the mystery it contains myself. If you choose to stop and help me to do so, you would confer a favour, and I will try to make it worth your while.'

'I will keep watch tonight, and see if I cannot give you some news tomorrow,' I thought. And then I opened Patty's letter—the best, dearest, sweetest letter any postman in all the world could have brought me.

If it had not been for what Lord Ladlow said about his sharing my undertaking, I should not have chosen that night for my vigil. I felt ill and languid—fancy, no doubt, to a great degree inducing these sensations. I had lost energy in a most unaccountable manner. The long, lonely days had told upon my spirits—the fidgety feeling which took me a hundred times in the twelve hours to look upon the open door, to close it, and to count how many steps I could take before it opened again, had tried my mental strength as a perpetual blister might have worn away my physical. In no sense was I fit for the task I had set myself, and yet I determined to go through with it. Why had I never before decided to watch in that mysterious chamber? Had I been at the bottom of my heart afraid? In the bravest of us there are depths of cowardice that lurk unsuspected till they engulf our courage.

The day wore on—the long, dreary day; evening approached—the night shadows closed over the Hall. The moon would not rise for a couple of hours more. Everything was still as death. The house had never before seemed to me so silent and so deserted.

I took a light, and went up to my accustomed room, moving about for a time as though preparing for bed; then I extinguished the candle, softly opened the door, turned the key, and put it in my pocket, slipped softly downstairs, across the hall, through the open door. Then I knew I had been afraid, for I felt a thrill of terror as in the dark I stepped over the threshold. I paused and listened—there was not a sound—the night was still and sultry, as though a storm were brewing. Not a leaf seemed moving—the very mice remained in their holes! Noiselessly I made my way to the other side of the room. There was an old-fashioned easy-chair between the bookshelves and the bed; I sat down in it, shrouded by the heavy curtain.

The hours passed—were ever hours so long? The moon rose, came and looked in at the windows, and then sailed away to the west; but not a sound, no, not even the cry of a bird. I seemed to myself a mere collection of nerves. Every part of my body appeared twitching. It was agony to remain still; the desire to move became a form of torture. Ah! a streak in the sky; morning at last, Heaven be praised! Had ever anyone before so welcomed the dawn? A thrush began to sing—was there ever heard such delightful music? It was the morning twilight, soon the sun would rise; soon that awful vigil would be over, and yet I was no nearer the mystery than before. Hush! what was that? *It had come.* After the hours of watching and waiting; after the long night and the long suspense, it came in a moment.

The locked door opened—so suddenly, so silently, that I had barely time to draw back behind the curtain, before I saw a woman in the room. She went straight across to the other door and closed it, securing it as I saw with bolt and lock. Then just glancing around, she made her way to the cabinet, and with a key she produced shot back the wards. I did not stir, I scarcely breathed, and yet she seemed uneasy. Whatever she wanted to do she evidently was in haste to finish, for she took out the drawers one by one, and placed them on the floor; then, as the light grew better, I saw her first kneel on the floor, and peer into every aperture, and subsequently repeat the same process, standing on a chair she drew forward for the purpose. A slight, lithe woman, not a lady, clad all in black—not a bit of white about her. What on earth could she want? In a moment it flashed upon me—THE WILL AND THE LETTER! SHE IS SEARCHING FOR THEM.

I sprang from my concealment—I had her in my grasp; but she tore herself out of my hands, fighting like a wild-cat: she bit, scratched, kicked, shifting her body as though she had not a bone in it, and at last slipped herself free, and ran wildly towards the door by which she had entered.

If she reached it, she would escape me. I rushed across the room and just caught her dress as she was on the threshold. My blood was up, and I dragged her back: she had the strength of twenty devils, I think, and struggled as surely no woman ever did before.

'I do not want to kill you,' I managed to say in gasps, 'but I will if you do not keep quiet.'

'Bah!' she cried; and before I knew what she was doing she had the revolver out of my pocket and fired.

She missed: the ball just glanced off my sleeve. I fell upon her—I can use no other expression, for it had become a fight for life, and no man can tell the ferocity there is in him till he is placed as I was then—fell upon her, and seized the weapon. She would not let it go, but I held her so tight she could not use it. She bit my face; with her disengaged hand she tore my hair. She turned and twisted and slipped about like a snake, but I did not feel pain or anything except a deadly horror lest my strength should give out.

Could I hold out much longer? She made one desperate plunge, I felt the grasp with which I held her slackening; she felt it too, and seizing her advantage tore herself free, and at the same instant fired again blindly, and again missed.

Suddenly there came a look of horror into her eyes—a frozen expression of fear.

'See!' she cried; and flinging the revolver at me, fled.

I saw, as in a momentary flash, that the door I had beheld locked stood wide—that there stood beside the table an awful figure, with uplifted hand—and then I saw no more. I was struck at last; as she threw the revolver at me she must have pulled the trigger, for I felt something like red-hot iron enter my shoulder, and I could but rush from the room before I fell senseless on the marble pavement of the hall.

When the postman came that morning, finding no one stirring, he looked through one of the long windows that flanked the door; then he ran to the farmyard and called for help.

'There is something wrong inside,' he cried. 'That young gentleman is lying on the floor in a pool of blood.'

As they rushed round to the front of the house they saw Lord Ladlow riding up the avenue, and breathlessly told him what had happened.

'Smash in one of the windows,' he said; 'and go instantly for a doctor.'

They laid me on the bed in that terrible room, and telegraphed for my father. For long I hovered between life and death, but at length I recovered sufficiently to be removed to the house Lord Ladlow owned on the other side of the Hollow.

Before that time I had told him all I knew, and begged him to make instant search for the will.

'Break up the cabinet if necessary,' I entreated, 'I am sure the papers are there.'

And they were. His lordship got his own, and as to the scandal and the crime, one was hushed up and the other remained unpunished. The dowager and her maid went abroad the very morning I lay on the marble pavement at Ladlow Hall—they never returned.

My lord made that one condition of his silence.

Not in Meadowshire, but in a fairer county still, I have a farm which I manage, and make both ends meet comfortably.

Patty is the best wife any man ever possessed—and I—well, I am just as happy if a trifle more serious than of old; but there are times when a great horror of darkness seems to fall upon me, and at such periods I cannot endure to be left alone.

The Captain of the 'Pole-star'

SIR ARTHUR CONAN DOYLE

[Being an extract from the journal of JOHN MCALISTER RAY, student of medicine, kept by him during the six months' voyage in the Arctic Seas, of the steam-whaler *Pole-star*, of Dundee, Captain Nicholas Craigie.]

September 11th. Lat. 81° 40′ N.; Long. 2° E.—Still lying-to amid enormous ice fields. The one which stretches away to the north of us, and to which our ice-anchor is attached, cannot be smaller than an English county. To the right and left unbroken sheets extend to the horizon. This morning the mate reported that there were signs of pack ice to the southward. Should this form of sufficient thickness to bar our return, we shall be in a position of danger, as the food, I hear, is already running somewhat short. It is late in the season and the nights are beginning to reappear. This morning I saw a star twinkling just over the fore-yard—the first since the beginning of May. There is considerable discontent among the crew, many of whom are anxious to get back home to be in time for the herring season, when labour always commands a high price upon the Scotch coast. As yet their displeasure is only signified by sullen countenances and black looks, but I heard from the second mate this afternoon that they contemplated sending a deputation to the Captain to explain their grievance. I much doubt how he will receive it, as he is a man of fierce temper, and very sensitive about anything approaching to an infringement of his rights. I shall venture after dinner to say a few words to him upon the subject. I have always found that he will tolerate from me what he would resent from any other member of the crew. Amsterdam Island, at the north-west corner of Spitzbergen, is visible upon our starboard quarter—a rugged line of volcanic rocks, intersected by white seams, which represent glaciers. It is curious to think that at the present moment there is probably no human being nearer to us than the Danish settlements in the south of Greenland—a good nine hundred miles as the crow flies. A captain takes a great responsibility upon himself when he risks his vessel under such circumstances. No whaler has ever remained in these latitudes till so advanced a period of the year.

9 P.M. I have spoken to Captain Craigie, and though the result has been hardly satisfactory, I am bound to say that he listened to what I

had to say very quietly and even deferentially. When I had finished he put on that air of iron determination which I have frequently observed upon his face, and paced rapidly backwards and forwards across the narrow cabin for some minutes. At first I feared that I had seriously offended him, but he dispelled the idea by sitting down again, and putting his hand upon my arm with a gesture which almost amounted to a caress. There was a depth of tenderness too in his wild dark eyes which surprised me considerably. 'Look here, Doctor', he said, 'I'm sorry I ever took you—I am indeed—and I would give fifty pounds this minute to see you standing safe upon the Dundee quay. It's hit or miss with me this time. There are fish to the north of us. How dare you shake your head, sir, when I tell you I saw them blowing from the masthead!'—this in a sudden burst of fury, though I was not conscious of having shown any signs of doubt. 'Two and twenty fish in as many minutes as I am a living man, and not one under ten foot.* Now, Doctor, do you think I can leave the country when there is only one infernal strip of ice between me and my fortune. If it came on to blow from the north tomorrow we could fill the ship and be away before the frost could catch us. If it came on to blow from the south—well, I suppose, the men are paid for risking their lives, and as for myself it matters but little to me, for I have more to bind me to the other world than to this one. I confess that I am sorry for *you*, though. I wish I had old Angus Tait who was with me last voyage, for he was a man that would never be missed, and you—you said once that you were engaged, did you not?'

'Yes,' I answered, snapping the spring of the locket which hung from my watch-chain, and holding up the little vignette of Flora.

'Blast you!' he yelled, springing out of his seat, with his very beard bristling with passion. 'What is your happiness to me? What have I to do with her that you must dangle her photograph before my eyes?' I almost thought that he was about to strike me in the frenzy of his rage, but with another imprecation he dashed open the door of the cabin and rushed out upon deck, leaving me considerably astonished at his extraordinary violence. It is the first time that he has ever shown me anything but courtesy and kindness. I can hear him pacing excitedly up and down overhead as I write these lines.

I should like to give a sketch of the character of this man, but it seems presumptuous to attempt such a thing upon paper, when the idea in my own mind is at best a vague and uncertain one. Several

* A whale is measured among whalers not by the length of its body, but by the length of its whalebone.

times I have thought that I grasped the clue which might explain it, but only to be disappointed by his presenting himself in some new light which would upset all my conclusions. It may be that no human eye but my own shall ever rest upon these lines, yet as a psychological study I shall attempt to leave some record of Captain Nicholas Craigie.

A man's outer case generally gives some indication of the soul within. The Captain is tall and well formed, with dark, handsome face, and a curious way of twitching his limbs, which may arise from nervousness, or be simply an outcome of his excessive energy. His jaw and whole cast of countenance is manly and resolute, but the eyes are the distinctive feature of his face. They are of the very darkest hazel, bright and eager, with a singular mixture of recklessness in their expression, and of something else which I have sometimes thought was more allied with horror than any other emotion. Generally the former predominated, but on occasions, and more particularly when he was thoughtfully inclined, the look of fear would spread and deepen until it imparted a new character to his whole countenance. It is at these times that he is most subject to tempestuous fits of anger, and he seems to be aware of it, for I have known him lock himself up so that no one might approach him until his dark hour was passed. He sleeps badly, and I have heard him shouting during the night, but his room is some little distance from mine, and I could never distinguish the words which he said.

This is one phase of his character, and the most disagreeable one. It is only through my close association with him, thrown together as we are day after day, that I have observed it. Otherwise he is an agreeable companion, well read and entertaining, and as gallant a seaman as ever trod a deck. I shall not easily forget the way in which he handled the ship when we were caught by a gale among the loose ice at the beginning of April. I have never seen him so cheerful, and even hilarious, as he was that night as he paced backwards and forwards upon the bridge amid the flashing of the lightning and the howling of the wind. He has told me several times that the thought of death was a pleasant one to him, which is a sad thing for a young man to say; he cannot be much more than thirty, though his hair and moustache are already slightly grizzled. Some great sorrow must have overtaken him and blighted his whole life. Perhaps I should be the same if I lost my Flora—God knows! I think if it were not for her that I should care very little whether the wind blew from the north or the south tomorrow. There, I hear him come down the companion and he has locked himself up in his room, which shows that he is still in an amiable mood. And so to bed, as old Pepys would say, for the candle is burning down (we have to use them now since the nights are closing in), and the steward has turned in, so there are no hopes of another one.

September 12th. Calm clear day, and still lying in the same position. What wind there is comes from the south-east, but it is very slight. Captain is in a better humour, and apologized to me at breakfast for his rudeness. He still looks somewhat *distrait*, however, and retains that wild look in his eyes which in a Highlander would mean that he was 'fey'—at least so our chief engineer remarked to me, and he has some reputation among the Celtic portion of our crew as a seer and expounder of omens.

It is strange that superstition should have obtained such mastery over this hard-headed and practical race. I could not have believed to what an extent it is carried had I not observed it for myself. We have had a perfect epidemic of it this voyage, until I have felt inclined to serve out rations of sedatives and nerve tonics with the Saturday allowance of grog. The first symptom of it was that shortly after leaving Shetland the men at the wheel used to complain that they heard plaintive cries and screams in the wake of the ship, as if something were following it and were unable to overtake it. This fiction has been kept up during the whole voyage, and on dark nights at the beginning of the seal-fishing it was only with great difficulty that men could be induced to do their spell. No doubt what they heard was either the creaking of the rudder-chains, or the cry of some passing sea-bird. I have been fetched out of bed several times to listen to it, but I need hardly say that I was never able to distinguish anything unnatural. The men, however, are so absurdly positive upon the subject that it is hopeless to argue with them. I mentioned the matter to the Captain once, but to my surprise he took it very gravely, and indeed appeared to be considerably disturbed by what I told him. I should have thought that he at least would have been above such vulgar delusions.

All this disquisition upon superstition leads me up to the fact that Mr Manson, our second mate, saw a ghost last night—or, at least, says that he did, which of course is the same thing. It is quite refreshing to have some new topic of conversation after the eternal routine of bears and whales which has served us for so many months. Manson swears the ship is haunted, and that he would not stay in her a day if he had any other place to go to. Indeed the fellow is honestly frightened, and I had to give him some chloral and bromide of potassium this morning to steady him down. He seemed quite indignant when I suggested that he had been having an extra glass the night before, and I was obliged to pacify him by keeping as grave a countenance as possible during his story, which he certainly narrated in a very straightforward and matter-of-fact way.

'I was on the bridge,' he said, 'about four bells in the middle watch, just when the night was at its darkest. There was a bit of a moon, but the clouds were blowing across it so that you couldn't see far from the ship. John McLeod, the harpooner, came aft from the foc'sle-head and reported a strange noise on the starboard bow. I went forrard and we both heard it, sometimes like a bairn crying and sometimes like a wench in pain. I've been seventeen years to the country and I never heard seal, old or young, make a sound like that. As we were standing there on the foc'sle-head the moon came out from behind a cloud, and we both saw a sort of white figure moving across the ice field in the same direction that we had heard the cries. We lost sight of it for a while, but it came back on the port bow, and we could just make it out like a shadow on the ice. I sent a hand aft for the rifles, and McLeod and I went down on to the pack, thinking that maybe it might be a bear. When we got on the ice I lost sight of McLeod, but I pushed on in the direction where I could still hear the cries. I followed them for a mile or maybe more, and then running round a hummock I came right on to the top of it standing and waiting for me seemingly. I don't know what it was. It wasn't a bear any way. It was tall and white and straight, and if it wasn't a man nor a woman, I'll stake my davy it was something worse. I made for the ship as hard as I could run, and precious glad I was to find myself aboard. I signed articles to do my duty by the ship, and on the ship I'll stay, but you don't catch me on the ice again after sundown.'

That is his story given as far as I can in his own words. I fancy what he saw must in spite of his denial, have been a young bear erect upon its hind legs, and attitude which they often assume when alarmed. In the uncertain light this would bear a resemblance to a human figure, especially to a man whose nerves were already somewhat shaken. Whatever it may have been, the occurrence is unfortunate, for it has produced a most unpleasant effect upon the crew. Their looks are more sullen than before and their discontent more open. The double grievance of being debarred from the herring fishing and of being detained in what they choose to call a haunted vessel, may lead them to do something rash. Even the harpooners, who are the oldest and steadiest among them, are joining in the general agitation.

Apart from this absurd outbreak of superstition, things are looking rather more cheerful. The pack which was forming to the south of us has partly cleared away, and the water is so warm as to lead me to believe that we are lying in one of those branches of the gulf-stream which run up between Greenland and Spitzbergen. There are numerous small Medusæ and sealemons about the ship, with abundance of shrimps, so that there is every possibility of 'fish' being sighted. Indeed one was seen blowing about dinner-time, but in such a position that it was impossible for the boats to follow it.

September 13th. Had an interesting conversation with the chief mate Mr Milne upon the bridge. It seems that our Captain is as great an enigma to the seamen, and even to the owners of the vessel, as he has been to me. Mr Milne tells me that when the ship is paid off, upon returning from a voyage, Captain Craigie disappears, and is not seen again until the approach of another season, when he walks quietly into the office of the company, and asks whether his services will be required. He has no friend in Dundee, nor does anyone pretend to be acquainted with his early history. His position depends entirely upon his skill as a seaman, and the name for courage and coolness which he had earned in the capacity of mate, before being entrusted with a separate command. The unanimous opinion seems to be that he is not a Scotchman, and that his name is an assumed one. Mr Milne thinks that he has devoted himself to whaling simply for the reason that it is the most dangerous occupation which he could select, and that he courts death in every possible manner. He mentioned several instances of this, one of which is rather curious, if true. It seems that on one occasion he did not put in an appearance at the office, and a substitute had to be selected in his place. That was at the time of the last Russian and Turkish war. When he turned up again next spring he had a puckered wound in the side of his neck which he used to endeavour to conceal with his cravat. Whether the mate's inference that he had been engaged in the war is true or not I cannot say. It was certainly a strange coincidence.

The wind is veering round in an easterly direction, but is still very slight. I think the ice is lying closer than it did yesterday. As far as the eye can reach on every side there is one wide expanse of spotless white, only broken by an occasional rift or the dark shadow of a hummock. To the south there is the narrow lane of blue water which is our sole means of escape, and which is closing up every day. The Captain is taking a heavy responsibility upon himself. I hear that the tank of potatoes has been finished, and even the biscuits are running short, but he preserves the same impossible countenance and spends the greater part of the day at the crow's nest, sweeping the horizon with his glass. His manner is very variable, and he seems to avoid my society, but there has been no repetition of the violence which he showed the other night.

7.30 P.M. My deliberate opinion is that we are commanded by a madman. Nothing else can account for the extraordinary vagaries of Captain Craigie. It is fortunate that I have kept this journal of our voyage, as it will serve to justify us in case we have to put him under any sort of restraint, a step which I should only consent to as a last resource. Curiously enough it was he himself who suggested lunacy and not mere eccentricity as the secret of his strange conduct. He was standing upon the bridge about an hour ago, peering as usual through his glass, while I was walking up and down the quarterdeck. The majority of the men were below at their tea, for the watches have not been regularly kept of late. Tired of walking, I leaned against the bulwarks, and admired the mellow glow cast by the sinking sun upon the great ice fields which surround us. I was suddenly aroused from the reverie into which I had fallen by a hoarse voice at my elbow, and starting round I found that the Captain had descended and was standing by my side. He was staring out over the ice with an expression in which horror, surprise, something approaching to joy were contending for the mastery. In spite of the cold, great drops of perspiration were coursing down his forehead and he was evidently fearfully excited. His limbs twitched like those of a man upon the verge of an epileptic fit, and the lines about his mouth were drawn and hard.

'Look!' he gasped, seizing me by the wrist, but still keeping his eyes upon the distant ice, and moving his head slowly in a horizontal direction, as if following some object which was moving across the field of vision. 'Look! There, man, there! Between the hummocks! Now coming out from behind the far one! You see her, you *must* see her! There still! Flying from me, by God, flying from me—and gone!'

He uttered the last two words in a whisper of concentrated agony which shall never fade from my remembrance. Clinging to the ratlines he endeavoured to climb up upon the top of the bulwarks as if in the hope of obtaining a last glance at the departing object. His strength was not equal to the attempt, however, and he staggered back against the saloon skylights, where he leaned panting and exhausted. His face was so livid that I expected him to become unconscious, so lost no time leading him down the companion, and stretching him upon one of the sofas in the cabin. I then poured him out some brandy which I held to his lips, and which had a wonderful effect upon him, bringing the blood back into his white face and steadying his poor shaking limbs. He rasied himself up upon his elbow, and looking round to see that we were alone, he beckoned to me to come and sit beside him.

'You saw it, didn't you?' he asked, still in the same subdued awesome tone so foreign to the nature of the man.

'No, I saw nothing.'

His head sank back again upon the cushions. 'No, he wouldn't without the glass,' he murmured. 'He couldn't. It was the glass that showed her to me, and then the eyes of love—the eyes of love. I say, Doc, don't let the steward in! He'll think I'm mad. Just bolt the door, will you!'

I rose and did what he had commanded.

He lay quiet for a little, lost in thought apparently, and then raised himself up upon his elbow again, and asked for some more brandy.

'You don't think I am, do you, Doc?' he asked as I was putting the bottle back into the after-locker. 'Tell me now, as man to man, do you think that I am mad?'

'I think you have something on your mind,' I answered, 'which is exciting you and doing you a good deal of harm.'

'Right there, lad!' he cried, his eyes sparkling from the effects of the brandy. 'Plenty on my mind—plenty! But I can work out the latitude and the longitude, and I can handle my sextant and manage my logarithms. You couldn't prove me mad in a court of law, could you, now?' It was curious to hear the man lying back and coolly arguing out the question of his own sanity.

'Perhaps not,' I said, 'but still I think you would be wise to get home as soon as you can and settle down to a quiet life for a while.'

'Get home, eh?' he muttered with a sneer upon his face. 'One word for me and two for yourself, lad. Settle down with Flora—pretty little Flora. Are bad dreams signs of madness?'

'Sometimes,' I answered.

'What else? what would be the first symptoms?'

'Pains in the head, noises in the ears, flashes before the eyes, delusions——'

'Ah! what about them?' he interrupted. 'What would you call a delusion?'

'Seeing a thing which is not there is a delusion.'

'But she *was* there!' he groaned to himself. 'She *was* there!' and rising, he unbolted the door and walked with slow and uncertain steps to his own cabin, where I have no doubt that he will remain until tomorrow morning. His system seems to have received a terrible shock, whatever it may have been that he imagined himself to have seen. The man becomes a greater mystery every day, though I fear that the solution which he has himself suggested is the correct one, and that his reason is affected. I do not think that a guilty conscience has anything to do with his behaviour. The idea is a popular one among the officers, and, I believe, the crew; but I have seen nothing to support it. He has not the air of a guilty man, but of one who has had terrible usage at the hands of fortune, and who should be regarded as a martyr rather than a criminal.

The wind is veering round to the south tonight. God help us if it blocks that narrow pass which is our only road to safety! Situated as we are on the edge of the main Arctic pack, or the 'barrier' as it is called by the whalers, any wind from the north has the effect of shredding out the ice around us and allowing our escape, while a wind from the south blows up all the loose ice behind us and hems us in between two packs. God help us, I say again!

September 14th.—Sunday, and a day of rest. My fears have been confirmed, and the thin strip of blue water has disappeared from the southward. Nothing but the great motionless ice fields around us, with their weird hummocks and fantastic pinnacles. There is a deathly silence over their wide expanse which is horrible. No lapping of the waves now, no cries of seagulls or straining of sails, but one deep universal silence in which the murmurs of the seamen, and the creak of their boots upon the white shining deck, seem discordant and out of place. Our only visitor was an Arctic fox, a rare animal upon the pack, though common enough upon the land. He did not come near the ship, however, but after surveying us from a distance fled rapidly across the ice. This was curious conduct, as they generally know nothing of man, and being of an inquisitive nature become so familiar that they are easily captured. Incredible as it may seem, even this little incident produced a bad effect upon the crew. 'Yon puir beastie kens mair, aye an' sees mair nor you nor me!' was the comment of one of the leading harpooners, and the others nodded their acquiescence. It is vain to attempt to argue against such puerile superstition. They have made up their minds that there is a curse upon the ship, and nothing will ever persuade them to the contrary.

The Captain remained in seclusion all day except for about half an hour in the afternoon, when he came out upon the quarterdeck. I observed that he kept his eye fixed upon the spot where the vision of yesterday had appeared, and was quite prepared for another outburst, but none such came. He did not seem to see me although I was standing close beside him. Divine service was read as usual by the chief engineer. It is a curious thing that in whaling vessels the Church of England Prayer-book is always employed, although there is never a member of that Church among either officers or crew. Our men are all Roman Catholics or Presbyterians, the former predominating. Since a ritual is used which is foreign to both, neither can complain that the other is preferred to them, and they listen with all attention and devotion, so that the system has something to recommend it.

A glorious sunset, which made the great fields of ice look like a lake of blood. I have never seen a finer and at the same time more ghastly effect. Wind is veering round. If it will blow twenty-four hours from the north all will yet be well.

September 15th. Today is Flora's birthday. Dear lass! it is well that she cannot see her boy, as she used to call me, shut up among the ice fields with a crazy captain and a few weeks' provisions. No doubt she scans the shipping list in the *Scotsman* every morning to see if we are reported from Shetland. I have to set an example to the men and look cheery and unconcerned; but God knows my heart is very heavy at times.

The thermometer is at nineteen Fahrenheit today. There is but little wind, and what there is comes from an unfavourable quarter. Captain is in an excellent humour; I think he imagines he has seen some other omen or vision, poor fellow, during the night, for he came into my room early in the morning, and stooping down over my bunk whispered, 'It wasn't a delusion, Doc, it's all right!' After breakfast he asked me to find out how much food was left, which the second mate and I proceeded to do. It is even less than we had expected. Forward they have half a tank full of biscuits, three barrels of salt meat, and a very limited supply of coffee beans and sugar. In the after-hold and lockers there are a good many luxuries such as tinned salmon, soups, haricot mutton, etc., but they will go a very short way among a crew of fifty men. There are two barrels of flour in the store-room, and an unlimited supply of tobacco. Altogether there is about enough to keep the men on half rations for eighteen or twenty days—certainly not more. When we reported the state of things to the Captain, he ordered all hands to be piped, and addressed them from the quarterdeck. I never saw him to better advantage. With his tall, well-knit figure and dark animated face, he seemed a man born to command, and he discussed the situation in a cool sailor-like way which showed that while appreciating the danger he had an eye for every loophole of escape.

'My lads,' he said, 'no doubt you think I brought you into this fix, if it is a fix, and maybe some of you feel bitter against me on account of it. But you must remember that for many a season no ship that comes to the country has brought in as much oil-money as the old *Pole-star*, and every one of you has had his share of it. You can leave your wives behind you in comfort while other poor fellows come back to find their lasses on the parish. If you have to thank me for the one you have to thank me for the other, and we may call it quits. We've tried a bold venture before this and succeeded, so now that we've tried one and failed we've no cause to cry out about it. If the worst comes to the worst, we can make the land across the ice, and lay in a stock of seals which will keep us alive until the spring. It won't come to that, though, for you'll see the Scotch coast again before three weeks are out. At present every man must go on half rations, share and share alike, and no favour to any. Keep up your hearts and you'll pull through this as you've pulled through many a danger before. These few simple words of his had a wonderful effect upon the crew. His former unpopularity was forgotten, and the old harpooner whom I have already mentioned for his superstition, led off three cheers, which were heartily joined in by all hands.

September 16th.—The wind has veered round to the north during the night, and the ice shows some symptoms of opening out. The men are in a good humour in spite of the short allowance upon which they have been placed. Steam is kept up in the engine-room, that there may be no delay should an opportunity for escape present itself. The Captain is in exuberant spirits, though he still retains that wild 'fey' expression which I have already remarked upon. This burst of cheerfulness puzzles me more than his former gloom. I cannot understand it. I think I mentioned in an early part of this journal that one of his oddities is that he never permits any person to enter his cabin, but insists upon making his own bed, such as it is, and performing every other office for himself. To my surprise he handed me the key today and requested me to go down there and take the time by his chronometer while he measured the altitude of the sun at noon. It is a bare little room containing a washing-stand and a few books, but little else in the way of luxury, except some pictures upon the walls. The majority of these are small cheap oleographs, but there was one water-colour sketch of the head of a young lady which arrested my attention. It was evidently a portrait, and not one of those fancy types of female beauty which sailors particularly affect. No artist could have evolved from his own mind such a curious mixture of character and weakness. The languid, dreamy eyes with their drooping lashes, and the broad, low brow unruffled by thought or care, were in strong contrast with the clean-cut, prominent jaw, and the resolute set of the lower lip. Underneath it in one of the corners was written 'M. B., æt. 19.' That any one in the short space of nineteen years of existence could develop such strength of will as was stamped upon her face seemed to me at the time to be well-nigh incredible. She must have been an extraordinary woman. Her features have thrown such a glamour over me that though I had but a fleeting glance at them, I could, were I a draughtsman, reproduce them line for line upon this page of the journal. I wonder what part she has played in our Captain's life. He has hung her picture at the end of his berth so that his eyes continually rest upon it. Were he a less reserved man I should make some remark upon the subject. Of the other things in his cabin there was nothing worthy of mention—uniform coats, a camp stool, small looking-glass, tobacco box and numerous pipes, including an oriental hookah—which by-the-bye gives some colour to Mr Milne's story about his participation in the war, though the connection may seem rather a distant one.

11.20 P.M. Captain just gone to bed after a long and interesting conversation on general topics. When he chooses he can be a most fascinating companion, being remarkably well read, and having the power of expressing his opinion forcibly without appearing to be dogmatic. I hate to have my intellectual toes trod upon. He spoke about the nature of the soul and sketched out the views of Aristotle and Plato upon the subject in a masterly manner. He seems to have a leaning for metempsychosis and the doctrines of Pythagoras. In discussing them we touched upon modern spiritualism, and I made some joking allusion to the impostures of Slade, upon which, to my surprise, he warned me most impressively against confusing the innocent with the guilty, and argued that it would be as logical to brand Christianity as an error, because Judas who professed that religion was a villain. He shortly afterwards bade me goodnight and retired to his room.

The wind is freshening up, and blows steadily from the north. The nights are as dark now as they are in England. I hope tomorrow may set us free from our frozen fetters.

September 17th. The Bogie again. Thank Heaven that I have strong nerves! The superstition of these poor fellows, and the circumstantial accounts which they give, with the utmost earnestness and self conviction, would horrify any man not accustomed to their ways. There are many versions of the matter, but the sum-total of them all is that something uncanny has been flitting round the ship all night, and that Sandie McDonald of Peterhead and 'lang' Peter Williamson of Shetland saw it, as also did Mr Milne on the bridge—so having three witnesses, they can make a better case of it than the second mate did. I spoke to Milne after breakfast and told him that he should be above such nonsense, and that as an officer he ought to set the men a better example. He shook his weatherbeaten head ominously, but answered with characteristic caution, 'Mebbe aye, mebbe na, Doctor,' he said; 'I didna ca' it a ghaist. I canna' say I preen my faith in sea bogles an' the like, though there's a mony as claims to ha' seen a' that and waur. I'm no easy feared, but may be your ain bluid would run a bit cauld, mun, if instead o' speerin' aboot it in daylicht ye were wi' me last night, an' seed an awfu' like shape, white an' gruesome, whiles here, whiles there, an' it greetin' and ca'ing in the darkness like a bit lambie that hae lost its mither. Ye would na' be sae ready to put it a' doon to auld wives' clavers then, I'm thinkin'.' I saw it was hopeless to reason with him, so contented myself with begging him as a personal favour to call me up the next time the spectre appeared—a request to which he acceded with many ejaculations expressive of his hopes that such an opportunity might never arise.

As I had hoped, the white desert behind us has become broken by many thin streaks of water which intersect it in all directions. Our latitude today was 80° 52′ N., which shows that there is a strong southerly drift upon the pack. Should the wind continue favourable it will break up as rapidly as it formed. At present we can do nothing but smoke and wait and hope for the best. I am rapidly becoming a fatalist. When dealing with such uncertain factors as wind and ice a man can be nothing else. Perhaps it was the wind and sand of the Arabian deserts which gave the minds of the original followers of Mahomet their tendency to bow to kismet.

These spectral alarms have a very bad effect upon the Captain. I feared that it might excite his sensitive mind, and endeavoured to conceal the absurd story from him, but unfortunately he overheard one of the men making an allusion to it, and insisted upon being informed about it. As I had expected, it brought out all his latent lunacy in an exaggerated form. I can hardly believe that this is the same man who discoursed philosophy last night with the most critical acumen, and coolest judgement. He is pacing backwards and forwards upon the quarterdeck like a caged tiger, stopping now and again to throw out his hands with a yearning gesture, and stare impatiently out over the ice. He keeps up a continual mutter to himself, and once he called out, 'But a little time, love—but a little time!' Poor fellow, it is sad to see a gallant seaman and accomplished gentleman reduced to such a pass, and to think that imagination and delusion can cow a mind to which real danger was but the salt of life. Was ever a man in such a position as I, between a demented captain and a ghost-seeing mate? I sometimes think I am the only really sane man aboard the vessel—except perhaps the second engineer, who is a kind of ruminant and would care nothing for all the fiends in the Red Sea, so long as they would leave him alone and not disarrange his tools.

The ice is still opening rapidly, and there is every probability of our being able to make a start tomorrow morning. They will think I am inventing when I tell them at home all the strange things that have befallen me.

12 P.M. I have been a good deal startled, though I feel steadier now, thanks to a stiff glass of brandy. I am hardly myself yet however, as this handwriting will testify. The fact is that I have gone through a very strange experience, and am beginning to doubt whether I was justified in branding everyone on board as madmen, because they professed to have seen things which did not seem reasonable to my understanding. Pshaw! I am a fool to let such a trifle unnerve me, and yet coming as it does after all these alarms, it has an additional significance, for I cannot doubt either Mr Manson's story or that of the mate, now that I have experienced that which I used formerly to scoff at.

After all it was nothing very alarming—a mere sound, and that was all. I cannot expect that anyone reading this, if anyone ever should read it, will sympathize with my feelings, or realize the effect which it produced upon me at the time. Supper was over and I had gone on deck to have a quiet pipe before turning in. The night was very dark—so dark that standing under the quarter boat, I was unable to see the officer upon the bridge. I think I have already mentioned the extraordinary silence which prevails in these frozen seas. In other parts of the world, be they ever so barren, there is some slight vibration of the air—some faint hum, be it from the distant haunts of men, or from the leaves of the trees, or the wings of the birds, or even the faint rustle of the grass that covers the ground. One may not actively perceive the sound, and yet if it were withdrawn it would be missed. It is only here in these Arctic seas that stark, unfathomable stillness obtrudes itself upon you in all its gruesome reality. You find your tympanum straining to catch some little murmur and dwelling eagerly upon every accidental sound within the vessel. In this state I was leaning against the bulwarks when there arose from the ice almost directly underneath me, a cry, sharp and shrill, upon the silent air of the night, beginning, as it seemed to me, at a note such as prima donna never reached, and mounting from that ever higher and higher until it culminated in a long wail of agony, which might have been the last cry of a lost soul. The ghastly scream is still ringing in my ears. Grief, unutterable grief, seemed to be expressed in it and a great longing, and yet through it all there was an occasional wild note of exultation. It seemed to come from close beside me, and yet as I glared into the darkness I could make out nothing. I waited some little time, but without hearing any repetition of the sound, so I came below, more shaken than I have ever been in my life before. As I came down the companion I met Mr Milne coming up to relieve the watch. 'Weel, Doctor,' he said, 'may be that's auld wives' clavers tae? Did ye no hear it skirling? Maybe that's a supersteetion? what d'ye think o't noo?' I was obliged to apologize to the honest fellow, and acknowledge that I was as puzzled by it as he was. Perhaps tomorrow things may look different. At present I dare hardly write all that I think. Reading it again in days to come, when I have shaken off all these associations, I should despise myself for having been so weak.

September 18th.—Passed a restless and uneasy night still haunted by that strange sound. The Captain does not look as if he had had much repose either, for his face is haggard and his eyes bloodshot. I have not told him of my adventure of last night, nor shall I. He is already restless and excited, standing up, sitting down, and apparently utterly unable to keep still.

A fine lead appeared in the pack this morning, as I had expected, and we were able to cast off our ice-anchor, and steam about twelve miles in a west-sou'-westerly direction. We were then brought to a halt by a great floe as massive as any which we have left behind us. It bars our progress completely, so we can do nothing but anchor again and wait until it breaks up, which it will probably do within twenty-four hours, if the wind holds. Several bladder-nosed seals were seen swimming in the water, and one was shot, an immense creature more than eleven feet long. They are fierce, pugnacious animals, and are said to be more than a match for a bear. Fortunately they are slow and clumsy in their movements, so that there is little danger in attacking them upon the ice.

The Captain evidently does not think we have seen the last of our troubles, though why he should take a gloomy view of the situation is more than I can fathom, since everyone else on board considers that we have had a miraculous escape, and are sure now to reach the open sea.

'I suppose you think it's all right now, Doctor?' he said as we sat together after dinner.

'I hope so,' I answered.

'We mustn't be too sure—and yet no doubt you are right. We'll all be in the arms of our own true loves before long, lad, won't we? But we mustn't be too sure—we mustn't be too sure.'

He sat silent a little, swinging his leg thoughtfully backwards and forwards. 'Look here,' he continued. 'It's a dangerous place this, even at its best—a treacherous, dangerous place. I have known men cut off very suddenly in a land like this. A slip would do it sometimes—a single slip, and down you go through a crack and only a bubble on the green water to show where it was that you sank. It's a queer thing,' he continued with a nervous laugh, 'but all the years I've been in this country I never once thought of making a will—not that I have anything to leave in particular, but still when a man is exposed to danger he should have everything arranged and ready—don't you think so?'

'Certainly,' I answered, wondering what on earth he was driving at.

'He feels better for knowing it's all settled,' he went on. 'Now if anything should ever befall me, I hope that you will look after things for me. There is very little in the cabin, but such as it is I should like it to be sold, and the money divided in the same proportion as the oil-money among the crew. The chronometer I wish you to keep yourself as some slight remembrance of our voyage. Of course all this is a mere precaution, but I thought I would take the opportunity of speaking to you about it. I suppose I might rely upon you if there were any necessity?'

'Most assuredly,' I answered; 'and since you are taking this step, I may as well——'

'You! you!' he interrupted. *You're* all right. What the devil is the matter with *you*? There, I didn't mean to be peppery, but I don't like to hear a young fellow, that has hardly began life, speculating about death. Go up on deck and get some fresh air into your lungs instead of talking nonsense in the cabin, and encouraging me to do the same.'

The more I think of this conversation of ours the less do I like it. Why should the man be settling his affairs at the very time when we seem to be emerging from all danger? There must be some method in his madness. Can it be that he contemplates suicide? I remember that upon one occasion he spoke in a deeply reverent manner of the heinousness of the crime of self-destruction. I shall keep my eye upon him however, and though I cannot obtrude upon the privacy of his cabin, I shall at least make a point of remaining on deck as long as he stays up.

Mr Milne pooh-poohs my fears, and says it is only the 'skipper's little way'. He himself takes a very rosy view of the situation. According to him we shall be out of the ice by the day after tomorrow, pass Jan Meyen two days after that, and sight Shetland in little more than a week. I hope he may not be too sanguine. His opinion may be fairly balanced against the gloomy precautions of the Captain, for he is an old and experienced seaman, and weights his words well before uttering them.

The long-impending catastrophe has come at least. I hardly know what to write about it. The Captain is gone. He may come back to us again alive, but I fear me—I fear me. It is now seven o'clock of the morning of the 19th of September. I have spent the whole night traversing the great ice-floe in front of us with a party of seamen in the hope of coming upon some trace of him, but in vain. I shall try to give some account of the circumstances which attended upon his disappearance. Should anyone ever chance to read the words which I put down, I trust they will remember that I do not write from conjecture or from hearsay, but that I, a sane and educated man, am describing accurately what actually occurred before my very eyes. My inferences are my own but I shall be answerable for the facts.

The Captain remained in excellent spirits after the conversation which I have recorded. He appeared to be nervous and impatient however, frequently changing his position, and moving his limbs in an aimless choreic way which is characteristic of him at times. In a quarter of an hour he went upon deck seven times, only to descend after a few hurried paces. I followed him each time, for there was something about his face which confirmed my resolution of not letting him out of my sight. He seemed to observe the effect which his movements had produced, for he endeavoured by an over-done hilarity, laughing boisterously at the very smallest of jokes, to quiet my apprehensions.

After supper he went on to the poop once more, and I with him. The night was dark and very still, save for the melancholy soughing of the wind among the spars. A thick cloud was coming up from the north-west, and the ragged tentacles which it threw out in front of it were drifting across the face of the moon, which only shone now and again through a rift in the wrack. The Captain paced rapidly backwards and forwards, and then seeing me still dogging him he came across and hinted that he thought I should be better below—which I need hardly say had the effect of strengthening my resolution to remain on deck.

I think he forgot about my presence after this, for he stood silently leaning over the taffrail, and peering out across the great desert of snow, part of which lay in shadow, while part glittered mistily in the moonlight. Several times I could see by his movements that he was referring to his watch, and once he muttered a short sentence of which I could only catch the one word 'ready'. I confess to having felt an eerie feeling creeping over me as I watched the loom of his tall figure through the darkness, and noted how completely he fulfilled the idea of a man who is keeping a tryst. A tryst with whom? Some vague perception began to dawn upon me as I pieced one fact with another, but I was utterly unprepared for the sequel.

By the sudden intensity of his attitude I felt that he saw something. I crept up behind him. He was staring with an eager questioning gaze at what seemed to be a wreath of mist, blown swiftly in a line with the ship. It was a dim nebulous body devoid of shape, sometimes more, sometimes less apparent, as the light fell on it. The moon was dimmed in its brilliancy at the moment by a canopy of thinnest cloud, like the coating of an anemone.

'Coming, lass, coming,' cried the skipper, in a voice of unfathomable tenderness and compassion, like one who soothes a beloved one by some favour long looked for, and as pleasant to bestow as to receive.

What followed, happened in an instant. I had no power to interfere. He gave one spring to the top of the bulwarks, and another which took him on to the ice, almost to the feet of the pale misty figure. He held out his hands as if to clasp it, and so ran into the darkness with outstretched arms and loving words. I still stood rigid and motionless, straining my eyes after his retreating form, until his voice died away in the distance. I never thought to see him again, but at that moment the moon shone out brilliantly through a chink in the cloudy heaven, and illuminated the great field of ice. Then I saw his dark figure already a very long way off, running with prodigious speed across the frozen plain. That was the last glimpse which we caught of him—perhaps the last we ever shall. A party was organized to follow him, and I accompanied them, but the men's hearts were not in the work, and nothing was found. Another will be formed within a few hours. I can hardly believe I have not been dreaming, or suffering from some hideous nightmare as I write these things down.

7.30 P.M.—Just returned dead beat and utterly tired out from a second unsuccessful search for the Captain. The floe is of enormous extent, for though we have traversed at least twenty miles of its surface, there has been no sign of its coming to an end. The frost has been so severe of late that the overlying snow is frozen as hard as granite, otherwise we might have had the footsteps to guide us. The crew are anxious that we should cast off and steam round the floe and so to the southward, for the ice has opened up during the night, and the sea is visible upon the horizon. They argue that Captain Craigie is certainly dead, and that we are all risking our lives to no purpose by remaining when we have an opportunity of escape. Mr Milne and I have had the greatest difficulty in persuading them to wait until tomorrow night, and have been compelled to promise that we will not under any circumstances delay our departure longer than that. We propose therefore to take a few hour's sleep and then to start upon a final search.

September 20th, evening.—I crossed the ice this morning with a party of men exploring the southern part of the floe, while Mr Milne went off in a northerly direction. We pushed on for ten or twelve miles without seeing a trace of any living thing except a single bird, which fluttered a great way over our heads, and which by its flight I should judge to have been a falcon. The southern extremity of the ice field tapered away into a long narrow spit which projected out into the sea. When we came to the base of this promontory, the men halted, but I begged them to continue to the extreme end of it that we might have the satisfaction of knowing that no possible chance had been neglected.

We had hardly gone a hundred yards before McDonald of Peterhead cried out that he saw something in front of us, and began to run. We all got a glimpse of it and ran too. At first it was only a vague darkness against the white ice, but as we raced along together it took the shape of a man, and eventually of the man of whom we were in search. He was lying face downwards upon a frozen bank. Many little crystals of ice and feathers of snow had drifted on to him as he lay, and sparkled upon his dark seaman's jacket. As we came up some wandering puff of wind caught these tiny flakes in its vortex, and they whirled up into the air, partially descended again, and then, caught once more in the current, sped rapidly away in the direction of the sea. To my eyes it seemed but a snow-drift, but many of my companions averred that it started up in the shape of a woman, stooped over the corpse and kissed it, and then hurried away across the floe. I have learned never to ridicule any man's opinion, however strange it may seem. Sure it is that Captain Nicholas Craigie had met with no painful end, for there was a bright smile upon his blue pinched features, and his hands were still outstretched as though grasping at the strange visitor which had summoned him away into the dim world that lies beyond the grave.

We buried him the same afternoon with the ship's ensign around him, and a thirty-two pound shot at his feet. I read the burial service, while the rough sailors wept like children, for these were many who owed much to his kind heart, and who showed now the affection which his strange ways had repelled during his lifetime. He went off the grating with a dull, sullen splash, and as I looked into the green water I saw him go down, down, down, until he was but a little flickering patch of white hanging upon the outskirts of eternal darkness. Then even that faded away and he was gone. There he shall lie, with his secret and his sorrows and his mystery all still buried in his breast, until that great day when the sea shall give up its dead, and Nicholas Craigie come out from among the ice with the smile upon his face, and his stiffened arms outstretched in greeting. I pray that his lot may be a happier one in that life than it has been in this.

I shall not continue my journal. Our road to home lies plain and clear before us, and the great ice field will soon be but a remembrance of the past. It will be some time before I get over the shock produced by recent events. When I began this record of our voyage I little thought of how I should be compelled to finish it. I am writing these final words in the lonely cabin, still starting at times and fancying I hear the quick nervous step of the dead man upon the deck above me. I entered his cabin tonight as was my duty, to make a list of his effects in order that they might be entered in the official log. All was as it had been upon my previous visit, save that the picture which I have described as having hung at the end of his bed had been cut out of its frame, as with a knife, and was gone. With this last link in a strange chain of evidence I close my diary of the voyage of the *Pole-star*.

[NOTE by Dr John McAlister Ray, senior. 'I have read over the strange events connected with the death of the Captain of the *Pole-star*, as narrated in the journal of my son. That everything occurred exactly as he describes it I have the fullest confidence, and, indeed, the most positive certainty, for I know him to be a strong-nerved and unimaginative man, with the strictest regard for veracity. Still, the story is, on the face of it, so vague and so improbable, that I was long opposed to its publication. Within the last few days, however, I have had independent testimony upon the subject which throws a new light upon it. I had run down to Edinburgh to attend a meeting of the British Medical Association, when I chanced to come across Dr P——, an old college chum of mine, now practising at Saltash, in Devonshire. Upon my telling him of this experience of my son's, he declared to me that he was familiar with the man, and proceeded, to my no small surprise, to give me a description of him, which tallied remarkably well with that given in the journal, except that he depicted him as a younger man. According to his account, he had been engaged to a young lady of singular beauty residing upon the Cornish coast. During his absence at sea his betrothed had died under circumstances of peculiar horror.]

The Body-Snatcher

ROBERT LOUIS STEVENSON

Every night in the year, four of us sat in the small parlour of the *George* at Debenham—the undertaker, and the landlord, and Fettes, and myself. Sometimes there would be more; but blow high, blow low, come rain or snow or frost, we four would be each planted in his own particular armchair. Fettes was an old drunken Scotsman, a man of education obviously, and a man of some property, since he lived in idleness. He had come to Debenham years ago, while still young, and by a mere continuance of living had grown to be an adopted townsman. His blue camlet cloak was a local antiquity, like the church-spire. His place in the parlour at the *George*, his absence from church, his old, crapulous, disreputable vices, were all things of course in Debenham. He had some vague Radical opinions and some fleeting infidelities, which he would now and again set forth and emphasize with tottering slaps upon the table. He drank rum—five glasses regularly every evening; and for the greater portion of his nightly visit to the *George* sat, with his glass in his right hand, in a state of melancholy alcoholic saturation. We called him the Doctor, for he was supposed to have some special knowledge of medicine, and had been known upon a pinch, to set a fracture or reduce a dislocation; but, beyond these slight particulars, we had no knowledge of his character and antecedents.

One dark winter night—it had struck nine some time before the landlord joined us—there was a sick man in the *George*, a great neighbouring proprietor suddenly struck down with apoplexy on his way to Parliament; and the great man's still greater London doctor had been telegraphed to his bedside. It was the first time that such a thing had happened in Debenham, for the railway was but newly open, and we were all proportionately moved by the occurrence.

'He's come,' said the landlord, after he had filled and lighted his pipe.

'He?' said I. 'Who?—not the doctor?'

'Himself,' replied our host.

'What is his name?'

'Dr Macfarlane,' said the landlord.

Fettes was far through his third tumbler, stupidly fuddled, now nodding over, now staring mazily around him; but at the last word he seemed to awaken, and repeated the name 'Macfarlane' twice, quietly enough the first time, but with sudden emotion at the second.

'Yes,' said the landlord, 'that's his name, Doctor Wolfe Macfarlane.'

Fettes became instantly sober: his eyes awoke, his voice became clear, loud, and steady, his language forcible and earnest. We were all startled by the transformation, as if a man had risen from the dead.

'I beg your pardon,' he said, 'I am afraid I have not been paying much attention to your talk. Who is this Wolfe Macfarlane?' And then, when he had heard the landlord out, 'It cannot be, it cannot be,' he added; 'and yet I would like well to see him face to face.'

'Do you know him, Doctor?' asked the undertaker, with a gasp.

'God forbid!' was the reply. 'And yet the name is a strange one; it were too much to fancy two. Tell me, landlord, is he old?'

'Well,' said the host, 'he's not a young man, to be sure, and his hair is white; but he looks younger than you.'

'He is older, though; years older. But,' with a slap upon the table, 'it's the rum you see in my face—rum and sin. This man, perhaps, may have an easy conscience and a good digestion. Conscience! Hear me speak. You would think I was some good, old, decent Christian, would you not? But no, not I; I never canted. Voltaire might have canted if he'd stood in my shoes; but the brains'—with a rattling fillip on his bald head—'the brains were clear and active, and I saw and made no deductions'.

'If you know this doctor,' I ventured to remark, after a somewhat awful pause, 'I should gather that you do not share the landlord's good opinion.'

Fettes paid no regard to me.

'Yes,' he said, with sudden decision, 'I must see him face to face.'

There was another pause, and then a door was closed rather sharply on the first floor, and a step was heard upon the stair.

'That's the doctor,' cried the landlord. 'Look sharp, and you can catch him.'

It was but two steps from the small parlour to the door of the old *George* inn; the wide oak staircase landed almost in the street; there was room for a Turkey rug and nothing more between the threshold and the last round of the descent; but this little space was every evening brilliantly lit up, not only by the light upon the stair and the great signal-lamp below the sign, but by the warm radiance of the bar-room window. The *George* thus brightly advertised itself to passers-by in the cold street. Fettes walked steadily to the spot, and we, who were hanging behind, beheld the two men meet, as one of them had phrased it, face to face. Dr Macfarlane was alert and vigorous. His white hair set off his pale and placid, although energetic, countenance. He was richly dressed in the finest of broadcloth and the whitest of linen, with a great gold watchchain, and studs and spectacles of the same precious material. He wore a broad-folded tie, white and speckled with lilac, and he carried on his arm a comfortable driving-coat of fur. There was no doubt but he became his years, breathing as he did, of wealth and consideration; and it was a surprising contrast to see our parlour sot—bald, dirty, pimpled, and robed in his old camlet cloak—confront him at the bottom of the stairs.

'Macfarlane!' he said somewhat loudly, more like a herald than a friend.

The great doctor pulled up short on the fourth step, as though the familiarity of the address surprised and somewhat shocked his dignity.

'Toddy Macfarlane!' repeated Fettes.

The London man almost staggered. He stared for the swiftest of seconds at the man before him, glanced behind him with a sort of scare, and then in a startled whisper, 'Fettes!' he said, 'you!'

'Ay,' said the other, 'me! Did you think I was dead too? We are not so easy shut of our acquaintance.'

'Hush, hush!' exclaimed the doctor. 'Hush, hush! this meeting is so unexpected—I can see you are unmanned. I hardly knew you, I confess, at first; but I am overjoyed—overjoyed to have this opportunity. For the present it must be how-d'ye-do and goodbye in one, for my fly is waiting, and I must not fail the train; but you shall—let me see—yes—you shall give me your address, and you can count on early news of me. We must do something for you, Fettes. I fear you are out at elbows; but we must see to that for auld lang syne, as once we sang at suppers.'

'Money!' cried Fettes; 'money from you! The money that I had from you is lying where I cast it in the rain.'

Dr Macfarlane had talked himself into some measure of superiority and confidence, but the uncommon energy of this refusal cast him back into his first confusion.

A horrible, ugly look came and went across his almost venerable countenance. 'My dear fellow,' he said, 'be it as you please; my last thought is to offend you. I would intrude on none. I will leave you my address, however——'

'I do not wish it—I do not wish to know the roof that shelters you,' interrupted the other. 'I heard your name; I feared it might be you; I wished to know if, after all, there were a God; I know now that there is none. Begone!'

He still stood in the middle of the rug, between the stair and the doorway; and the great London physician, in order to escape, would be forced to step to one side. It was plain that he hesitated before the thought of this humiliation. White as he was, there was a dangerous glitter in his spectacles; but while he still paused uncertain, he became aware that the driver of his fly was peering in from the street at this unusual scene and caught a glimpse at the same time of our little body from the parlour, huddled by the corner of the bar. The presence of so many witnesses decided him at once to flee. He crouched together, brushing on the wainscot, and made a dart like a serpent, striking for the door. But his tribulation was not yet entirely at an end, for even as he was passing Fettes clutched him by the arm and these words came in a whisper, and yet painfully distinct, 'Have you seen it again?'

The great rich London doctor cried out aloud with a sharp, throttling cry; he dashed his questioner across the open space, and, with his hands over his head, fled out of the door like a detected thief. Before it had occurred to one of us to make a movement, the fly was already rattling toward the station. The scene was over like a dream, but the dream had left proofs and traces of its passage. Next day the servant found the find gold spectacles broken on the threshold, and that very night we were all standing breathless by the bar-room window, and Fettes at our side, sober, pale, and resolute in look.

'God protect us, Mr Fettes!' said the landlord, coming first into possession of his customary senses. 'What in the universe is all this? These are strange things you have been saying.'

Fettes turned toward us; he looked us each in succession in the face. 'See if you can hold your tongues,' said he. 'That man Macfarlane is not safe to cross; those that have done so already have repented it too late.'

And then, without so much as finishing his third glass, far less waiting for the other two, he bade us goodbye and went forth, under the lamp of the hotel, into the black night.

We three turned to our places in the parlour, with the big red fire and four clear candles; and as we recapitulated what had passed the first chill of our surprise soon changed into a glow of curiosity. We sat late; it was the latest session I have known in the old *George*. Each man, before we parted, had his theory that he was bound to prove; and none of us had any nearer business in this world than to track out the past of our condemned companion, and surprise the secret that he shared with the great London doctor. It is no great boast, but I believe I was a better hand at worming out a story than either of my fellows at the

George; and perhaps there is now no other man alive who could narrate to you the following foul and unnatural events.

In his young days Fettes studied medicine in the schools of Edinburgh. He had talent of a kind, the talent that picks up swiftly what it hears and readily retails it for its own. He worked little at home; but he was civil, attentive, and intelligent in the presence of his masters. They soon picked him out as a lad who listened closely and remembered well; nay, strange as it seemed to me when I first heard it, he was in those days well favoured, and pleased by his exterior. There was, at that period, a certain extramural teacher of anatomy, whom I shall here designate by the letter K. His name was subsequently too well known. The man who bore it skulked through the streets of Edinburgh in disguise, while the mob that applauded at the execution of Burke called loudly for the blood of his employer. But Mr K—— was then at the top of his vogue; he enjoyed a popularity due partly to his own talent and address, partly to the incapacity of his rival, the university professor. The students, at least, swore by his name, and Fettes believed himself, and was believed by others, to have laid the foundations of success when he had acquired the favour of this meteorically famous man. Mr K—— was a *bon vivant* as well as an accomplished teacher; he liked a sly allusion no less than a careful preparation. In both capacities Fettes enjoyed and deserved his notice, and by the second year of his attendance he held the half-regular position of second demonstrator or sub-assistant in his class.

In this capacity, the charge of the theatre and lecture-room devolved in particular upon his shoulders. He had to answer for the cleanliness of the premises and the conduct of the other students, and it was a part of his duty to supply, receive, and divide the various subjects. It was with a view to this last—at that time very delicate—affair that he was lodged by Mr K—— in the same wynd, and at last in the same building, with the dissecting-rooms. Here, after a night of turbulent pleasures, his hand still tottering, his sight still misty and confused, he would be called out of bed in the black hours before the winter dawn by the unclean and desperate interlopers who supplied the table. He would open the door to these men, since infamous throughout the land. He would help them with their tragic burthen, pay them their sordid price, and remain alone, when they were gone, with the unfriendly relics of humanity. From such a scene he would return to snatch another hour or two of slumber, to repair the abuses of the night, and refresh himself for the labours of the day.

Few lads could have been more insensible to the impressions of a life thus passed among the ensigns of mortality. His mind was closed against all general considerations. He was incapable of interest in the fate and fortunes of another, the slave of his own desires and low ambitions. Cold, light, and selfish in the last resort, he had that modicum of prudence, miscalled morality, which keeps a man from inconvenient drunkenness or punishable theft. He coveted, besides, a measure of consideration from his masters and his fellow-pupils, and he had no desire to fail conspicuously in the external parts of life. Thus he made it his pleasure to gain some distinction in his studies, and day after day rendered unimpeachable eye-service to his employer, Mr K——. For his day of work he indemnified himself by nights of roaring, blackguardly enjoyment; and when that balance had been struck, the organ that he called his conscience declared itself content.

The supply of subjects was a continual trouble to him as well as to his master. In that large and busy class, the raw material of the anatomists kept perpetually running out; and the business thus rendered necessary was not only unpleasant in itself, but threatened dangerous consequences to all who were concerned. It was the policy of Mr K—— to ask no questions in his dealings with the trade. 'They bring the boy, and we pay the price,' he used to say, dwelling on the alliteration—*quid pro quo.* And, again, and somewhat profanely, 'Ask no questions,' he would tell his assistants, 'for conscience' sake.' There was no understanding that the subjects were provided by the crime of murder. Had that idea been broached to him in words, he would have recoiled in horror; but the lightness of his speech upon so grave a matter was, in itself, an offence against good manners, and a temptation to the men with whom he dealt. Fettes, for instance, had often remarked to himself upon the singular freshness of the bodies. He had been struck again and again by the hang-dog, abominable looks of the ruffians who came to him before the dawn; and, putting things together clearly in his private thoughts, he perhaps attributed a meaning too immoral and too categorical to the unguarded counsels of his master. He understood his duty, in short, to have three branches: to take what was brought to pay the price, and to avert the eye from any evidence of crime.

One November morning this policy of silence was put sharply to the test. He had been awake all night with a racking toothache—pacing his room like a caged beast or throwing himself in fury on his bed—and had fallen at last into that profound, uneasy slumber that so often follows on a night of pain, when he was awakened by the third or fourth angry repetition of the concerted signal. There was a thin, bright moonshine: it was bitter cold, windy, and frosty; the town had not yet awakened, but an indefinable stir already preluded the noise and business of the day. The ghouls had come later than usual, and they seemed more than usually eager to be gone. Fettes, sick with sleep, lighted them upstairs. He heard their grumbling Irish voices through a dream; and as they stripped the sack from their sad merchandise he leaned dozing, with his shoulder propped against the wall; he had to shake himself to find the men their money. As he did so his eyes lighted on the dead face. He started; he took two steps nearer, with the candle raised.

'God Almighty!' he cried. 'That is Jane Galbraith!'

The men answered nothing, but they shuffled nearer the door.

'I know her, I tell you,' he continued. 'She was alive and hearty yesterday. It's impossible she can be dead; it's impossible you should have got this body fairly.'

'Sure, sir, you're mistaken entirely,' said one of the men.

But the other looked Fettes darkly in the eyes, and demanded the money on the spot.

It was impossible to misconceive the threat or to exaggerate the danger. The lad's heart failed him. He stammered some excuses, counted out the sum, and saw his hateful visitors depart. No sooner were they gone than he hastened to confirm his doubts. By a dozen unquestionable marks he identified the girl he had jested with the day before. He saw, with horror, marks upon her body that might well betoken violence. A panic seized him, and he took refuge in his room. There he reflected at length over the discovery that he had made; considered soberly the bearing of Mr K——'s instructions and the danger to himself of interference in so serious a business, and at last, in sore perplexity, determined to wait for the advice of his immediate superior, the class assistant.

This was a young doctor, Wolfe Macfarlane, a high favourite among all the reckless students, clever, dissipated, and unscrupulous to the last degree. He had travelled and studied abroad. His manners were agreeable and a little forward. He was an authority on the stage, skilful on the ice or the links with skate or golf-club; he dressed with nice audacity, and, to put the finishing touch upon his glory, he kept a gig and a strong trotting-horse. With Fettes he was on terms of intimacy; indeed their relative positions called for some community of life; and when subjects were scarce the pair would drive far into the country in Macfarlane's gig, visit and desecrate some lonely graveyard, and return before dawn with their booty to the door of the dissecting-room.

On that particular morning Macfarlane arrived somewhat earlier than his wont. Fettes heard him, and met him on the stairs, told him his story, and showed him the cause of his alarm. Macfarlane examined the marks on her body.

'Yes', he said with a nod, 'it looks fishy.'

'Well, what should I do?' asked Fettes.

'Do?' repeated the other. 'Do you want to do anything? Least said soonest mended, I should say.'

'Someone else might recognize her,' objected Fettes. 'She was as well known as the Castle Rock.'

'We'll hope not,' said Macfarlane, 'and if anybody does—well, you didn't, don't you see, and there's an end. The fact is, this has been going on too long. Stir up the mud, and you'll get K—— into the most unholy trouble; you'll be in a shocking box yourself. So will I, if you come to that. I should like to know how any one of us would look, or what the devil we should have to say for ourselves, in any Christian witness-box. For me, you know there's one thing certain—that, practically speaking, all our subjects have been murdered.'

'Macfarlane!' cried Fettes.

'Come now!' sneered the other. 'As if you hadn't suspected it yourself!'

'Suspecting is one thing——'

'And proof another. Yes, I know; and I'm as sorry as you are this should have come here,' tapping the body with his cane. 'The next best thing for me is not to recognize it; and,' he added coolly, 'I don't. You may, if you please. I don't dictate, but I think a man of the world would do as I do; and I may add, I fancy that is what K—— would look for at our hands. The question is, Why did he choose us two for his assistants? And I answer, because he didn't want old wives.'

This was the tone of all others to affect the mind of a lad like Fettes. He agreed to imitate Macfarlane. The body of the unfortunate girl was duly dissected, and no one remarked or appeared to recognize her.

One afternoon, when his day's work was over, Fettes dropped into a popular tavern and found Macfarlane sitting with a stranger. This was a small man, very pale and dark, with coal-black eyes. The cut of his features gave a promise of intellect and refinement which was but feebly realized in his manners, for he proved, upon a nearer acquaintance, coarse, vulgar, and stupid. He exercised, however, a very remarkable control over Macfarlane; issued orders like the Great Bashaw; became inflamed at the least discussion or delay, and commented rudely on the servility with which he was obeyed. This most offensive person took a fancy to Fettes on the spot, plied him with drinks, and honoured him with unusual confidences on his past career. If a tenth part of what he confessed were true, he was a very loathsome rogue;

and the lad's vanity was tickled by the attention of so experienced a man.

'I'm a pretty bad fellow myself,' the stranger remarked, 'but Macfarlane is the boy—Toddy Macfarlane I call him. Toddy, order your friend another glass.' Or it might be, 'Toddy, you jump up and shut the door.' 'Toddy hates me,' he said again. 'Oh, yes, Toddy, you do!'

'Don't you call me that confounded name,' growled Macfarlane.

'Hear him! Did you ever see the lads play knife? He would like to do that all over my body,' remarked the stranger.

'We medicals have a better way than that,' said Fettes. 'When we dislike a dead friend of ours, we dissect him.'

Macfarlane looked up sharply, as though this jest was scarcely to his mind.

The afternoon passed. Gray, for that was the stranger's name, invited Fettes to join them at dinner, ordered a feast so sumptuous that the tavern was thrown in commotion, and when all was done commanded Macfarlane to settle the bill. It was late before they separated; the man Gray was incapably drunk. Macfarlane, sobered by his fury, chewed the cud of the money he had been forced to squander and the slights he had been obliged to swallow. Fettes, with various liquors singing in his head, returned home with devious footsteps and a mind entirely in abeyance. Next day Macfarlane was absent from the class, and Fettes smiled to himself as he imagined him still squiring the intolerable Gray from tavern to tavern. As soon as the hour of liberty had struck he posted from place to place in quest of his last night's companions. He could find them, however, nowhere; so returned early to his rooms, went early to bed, and slept the sleep of the just.

At four in the morning he was awakened by the well-known signal. Descending to the door, he was filled with astonishment to find Macfarlane with his gig, and in the gig one of those long and ghastly packages with which he was so well acquainted.

'What?' he cried. 'Have you been out alone? How did you manage?'

But Macfarlane silenced him roughly, bidding him turn to business. When they had got the body upstairs and laid it on the table, Macfarlane made at first as if he were going away. Then he paused and seemed to hesitate; and then, 'You had better look at the face,' said he, in tones of some constraint. 'You had better,' he repeated, as Fettes only stared at him in wonder.

'But where, and how, and when did you come by it?' cried the other.

'Look at the face,' was the only answer.

Fettes was staggered; strange doubts assailed him. He looked from the young doctor to the body, and then back again. At last, with a start, he did as he was bidden. He had almost expected the sight that met his eyes, and yet the shock was cruel. To see, fixed in the rigidity of death and naked on that coarse layer of sack-cloth, the man whom he had left well-clad and full of meat and sin upon the threshold of a tavern, awoke, even in the thoughtless Fettes, some of the terrors of the conscience. It was a *cras tibi* which re-echoed in his soul, that two whom he had known should have come to lie upon these icy tables. Yet these were only secondary thoughts. His first concern regarded Wolfe. Unprepared for a challenge so momentous, he knew not how to look his comrade in the face. He durst not meet his eye, and he had neither words nor voice at his command.

It was Macfarlane himself who made the first advance. He came up quietly behind and laid his hand gently but firmly on the other's shoulder.

'Richardson,' said he, 'may have the head.'

Now Richardson was a student who had long been anxious for that portion of the human subject to dissect. There was no answer, and the murderer resumed: 'Talking of business, you must pay me; your accounts, you see, must tally.'

Fettes found a voice, the ghost of his own: 'Pay you!' he cried, 'Pay you for that?'

'Why, yes, of course you must. By all means and on every possible account, you must,' returned the other. 'I dare not give it for nothing, you dare not take it for nothing; it would compromise us both. This is another case like Jane Galbraith's. The more things are wrong the more we must act as if all were right. Where does old K—— keep his money?'

'There,' answered Fettes hoarsely, pointing to a cupboard in the corner.

'Give me the key, then,' said the other, calmly, holding out his hand.

There was an instant's hesitation, and the die was cast. Macfarlane could not suppress a nervous twitch, the infinitesimal mark of an immense relief, as he felt the key between his fingers. He opened the cupboard, brought out pen and ink and a paper-book that stood in one compartment, and separated from the funds in a drawer a sum suitable to the occasion.

'Now, look here,' he said, 'there is the payment made—first proof of your good faith: first step to your security. You have now to clinch it by a second. Enter the payment in your book, and then you for your part may defy the devil.'

The next few seconds were for Fettes an agony of thought; but in balancing his terrors it was the most immediate that triumphed. Any future difficulty seemed almost welcome if he could avoid a present quarrel with Macfarlane. He set down the candle which he had been carrying all the time, and with a steady hand entered the date, the nature, and the amount of the transaction.

'And now,' said Macfarlane, 'it's only fair that you should pocket the lucre. I've had my share already. By-the-by, when a man of the world falls into a bit of luck, has a few shillings extra in his pocket—I'm ashamed to speak of it, but there's a rule of conduct in the case. No treating, no purchase of expensive class-books, no squaring of old debts; borrow, don't lend.'

'Macfarlane,' began Fettes, still somewhat hoarsely, 'I have put my neck in a halter to oblige you.'

'To oblige me?' cried Wolfe. 'Oh, come! You did, as near as I can see the matter, what you downright had to do in self-defence. Suppose I got into trouble, where would you be? This second little matter flows clearly from the first. Mr Gray is the continuation of Miss Galbraith. You can't begin and then stop. If you begin, you must keep on beginning; that's the truth. No rest for the wicked.'

A horrible sense of blackness and the treachery of fate seized hold upon the soul of the unhappy student.

'My God!' he cried, 'but what have I done? and when did I begin? To be made a class assistant—in the name of reason, where's the harm in that? Service wanted the position; Service might have got it. Would *he* have been where *I* am now?'

'My dear fellow,' said Macfarlane, 'what a boy you are! What harm *has* come to you? What harm *can* come to you if you hold your tongue? Why, man, do you know what this life is? There are two squads of us—the lions and the lambs. If you're a lamb, you'll come to lie upon these tables like Gray or Jane Galbraith; if you're a lion, you'll live and drive a horse like me, like K——, like all the world with any wit or courage. You're staggered at the first. But look at K——! My dear fellow, you're clever, you have pluck. I like you, and K—— likes you. You were born to lead the hunt; and I tell you, on my honour and my experience of life, three days from now you'll laugh at all these scarecrows like a high-school boy at a farce.'

And with that Macfarlane took his departure and drove off up the wynd in his gig to get under cover before daylight. Fettes was thus left alone with his regrets. He saw the miserable peril in which he stood involved. He saw, with inexpressible dismay, that there was no limit to his weakness, and that, from concession to concession, he had fallen from the arbiter of Macfarlane's destiny to his paid and helpless accomplice. He would have given the world to have been a little braver at the time, but it did not occur to him that he might still be brave. The secret of Jane Galbraith and the cursed entry in the day-book closed his mouth.

Hours passed; the class began to arrive; the members of the unhappy Gray were dealt out to one and to another, and received without remark. Richardson was made happy with the head; and before the hour of freedom rang Fettes trembled with exultation to perceive how far they had already gone toward safety.

For two days he continued to watch, with increasing joy, the dreadful process of disguise.

On the third day Macfarlane made his appearance. He had been ill, he said; but he made up for lost time by the energy with which he directed the students. To Richardson in particular he extended the most valuable assistance and advice, and that student, encouraged by the praise of the demonstrator, burned high with ambitious hopes, and saw the medal already in his grasp.

Before the week was out Macfarlane's prophecy had been fulfilled. Fettes had outlived his terrors and had forgotten his baseness. He began to plume himself upon his courage, and had so arranged the story in his mind that he could look back on these events with an unhealthy pride. Of his accomplice he saw but little. They met, of course, in the business of the class; they received their orders together from Mr K——. At times they had a word or two in private, and Macfarlane was from first to last particularly kind and jovial. But it was plain that he avoided any reference to their common secret; and even when Fettes whispered to him that he had cast in his lot with the lions and forsworn the lambs, he only signed to him smilingly to hold his peace.

At length an occasion arose which threw the pair once more into a closer union. Mr K—— was again short of subjects; pupils were eager, and it was a part of this teacher's pretensions to be always well supplied. At the same time there came the news of a burial in the rustic graveyard of Glencorse. Time has little changed the place in question. It stood then, as now, upon a crossroad, out of call of human habitations, and buried fathom deep in the foliage of six cedar trees. The cries of the sheep upon the neighbouring hills, the streamlets upon either hand, one loudly singing among pebbles, the other dripping furtively from pond to pond, the stir of the wind in mountainous old flowering chestnuts, and once in seven days the voice of the bell and the old tunes of the precentor, were the only sounds that disturbed the silence around the rural church. The Resurrection Man—to use a by-name of the period—was not to be deterred by any of the sanctities of customary piety. It was part of his trade to despise and desecrate the scrolls and trumpets of old tombs, the paths worn by the feet of worshippers and mourners, and the offerings and the inscriptions of bereaved affection. To rustic neighbourhoods, where love is more than commonly tenacious, and where some bonds of blood or fellowship unite the entire society of a parish, the body-snatcher, far from being repelled by natural respect, was attracted by the ease and safety of the task. To bodies that had been laid in earth, in joyful expectation of a far different awakening, there came that hasty, lamp-lit, terror-haunted resurrection of the spade and mattock. The coffin was forced, the cerements torn, and the melancholy relics, clad in sackcloth, after being rattled for hours on moonless by-ways, were at length exposed to uttermost indignities before a class of gaping boys.

Somewhat as two vultures may swoop upon a dying lamb, Fettes and Macfarlane were to be let loose upon a grave in that green and quiet resting-place. The wife of a farmer, a woman who had lived for sixty years, and been known for nothing but good butter and a godly conversation, was to be rooted from her grave at midnight and carried, dead and naked, to that far-away city that she had always honoured with her Sunday best; the place beside her family was to be empty till the crack of doom; her innocent and almost venerable members to be exposed to that last curiosity of the anatomist.

Late one afternoon the pair set forth, well wrapped in cloaks and furnished with a formidable bottle. It rained without remission—a cold, dense, lashing rain. Now and again there blew a puff of wind, but these sheets of falling water kept it down. Bottle and all, it was a sad and silent drive as far as Penicuik, where they were to spend the evening. They stopped once, to hide their implements in a thick bush not far from the churchyard, and once again at the Fisher's Tryst, to have a toast before the kitchen fire and vary their nips of whisky with a glass of ale. When they reached their journey's end the gig was housed, the horse was fed and comforted, and the two young doctors in a private room sat down to the best dinner and the best wine the house afforded. The lights, the fire, the beating rain upon the window, the cold, incongruous work that lay before them, added zest to their enjoyment of the meal. With every glass their cordiality increased. Soon Macfarlane handed a little pile of gold to his companion.

'A compliment,' he said. 'Between friends these little d——d accommodations ought to fly like pipe-lights.'

Fettes pocketed the money, and applauded the sentiment to the echo. 'You are a philosopher,' he cried. 'I was an ass till I knew you. You and K—— between you, by the Lord Harry! but you'll make a man of me.'

'Of course we shall,' applauded Macfarlane. 'A man? I tell you, it required a man to back me up the other morning. There are some big, brawling, forty-year-old cowards who would have turned sick at the look of the d——d thing; but not you—you kept your head. I watched you.'

'Well, and why not?' Fettes thus vaunted himself. 'It was no affair of mine. There was nothing to gain on the one side but disturbance, and on the other I could count on your gratitude, don't you see?' And he slapped his pocket till the gold pieces rang.

Macfarlane somehow felt a certain touch of alarm at these unpleasant words. He may have regretted that he had taught his young companion so successfully, but he had no time to interfere, for the other noisily continued in this boastful strain:

'The great thing is not to be afraid. Now, between you and me, I don't want to hang—that's practical; but for all cant, Macfarlane, I was born with a contempt. Hell, God, Devil, right, wrong, sin, crime, and all the old gallery of curiosities—they may frighten boys, but men of the world, like you and me, despise them. Here's to the memory of Gray!'

It was by this time growing somewhat late. The gig, according to order, was brought round to the door with both lamps brightly shining, and the young men had to pay their bill and take the road. They announced that they were bound for Peebles, and drove in that direction till they were clear of the last houses of the town; then, extinguishing the lamps, returned upon their course, and followed a by-road toward Glencorse. There was no sound but that of their own passage, and the incessant, strident pouring of the rain. It was pitch dark; here and there a white gate or a white stone in the wall guided them for a short space across the night; but for the most part it was at a foot pace, and almost groping, that they picked their way through that resonant blackness to their solemn and isolated destination. In the sunken woods that traverse the neighbourhood of the burying-ground the last glimmer failed them, and it became necessary to kindle a match and reillumine one of the lanterns of the gig. Thus, under the dripping trees, and environed by huge and moving shadows, they reached the scene of their unhallowed labours.

They were both experienced in such affairs, and powerful with the spade; and they had scarce been twenty minutes at their task before they were rewarded by a dull rattle on the coffin lid. At the same moment Macfarlane, having hurt his hand upon a stone, flung it carelessly above his head. The grave, in which they now stood almost to the shoulders, was close to the edge of the plateau of the graveyard; and the gig lamp had been propped, the better to illuminate their labours, against a tree, and on the immediate verge of the steep bank descending to the stream. Chance had taken a sure aim with the stone. Then came a clang of broken glass; night fell upon them; sounds alternately dull and ringing announced the bounding of the lantern down the bank, and its occasional collision with the trees. A stone or two, which it had dislodged in its descent, rattled behind it into the profundities of the glen; and then silence, like night, resumed its sway; and they might bend their hearing to its utmost pitch, but naught was to be heard except the rain, now marching to the wind, now steadily falling over miles of open country.

They were so nearly at an end of their abhorred task that they judged it wisest to complete it in the dark. The coffin was exhumed and broken open; the body inserted in the dripping sack and carried between them to the gig; one mounted to keep it in its place, and the other, taking the horse by the mouth, groped along by wall and bush until they reached the wider road by the Fisher's Tryst. Here was a faint, diffused radiancy, which they hailed like daylight; by that they pushed the horse to a good pace and began to rattle along merrily in the direction of the town.

They had both been wetted to the skin during their operations, and now, as the gig jumped among the deep ruts, the thing that stood propped between them fell now upon one and now upon the other. At every repetition of the horrid contact each instinctively repelled it with greater haste; and the process, natural although it was, began to tell upon the nerves of the companions. Macfarlane made some ill-favoured jest about the farmer's wife, but it came hollowly from his lips, and was allowed to drop in silence. Still their unnatural burthen bumped from side to side; and now the head would be laid, as if in confidence, upon their shoulders, and now the drenching sackcloth would flap icily about their faces. A creeping chill began to possess the soul of Fettes. He peered at the bundle, and it seemed somehow larger than at first. All over the countryside, and from every degree of distance, the farm dogs accompanied their passage with tragic ululations; and it grew and grew upon his mind that some unnatural miracle had been accomplished, that some nameless change had befallen the dead body, and that it was in fear of their unholy burden that the dogs were howling.

'For God's sake,' said he, making a great effort to arrive at speech, 'for God's sake, let's have a light!'

Seemingly Macfarlane was affected in the same direction; for though he made no reply, he stopped the horse, passed the reins to his companion, got down, and proceeded to kindle the remaining lamp. They had by that time got no further than the cross-road down to

Auchendinny. The rain still poured as though the deluge were returning, and it was no easy matter to make a light in such a world of wet and darkness. When at last the flickering blue flame had been transferred to the wick and began to expand and clarify, and shed a wide circle of misty brightness round the gig, it became possible for the two young men to see each other and the thing they had along with them. The rain had moulded the rough sacking to the outlines of the body underneath; the head was distinct from the trunk, the shoulders plainly modelled; something at once spectral and human riveted their eyes upon the ghastly comrade of their drive.

For some time Macfarlane stood motionless, holding up the lamp. A nameless dread was swathed, like a wet sheet, about the body, and tightened the white skin upon the face of Fettes; a fear that was meaningless, a horror of what could not be, kept mounting to his brain. Another beat of the watch, and he had spoken. But his comrade forestalled him.

'That is not a woman,' said Macfarlane, in a hushed voice.'

'It was a woman when we put her in,' whispered Fettes.

'Hold that lamp,' said the other. 'I must see her face.'

And as Fettes took the lamp his companion untied the fastenings of the sack and drew down the cover from the head. The light fell very clear upon the dark, well-moulded features and smooth-shaven cheeks of a too familiar countenance, often beheld in dreams of both of these young men. A wild yell rang up into the night; each leaped from his own side into the roadway; the lamp fell, broke, and was extinguished; and the horse, terrified by this unusual commotion, bounded and went off toward Edinburgh at a gallop, bearing along with it, sole occupant of the gig, the body of the dead and long-dissected Gray.

The Story of the Rippling Train

MARY LOUISA MOLESWORTH

'Let's tell ghost stories then,' said Gladys.

'Aren't you tired of them? One hears nothing else nowadays. And they're all "authentic," really vouched for, only you never see the person who saw or heard or felt the ghost. It is always somebody's sister or cousin, or friend's friend,' objected young Mrs Snowdon, another of the guests at the Quarries.

'I don't know that that is quite a reasonable ground for discrediting them *en masse*,' said her husband. 'It is natural enough, indeed inevitable, that the principal or principals in such cases should be much more rarely come across than the stories themselves. A hundred people can repeat the story, but the author, or rather hero, of it, can't be in a hundred places at once. You don't disbelieve in any other statement or narrative merely because you have never seen the prime mover in it?'

'But I didn't say I discredited them on that account,' said Mrs Snowdon. 'You take one up so, Archie. I'm not logical and reasonable—I don't pretend to be. If I meant anything, it was that a ghost story would have a great pull over other ghost stories if one could see the person it happened to. One does get rather provoked at *never* coming across him or her,' she added, a little petulantly.

She was tired; they were all rather tired, for it was the first evening since the party had assembled at the large country house known as 'The Quarries', on which there was not to be dancing, with the additonal fatigue of 'ten miles there and ten back again'; and three or four evenings of such doings without intermission tell, even on the young and vigorous.

Tonight, various less energetic ways of passing the evening had been proposed. Music, games, reading aloud, recitation—none had found favour in everybody's sight, and now Gladys Lloyd's proposal that they should 'tell ghost stories', seemed likely to fall flat also.

For a moment or two no one answered Mrs Snowdon's last remarks. Then, somewhat to everybody's surprise, the young daughter of the house turned to her mother.

'Mamma,' she said, 'don't be vexed with me—I know you warned me once to be careful how I spoke of it; but *wouldn't* it be nice if Uncle Paul would tell us his ghost story? And then, Mrs Snowdon,' she went on, 'you could always say you had heard *one* ghost story at or from—which should I say?—headquarters.'

Lady Denholme glanced round half nervously before she replied.

'Locally speaking, it would not be *at* headquarters, Nina', she said. 'The Quarries was not the scene of your uncle's ghost story. But I almost think it is better not to speak about it—I am not sure that he would like it mentioned, and he will be coming in a moment. He had only a note to write.'

'I do wish he would tell it to us,' said Nina regretfully. 'Don't you think, mamma, I might just run to the study and ask him, and if he did not like the idea he might say so to me, and no one would seem to know anything about it? Uncle Paul is so kind—I'm never afraid of asking him any favour.'

'Thank you, Nina, for your good opinion of me; you see there is no rule without exceptions; listeners do sometimes hear pleasant things of themselves,' said Mr Marischal, as he at that moment came round the screen which half concealed the doorway. 'What is the special favour you were thinking of asking me?'

Nina looked rather taken aback.

'How softly you opened the door, Uncle Paul,' she said. 'I would not have spoken of you if I had known you were there.'

'But after all you were saying no harm,' observed her brother Michael. 'And for my part I don't believe Uncle Paul would mind our asking him what we were speaking of.'

'What was it?' asked Mr Marischal. 'I think, as I have heard so much, you may as well tell me the whole.'

'It was only——,' began Nina, but her mother interrupted her.

'I have told Nina not to speak of it, Paul,' she said anxiously; 'but—it was only that all these young people are talking about ghost stories, and they want you to tell them your own strange experience. You must not be vexed with them.'

'Vexed,' said Mr Marischal; 'not in the least.' But for a moment or two he said no more, and even pretty spoilt Mrs Snowdon looked a little uneasy.

'You shouldn't have persisted, Nina,' she whispered.

Mr Marischal must have had unusually quick ears. He looked up and smiled.

'I really don't mind telling you all there is to hear,' he said. 'At one time I had a sort of dislike to mentioning the story, for the sake of others. The details would have led to its being recognized—and it might have been painful. But there is no one now living to whom it would matter—you know,' he added, turning to his sister, 'her husband is dead, too.'

Lady Denholme shook her head.

'No,' she said, 'I did not hear.'

'Yes,' said her brother, 'I saw his death in the papers last year. He had married again, I believe. There is not now, therefore, any reason why I should not tell the story, if it will interest you,' he went on, turning to the others. 'And there is not very much to tell. Not worth making such a preamble about. It was—let me see—yes, it must be nearly fifteen years ago.'

'Wait a moment, Uncle Paul,' said Nina. 'Yes, that's all right, Gladys. You and I will hold each other's hands, and pinch hard if we get very frightened.'

'Thank you,' Miss Lloyd replied. 'On the whole I should prefer for you not to hold my hand.'

'But I won't pinch you so as to hurt,' said Nina, reassuringly; 'and it isn't as if we were in the dark.'

'Shall I turn down the lamps?' asked Mr Snowdon.

'No, no,' exclaimed his wife.

'There really is nothing frightening—scarcely even "creepy", in my story at all,' said Mr Marischal, half apologetically. 'You make me feel like an impostor.'

'Oh no, Uncle Paul, don't say that. It is all my fault for interrupting,' said Nina. 'Now go on, please. I have Gladys's hand all the same,' she added, *sotto voce*; 'it's just as well to be prepared.'

'Well then,' began Mr Marischal once more, 'it must be nearly fifteen years ago. And I had not seen her for fully ten years before that again! I was not thinking of her in the least; in a sense I had really forgotten her: she had quite gone out of my life—that has always struck me as a very curious point in the story,' he added parenthetically.

'Won't you tell us who "she" was, Uncle Paul?' asked Nina, half shyly.

'Oh yes, I was going to do so. I am not skilled in story-telling, you see. She was, at the time I first knew her—at the only time indeed that I knew her—a very sweet and attractive girl, named Maud Bertram. She was very pretty—more than pretty, for she had remarkably regular features—her profile was always admired, and a tall and graceful figure. And she was a bright and happy creature too; that perhaps was almost her greatest charm. You will wonder—I see the question hovering on your lips, Miss Lloyd, and on yours too, Mrs Snowdon—

why if I admired and liked her so much I did not go further. And I will tell you frankly that I did not because I dared not. I had then no prospect of being able to marry for years to come, and I was not very young. I was already nearly thirty, and Maud was quite ten years younger. I was wise enough and old enough to realize the situation thoroughly, and to be on my guard.'

'And Maud?' asked Mrs Snowdon.

'She was surrounded by admirers—it seemed to me then that it would have been insufferable conceit to have even asked myself if it could matter to her. It was only in the light of after events that the possibility of my having been mistaken occurred to me. And I don't even now see that I could have acted otherwise——' here Uncle Paul sighed a little. 'We were the best of friends. She knew that I admired her, and she seemed to take a frank pleasure in its being so. I had always hoped that she really liked and trusted me as a friend, but no more. The last time I saw her was just before I started for Portugal, where I remained three years. When I returned to London Maud had been married for two years, and had gone straight out to India on her marriage, and except by some few friends who had known us both intimately I seldom heard her mentioned. And time passed—I cannot say I had exactly forgotten her, but she was not much or often in my thoughts. I was a busy and much absorbed man, and life had proved a serious matter to me. Now and then some passing resemblance would recall her to my mind—once especially when I had been asked to look in to see the young wife of one of my cousins in her court-dress, something in her figure and bearing brought back Maud to my memory, for it was thus, in full dress, that I had last seen her, and thus, perhaps unconsciously, her image had remained photographed on my brain. But as far as I can recollect at the time when the occurrence I am going to relate to you happened, I had not been thinking of Maud Bertram for months. I was in London just then, staying with my brother, my eldest brother, who had been married for several years and lived in our own old town house in —— Square. It was in April, a clear spring day, with no fog or half-lights about, and it was not yet four o'clock in the afternoon—not very ghost-like circumstances, you will admit. I had come home early from my club—it was a sort of holiday time with me just then for a few weeks—intending to get some letters written which had been on my mind for some days, and I had sauntered into the library, a pleasant, fair-sized room lined with books, on the first-floor. Before setting to work I sat down for a moment or two in an easy-chair by the fire, for it was still cool enough weather to make a fire desirable, and began thinking over my letters. No thought, no shadow of a thought of my old friend Miss Bertram was present with me, of that I am perfectly certain. The door was on the same side of the room as the fireplace; as I sat there, half facing the fire, I also half faced the door. I had not shut it properly on coming in, I had only closed it without turning the handle, and I did not feel surprised when it slowly and noiselessly swung open, till it stood right out into the room, concealing the actual doorway from my view. You will perhaps understand the position better if you think of the door as just then acting like a screen to the doorway. From where I sat I could not have seen anyone entering the room till he or she had got beyond the door itself. I glanced up, half expecting to see someone come in, but there was no one; the door had swung open of itself. For the moment I sat on, with only the vague thought passing through my mind, "I must shut it before I begin to write."

'But suddenly I found my eyes fixing themselves on the carpet; something had come within their range of vision, compelling their attention in a mechanical sort of way. What was it?

'"Smoke," was my first idea. "Can there be anything on fire?" But I dismissed the notion almost as soon as it suggested itself. The something, faint and shadowy, that came slowly rippling itself in as it were, beyond the dark wood of the open door, was yet too material for "smoke". My next idea was a curious one: "It looks like soapy water," I said to myself; "can one of the housemaids have been scrubbing, and upset a pail on the stairs?" For the stair to the next floor almost faced the library door. But—no, I rubbed my eyes and looked again—the soapy water theory gave way. The wavy something that kept gliding, rippling in, gradually assumed a more substantial appearance. It was—yes, I suddenly became convinced of it, it was ripples of soft silken stuff, creeping in as if in some mysterious way unfolded or unrolled, not jerkily or irregularly, but glidingly and smoothly, like little wavelets on the sea-shore.

'And I sat there and gazed. "Why did you not jump up and look behind the door to see what it was?" you may reasonably ask. That question I cannot answer. Why I sat still, as if bewitched, or under some irresistible influence, I cannot tell, but so it was.

'And it—came always rippling in, till at last it began to rise as it still came on, and I saw that a figure, a tall graceful woman's figure, was slowly advancing, backwards of course, into the room, and that the waves of pale silk—a very delicate shade of pearly grey I think it must have been—were in fact the lower portion of a long court-train, the upper part of which hung in deep folds from the lady's waist. She moved in—I cannot describe the motion, it was not like ordinary walking or stepping backwards—till the whole of her figure and the clear profile of her face and head were distinctly visible, and when at last she stopped and stood there full in my view just, but only just beyond the door, I saw—it came upon me like a flash, that she was no stranger to me, this mysterious visitant! I recognized, unchanged it seemed to me since the day, ten years ago, when I had last seen her, the beautiful features of Maud Bertram.'

Mr Marischal stopped a moment. Nobody spoke. Then he went on again.

'I should not have said "unchanged". There was one great change in the sweet face. You remember my telling you that one of my girl-friend's greatest charms was her bright sunny happiness—she never seemed gloomy or depressed or dissatisfied, seldom even pensive. But in this respect the face I sat there gazing at was utterly unlike Maud Bertram's. Its expression, as she—or 'it'—stood there looking, not towards me, but out beyond, as if at someone or something outside the doorway, was of the profoundest sadness. Anything *so* sad I have never seen in a human face, and I trust I never may. But I sat on, as motionless almost as she, gazing at her fixedly, with no desire, no power perhaps, to move or approach more nearly to the phantom. I was not in the least frightened. I knew it *was* a phantom, but I felt paralysed and as if I myself had somehow got outside of ordinary conditions. And there I sat—staring at Maud, and there she stood, gazing before her with that terrible, unspeakable sadness in her face, which, even though I felt no *fear*, seemed to freeze me with a kind of unutterable pity.

'I don't know how long I had sat thus, or how long I might have continued to sit there, almost as if in a trance, when suddenly I heard the front-door bell ring. It seemed to awaken me. I started up and glanced round, half-expecting that I should find the vision dispelled. But no; she was still there, and I sank back into my seat just as I heard my brother coming quickly upstairs. He came towards the library, and seeing the door wide open walked in, and I, still gazing, saw his figure *pass through that of the woman in the doorway* as you may walk through a wreath of mist or smoke—only, don't misunderstand me, the figure of Maud till that moment had had nothing unsubstantial about it. She had looked to me, as she stood there, literally and exactly like a living woman—the shade of her dress, the colour of her hair, the few ornaments she wore, all were as defined and clear as yours, Nina, at the present moment, and remained so, or perhaps became so again as soon as my brother was well within the room. He came forward, addressing me by name, but I answered him in a whisper, begging him to be silent and to sit down on the seat opposite me for a moment or two. He did so, though he was taken aback by my strange manner, for I still kept my eyes fixed on the door. I had a queer consciousness that if I looked away *it* would fade, and I wanted to keep cool and see what would happen. I asked Herbert in a low voice if *he* saw nothing, but though he mechanically followed the direction of my eyes, he shook his head in bewilderment. And for a moment or two he remained thus. Then I began to notice that the figure was growing less clear, as if it were receding, yet without growing smaller to the sight; it grew fainter and vaguer, the colours grew hazy. I rubbed my eyes once or twice with a half idea that my long watching was making them misty, but it was not so. My eyes were not at fault—slowly but surely Maud Bertram, or her ghost, melted away, till all trace of her had gone. I saw again the familiar pattern of the carpet where she had stood and the objects of the room that had been hidden by her draperies—all again in the most commonplace way: but she was gone, quite gone.

'Then Herbert, seeing me relax my intense gaze, began to question me. I told him exactly what I have told you. He answered, as every common-sensible person of course would, that it was strange, but that such things did happen sometimes, and were classed by the wise under the head of "optical delusions". I was not well, perhaps, he suggested. Been overworking? Had I not better see a doctor? But I shook my head. I was quite well, and I said so. And perhaps he was right, it might be an optical delusion only. I had never had any experience of such things.

'"All the same," I said, "I shall mark down the date."

'Herbert laughed and said that was what people always did in such cases. If he knew where Mrs —— then was, he would write to her, just for the fun of the thing, and ask her to be so good as to look up in her diary, if she kept one, and let us know what she had been doing on that particular day—"the 6th of April, isn't it?" he said—when I would have it her wraith had paid me a visit. I let him talk. It seemed to remove the strange, painful impression—painful because of that terrible sadness in the sweet face. But we neither of us knew where she was, we scarcely remembered her married name! And so there was nothing to be done—except, what I did at once in spite of Herbert's rallying—to mark down the day and hour with scrupulous exactness in *my* diary.

'Time passed. I had not forgotten my strange experience, but of course the impression of it lessened by degrees till it seemed more like a curious dream than anything more real, when one day I *did* hear of poor Maud again. "Poor" Maud I cannot help calling her. I heard of her indirectly, and probably, but for the sadness of her story, I should never have heard it at all. It was a friend of her husband's family who had mentioned the circumstances in the hearing of a friend of mine, and one day something brought round the conversation to old times, and he startled me by suddenly enquiring if I remembered Maud Bertram. I said, of course, I did. Did he know anything of her? And then he told me.

'She was dead—she had died some months ago after a long and trying illness, the result of a terrible accident. She had caught fire one evening when dressed for some grand entertainment or other, and though her injuries did not seem likely to be fatal at the time, she had never recovered the shock.

' "She was so pretty," my friend said, "and one of the saddest parts of it was that I hear she was terrifically disfigured, and she took this most sadly to heart. The right side of her face was utterly ruined, and the sight of the right eye lost, though, strange to say, the left side entirely escaped, and seeing her in profile one would have had no notion of what had happened. Was it not sad? She was such a sweet bright creature."

'I did not tell him *my* story, for I did not want it chattered about, but a strange sort of shiver ran through me at his words. *It was the left side of her face only* that the wraith of my poor friend had allowed me to see.'

'Oh, Uncle Paul!' exclaimed Nina.

'And—as to the dates?' enquired Mr Snowdon.

'I never knew the exact date of the accident,' said Mr Marischal, 'but that of her death was fully six months after I had seen her. And in my own mind, I have never made any doubt that it was at, or about, probably a short time after, the accident that she came to me. It seemed a kind of appeal for sympathy—and—a farewell also, poor child.'

They all sat silent for some little time, and then Mr Marischal got up and went off to his own quarters, saying something vaguely about seeing if his letters had gone.

'What a touching story,' said Gladys Lloyd. 'I am afraid, after all, it has been more painful than he realized for Mr Marischal to tell it. Did you know anything of Maud's husband, dear Lady Denholme? Was he kind to her? Was she happy?'

'We never heard much about her married life,' her hostess replied. 'But I have no reason to think she was unhappy. Her husband married again two or three years after her death, but that says nothing.'

'N—no,' said Nina. 'All the same, mamma, I am sure she really did love Uncle Paul very much—much more than he had any idea of. Poor Maud!'

'And he has never married,' added Gladys.

'No,' said Lady Denholme; 'but there have been many practical difficulties in the way of his doing so. He has had a most absorbingly busy life, and now that he is more at leisure he feels himself too old to form new ties.'

'But,' persisted Nina, 'if he had had any idea at the time, that Maud cared for him so?'

'Ah well,' Lady Denholme allowed, 'in that case, in spite of the practical difficulties, things would probably have been different.'

And again Nina repeated softly,

'Poor Maud!'

At the End of the Passage

RUDYARD KIPLING

The sky is lead and our faces are red,
  And the gates of Hell are opened and riven,
  And the winds of Hell are loosened and driven,
And the dust flies up in the face of Heaven,
  And the clouds come down in a fiery sheet,
Heavy to raise and hard to be borne.
  And the soul of man is turned from his meat,
Turned from the trifles for which he has striven
  Sick in his body, and heavy hearted,
  And his soul flies up like the dust in the sheet
  Breaks from his flesh and is gone and departed,
As the blasts they blow on the cholera-horn.

Himalayan

Four men, each entitled to 'life, liberty, and the pursuit of happiness', sat at a table playing whist. The thermometer marked—for them—one hundred and one degrees of heat. The room was darkened till it was only just possible to distinguish the pips of the cards and the very white faces of the players. A tattered, rotten punkah of whitewashed calico was puddling the hot air and whining dolefully at each stroke. Outside lay gloom of a November day in London. There was neither sky, sun, nor horizon—nothing but a brown purple haze of heat. It was as though the earth were dying of apoplexy.

From time to time clouds of tawny dust rose from the ground without wind or warning, flung themselves tablecloth-wise among the tops of the parched trees, and came down again. Then a-whirling dust-devil would scutter across the plain for a couple of miles, break, and fall outward, though there was nothing to check its flight save a long low line of piled railway-sleepers white with the dust, a cluster of huts made of mud, condemned rails, and canvas, and the one squat four-roomed bungalow that belonged to the assistant engineer in charge of a section of the Gaudhari State line then under construction.

The four, stripped to the thinnest of sleeping-suits, played whist crossly, with wranglings as to leads and returns. It was not the best kind of whist, but they had taken some trouble to arrive at it. Mottram of the Indian Survey had ridden thirty and railed one hundred miles from his lonely post in the desert since the night before; Lowndes of the Civil Service, on special duty in the political department, had come as far to escape for an instant the miserable intrigues of an impoverished native State whose king alternately fawned and blustered for more money from the pitiful revenues contributed by hard-wrung peasants and despairing camel-breeders; Spurstow, the doctor of the line, had left a cholera-stricken camp of coolies to look after itself for forty-eight hours while he associated with white men once more. Hummil, the assistant engineer, was the host. He stood fast and received his friends thus every Sunday if they could come in. When one of them failed to appear, he would send a telegram to his last address, in order that he might know whether the defaulter were dead or alive. There are very many places in the East where it is not good or kind to let your acquaintances drop out of sight even for one short week.

The players were not conscious of any special regard for each other. They squabbled whenever they met; but they ardently desired to meet, as men without water desire to drink. They were lonely folk who understood the dread meaning of loneliness. They were all under thirty years of age—which is too soon for any man to possess that knowledge.

'Pilsener?' said Spurstow, after the second rubber, mopping his forehead.

'Beer's out, I'm sorry to say, and there's hardly enough soda-water for tonight,' said Hummil.

'What filthy bad management!' Spurstow snarled.

'Can't help it. I've written and wired; but the trains don't come through regularly yet. Last week the ice ran out—as Lowndes knows.'

'Glad I didn't come. I could ha' sent you some if I had known, though. Phew! it's too hot to go on playing bumblepuppy.' This with a savage scowl at Lowndes, who only laughed. He was a hardened offender.

Mottram rose from the table and looked out of a chink in the shutters.

'What a sweet day!' said he.

The company yawned all together and betook themselves to an aimless investigation of all Hummil's possessions—guns, tattered novels, saddlery, spurs, and the like. They had fingered them a score of times before, but there was really nothing else to do.

'Got anything fresh?' said Lowndes.

'Last week's *Gazette of India*, and a cutting from a home paper. My father sent it out. It's rather amusing.'

'One of those vestrymen that call 'emselves M.P.s again, is it?' said Spurstow, who read his newspapers when he could get them.

'Yes. Listen to this. It's to your address, Lowndes. The man was making a speech to his constituents, and he piled it on. Here's a sample, "And I assert unhesitatingly that the Civil Service in India is the preserve—the pet preserve—of the aristocracy of England. What does the democracy—what do the masses—get from that country, which we have step by step fraudulently annexed? I answer, nothing whatever. It is farmed with a single eye to their own interests by the scions of the aristocracy. They take good care to maintain their lavish scale of incomes, to avoid or stifle any inquiries into the nature and conduct of their administration, while they themselves force the unhappy peasant to pay with the sweat of his brow for all the luxuries in which they are lapped."' Hummil waved the cutting above his head. ''Ear! 'ear!' said his audience.

Then Lowndes, meditatively, 'I'd give—I'd give three months' pay to have that gentleman spend one month with me and see how the free and independent native prince works things. Old Timbersides'—this was his flippant title for an honoured and decorated feudatory prince—'has been wearing my life out this week past for money. By Jove, his latest performance was to send me one of his women as a bribe!'

'Good for you! Did you accept it?' said Mottram.

'No. I rather wish I had, now. She was a pretty little person, and she yarned away to me about the horrible destitution among the king's women-folk. The darlings haven't had any new clothes for nearly a month, and the old man wants to buy a new drag from Calcutta—solid silver railings and silver lamps, and trifles of that kind. I've tried to make him understand that he has played the deuce with the revenues for the last twenty years and must go slow. He can't see it.'

'But he has the ancestral treasure-vaults to draw on. There must be three millions at least in jewels and coin under his palace,' said Hummil.

'Catch a native king disturbing the family treasure! The priests forbid it except as the last resort. Old Timbersides has added something like a quarter of a million to the deposit in his reign.'

'Where the mischief does it all come from?' said Mottram.

'The country. The state of the people is enough to make you sick. I've known the taxmen wait by a milch-camel till the foal was born and then hurry off the mother for arrears. And what can I do? I can't get the court clerks to give me any accounts; I can't raise anything more than a fat smile from the commander-in-chief when I find out the troops are three months in arrears; and old Timbersides begins to weep when I speak to him. He has taken to the King's Peg heavily, liqueur brandy for whisky, and Heidsieck for soda-water.'

'That's what the Rao of Jubela took to. Even a native can't last long at that,' said Spurstow. 'He'll go out.'

'And a good thing, too. Then I suppose we'll have a council of regency, and a tutor for the young prince, and hand him back his kingdom with ten years' accumulations.'

'Whereupon that young prince, having been taught all the vices of the English, will play ducks and drakes with the money and undo ten years' work in eighteen months. I've seen that business before,' said Spurstow. 'I should tackle the king with a light hand if I were you, Lowndes. They'll hate you quite enough under any circumstances.

'That's all very well. The man who looks on can talk about the light hand; but you can't clean a pig-sty with a pen dipped in rose-water. I know my risks; but nothing has happened yet. My servant's an old Pathan, and he cooks for me. They are hardly likely to bribe him, and I don't accept food from my true friends, as they call themselves. Oh, but it's weary work! I'd sooner be with you, Spurstow. There's shooting near your camp.'

'Would you? I don't think it. About fifteen deaths a day don't incite a man to shoot anything but himself. And the worst of it is that the poor devils look at you as though you ought to save them. Lord knows, I've tried everything. My last attempt was empirical, but it pulled an old man through. He was brought to me apparently past hope, and I gave him gin and Worcester sauce with cayenne. It cured him; but I don't recommend it.'

'How do the cases run generally?' said Hummil.

'Very simply indeed. Chlorodyne, opium pill, chlorodyne, collapse, nitre, bricks to the feet, and then—the burning-ghaut. The last seems to be the only thing that stops the trouble. It's black cholera, you know. Poor devils! But, I will say, little Bunsee Lal, my apothecary, works like a demon. I've recommended him for promotion if he comes through it all alive.'

'And what are your chances, old man?' said Mottram.

'Don't know; don't care much; but I've sent the letter in. What are you doing with yourself generally?'

'Sitting under a table in the tent and spitting on the sextant to keep it cool,' said the man of the survey. 'Washing my eyes to avoid ophthalmia, which I shall certainly get, and trying to make a sub-surveyor understand that an error of five degrees in an angle isn't quite so small as it looks. I'm altogether alone, y' know, and shall be till the end of the hot weather.'

'Hummil's the lucky man,' said Lowndes, flinging himself into a long chair. 'He has an actual roof—torn as to the ceiling-cloth, but still a roof—over his head. He sees one train daily. He can get beer and soda-water and ice 'em when God is good. He has books, pictures—they were torn from the *Graphic*—and the society of the excellent sub-contractor Jevins, besides the pleasure of receiving us weekly.'

Hummil smiled grimly. 'Yes, I'm the lucky man, I suppose. Jevins is luckier.'

'How? Not——'

'Yes. Went out. Last Monday.'

'By his own hand?' said Spurstow quickly, hinting the suspicion that was in everybody's mind. There was no cholera near Hummil's section. Even fever gives a man at least a week's grace, and sudden death generally implied self-slaughter.

'I judge no man this weather,' said Hummil. 'He had a touch of the sun, I fancy; for last week, after you fellows had left, he came into the verandah and told me that he was going home to see his wife, in Market Street, Liverpool, that evening.

'I got the apothecary in to look at him, and we tried to make him lie down. After an hour or two he rubbed his eyes and said he believed he had had a fit, hoped he hadn't said anything rude. Jevins had a great idea of bettering himself socially. He was very like Chucks in his language.'

'Well?'

'Then he went to his own bungalow and began cleaning a rifle. He told the servant that he was going to shoot buck in the morning. Naturally he fumbled with the trigger, and shot himself through the head—accidentally. The apothecary sent in a report to my chief, and Jevins is buried somewhere out there. I'd have wired to you, Spurstow, if you could have done anything.'

'You're a queer chap,' said Mottram. 'If you'd killed the man yourself you couldn't have been more quiet about the business.'

'Good Lord! what does it matter?' said Hummil calmly. 'I've got to do a lot of his overseeing work in addition to my own. I'm the only person that suffers. Jevins is out of it, by pure accident, of course, but out of it. The apothecary was going to write a long screed on suicide. Trust a babu to drivel when he gets the chance.'

'Why didn't you let it go in as suicide?' said Lowndes.

'No direct proof. A man hasn't many privileges in his country, but he might at least be allowed to mishandle his own rifle. Besides, some day I may need a man to smother up an accident to myself. Live and let live. Die and let die.'

'You take a pill,' said Spurstow, who had been watching Hummil's white face narrowly. 'Take a pill, and don't be an ass. That sort of talk is skittles. Anyhow, suicide is shirking your work. If I were Job ten times over, I should be so interested in what was going to happen next that I'd stay on and watch.'

'Ah! I've lost that curiosity,' said Hummil.

'Liver out of order?' said Lowndes feelingly.

'No. Can't sleep. That's worse.'

'By Jove, it is!' said Mottram. 'I'm that way every now and then, and the fit has to wear itself out. What do you take for it?'

'Nothing. What's the use? I haven't had ten minutes' sleep since Friday morning.'

'Poor chap! Spurstow, you ought to attend to this,' said Mottram. 'Now you mention it, your eyes are rather gummy and swollen.'

Spurstow, still watching Hummil, laughed lightly. 'I'll patch him up, later on. Is it too hot, do you think, to go for a ride?'

'Where to?' said Lowndes wearily. 'We shall have to go away at eight, and there'll be riding enough for us then. I hate a horse when I have to use him as a necessity. Oh, heavens! what is there to do?'

'Begin whist again, at chick points ['a chick' is supposed to be eight shillings] and a gold mohur on the rub,' said Spurstow promptly.

'Poker. A month's pay all round for the pool—no limit—and fifty-rupee raises. Somebody would be broken before we got up,' said Lowndes.

'Can't say that it would give me any pleasure to break any man in this company,' said Mottram. 'There isn't enough excitement in it, and it's foolish.' He crossed over to the worn and battered little camp-piano—wreckage of a married household that had once held the bungalow—and opened the case.

'It's used up long ago,' said Hummil. 'The servants have picked it to pieces.'

The piano was indeed hopelessly out of order, but Mottram managed to bring the rebellious notes into a sort of agreement, and there rose from the ragged keyboard something that might once have been the ghost of a popular music-hall song. The men in the long chairs turned with evident interest as Mottram banged the more lustily.

'That's good!' said Lowndes. 'By Jove! the last time I heard that song was in '79, or thereabouts, just before I came out.'

'Ah!' said Spurstow with pride, 'I was home in '80.' And he mentioned a song of the streets popular at that date.

Mottram executed it roughly. Lowndes criticized and volunteered emendations. Mottram dashed into another ditty, not of the music-hall character, and made as if to rise.

'Sit down,' said Hummil. 'I didn't know that you had any music in your composition. Go on playing until you can't think of anything more. I'll have that piano tuned up before you come again. Play something festive.'

Very simple indeed were the tunes to which Mottram's art and the limitations of the piano could give effect, but the men listened with pleasure, and in the pauses talked all together of what they had seen or heard when they were last at home. A dense dust-storm sprung up outside, and swept roaring over the house, enveloping it in the choking darkness of midnight, but Mottram continued unheeding, and the crazy tinkle reached the ears of the listeners above the flapping of the tattered ceiling-cloth.

In the silence after the storm he glided from the more directly personal songs of Scotland, half humming them as he played, into the Evening Hymn.

'Sunday,' said he, nodding his head.

'Go on. Don't apologize for it,' said Spurstow.

Hummil laughed long and riotously. 'Play it, by all means. You're full of surprises today. I didn't know you had such a gift of finished sarcasm. How does that thing go?'

Mottram took up the tune.

'Too slow by half. You miss the note of gratitude,' said Hummil. 'It ought to go to the "Grasshopper's Polka"—this way.' And he chanted, *prestissimo*,

> 'Glory to thee, my God, this night,
> For all the blessings of the light.

That shows we really feel our blessings. How does it go on?—

> If in the night I sleepless lie,
> My soul with sacred thoughts supply;
> May no ill dreams disturb my rest,—

Quicker, Mottram!—

> Or powers of darkness me molest!'

'Bah! what an old hypocrite you are!'

'Don't be an ass,' said Lowndes. 'You are at full liberty to make fun of anything else you like, but leave that hymn alone. It's associated in my mind with the most sacred recollections——'

'Summer evenings in the country, stained-glass window, light going out, and you and she jamming your heads together over one hymnbook,' said Mottram.

'Yes, and a fat old cockchafer hitting you in the eye when you walked home. Smell of hay, and a moon as big as a bandbox sitting on the top of a haycock; bats, roses, milk and midges,' said Lowndes.

'Also mothers. I can just recollect my mother singing me to sleep with that when I was a little chap,' said Spurstow.

The darkness had fallen on the room. They could hear Hummil squirming in his chair.

'Consequently,' said he testily, 'you sing it when you are seven fathom deep in Hell! It's an insult to the intelligence of the Deity to pretend we're anything but tortured rebels.'

'Take *two* pills,' said Spurstow; 'that's tortured liver.'

'The usually placid Hummil is in a vile bad temper. I'm sorry for his coolies tomorrow,' said Lowndes, as the servants brought in the lights and prepared the table for dinner.

As they were settling into their places about the miserable goat-chops, and the smoked tapioca pudding, Spurstow took occasion to whisper to Mottram, 'Well done, David!'

'Look after Saul, then,' was the reply.

'What are you two whispering about?' said Hummil suspiciously.

'Only saying that you are a damned poor host. This fowl can't be cut,' returned Spurstow with a sweet smile. 'Call this a dinner?'

'I can't help it. You don't expect a banquet, do you?'

Throughout that meal Hummil contrived laboriously to insult directly and pointedly all his guests in succession, and at each insult Spurstow kicked the aggrieved persons under the table; but he dared not exchange a glance of intelligence with either of them. Hummil's face was white and pinched, while his eyes were unnaturally large. No man dreamed for a moment of resenting his savage personalities, but as soon as the meal was over they made haste to get away.

'Don't go. You're just getting amusing, you fellows. I hope I haven't said anything that annoyed you. You're such touchy devils.' Then, changing the note into one of almost abject entreaty, Hummil added, 'I say, you surely aren't going?'

'In the language of the blessed Jorrocks, where I dines I sleeps,' said Spurstow. 'I want to have a look at your coolies tomorrow, if you don't mind. You can give me a place to lie down in, I suppose?'

The others pleaded the urgency of their several duties next day, and, saddling up, departed together, Hummil begging them to come next Sunday. As they jogged off, Lowndes unbosomed himself to Mottram—

'. . . And I never felt so like kicking a man at his own table in my life. He said I cheated at whist, and reminded me I was in debt! 'Told you you were as good as a liar to your face! You aren't half indignant enough over it.'

'Not I,' said Mottram. 'Poor devil! Did you ever know old Hummy behave like that before or within a hundred miles of it?'

'That's no excuse. Spurstow was hacking my shin all the time, so I kept a hand on myself. Else I should have——'

'No, you wouldn't. You'd have done as Hummy did about Jevins; judge no man this weather. By Jove! the buckle of my bridle is hot in my hand! Trot out a bit, and 'ware rat-holes.'

Ten minutes' trotting jerked out of Lowndes one very sage remark when he pulled up, sweating from every pore—

''Good thing Spurstow's with him tonight.'

'Ye-es. Good man, Spurstow. Our roads turn here. See you again next Sunday, if the sun doesn't bowl me over.'

'S'pose so, unless old Timbersides' finance minister manages to dress some of my food. Goodnight, and—God bless you!'

'What's wrong now?'

'Oh, nothing.' Lowndes gathered up his whip, and, as he flicked Mottram's mare on the flank, added, 'You're not a bad little chap, that's all.' And the mare bolted half a mile across the sand, on the word.

In the assistant engineer's bungalow Spurstow and Hummil smoked the pipe of silence together, each narrowly watching the other. The capacity of a bachelor's establishment is as elastic as its arrangements are simple. A servant cleared away the dining-room table, brought in a couple of rude native bedsteads made of tape strung on a light wood frame, flung a square of cool Calcutta matting over each, set them side by side, pinned two towels to the punkah so that their fringes should just sweep clear of the sleeper's nose and mouth, and announced that the couches were ready.

The men flung themselves down, ordering the punkah-coolies by all the powers of Hell to pull. Every door and window was shut, for the outside air was that of an oven. The atmosphere within was only 104°, as the thermometer bore witness, and heavy with the foul smell of badly-trimmed kerosene lamps; and this stench, combined with that of native tobacco, baked brick, and dried earth, sends the heart of many a strong man down to his boots, for it is the smell of the Great Indian

Empire when she turns herself for six months into a house of torment. Spurstow packed his pillows craftily so that he reclined rather than lay, his head at a safe elevation above his feet. It is not good to sleep on a low pillow in the hot weather if you happen to be of thick-necked build, for you may pass with lively snores and gugglings from natural sleep into the deep slumber of heat-apoplexy.

'Pack your pillows,' said the doctor sharply, as he saw Hummil preparing to lie down at full length.

The night-light was trimmed; the shadow of the punkah wavered across the room, and the '*flick*' of the punkah-towel and the soft whine of the rope through the wall-hole followed it. Then the punkah flagged, almost ceased. The sweat poured from Spurstow's brow. Should he go out and harangue the coolie? It started forward again with a savage jerk, and a pin came out of the towels. When this was replaced, a tomtom in the coolie-lines began to beat with the steady throb of a swollen artery inside some brain-fevered skull. Spurstow turned on his side and swore gently. There was no movement on Hummil's part. The man had composed himself as rigidly as a corpse, his hands clinched at his sides. The respiration was too hurried for any suspicion of sleep. Spurstow looked at the set face. The jaws were clinched, and there was a pucker round the quivering eyelids.

'He's holding himself as tightly as ever he can,' thought Spurstow. 'What in the world is the matter with him?—Hummil!'

'Yes,' in a thick constrained voice.

'Can't you get to sleep?'

'No.'

'Head hot? Throat feeling bulgy? or how?'

'Neither, thanks. I don't sleep much, you know.'

' 'Feel pretty bad?'

'Pretty bad, thanks. There is a tomtom outside, isn't there? I thought it was my head at first. . . . Oh, Spurstow, for pity's sake give me something that will put me asleep, sound asleep, if it's only for six hours!' He sprang up, trembling from head to foot. 'I haven't been able to sleep naturally for days, and I can't stand it! I can't stand it!'

'Poor old chap!'

'That's no use. Give me something to make me sleep. I tell you I'm nearly mad. I don't know what I say half my time. For three weeks I've had to think and spell out every word that has come through my lips before I dared say it. Isn't that enough to drive a man mad? I can't see things correctly now, and I've lost my sense of touch. My skin aches—my skin aches! Make me sleep. Oh, Spurstow, for the love of

God make me sleep sound. It isn't enough merely to let me dream. Let me sleep!'

'All right, old man, all right. Go slow; you aren't half as bad as you think.'

The flood-gates of reserve once broken, Hummil was clinging to him like a frightened child. 'You're pinching my arm to pieces.'

'I'll break your neck if you don't do something for me. No, I didn't mean that. Don't be angry, old fellow.' He wiped the sweat off himself as he fought to regain composure. 'I'm a bit restless and off my oats, and perhaps you could recommend some sort of sleeping mixture—bromide of potassium.'

'Bromide of skittles! Why didn't you tell me this before? Let go of my arm, and I'll see if there's anything in my cigarette-case to suit your complaint.' Spurstow hunted among his day-clothes, turned up the lamp, opened a little silver cigarette-case, and advanced on the expectant Hummil with the daintiest of fairy squirts.

'The last appeal of civilization,' said he, 'and a thing I hate to use. Hold out your arm. Well, your sleeplessness hasn't ruined your muscle; and what a thick hide it is! Might as well inject a buffalo subcutaneously. Now in a few minutes the morphia will begin working. Lie down and wait.'

A smile of unalloyed and idiotic delight began to creep over Hummil's face. 'I think,' he whispered,—'I think I'm going off now. Gad! it's positively heavenly! Spurstow, you must give me that case to keep; you——' The voice ceased as the head fell back.

'Not for a good deal,' said Spurstow to the unconscious form. 'And now, my friend, sleeplessness of your kind being very apt to relax the moral fibre in little matters of life and death, I'll just take the liberty of spiking your guns.'

He paddled into Hummil's saddle-room in his bare feet and uncased a twelve-bore rifle, an express, and a revolver. Of the first he unscrewed the nipples and hid them in the bottom of a saddlery-case; of the second he abstracted the lever, kicking it behind a big wardrobe. The third he merely opened, and knocked the doll-head bolt of the grip up with the heel of a riding-boot.

'That's settled,' he said, as he shook the sweat off his hands. 'These little precautions will at least give you time to turn. You have too much sympathy with gun-room accidents.'

And as he rose from his knees, the thick muffled voice of Hummil cried in the doorway, 'You fool!'

Such tones they use who speak in the lucid intervals of delirium to their friends a little before they die.

Spurstow started, dropping the pistol. Hummil stood in the doorway, rocking with helpless laughter.

'That was awf'ly good of you, I'm sure,' he said, very slowly, feeling for his words. 'I don't intend to go out by my own hand at present. I say, Spurstow, that stuff won't work. What shall I do? What shall I do?' And panic terror stood in his eyes.

'Lie down and give it a chance. Lie down at once.'

'I daren't. It will only take me half-way again, and I shan't be able to get away this time. Do you know it was all I could do to come out just now? Generally I am as quick as lightning; but you had clogged my feet. I was nearly caught.'

'Oh yes, I understand. Go and lie down.'

'No, it isn't delirium; but it was an awfully mean trick to play on me. Do you know I might have died?'

As a sponge rubs a slate clean, so some power unknown to Spurstow had wiped out of Hummil's face all that stamped it for the face of a man, and he stood at the doorway in the expression of his lost innocence. He had slept back into terrified childhood.

'Is he going to die on the spot?' thought Spurstow. Then, aloud, 'All right, my son. Come back to bed, and tell me all about it. You couldn't sleep; but what was all the rest of the nonsense?'

'A place, a place down there,' said Hummil, with simple sincerity. The drug was acting on him by waves, and he was flung from the fear of a strong man to the fright of a child as his nerves gathered sense or were dulled.

'Good God! I've been afraid of it for months past, Spurstow. It has made every night hell to me; and yet I'm not conscious of having done anything wrong.'

'Be still, and I'll give you another dose. We'll stop your nightmares, you unutterable idiot!'

'Yes, but you must give me so much that I can't get away. You must make me quite sleepy, not just a little sleepy. It's so hard to run then.'

'I know it; I know it. I've felt it myself. The symptoms are exactly as you describe.'

'Oh, don't laugh at me, confound you! Before this awful sleeplessness came to me I've tried to rest on my elbow and put a spur in the bed to sting me when I fell back. Look!'

'By Jove! the man has been rowelled like a horse! Ridden by the nightmare with a vengeance! And we all thought him sensible enough. Heaven send us understanding! You like to talk, don't you?'

'Yes, sometimes. Not when I'm frightened. *Then* I want to run. Don't you?'

'Always. Before I give you your second dose try to tell me exactly what your trouble is.'

Hummil spoke in broken whispers for nearly ten minutes, whilst Spurstow looked into the pupils of his eyes and passed his hand before them once or twice.

At the end of the narrative the silver cigarette-case was produced, and the last words that Hummil said as he fell back for the second time were, 'Put me quite to sleep; for if I'm caught I die, I die!'

'Yes, yes; we all do that sooner or later, thank Heaven who has set a term to our miseries,' said Spurstow, settling the cushions under the head. 'It occurs to me that unless I drink something I shall go out before my time. I've stopped sweating, and—I wear a seventeen-inch collar.' He brewed himself scalding hot tea, which is an excellent remedy against heat-apoplexy if you take three or four cups of it in time. Then he watched the sleeper.

'A blind face that cries and can't wipe its eyes, a blind face that chases him down corridors! H'm! Decidedly, Hummil ought to go on leave as soon as possible; and, sane or otherwise, he undoubtedly did rowel himself most cruelly. Well, Heaven send us understanding!'

At mid-day Hummil rose, with an evil taste in his mouth, but an unclouded eye and a joyful heart.

'I was pretty bad last night, wasn't I?' said he.

'I have seen healthier men. You must have had a touch of the sun. Look here: if I write you a swingeing medical certificate, will you apply for leave on the spot?'

'No.'

'Why not? You want it.'

'Yes, but I can hold on till the weather's a little cooler.'

'Why should you, if you can get relieved on the spot?'

'Burkett is the only man who could be sent; and he's a born fool.'

'Oh, never mind about the line. You aren't so important as all that. Wire for leave, if necessary.'

Hummil looked very uncomfortable.

'I can hold on till the Rains,' he said evasively.

'You can't. Wire to headquarters for Burkett.'

'I won't. If you want to know why, particularly, Burkett is married, and his wife's just had a kid, and she's up at Simla, in the cool, and Burkett has a very nice billet that takes him into Simla from Saturday to Monday. That little woman isn't at all well. If Burkett was transferred she'd try to follow him. If she left the baby behind she'd fret herself to death. If she came—and Burkett's one of those selfish little beasts who are always talking about a wife's place being with her husband—she'd die. It's murder to bring a woman here just now. Burkett hasn't the physique of a rat. If he came here he'd go out; and I know she hasn't any money, and I'm pretty sure she'd go out too. I'm salted in a sort of way, and I'm not married. Wait till the Rains, and then Burkett can get thin down here. It'll do him heaps of good.'

'Do you mean to say that you intend to face—what you have faced, till the Rains break?'

'Oh, it won't be so bad, now you've shown me a way out of it. I can always wire to you. Besides, now I've once got into the way of sleeping, it'll be all right. Anyhow, I shan't put in for leave. That's the long and the short of it.'

'My great Scott! I thought all that sort of thing was dead and done with.'

'Bosh! You'd do the same yourself. I feel a new man, thanks to that cigarette-case. You're going over to camp now, aren't you?'

'Yes; but I'll try to look you up every other day, if I can.'

'I'm not bad enough for that. I don't want you to bother. Give the coolies gin and ketchup.'

'Then you feel all right?'

'Fit to fight for my life, but not to stand out in the sun talking to you. Go along, old man, and bless you!'

Hummil turned on his heel to face the echoing desolation of his bungalow, and the first thing he saw standing in the verandah was the figure of himself. He had met a similar apparition once before, when he was suffering from overwork and the strain of the hot weather.

'This is bad—already,' he said, rubbing his eyes. 'If the thing slides away from me all in one piece, like a ghost, I shall know it is only my eyes and stomach that are out of order. If it walks—my head is going.'

He approached the figure, which naturally kept at an unvarying distance from him, as is the use of all spectres that are born of overwork. It slid through the house and dissolved into swimming specks within the eyeball as soon as it reached the burning light of the garden. Hummil went about his business till even. When he came in to dinner he found himself sitting at the table. The vision rose and walked out hastily. Except that it cast no shadow it was in all respects real.

No living man knows what that week held for Hummil. An increase of the epidemic kept Spurstow in camp among the coolies, and all he could do was to telegraph to Mottram, bidding him go to the bungalow and sleep there. But Mottram was forty miles away from the nearest telegraph, and knew nothing of anything save the needs of the survey till he met, early on Sunday morning, Lowndes and Spurstow heading towards Hummil's for the weekly gathering.

'Hope the poor chap's in a better temper,' said the former, swinging himself off his horse at the door. 'I suppose he isn't up yet.'

'I'll just have a look at him,' said the doctor. 'If he's asleep there's no need to wake him.'

And an instant later, by the tone of Spurstow's voice calling upon them to enter, the men knew what had happened. There was no need to wake him.

The punkah was still being pulled over the bed, but Hummil had departed this life at least three hours.

The body lay on its back, hands clinched by the side, as Spurstow had seen it lying seven nights previously. In the staring eyes was written terror beyond the expression of any pen.

Mottram, who had entered behind Lowndes, bent over the dead and touched the forehead lightly with his lips. 'Oh, you lucky, lucky devil!' he whispered.

But Lowndes had seen the eyes, and withdrew shuddering to the other side of the room.

'Poor chap! poor old chap! And the last time I met him I was angry. Spurstow, we should have watched him. Has he——?'

Deftly Spurstow continued his investigations, ending by a search round the room.

'No, he hasn't,' he snapped. 'There's no trace of anything. Call the servants.'

They came, eight or ten of them, whispering and peering over each other's shoulders.

'When did your Sahib go to bed?' said Spurstow.

'At eleven or ten, we think,' said Hummil's personal servant.

'He was well then? But how should you know?'

'He was not ill, as far as our comprehension extended. But he had slept very little for three nights. This I know, because I saw him walking much, and specially in the heart of the night.'

As Spurstow was arranging the sheet, a big straight-necked hunting-spur tumbled on the ground. The doctor groaned. The personal servant peeped at the body.

'What do you think, Chuma?' said Spurstow, catching the look on the dark face.

'Heaven-born, in my poor opinion, this that was my master has descended into the Dark Places, and there has been caught because he was not able to escape with sufficient speed. We have the spur for evidence that he fought with Fear. Thus have I seen men of my race do with thorns when a spell was laid upon them to overtake them in their sleeping hours and they dared not sleep.'

'Chuma, you're a mud-head. Go out and prepare seals to be set on the Sahib's property.'

'God has made the Heaven-born. God has made me. Who are we, to enquire into the dispensations of God? I will bid the other servants hold aloof while you are reckoning the tale of the Sahib's property. They are all thieves, and would steal.'

'As far as I can make out, he died from—oh, anything; stoppage of the heart's action, heat-apoplexy, or some other visitation,' said Spurstow to his companions. 'We must make an inventory of his effects, and so on.'

'He was scared to death,' insisted Lowndes. 'Look at those eyes! For pity's sake don't let him be buried with them open!'

'Whatever it was, he's clear of all the trouble now,' said Mottram softly.

Spurstow was peering into the open eyes.

'Come here,' said he. 'Can you see anything there?'

'I can't face it!' whimpered Lowndes. 'Cover up the face! Is there any fear on earth that can turn a man into that likeness? It's ghastly. Oh, Spurstow, cover it up!'

'No fear—on earth,' said Spurstow. Mottram leaned over his shoulder and looked intently.

'I see nothing except some grey blurs in the pupil. There can be nothing there, you know.'

'Even so. Well, let's think. It'll take half a day to knock up any sort of coffin; and he must have died at midnight. Lowndes, old man, go out and tell the coolies to break ground next to Jevins's grave. Mottram, go round the house with Chuma and see that the seals are put on things. Send a couple of men to me here, and I'll arrange.'

The strong-armed servants when they returned to their own kind told a strange story of the doctor Sahib vainly trying to call their master back to life by magic arts—to wit, the holding of a little green box that clicked to each of the dead man's eyes, and of a bewildered muttering on the part of the doctor Sahib, who took the little green box away with him.

The resonant hammering of a coffin-lid is no pleasant thing to hear, but those who have experience maintain that much more terrible is the soft swish of the bed-linen, the reeving and unreeving of the bed-tapes, when he who has fallen by the roadside is apparelled for burial, sinking gradually as the tapes are tied over, till the swaddled shape touches the floor and there is no protest against the indignity of hasty disposal.

At the last moment Lowndes was seized with scruples of conscience. 'Ought you to read the service, from beginning to end?' said he to Spurstow.

'I intend to. You're my senior as a civilian. You can take it if you like.'

'I didn't mean that for a moment. I only thought if we could get a chaplain from somewhere, I'm willing to ride anywhere, and give poor Hummil a better chance. That's all.'

'Bosh!' said Spurstow, as he framed his lips to the tremendous words that stand at the head of the burial service.

After breakfast they smoked a pipe in silence to the memory of the dead. Then Spurstow said absently—

' 'Tisn't medical science.'

'What?'

'Things in a dead man's eye.'

'For goodness' sake leave that horror alone!' said Lowndes. 'I've seen a native die of pure fright when a tiger chivied him. I know what killed Hummil.'

'The deuce you do! I'm going to try to see.' And the doctor retreated into the bathroom with a Kodak camera. After a few minutes there was the sound of something being hammered to pieces, and he emerged, very white indeed.

'Have you got a picture?' said Mottram. 'What does the thing look like?'

'It was impossible, of course. You needn't look, Mottram. I've torn up the films. There was nothing there. It was impossible.'

'That,' said Lowndes, very distinctly, watching the shaking hand striving to relight the pipe, 'is a damned lie.'

Mottram laughed uneasily. 'Spurstow's right,' he said. 'We're all in such a state now that we'd believe anything. For pity's sake let's try to be rational.'

There was no further speech for a long time. The hot wind whistled without, and the dry trees sobbed. Presently the daily train, winking brass, burnished steel, and spouting steam, pulled up panting in the intense glare. 'We'd better go on that,' said Spurstow. 'Go back to work. I've written my certificate. We can't do any more good here, and work'll keep our wits together. Come on.'

No one moved. It is not pleasant to face railway journeys at mid-day in June. Spurstow gathered up his hat and whip, and, turning in the doorway, said—

> 'There may be Heaven—there must be Hell.
> Meantime, there is our life here. We-ell?'

Neither Mottram nor Lowndes had any answer to the question.

'To Let'

B. M. CROKER

Some years ago, when I was a slim young spin, I came out to India to live with my brother Tom: he and I were members of a large and somewhat impecunious family, and I do not think my mother was sorry to have one of her four grown-up daughters thus taken off her hands. Tom's wife, Aggie, had been at school with my eldest sister; we had known and liked her all our lives. She was quite one of ourselves, and as she and the children were at home when Tom's letter was received, and his offer accepted, she helped me to choose my slender outfit with judgement, zeal, and taste; endowed me with several pretty additions to my wardrobe; superintended the fitting of my gowns and the trying on of my hats, with most sympathetic interest, and finally escorted me out to Lucknow, under her own wing, and installed me in the only spare room in her comfortable bungalow in Dilkongha.

My sister-in-law is a pretty little brunette, rather pale, with dark hair, brilliant black eyes, a resolute mouth and a bright, intelligent expression. She is orderly, trim and feverishly energetic, and seems to live every moment of her life. Her children, her wardrobe, her house, her servants, and last, not least, her husband, are all models in their way; and yet she has plenty of time for tennis and dancing, and talking and walking. She is, undoubtedly, a remarkably talented little creature, and especially prides herself on her nerve and her power of will, or will power. I suppose they are the same thing? and I am sure they are all the same to Tom, who worships the sole of her small slipper. Strictly between ourselves she is the ruling member of the family, and turns her lord and master round her little finger. Tom is big and fair, of course, the opposite to his wife, quiet, rather easy-going and inclined to be indolent, but Aggie rouses him up, and pushes him to the front, and keeps him there. She knows all about his department, his prospects of promotion, his prospects of furlough, of getting acting appointments, and so on, even better than he does himself. The chief of Tom's department—have I said that Tom is in the Irritation Office?—has placed it solemnly on record that he considers little Mrs Shandon a surprisingly clever woman. The two children, Bob and Tor, are merry, oppressively active monkeys, aged three and five years respectively. As for myself I am tall and fair, and I wish I could add pretty; but this is a true story. My eyes are blue, my teeth are white, my hair is red—alas, a blazing red; and I was, at this period, nineteen years of age; and now I think I have given a sufficient outline of the whole family.

We arrived at Lucknow in November, when the cold weather is delightful, and everything was delightful to me. The bustle and life of a great Indian station, the novelty of my surroundings, the early morning rides, picnics down the river, and dances at the 'Chutter Munzil', made me look upon Lucknow as a paradise on earth; and in this light I still regarded it, until a great change came over the temperature, and the month of April introduced me to red-hot winds, sleepless nights, and the intolerable 'brain fever' bird. Aggie had made up her mind definitely on one subject: we were not to go away to the hills until the rains. Tom could only get two months' leave (July and August), and she did not intend to leave him to grill on the plains alone. As for herself and the children—not to speak of me—we had all come out from home so recently we did not require a change. The trip to Europe had made a vast hole in the family stocking, and she wished to economize; and who can economize with two establishments in full swing? Tell me this, ye Anglo-Indian matrons. With a large, cool bungalow, plenty of punkhas, khuskhus tatties, ice, and a thermantidote, surely we could manage to brave May and June—at any rate the attempt was made. Gradually the hills drained Lucknow week by week; family after family packed up, warned us of our folly in remaining on the plains, offered to look for houses for us, and left by the night mail. By the middle of May, the place was figuratively empty. Nothing can be more dreary than a large station in the hot weather, unless it is an equally forsaken hill station in the depths of winter, when the mountains are covered with snow: the mall no longer resounds with gay voices and the tramp of Jampanies, but is visited by bears and panthers, and the houses are closed, and, as it were, put to bed in straw! As for Lucknow in the summer, it was a melancholy spot; the public gardens were deserted, the chairs at the Chutter Munzil stood empty, the very bands had gone to the hills!, the shops were shut, the baked white roads, no longer thronged with carriages and bamboo carts, gave ample room to the humble ekka, or a Dhobie's meagre donkey shuffling along in the dust.

Of course we were not the *only* people remaining in the place, grumbling at the heat and dust and life in general; but there can be no sociability with the thermometer above 100 in the shade. Through the long, long Indian day we sat and gasped, in darkened rooms, and consumed quantities of 'Nimbo pegs', i.e. limes and soda water, and listened to the fierce hot winds roaring along the road and driving the roasted leaves before it; and in the evening, when the sun had set, we went for a melancholy drive through the Wingfield Park, or round by Martiniere College, and met our friends at the library and compared sensations and thermometers. The season was exceptionally bad, but people say that every year, and presently Bobby and Tor began to fade: their little white faces and listless eyes appealed to Aggie as Tom's anxious expostulations had never done. 'Yes, they must go to the hills with *me*.' But this idea I repudiated at once; I refused to undertake the responsibility—I, who could scarcely speak a word to the servants—who had no experience! Then Bobbie had a bad go of fever—intermittent fever; the beginning of the end to his alarmed mother; the end being represented by a large gravestone! She now became as firmly determined to go as she had previously been resolved to stay; but it was so late in the season to take a house. Alas, alas, for the beautiful tempting advertisements in the *Pioneer*, which we had seen and scorned! Aggie wrote to a friend in a certain hill station, called for this occasion only 'Kantia', and Tom wired to a house agent, who triumphantly replied by letter that there was not *one* unlet bungalow on his books. This missive threw us into the depths of despair; there seemed no alternative but a hill hotel, and the usual quarters that await the last comers, and the proverbial welcome for children and dogs (we had only four); but the next day brought us good news from Aggie's friend Mrs Chalmers.

Dear Mrs Shandon—she said—

I received your letter, and went at once to Cursitjee, the agent. Every hole and corner up here seems full, and he had not a single house to let. Today I had a note from him, saying that Briarwood is vacant; the people who took it are not coming up, they have gone to Naini Tal. You *are* in luck. I have just been out to see the house, and have secured it for you. It is a mile and a half from the club, but I know that you and your sister are capital walkers. I envy you. Such a charming place—two sitting-rooms, four bedrooms, four bathrooms, a hall, servants' go-downs, stabling, and a splendid view from a very pretty garden, and only Rs. 800 for the season! Why, I am paying Rs. 1,000 for a *very* inferior house, with scarcely a stick of furniture and no view. I feel so proud of myself, and I am longing to show you my treasure trove. Telegraph when you start, and I shall have a milk man in waiting and fires in all the rooms.

Yours sincerely,
Edith Chalmers.

We now looked upon Mrs Chalmers as our best and dearest friend, and began to get under way at once. A long journey in India is a serious business when the party comprises two ladies, two children, two ayahs and five other servants, three fox terriers, a mongoose and a Persian cat—all these animals going to the hills for the benefit of their health—not to speak of a ton of luggage, including crockery and lamps, a cottage piano, a goat and a pony. Aggie and I, the children, one ayah, two terriers, the cat and mongoose, our bedding and pillows, the tiffin basket and ice basket, were all stowed into one compartment, and I must confess that the journey was truly miserable. The heat was stifling, despite the water tatties. One of the terriers had a violent dispute with the cat, and the cat had a difference with the mongoose, and Bob and Tor had a pitched battle more than once. I actually wished myself back in Lucknow. I was most truly thankful to wake one morning to find myself under the shadow of the Himalayas—not a mighty, snow-clad range of everlasting hills, but merely the spurs—the moderate slopes, covered with scrub and loose shale and jungle, and deceitful little trickling watercourses. We sent the servants on ahead, whilst we rested at the Dak bungalow near the railway station, and then followed them at our leisure. We accomplished the ascent in dandies—open kind of boxes, half box half chair, carried on the shoulders of four men. This was an entirely novel sensation to me, and at first an agreeable one, so long as the slopes were moderate and the paths wide; but the higher we went, the narrower became the path, the steeper the naked precipice; and as my coolies would walk at the extreme edge, with the utmost indifference to my frantic appeals to 'Bector! Bector!'—and would change poles at the most agonizing corners—my feelings were very mixed, especially when droves of loose pack ponies came thundering down hill, with no respect for the rights of the road. Late at night we passed through Kantia, and arrived at Briarwood far too weary to be critical. Fires were blazing, supper was prepared, and we dispatched it in haste, and most thankfully went to bed and slept soundly, as anyone would do who had spent thirty-six hours in a crowded compartment and ten in a cramped wooden case.

The next morning, rested and invigorated, we set out on a tour of inspection; and it is almost worth while to undergo a certain amount of baking on the sweltering heat of the plains, in order to enjoy those deep first draughts of cool hill air, instead of a stifling, dust-laden atmosphere, and to appreciate the green valleys and blue hills by force of contrast to the far-stretching, eye-smarting, white glaring roads that intersect the burnt-up plains—roads and plains that even the pariah abandons, salamander though he be!

To our delight and surprise, Mrs Chalmers had by no means

overdrawn the advantages of our new abode. The bungalow was solidly built of stone, two storied, and ample in size. It stood on a kind of shelf, cut out of the hillside, and was surrounded by a pretty flower garden, full of roses, fuchsias, carnations. The high road passed the gate, from which the avenue descended direct to the entrance door, which was at the end of the house, and from whence ran a long passage. Off this passage three rooms opened to the right, all looking south, and all looking into a deep, delightful, flagged verandah. The stairs were very steep. At the head of them, the passage and rooms were repeated. There were small nooks, and dressing-rooms, and convenient out-houses, and plenty of good water; but the glory of Briarwood was undoubtedly its verandah: it was fully twelve feet wide, roofed with zinc, and overhung a precipice of a thousand feet—not a startlingly sheer khud, but a tolerably straight descent of grey-blue shale rocks and low jungle. From it there was a glorious view, across a valley, far away, to the snowy range. It opened at one end into the avenue, and was not inclosed; but at the side next the precipice there was a stout wooden railing, with netting at the bottom, for the safety of too enterprising dogs or children. A charming spot, despite its rather bold situation; and as Aggie and I sat in it, surveying the scenery and inhaling the pure hill air, and watching Bob and Tor tearing up and down playing horses, we said to one another that 'the verandah alone was worth half the rent'.

'It's absurdly cheap,' exclaimed my sister-in-law complacently. 'I wish you saw the hovel *I* had, at Simla, for the same rent. I wonder if it is feverish, or badly drained, or what?'

'Perhaps it has a ghost,' I suggested facetiously; and at such an absurd idea we both went into peals of laughter.

At this moment Mrs Chalmers appeared, brisk, rosy, and breathlessly benevolent, having walked over from Kantia.

'So you have found it,' she said as we shook hands. 'I said nothing about this delicious verandah! I thought I would keep it as a surprise. I did not say a word too much for Briarwood, did I?'

'Not half enough,' we returned rapturously; and presently we went in a body, armed with a list from the agent, and proceeded to go over the house and take stock of its contents.

'It's not a bit like a *hill* furnished house,' boasted Mrs Chalmers, with a glow of pride, as she looked round the drawing-room; 'carpets, curtains, solid, *very* solid chairs, and Berlin wool worked screens, a card-table, and any quantity of pictures.'

'Yes, don't they look like family portraits?' I suggested, as we gazed at them. There was one of an officer in faded water colours, another of his wife, two of a previous generation in oils and amply gilded frames, two sketches of an English country house, and some framed photographs, groups of grinning cricketers or wedding guests. All the rooms were well, almost handsomely, furnished in an old-fashioned style. There was no scarcity of wardrobes, looking-glasses, or even armchairs, in the bedrooms, and the pantry was fitted out—a most singular circumstance—with a large supply of handsome glass and china, lamps, old moderators, coffee and tea pots, plated side dishes and candlesticks, cooking utensils and spoons and forks, wine coasters, and a cake-basket. These articles were all let with the house, much to our amazement, provided we were responsible for the same. The china was Spode, the plate old family heirlooms, with a crest—a winged horse—on everything, down to the very mustard spoons.

'The people who own this house must be lunatics,' remarked Aggie as she peered round the pantry; 'fancy hiring out one's best family plate and good old china! And I saw some ancient music books in the drawing-room, and there is a side saddle in the bottle khana.'

'My dear, the people who owned this house are dead,' explained Mrs Chalmers. 'I heard all about them last evening from Mrs Starkey.'

'Oh, is *she* up there?' exclaimed Aggie somewhat fretfully.

'Yes, her husband is cantonment magistrate. This house belonged to an old retired colonel and his wife. They and his niece lived here. These were all their belongings. They died within a short time of one another, and the old man left a queer will, to say that the house was to remain precisely as they left it for twenty years, and at the end of that time it was to be sold and all the property dispersed. Mrs Starkey says she is sure that he never intended it to be *let*, but the heir-at-law insists on that, and is furious at the terms of the will.'

'Well, it is a very good thing for us,' remarked Aggie; 'we are as comfortable here as if we were in our own house: there is a stove in the kitchen; there are nice boxes for firewood in every room, clocks, real hair mattresses—in short, it is as you said, a treasure trove.'

We set to work to modernize the drawing-room with phoolkaries, Madras muslin curtains, photograph screens and frames, and such like portable articles. We placed the piano across a corner, arranged flowers in some handsome Dresden china vases, and entirely altered and improved the character of the room. When Aggie had dispatched a most glowing description of our new quarters to Tom, and we had had tiffin, we set off to walk into Kantia to put our names down at the library and to enquire for letters at the post office. Aggie met a good many acquaintances—who does not who has lived five years in India in the same district?

Among them Mrs Starkey, an elderly lady with a prominent nose and goggle eyes, who greeted her loudly across the reading-room table in this agreeable fashion:

'And so you have come up after *all*, Mrs Shandon. Someone told me that you meant to remain below, but I knew you never could be so wicked as to keep your poor little children in that heat.' Then coming round and dropping into a chair beside her she said, 'And I suppose this young lady is your sister-in-law?'

Mrs Starkey eyed me critically, evidently appraising my chances in the great marriage market. She herself had settled her own two daughters most satisfactorily, and had now nothing to do but interest herself in these people's affairs.

'Yes,' acquiesced Aggie, 'Miss Shandon—Mrs Starkey.'

'And so you have taken Briarwood?'

'Yes, we have been most lucky to get it.'

'I hope you will think so at the end of three months,' observed Mrs Starkey with a significant pursing of her lips. 'Mrs Chalmers is a stranger up here, or she would not have been in such a hurry to jump at it.'

'Why, what is the matter with it?' enquired Aggie. 'It is well built, well furnished, well situated, and very cheap.'

'That's just it—*suspiciously* cheap. Why, my dear Mrs Shandon, if there was not something against it, it would let for two hundred rupees a month.'

'And what is against it?'

'It's haunted! There you have the reason in two words.'

'Is that all? I was afraid it was the drains. I don't believe in ghosts and haunted houses. What are we supposed to see?'

'Nothing,' retorted Mrs Starkey, who seemed a good deal nettled at our smiling incredulity.

'Nothing!' with an exasperating laugh.

'No, but you will make up for it in hearing. Not now—you are all right for the next six weeks—but after the monsoon breaks I give you a week at Briarwood. No one would stand it longer, and indeed you might as well bespeak your rooms at Cooper's Hotel *now*. There is always a rush up here in July by the two month's leave people, and you will be poked into some wretched go-down.'

Aggie laughed rather a careless ironical little laugh and said, 'Thank you, Mrs Starkey; but I think we will stay on where we are; at any rate for the present.'

'Of course it will be as *you* please. What do you think of the verandah?' she enquired with a curious smile.

'I think, as I was saying to Susan, that it is worth half the rent of the house.'

'And in *my* opinion the house is worth double rent without it,' and with this enigmatic remark she rose and sailed away.

'Horrid old frump,' exclaimed Aggie as we walked home in the starlight. 'She is jealous and angry that she did not get Briarwood *herself*—I know her so well. She is always hinting and repeating stories about the nicest people—always decrying your prettiest dress or your best servant.'

We soon forgot all about Mrs Starkey and her dismal prophecy, being too gay and too busy to give her, or it, a thought. We had so many engagements—tennis parties and tournaments, picnics, concerts, dances and little dinners. We ourselves gave occasional afternoon teas in the verandah, using the best Spode cups and saucers and the old silver cake-basket, and were warmly complimented on our good fortune in securing such a charming house and garden. One day the children discovered to their great joy that the old chowkidar belonging to the bungalow possessed an African grey parrot—a rare bird indeed in India; he had a battered Europe cage, doubtless a remnant of better days, and swung on his ring, looking up at us enquiringly out of his impudent little black eyes.

The parrot had been the property of the former inmates of Briarwood, and as it was a long-lived creature, had survived its master and mistress, and was boarded out with the chowkidar, at one rupee per month.

The chowkidar willingly carried the cage into the verandah, where the bird seemed perfectly at home.

We got a little table for its cage, and the children were delighted with him, as he swung to and fro, with a bit of cake in his wrinkled claw.

Presently he startled us all by suddenly calling 'Lucy', in a voice that was as distinct as if it had come from a human throat. 'Pretty Lucy—Lu—cy.'

'That must have been the niece,' said Aggie. 'I expect she was the original of that picture over the chimney-piece in your room; she looks like a Lucy.'

It was a large framed half-length photograph of a very pretty girl, in a white dress, with gigantic open sleeves. The ancient parrot talked incessantly now that he had been restored to society; he whistled for the dogs, and brought them flying to his summons, to his great satisfaction and their equally great indignation. He called 'Qui hye' so naturally, in a lady's shrill soprano, or a gruff male bellow, that I have no doubt our servants would have liked to have wrung his neck. He coughed and expectorated like an old gentleman, and whined like a puppy, and mewed like a cat, and I am sorry to add, sometimes swore like a trooper; but his most constant cry was, 'Lucy, where are you, pretty Lucy—Lucy—Lu—cy?'

Aggie and I went to various picnics, but to that given by the Chalmers (in honour of Mr Chalmers's brother Charlie, a captain in a Gurkha regiment, just come up to Kantia on leave) Aggie was unavoidably absent. Tor had a little touch of fever, and she did not like to leave him; but I went under my hostess's care, and expected to enjoy myself immensely. Alas! on that self-same afternoon the long expected monsoon broke, and we were nearly drowned! We rode to the selected spot, five miles from Kantia, laughing and chattering, indifferent to the big blue-black clouds that came slowly, but surely, sailing up from below; it was a way they had had for days and nothing had come of it. We spread the tablecloth, boiled the kettle, unpacked the hampers, in spite of sharp gusts of wind and warning rumbling thunder. Just as we had commenced to reap the reward of our exertions, there fell a few huge drops, followed by a vivid flash, and then a tremendous crash of thunder, like a whole park of artillery, that seemed to shake the mountains, and after this the deluge. In less than a minute we were soaked through; we hastily gathered up the tablecloth by its four ends, gave it to the coolies and fled. It was all I could do to stand against the wind; only for Captain Chalmers I believe I would have been blown away; as it was I lost my hat, it was whirled into space. Mrs Chalmers lost her boa, and Mrs Starkey, not merely her bonnet, but some portion of her hair. We were truly in a wretched plight, the water streaming down our faces and squelching in our boots; the little trickling mountain rivulets were now like racing seas of turbid water; the lightning was almost blinding; the trees rocked dangerously and lashed one another with their quivering branches. I had never been out in such a storm before, and I hope I never may again. We reached Kantia more dead than alive, and Mrs Chalmers sent an express to Aggie, and kept me till the next day. After raining as it only *can* rain in the Himalayas, the weather cleared, the sun shone, and I rode home in borrowed plumes, full of my adventures and in the highest spirits. I found Aggie sitting over the fire in the drawing-room, looking ghastly white: that was nothing uncommon; but terribly depressed, which was most unusual. 'I am afraid you have neuralgia?' I said as I kissed her; she nodded and made no reply.

'How is Tor?' I enquired as I drew a chair up to the fire.

'Better—quite well.'

'Any news—any letter?'

'Not a word—not a line.'

'Has anything happened to Pip'—Pip was a fox terrier, renowned for having the shortest tail and being the most impertinent dog in Lucknow—'or the mongoose?'

'No, you silly girl! Why do you ask such questions?'

'I was afraid something was amiss; you seem rather down on your luck.' Aggie shrugged her shoulders and then said:

'What put such an absurd idea into your head? Tell me all about the picnic,' and she began to talk rapidly and to ask me various questions; but I observed that once she had set me going—no difficult task—her attention flagged, her eyes wandered from my face to the fire. She was not listening to half I said, and my most thrilling descriptions were utterly lost on this indifferent, abstracted little creature! I noticed from this time that she had become strangely nervous for her. She invited herself to the share of half my bed; she was restless, *distrait*, and even irritable; and when I was asked out to spend the day, dispensed with my company with an alacrity that was by no means flattering. Formerly, of an evening she used to herd the children home at sundown, and tear me away from the delights of the reading-room at seven o'clock; now she hung about the library until almost the last moment, until it was time to put out the lamps, and kept the children with her, making transparent pretexts for their company. Often we did not arrive at home till half-past eight o'clock. I made no objections to these late hours, neither did Charlie Chalmers, who often walked back with us and remained to dinner. I was amazed to notice that Aggie seemed delighted to have his company, for she had always expressed a rooted aversion to what she called 'tame young men', and here was this new acquaintance dining with us at least thrice a week! About a month after the picnic we had a spell of dreadful weather—thunderstorms accompanied by torrents. One pouring afternoon, Aggie and I were sitting over the drawing-room fire, whilst the rain came fizzing down among the logs and ran in rivers off the roof and out of the spouts. There had been no going out that day, and we were feeling rather flat and dull, as we sat in a kind of ghostly twilight, with all outdoor objects swallowed up in mist, listening to the violent battering of the rain on the zinc verandah, and the storm which was growling round the hills. 'Oh, for a visitor!' I exclaimed; 'but no one but a fish or a lunatic would be out on such an evening.'

'No one, indeed,' echoed Aggie in a melancholy tone. 'We may as well draw the curtains and have in the lamp and tea to cheer us up.'

She had scarcely finished speaking when I heard the brisk trot of a horse along the road. It stopped at the gate and came rapidly down our avenue. I heard the wet gravel crunching under his hoofs and—yes—a man's cheery whistle. My heart jumped, and I half rose from my chair. It must be Charlie Chalmers braving the elements to see *me!*—such, I must confess, was my incredible vanity! He did not stop at the front door as usual, but rode straight into the verandah, which afforded ample room and shelter for half-a-dozen mounted men.

'Aggie,' I said eagerly, 'do you hear? It must be——'

I paused—my tongue silenced by the awful pallor of her face and the expression of her eyes as she sat with her little hands clutching the arms of her chair, and her whole figure bent forward in an attitude of listening—an attitude of terror.

'What is it, Aggie?' I said, 'Are you ill?'

As I spoke the horse's hoofs made a loud clattering noise on the stone-paved verandah outside and a man's voice—a young man's eager voice—called, 'Lucy'.

Instantly a chair near the writing-table was pushed back and some-one went quickly to the window—a French one—and bungled for a moment with the fastening—I always had a difficulty with that window *myself*. Aggie and I were within the bright circle of the firelight, but the rest of the room was dim, and outside the streaming grey sky was spasmodically illuminated by occasional vivid flashes that lit up the surrounding hills as if it were daylight. The trampling of impatient hoofs and the rattling of a door handle were the only sounds that were audible for a few breathless seconds; but during those seconds Pip, bristling like a porcupine and trembling violently in every joint, had sprung off my lap and crawled abjectly under Aggie's chair, seemingly in a transport of fear. The door was opened audibly, and a cold, icy blast swept in, that seemed to freeze my very heart and made me shiver from head to foot. At this moment there came with a sinister blue glare the most vivid flash of lightning I ever saw. It lit up the whole room, which was empty save for ourselves, and was instantly followed by a clap of thunder that caused my knees to knock together and that terrified me and filled me with horror. It evidently terrified the horse too; there was a violent plunge, a clattering of hoofs on the stones, a sudden loud crash of smashing timber, a woman's long, loud, piercing shriek, which stopped the very beating of my heart, and then a frenzied struggle in the cruel, crumbling, treacherous shale, the rattle of loose stones and the hollow roar of something sliding down the precipice.

I rushed to the door and tore it open, with that awful despairing cry still ringing in my ears. The verandah was empty; there was not a soul to be seen or a sound to be heard, save the rain on the roof.

'Aggie,' I screamed, 'come here! Someone has gone over the verandah and down the khud! You heard him.'

'Yes,' she said, following me out; 'but come in—come in.'

'I believe it was Charlie Chalmers'—shaking her as I spoke. 'He has been killed—killed—killed! And you stand and do nothing. Send people! Let us go ourselves! Bearer! Ayah! Khidmatgar!' I cried, raising my voice.

'Hush! It was *not* Charlie Chalmers,' she said, vainly endeavouring to draw me into the drawing-room. 'Come in—come in.'

'No, no!'—pushing her away and wringing my hands. 'How cruel you are! How inhuman! There is a path. Let us go at once—at once!'

'You need not trouble yourself, Susan,' she interrupted; 'and you need not cry and tremble—*they* will bring him up. What you heard was supernatural; it was not real.'

'No—no—no! It was all real. Oh! that scream is in my ears still.'

'I will convince you,' said Aggie, taking my hand as she spoke. 'Feel all along the verandah. Are the railings broken?'

I did as she bade me. No, though very wet and clammy, the railing was intact.

'Where is the broken place?' she asked.

Where, indeed?

'Now,' she continued, 'since you will not come in, look over, and you will see something more presently.'

Shivering with fear and cold, drifting rain, I gazed down as she bade me, and there far below I saw lights moving rapidly to and fro, evidently in search of something. After a little delay they congregated in one place. There was a low, booming murmur—they had found him—and presently they commenced to ascend the hill, with the 'hum-hum' of coolies carrying a burden. Nearer and nearer the lights and sounds came up to the very brink of the khud, past the end of the verandah. Many steps and many torches—faint blue torches held by invisible hands—invisible but heavy-footed bearers carried their burden slowly upstairs and along the passage, and deposited it with a dump in Aggie's bedroom! As we stood clasped in one another's arms and shaking all over, the steps descended, the ghostly lights passed up the avenue and disappeared in the gathering darkness. The repetition of the tragedy was over for that day.

'Have you heard it before?' I asked with chattering teeth, as I bolted the drawing-room window.

'Yes, the evening of the picnic and twice since. That is the reason I have always tried to stay out till late and to keep you out. I was hoping and praying you might never hear it. It always happens just before dark. I am afraid you have thought me very queer of late. I have told no end of stories to keep you and the children from harm—I have——'

'I think you have been very kind,' I interrupted. 'Oh, Aggie, shall you ever get that crash and that awful cry out of your head?'

'*Never!*' hastily lighting the candles as she spoke.

'Is there anything more?' I asked tremulously.

'Yes; sometimes at night the most terrible weeping and sobbing in my bedroom,' and she shuddered at the mere recollection.

'Do the servants know?' I asked anxiously.

'The ayah Mumà has heard it, and the khánsámáh says his mother is sick and he must go, and the bearer wants to attend his brother's wedding. They will *all* leave.'

'I suppose most people know too?' I suggested dejectedly.

'Yes, don't you remember Mrs Starkey's warnings and her saying that without the verandah the house was worth double rent? We understand that dark speech of hers *now*, and we have not come to Cooper's Hotel yet.'

'No, not *yet*. I wish we *had*. I wonder what Tom will say? He will be here in another fortnight. Oh, I wish he was here now.'

In spite of our heart-shaking experience, we managed to eat and drink and sleep, yea, to play tennis—somewhat solemnly, it is true—and go to the club, where we remained to the very *last* moment; needless to mention that I now entered into Aggie's manœuvre *con amore*. Mrs Starkey evidently divined the reason of our loitering in Kantia, and said in her most truculent manner, as she squared up to us:

'You keep your children out very late, Mrs Shandon.'

'Yes, but we like to have them with us,' rejoined Aggie in a meek apologetic voice.

'Then why don't you go home earlier?'

'Because it is so stupid and lonely,' was the mendacious answer.

'Lonely is not the word *I* should use. I wonder if you are as wise as your neighbours now? Come now, Mrs Shandon.'

'About what?' said Aggie with ill-feigned innocence.

'About Briarwood. Haven't you heard it yet? The ghastly precipice and horse affair?'

'Yes, I suppose we may as well confess that we *have*.'

'Humph! you are a brave couple to stay on. The Tombs tried it last year for three weeks. The Paxtons took it the year before, and then sub-let it, not that *they* believed in ghosts—oh, dear no,' and she laughed ironically.

'And what is the story?' I enquired eagerly.

'Well, the story is this. An old retired officer and his wife and their pretty niece lived at Briarwood a good many years ago. The girl was engaged to be married to a fine young fellow in the Guides. The day before the wedding what you know of happened, and has happened every monsoon ever since. The poor girl went out of her mind and destroyed herself, and the old colonel and his wife did not long survive her. The house is uninhabitable in the monsoon, and there seems nothing for it but to auction off the furniture and pull it down; it will always be the same as long as it stands. Take my advice and come into Cooper's Hotel. I believe you can have that small set of rooms at the back. The sitting-room smokes, but beggars can't be choosers.'

'That will only be our very last resource,' said Aggie hotly.

'It's not very grand, I grant you, but any port in a storm.'

Tom arrived, was doubly welcome, and was charmed with Briarwood. Chaffed us unmercifully and derided our fears until *he* himself had a similar experience, and he heard the phantom horse plunging in the verandah and that wild, unearthly and utterly appalling shriek. No, he could not laugh *that* away, and seeing that we had now a mortal abhorrence of the place, that the children had to be kept abroad in the damp till long after dark, that Aggie was a mere hollow-eyed spectre, and that we had scarcely a servant left, that—in short, one day we packed up precipitately and fled in a body to Cooper's Hotel. But we did not basely endeavour to sub-let, nor advertise Briarwood as 'a delightfully situated pucka built house, containing all the requirements of a gentleman's family'. No, no. Tom bore the loss of the rent and—a more difficult feat—Aggie bore Mrs Starkey's insufferable, 'I told you so.'

Aggie was at Kantia again last season. She walked out early one morning to see our former abode. The chowkidar and parrot are still in possession, and are likely to remain the sole tenants on the premises. The parrot suns and dusts his ancient feathers in the empty verandah, which re-echoes with his cry of 'Lucy, where are you, pretty Lucy?' The chowkidar inhabits a secluded go-down at the back, where he passes most of the day in sleeping, or smoking the soothing 'huka'. The place has a forlorn, uncared-for appearance now. The flowers are nearly all gone; the paint has peeled off the doors and windows; the avenue is grass-grown. Briarwood appears to have resigned itself to emptiness, neglect and decay, although outside the gate there still hangs a battered board on which, if you look very closely you can decipher the words '*To Let*'.

John Charrington's Wedding

E. NESBIT

No one ever thought that May Forster would marry John Charrington; but he thought differently, and things which John Charrington intended had a queer way of coming to pass. He asked her to marry him before he went up to Oxford. She laughed and refused him. He asked her again next time he came home. Again she laughed, tossed her dainty blonde head, and again refused. A third time he asked her; she said it was becoming a confirmed bad habit, and laughed at him more than ever.

John was not the only man who wanted to marry her: she was the belle of our village coterie, and we were all in love with her more or less; it was a sort of fashion, like masher collars or Inverness capes. Therefore we were as much annoyed as surprised when John Charrington walked into our little local Club—we held it in a loft over the saddler's, I remember—and invited us all to his wedding.

'Your wedding?'

'You don't mean it?'

'Who's the happy fair? When's it to be?'

John Charrington filled his pipe and lighted it before he replied. Then he said—

'I'm sorry to deprive you fellows of your only joke—but Miss Forster and I are to be married in September.'

'You don't mean it?'

'He's got the mitten again, and its turned his head.'

'No,' I said, rising, 'I see it's true. Lend me a pistol someone—or a first-class fare to the other end of Nowhere. Charrington has bewitched the only pretty girl in our twenty-mile radius. Was it mesmerism, or a love-potion, Jack?'

'Neither, sir, but a gift you'll never have—perseverance—and the best luck a man ever had in this world.'

There was something in his voice that silenced me, and all chaff of the other fellows failed to draw him further.

The queer thing about it was that when we congratulated Miss Forster, she blushed and smiled, and dimpled, for all the world as though she were in love with him, and had been in love with him all the time. Upon my word, I think she had. Women are strange creatures.

We were all asked to the wedding. In Brixham everyone who was anybody knew everybody else who was anyone. My sisters were, I truly believe, more interested in the *trousseau* than the bride herself, and I was to be best man. The coming marriage was much canvassed at afternoon tea-tables, and at our little Club over the saddler's, and the question was always asked: 'Does she care for him?'

I used to ask that question myself in the early days of their engagement, but after a certain evening in August I never asked it again. I was coming home from the Club through the churchyard. Our church is on a thyme-grown hill, and the turf about it is so thick and soft that one's footsteps are noiseless.

I made no sound as I vaulted the low lichened wall, and threaded my way between the tombstones. It was at the same instant that I heard John Charrington's voice, and saw her face. May was sitting on a low flat gravestone with the full splendour of the western sun upon her *mignonne* face. Its expression ended, at once and for ever, any question of her love for him; it was transfigured to a beauty I should not have believed possible, even to that beautiful little face.

John lay at her feet, and it was his voice that broke the stillness of the golden August evening.

'My dear, my dear, I believe I should come back from the dead if you wanted me!'

I coughed at once to indicate my presence, and passed on into the shadow fully enlightened.

The wedding was to be early in September. Two days before I had to run up to town on business. The train was late, of course, for we are on the South-Eastern, and as I stood grumbling with my watch in my hand, whom should I see but John Charrington and May Forster. They were walking up and down the unfrequented end of the platform, arm in arm, looking into each other's eyes, careless of the sympathetic interest of the porters.

Of course I knew better than to hesitate a moment before burying myself in the booking-office, and it was not till the train drew up at the platform, that I obtrusively passed the pair with my Gladstone, and took the corner in a first-class smoking-carriage. I did this with as good an air of not seeing them as I could assume. I pride myself on my discretion, but if John were travelling alone I wanted his company. I had it.

'Hullo, old man,' came his cheery voice as he swung his bag into my carriage; 'here's luck; I was expecting a dull journey!'

'Where are you off to?' I asked, discretion still bidding me turn my eyes away, though I saw, without looking, that hers were red-rimmed.

'To old Branbridge's,' he answered, shutting the door and leaning out for a last word with his sweetheart.

'Oh, I wish you wouldn't go, John,' she was saying in a low, earnest voice. 'I feel certain something will happen.'

'Do you think I should let anything happen to keep me, and the day after tomorrow our wedding-day?'

'Don't go,' she answered, with a pleading intensity which would have sent my Gladstone on to the platform and me after it. But she wasn't speaking to me. John Charrington was made differently; he rarely changed his opinions, never his resolutions.

He only stroked the little ungloved hands that lay on the carriage door.

'I must, May. The old boy's been awfully good to me, and now he's dying I must go and see him, but I shall come home in time for——' the rest of the parting was lost in a whisper and in the rattling lurch of the starting train.

'You're sure to come?' she spoke as the train moved.

'Nothing shall keep me,' he answered; and we steamed out. After he had seen the last of the little figure on the platform he leaned back in his corner and kept silence for a minute.

When he spoke it was to explain to me that his godfather, whose heir he was, lay dying at Peasmarsh Place, some fifty miles away, and had sent for John, and John had felt bound to go.

'I shall be surely back tomorrow,' he said, 'or, if not, the day after, in heaps of time. Thank Heaven, one hasn't to get up in the middle of the night to get married nowadays!'

'And suppose Mr Branbridge dies?'

'Alive or dead I mean to be married on Thursday!' John answered, lighting a cigar and unfolding *The Times*.

At Peasmarsh station we said 'goodbye', and he got out, and I saw him ride off; I went on to London, where I stayed the night.

When I got home the next afternoon, a very wet one, by the way, my sister greeted me with—

'Where's Charrington?'

'Goodness knows,' I answered testily. Every man, since Cain, has resented that kind of question.

'I thought you might have heard from him,' she went on, 'as you're to give him away tomorrow.'

'Isn't he back?' I asked, for I had confidently expected to find him at home.

'No, Geoffrey,'—my sister Fanny always had a way of jumping to conclusions, expecially such conclusions as were least favourable to her fellow-creatures—'he has not returned, and, what is more, you may depend upon it he won't. You mark my words, there'll be no wedding tomorrow.'

My sister Fanny has a power of annoying me which no other human being possesses.

'You mark my words,' I retorted with asperity, 'you had better give up making such a thundering idiot of yourself. There'll be more wedding tomorrow than ever you'll take the first part in.' A prophecy which, by the way, came true.

But though I could snarl confidently to my sister, I did not feel so comfortable when, late that night, I, standing on the doorstep of John's house, heard that he had not returned. I went home gloomily through the rain. Next morning brought a brilliant blue sky, gold sun, and all such softness of air and beauty of cloud as go to make up a perfect day. I woke with a vague feeling of having gone to bed anxious, and of being rather averse to facing that anxiety in the light of full wakefulness.

But with my shaving-water came a note from John which relieved my mind and sent me up to the Forsters with a light heart.

May was in the garden. I saw her blue gown through the hollyhocks as the lodge gates swung to behind me. So I did not go up to the house, but turned aside down the turfed path.

'He's written to you too,' she said, without preliminary greeting, when I reached her side.

'Yes, I'm to meet him at the station at three, and come straight on to the church.'

Her face looked pale, but there was a brightness in her eyes, and a tender quiver about the mouth that spoke of renewed happiness.

'Mr Branbridge begged him so to stay another night that he had not the heart to refuse,' she went on. 'He is so kind, but I wish he hadn't stayed.'

I was at the station at half-past two. I felt rather annoyed with John. It seemed a sort of slight to the beautiful girl who loved him, that he should come as it were out of breath, and with the dust of travel upon him to take her hand, which some of us would have given the best years of our lives to take.

But when the three o'clock train glided in, and glided out again having brought no passengers to our little station, I was more than annoyed. There was no other train for thirty-five minutes; I calculated that, with much hurry, we might just get to the church in time for the ceremony; but, oh, what a fool to miss that first train! What other man could have done it?

That thirty-five minutes seemed a year, as I wandered round the station reading the advertisements and the time-tables, and the company's bye-laws, and getting more and more angry with John Charrington. This confidence in his own power of getting everything he wanted the minute he wanted it was leading him too far. I hate waiting. Everyone does, but I believe I hate it more than anyone else. The three thirty-five was late, of course.

I ground my pipe between my teeth and stamped with impatience as I watched the signals. Click. The signal went down. Five minutes later I flung myself into the carriage that I had brought for John.

'Drive to the church!' I said, as someone shut the door. 'Mr Charrington hasn't come by this train.'

Anxiety now replaced anger. What had become of the man? Could he have been taken suddenly ill? I had never known him have a day's illness in his life. And even so he might have telegraphed. Some awful accident must have happened to him. The thought that he had played her false never—no, not for a moment, entered my head. Yes, something terrible had happened to him, and on me lay the task of telling his bride. I tell you, I almost wished the carriage would upset and break my head so that someone else might tell her, not I, who—but that's nothing to do with the story.

It was five minutes to four as we drew up at the churchyard gate. A double row of eager onlookers lined the path from lych-gate to porch. I sprang from the carriage and passed up between them. Our gardener had a good front place near the door. I stopped.

'Are they waiting still, Byles?' I asked, simply to gain time, for of course I knew they were by the waiting crowd's attentive attitude.

'Waiting, sir? No no, sir, why it must be over by now.'

'Over! Then Mr Charrington's come?'

'To the minute, sir; must have missed you somehow, and, I say, sir,' lowering his voice, 'I never see Mr John the least bit so afore, but my opinion is he's been drinking pretty free. His clothes was all dusty and his face like a sheet. I tell you I didn't like the looks of him at all, and the folks inside are saying all sorts of things. You'll see, something's gone very wrong with Mr John, and he's tried liquor. He looked like a ghost, and in he went with his eyes straight before him, with never a look or a word for none of us; him that was always such a gentleman!'

I had never heard Byles make so long a speech. The crowd in the churchyard were talking in whispers and getting ready rice and slippers to throw at the bride and bridegroom. The ringers were ready with their hands on the ropes to ring out the merry peal as the bride and bridegroom should come out.

A murmur from the church announced them; out they came, Byles was right. John Charrington did not look himself. There was dust on his coat, his hair was disarranged. He seemed to have been in some row, for there was a black mark above his eyebrow. He was deathly pale. But his pallor was not greater than that of the bride, who might have been carved in ivory—dress, veil, orange blossoms and all.

As they passed out the ringers stooped—there were six of them—and then, on the ears expecting the gay wedding peal, came the slow tolling of the passing bell.

A thrill of horror at so foolish a jest from the ringers passed through us all. But the ringers themselves dropped the ropes and fled like rabbits down the belfry stairs. The bride shuddered, and grey shadows came about her mouth, but the bridegroom led her on down the path where the people stood with the handfuls of rice; but the handfuls were never thrown, and the wedding-bells never rang. In vain the ringers were urged to remedy their mistake: they protested with many whispered expletives that they would see themselves further first.

In a hush like the hush in the chamber of death the bridal pair passed into their carriage and its door slammed behind them.

Then the tongues were loosed. A babel of anger, wonder, conjecture from the guests and the spectators.

'If I'd seen his condition, sir,' said old Forster to me as we drove off, 'I would have stretched him on the floor of the church, sir, by Heaven I would, before I'd have let him marry my daughter!'

Then he put his head out of the window.

'Drive like fury,' he cried to the coachman; 'don't spare the horses.'

He was obeyed. We passed the bride's carriage. I forbore to look at it, and old Forster turned his head away and swore. We reached home before it.

We stood in the hall doorway, in the blazing afternoon sun, and in about half a minute we heard wheels crunching the gravel. When the carriage stopped in front of the steps old Forster and I ran down.

'Great Heaven, the carriage is empty! And yet——'

I had the door open in a minute, and this is what I saw—

No sign of John Charrington; and of May, his wife only a huddled heap of white satin lying half on the floor of the carriage and half on the seat.

'I drove straight here, sir,' said the coachman, as the bride's father lifted her out; 'and I'll swear no one got out of the carriage.'

We carried her into the house in her bridal dress and drew back her veil. I saw her face. Shall I ever forget it? white, white and drawn with agony and horror, bearing such a look of terror as I have never seen since except in dreams. And her hair, her radiant blonde hair, I tell you it was white like snow.

As we stood, her father and I, half mad with the horror and mystery of it, a boy came up the avenue—a telegraph boy. They brought the orange envelope to me. I tore it open.

'Mr Charrington was thrown from his horse on his way to the station at half-past one. Killed on the spot!'

And he was married to May Forster in our parish church at half-past three, in presence of half the parish.

'I shall be married, dead or alive!'

What had passed in that carriage on the homeward drive? No one knows—no one will ever know. Oh, May! oh, my dear!

Before a week was over they laid her beside her husband in our little churchyard on the thyme-covered hill—the churchyard where they had kept their love-trysts.

Thus was accomplished John Charrington's wedding.

The Haunted Organist of Hurly Burly

ROSA MULHOLLAND

There had been a thunderstorm in the village of Hurly Burly. Every door was shut, every dog in his kennel, every rut and gutter a flowing river after the deluge of rain that had fallen. Up at the great house, a mile from the town, the rooks were calling to one another about the fright they had been in, the fawns in the deer-park were venturing their timid heads from behind the trunks of trees, and the old woman at the gate-lodge had risen from her knees, and was putting back her prayer-book on the shelf. In the garden, July roses, unwieldy with their full-blown richness, and saturated with rain, hung their heads heavily to the earth; others, already fallen, lay flat upon their blooming faces on the path, where Bess, Mistress Hurly's maid, would find them, when going on her morning quest of rose-leaves for her lady's pot-pourri. Ranks of white lilies, just brought to perfection by today's sun, lay dabbled in the mire of flooded mould. Tears ran down the amber cheeks of the plums on the south wall, and not a bee had ventured out of the hives, though the scent of the air was sweet enough to tempt the laziest drone. The sky was still lurid behind the boles of the upland oaks, but the birds had begun to dive in and out of the ivy that wrapped up the home of the Hurlys of Hurly Burly.

This thunderstorm took place more than half a century ago, and we must remember that Mistress Hurly was dressed in the fashion of that time as she crept out from behind the squire's chair, now that the lightning was over, and, with many nervous glances towards the window, sat down before her husband, the tea-urn, and the muffins. We can picture her fine lace cap, with its peachy ribbons, the frill on the hem of her cambric gown just touching her ankles, the embroidered clocks on her stockings, the rosettes on her shoes, but not so easily the lilac shade of her mild eyes, the satin skin, which still kept its delicate bloom, though wrinkled with advancing age, and the pale, sweet, puckered mouth, that time and sorrow had made angelic while trying vainly to deface its beauty.

The squire was as rugged as his wife was gentle, his skin as brown as hers was white, his grey hair as bristling as hers was glossed; the years had ploughed his face into ruts and channels; a bluff, choleric, noisy man he had been; but of late a dimness had come on his eyes, a hush on his loud voice, and a check on the spring of his hale step. He looked at his wife often, and very often she looked at him. She was not a tall woman, and he was only a head higher. They were a quaintly well-matched couple, despite their differences. She turned to you with nervous sharpness and revealed her tender voice and eye; he spoke and glanced roughly, but the turn of his head was courteous. Of late they fitted one another better than they had ever done in the heyday of their youthful love. A common sorrow had developed a singular likeness between them. In former years the cry from the wife had been, 'Don't curb my son too much!' and from the husband, 'You ruin the lad with softness.' But now the idol that had stood between them was removed, and they saw each other better.

The room in which they sat was a pleasant old-fashioned drawing-room, with a general spider-legged character about the fittings; spinnet and guitar in their places, with a great deal of copied music beside them; carpet, tawny wreaths on the pale blue; blue flutings on the walls, and faint gilding on the furniture. A huge urn, crammed with roses, in the open bay-window, through which came delicious airs from the garden, the twittering of birds settling to sleep in the ivy close by, and occasionally the pattering of a flight of rain-drops, swept to the ground as a bough bent in the breeze. The urn on the table was ancient silver, and the china rare. There was nothing in the room for luxurious ease of the body, but everything of delicate refinement for the eye.

There was a great hush all over Hurly Burly, except in the neighbourhood of the rooks. Every living thing had suffered from heat for the past month, and now, in common with all Nature, was receiving the boon of refreshed air in silent peace. The mistress and master of Hurly Burly shared the general spirit that was abroad, and were not talkative over their tea.

'Do you know,' said Mistress Hurly, at last, 'when I heard the first of the thunder beginning I thought it was—it was——'

The lady broke down, her lips trembling, and the peachy ribbons of her cap stirring with great agitation.

'Pshaw!' cried the old squire, making his cup suddenly ring upon the saucer, 'we ought to have forgotten that. Nothing has been heard for three months.'

At this moment a rolling sound struck upon the ears of both. The lady rose from her seat trembling, and folded her hands together, while the tea-urn flooded the tray.

'Nonsense, my love,' said the squire; 'that is the noise of wheels. Who can be arriving?'

'Who, indeed?' murmured the lady, reseating herself in agitation.

Presently pretty Bess of the rose-leaves appeared at the door in a flutter of blue ribbons.

'Please, madam, a lady has arrived, and says she is expected. She asked for her apartment, and I put her into the room that was got ready for Miss Calderwood. And she sends her respects to you, madam, and she'll be down with you presently.'

The squire looked at his wife, and his wife looked at the squire.

'It is some mistake,' murmured madam. 'Some visitor for Calderwood or the Grange. It is very singular.'

Hardly had she spoken when the door again opened, and the stranger appeared—a small creature, whether girl or woman it would be hard to say—dressed in a scanty black silk dress, her narrow shoulders covered with a white muslin pelerine. Her hair was swept up to the crown of her head, all but a little fringe hanging over her low forehead within an inch of her brows. Her face was brown and thin, eyes black and long, with blacker settings, mouth large, sweet, and melancholy. She was all head, mouth, and eyes; her nose and chin were nothing.

This visitor crossed the floor hastily, dropped a courtesy in the middle of the room, and approached the table, saying abruptly, with a soft Italian accent:

'Sir and madam, I am here. I am come to play your organ.'

'The organ!' gasped Mistress Hurly.

'The organ!' stammered the squire.

'Yes, the organ,' said the little stranger lady, playing on the back of a chair with her fingers, as if she felt notes under them. 'It was but last week that the handsome signor, your son, came to my little house, where I have lived teaching music since my English father and my Italian mother and brothers and sisters died and left me so lonely.'

Here the fingers left off drumming, and two great tears were brushed off, one from each eye with each hand, child's fashion. But the next moment the fingers were at work again, as if only whilst they were moving the tongue could speak.

'The noble signor, your son,' said the little woman, looking trustfully from one to the other of the old couple, while a bright blush shone through her brown skin, 'he often came to see me before that, always in the evening, when the sun was warm and yellow all through my little studio, and the music was swelling my heart, and I could play out grand with all my soul; then he used to come and say, "Hurry, little Lisa, and play better, better still. I have work for you to do by-and-by." Sometimes he said, "Brava!" and sometimes he said "Eccellentissima!" but one night last week he came to me and said, "It is enough. Will you swear to do my bidding, whatever it may be?" Here the black eyes fell. And I said, "Yes". And he said, "Now you are my betrothed". And I said, "Yes". And he said, "Pack up your music, little Lisa, and go off to England to my English father and mother, who have an organ in their house which must be played upon. If they refuse to let you play, tell them I sent you, and they will give you leave. You must play all day, and you must get up in the night and play. You must never tire. You are my betrothed, and you have sworn to do my work." I said, "Shall I see you there, signor?" And he said, "Yes, you shall see me there." I said, "I will keep my vow, signor." And so, sir and madam, I am come.'

The soft foreign voice left off talking, the fingers left off thrumming on the chair, and the little stranger gazed in dismay at her auditors, both pale with agitation.

'You are deceived. You make a mistake,' said they in one breath.

'Our son——' began Mistress Hurly, but her mouth twitched, her voice broke, and she looked piteously towards her husband.

'Our son,' said the squire, making an effort to conquer the quavering in his voice, 'our son is long dead.'

'Nay, nay,' said the little foreigner. 'If you have thought him dead have good cheer, dear sir and madam. He is alive; he is well, and strong, and handsome. But one, two, three, four, five' (on the fingers) 'days ago he stood by my side.'

'It is some strange mistake, some wonderful coincidence!' said the mistress and master of Hurly Burly.

'Let us take her to the gallery,' murmured the mother of this son who was thus dead and alive. 'There is yet light to see the pictures. She will not know his portrait.'

The bewildered wife and husband led their strange visitor away to a long gloomy room at the west side of the house, where the faint gleams from the darkening sky still lingered on the portraits of the Hurly family.

'Doubtless he is like this,' said the squire, pointing to a fair-haired young man with a mild face, a brother of his own who had been lost at sea.

But Lisa shook her head, and went softly on tiptoe from one picture to another, peering into the canvas, and still turning away troubled. But at last a shriek of delight startled the shadowy chamber.

'Ah, here he is! See, here he is, the noble signor, the beautiful signor, not half so handsome as he looked five days ago, when talking to poor little Lisa! Dear sir and madam, you are now content. Now take me to the organ, that I may commence to do his bidding at once.'

The mistress of Hurly Burly clung fast by her husband's arm.

'How old are you, girl?' she said faintly.

'Eighteen,' said the visitor impatiently, moving towards the door.

'And my son has been dead for twenty years!' said his mother, and swooned on her husband's breast.

'Order the carriage at once,' said Mistress Hurly, recovering from her swoon; 'I will take her to Margaret Calderwood. Margaret will tell her the story. Margaret will bring her to reason. No, not tomorrow; I cannot bear tomorrow, it is so far away. We must go tonight.'

The little signora thought the old lady mad, but she put on her cloak again obediently, and took her seat beside Mistress Hurly in the Hurly family coach. The moon that looked in at them through the pane as they lumbered along was not whiter than the aged face of the squire's wife, whose dim faded eyes were fixed upon it in doubt and awe too great for tears or words. Lisa, too, from her corner gloated upon the moon, her black eyes shining with passionate dreams.

A carriage rolled away from the Calderwood door as the Hurly coach drew up at the steps. Margaret Calderwood had just returned from a dinner-party, and at the open door a splendid figure was standing, a tall woman dressed in brown velvet, the diamonds on her bosom glistening in the moonlight that revealed her, pouring, as it did, over the house from eaves to basement. Mistress Hurly fell into her outstretched arms with a groan, and the strong woman carried her aged friend, like a baby, into the house. Little Lisa was overlooked, and sat down contentedly on the threshold to gloat awhile longer on the moon, and to thrum imaginary sonatas on the doorstep.

There were tears and sobs in the dusk, moonlit room into which Margaret Calderwood carried her friend. There was a long consultation, and then Margaret, having hushed away the grieving woman into some quiet corner, came forth to look for the little dark-faced stranger, who had arrived, so unwelcome, from beyond the seas, with such wild communication from the dead.

Up the grand staircase of handsome Calderwood the little woman followed the tall one into a large chamber where a lamp burned, showing Lisa, if she cared to see it, that this mansion of Calderwood was fitted with much greater luxury and richness than was that of Hurly Burly. The appointments of this room announced it the sanctum of a woman who depended for the interest of her life upon resources of intellect and taste. Lisa noticed nothing but a morsel of biscuit that was lying on a plate.

'May I have it?' said she eagerly. 'It is so long since I have eaten. I am hungry.'

Margaret Calderwood gazed at her with a sorrowful, motherly look, and, parting the fringing hair on her forehead, kissed her. Lisa, staring at her in wonder, returned the caress with ardour. Margaret's large fair shoulders, Madonna face, and yellow braided hair, excited a rapture within her. But when food was brought her, she flew to it and ate.

'It is better than I have ever eaten at home!' she said gratefully. And Margaret Calderwood murmured, 'She is physically healthy, at least.'

'And now, Lisa,' said Margaret Calderwood, 'come and tell me the whole history of the grand signor who sent you to England to play the organ.'

Then Lisa crept in behind a chair, and her eyes began to burn and her fingers to thrum, and she repeated word for word her story as she had told it at Hurly Burly.

When she had finished, Margaret Calderwood began to pace up and down the floor with a very troubled face. Lisa watched her, fascinated, and, when she bade her listen to a story which she would relate to her, folded her restless hands together meekly, and listened.

'Twenty years ago, Lisa, Mr and Mrs Hurly had a son. He was handsome, like that portrait you saw in the gallery, and he had brilliant talents. He was idolized by his father and mother, and all who knew him felt obliged to love him. I was then a happy girl of twenty. I was an orphan, and Mrs Hurly, who had been my mother's friend, was like a mother to me. I, too, was petted and caressed by all my friends, and I was very wealthy; but I only valued admiration, riches—every good gift that fell to my share—just in proportion as they seemed of worth in the eyes of Lewis Hurly. I was his affianced wife, and I loved him well.

'All the fondness and pride that were lavished on him could not keep him from falling into evil ways, nor from becoming rapidly more and more abandoned to wickedness, till even those who loved him best despaired of seeing his reformation. I prayed him with tears, for my sake, if not for that of his grieving mother, to save himself before it was too late. But to my horror I found that my power was gone, my words did not even move him; he loved me no more. I tried to think that this was some fit of madness that would pass, and still clung to hope. At last his own mother forbade me to see him.'

Here Margaret Calderwood paused, seemingly in bitter thought, but resumed:

'He and a party of his boon companions, named by themselves the "Devil's Club", were in the habit of practising all kinds of unholy pranks in the country. They had midnight carousings on the tombstones in the village graveyard; they carried away helpless old men and children, whom they tortured by making believe to bury them alive; they raised the dead and placed them sitting round the tombstones at a mock feast. On one occasion there was a very sad funeral from the village. The corpse was carried into the church, and prayers were read over the coffin, the chief mourner, the aged father of the dead man, standing weeping by. In the midst of this solemn scene the organ suddenly pealed forth a profane tune, and a number of voices shouted a drinking chorus. A groan of execration burst from the crowd, the clergyman turned pale and closed his book, and the old man, the father of the dead, climbed the altar steps, and, raising his arms above his head, uttered a terrible curse. He cursed Lewis Hurly to all eternity, he cursed the organ he played, that it might be dumb henceforth, except under the fingers that had now profaned it, which, he prayed, might be forced to labour upon it till they stiffened in death. And the curse seemed to work, for the organ stood dumb in the church from that day, except when touched by Lewis Hurly.

'For a bravado he had the organ taken down and conveyed to his father's house, where he had it put up in the chamber where it now stands. It was also for a bravado that he played on it every day. But, by-and-by, the amount of time which he spent at it daily began to increase rapidly. We wondered long at this whim, as we called it, and his poor mother thanked God that he had set his heart upon an occupation which would keep him out of harm's way. I was the first to suspect that it was not his own will that kept him hammering at the organ so many laborious hours, while his boon companions tried vainly to draw him away. He used to lock himself up in the room with the organ, but one day I hid myself among the curtains, and saw him writhing on his seat, and heard him groaning as he strove to wrench his hands from the keys, to which they flew back like a needle to a magnet. It was soon plainly to be seen that he was an involuntary slave to the organ; but whether through a madness that had grown within himself, or by some supernatural doom, having its cause in the old man's curse, we did not dare to say. By-and-by there came a time when we were wakened out of our sleep at nights by the rolling of the organ. He wrought now night and day. Food and rest were denied him. His face got haggard, his beard grew long, his eyes started from their sockets. His body became wasted, and his cramped fingers like the claws of a bird. He groaned piteously as he stooped over his cruel toil. All save his mother and I were afraid to go near him. She, poor, tender woman, tried to put wine and food between his lips, while the tortured fingers crawled over the keys; but he only gnashed his teeth at her with curses, and she retreated from him in terror, to pray. At last, one dreadful hour, we found him a ghastly corpse on the ground before the organ.

'From that hour the organ was dumb to the touch of all human fingers. Many, unwilling to believe the story, made persevering endeavours to draw sound from it, in vain. But when the darkened empty room was locked up and left, we heard as loud as ever the well-known sounds humming and rolling through the walls. Night and day the tones of the organ boomed on as before. It seemed that the doom of the wretched man was not yet fulfilled, although his tortured body had been worn out in the terrible struggle to accomplish it. Even his own mother was afraid to go near the room then. So the time went on, and the curse of this perpetual music was not removed from the house. Servants refused to stay about the place. Visitors shunned it. The squire and his wife left their home for years, and returned; left it, and returned again, to find their ears still tortured and their hearts wrung by the unceasing persecution of terrible sounds. At last, but a few months ago, a holy man was found, who locked himself up in the cursed chamber for many days, praying and wrestling with the demon. After he came forth and went away the sounds ceased, and the organ was heard no more. Since then there has been peace in the house. And now, Lisa, your strange appearance and your strange story convince us that you are a victim of a ruse of the Evil One. Be warned in time, and place yourself under the protection of God, that you may be saved from the fearful influences that are at work upon you. Come——'

Margaret Calderwood turned to the corner where the stranger sat, as she had supposed, listening intently. Little Lisa was fast asleep, her hands spread before her as if she played an organ in her dreams.

Margaret took the soft brown face to her motherly breast, and kissed the swelling temples, too big with wonder and fancy.

'We will save you from a horrible fate!' she murmured, and carried the girl to bed.

In the morning Lisa was gone. Margaret Calderwood, coming early from her own chamber, went into the girl's room and found the bed empty.

'She is just such a wild thing,' thought Margaret, 'as would rush out at sunrise to hear the larks!' and she went forth to look for her in the meadows, behind the beech hedges and in the home park. Mistress Hurly, from the breakfast-room window, saw Margaret Calderwood, large and fair in her white morning gown, coming down the garden-path between the rose bushes, with her fresh draperies dabbled by the dew, and a look of trouble on her calm face. Her quest had been unsuccessful. The little foreigner had vanished.

A second search after breakfast proved also fruitless, and towards evening the two women drove back to Hurly Burly together. There all was panic and distress. The squire sat in his study with the doors shut, and his hands over his ears. The servants, with pale faces, were huddled together in whispering groups. The haunted organ was pealing through the house as of old.

Margaret Calderwood hastened to the fatal chamber, and there, sure enough, was Lisa, perched upon the high seat before the organ, beating the keys with her small hands, her slight figure swaying, and the evening sunshine playing about her weird head. Sweet unearthly music she wrung from the groaning heart of the organ—wild melodies, mounting to rapturous heights and falling to mournful depths. She wandered from Mendelssohn to Mozart, and from Mozart to Beethoven. Margaret stood fascinated awhile by the ravishing beauty of the sounds she heard, but, rousing herself quickly, put her arms round the musician and forced her away from the chamber. Lisa returned next day, however, and was not so easily coaxed from her post again. Day after day she laboured at the organ, growing paler and thinner and more weird-looking as time went on.

'I work so hard,' she said to Mrs Hurly. 'The signor, your son, is he pleased? Ask him to come and tell me himself if he is pleased.'

Mistress Hurly got ill and took to her bed. The squire swore at the young foreign baggage, and roamed abroad. Margaret Calderwood was the only one who stood by to watch the fate of the little organist. The curse of the organ was upon Lisa; it spoke under her hand, and her hand was its slave.

At last she announced rapturously that she had had a visit from the brave signor, who had commended her industry, and urged her to work yet harder. After that she ceased to hold any communication with the living. Time after time Margaret Calderwood wrapped her arms about the frail thing, and carried her away by force, locking the door of the fatal chamber. But locking the chamber and burying the key were of no avail. The door stood open again, and Lisa was labouring on her perch.

One night, wakened from her sleep by the well-known humming and moaning of the organ, Margaret dressed hurriedly and hastened to the unholy room. Moonlight was pouring down the staircase and passages of Hurly Burly. It shone on the marble bust of the dead Lewis Hurly, that stood in the niche above his mother's sitting-room door. The organ room was full of it when Margaret pushed open the door and entered—full of the pale green moonlight from the window, mingled with another light, a dull lurid glare which seemed to centre round a dark shadow, like the figure of a man standing by the organ, and throwing out in fantastic relief the slight form of Lisa writhing, rather than swaying, back and forward, as if in agony. The sounds that came from the organ were broken and meaningless, as if the hands of the player lagged and stumbled on the keys. Between the intermittent chords low moaning cries broke from Lisa, and the dark figure bent towards her with menacing gestures. Trembling with the sickness of supernatural fear, yet strong of will, Margaret Calderwood crept forward within the lurid light, and was drawn into its influence. It grew and intensified upon her, it dazzled and blinded her at first; but presently, by a daring effort of will, she raised her eyes, and beheld Lisa's face convulsed with torture in the burning glare, and bending over her the figure and the features of Lewis Hurly! Smitten with horror, Margaret did not even then lose her presence of mind. She wound her strong arms around the wretched girl and dragged her from her seat and out of the influence of the lurid light, which immediately paled away and vanished. She carried her to her own bed, where Lisa lay, a wasted wreck, raving about the cruelty of the pitiless signor who would not see that she was labouring her best. Her poor cramped hands kept beating the coverlet, as though she were still at her agonizing task.

Margaret Calderwood bathed her burning temples, and placed fresh flowers upon her pillow. She opened the blinds and windows, and let in the sweet morning air and sunshine, and then, looking up at the newly awakened sky with its fair promise of hope for the day, and down at the dewy fields, and afar off at the dark green woods with the purple mists still hovering about them, she prayed that a way might be shown her by which to put an end to this curse. She prayed for Lisa, and then, thinking that the girl rested somewhat, stole from the room. She thought that she had locked the door behind her.

She went downstairs with a pale, resolved face, and, without consulting anyone, sent to the village for a bricklayer. Afterwards she sat by Mistress Hurly's bedside, and explained to her what was to be done. Presently she went to the door of Lisa's room, and hearing no sound, thought the girl slept, and stole away. By-and-by she went downstairs, and found that the bricklayer had arrived and already begun his task of building up the organ-room door. He was a swift workman, and the chamber was soon sealed safely with stone and mortar.

Having seen this work finished, Margaret Calderwood went and listened again at Lisa's door; and still hearing no sound, she returned, and took her seat at Mrs Hurly's bedside once more. It was towards evening that she at last entered her room to assure herself of the comfort of Lisa's sleep. But the bed and room were empty. Lisa had disappeared.

Then the search began, upstairs and downstairs, in the garden, in the grounds, in the fields and meadows. No Lisa. Margaret Calderwood ordered the carriage and drove to Calderwood to see if the strange little Will-o'-the-wisp might have made her way there; then to the village, and to many other places in the neighbourhood which it was not possible she could have reached. She made enquiries everywhere; she pondered and puzzled over the matter. In the weak, suffering state that the girl was in, how far could she have crawled?

After two days' search, Margaret returned to Hurly Burly. She was sad and tired, and the evening was chill. She sat over the fire wrapped in her shawl when little Bess came to her, weeping behind her muslin apron.

'If you'd speak to Mistress Hurly about it, please, ma'am,' she said. 'I love her dearly, and it breaks my heart to go away, but the organ haven't done yet, ma'am, and I'm frightened out of my life, so I can't stay.'

'Who has heard the organ, and when?' asked Margaret Calderwood, rising to her feet.

'Please, ma'am, I heard it the night you went away—the night after the door was built up!'

'And not since?'

'No, ma'am,' hesitatingly, 'not since. Hist! hark, ma'am! Is not that like the sound of it now?'

'No,' said Margaret Calderwood; 'it is only the wind.' But pale as death she flew down the stairs and laid her ear to the yet damp mortar of the newly built wall. All was silent. There was no sound but the monotonous sough of the wind in the trees outside. Then Margaret began to dash her soft shoulder against the strong wall, and to pick the mortar away with her white fingers, and to cry out for the bricklayer who had built up the door.

It was midnight, but the bricklayer left his bed in the village, and obeyed the summons to Hurly Burly. The pale woman stood by and watched him undo all his work of three days ago, and the servants gathered about in trembling groups, wondering what was to happen next.

What happened next was this: When an opening was made the man entered the room with a light, Margaret Calderwood and others following. A heap of something dark was lying on the ground at the foot of the organ. Many groans arose in the fatal chamber. Here was little Lisa dead!

When Mistress Hurly was able to move, the squire and his wife went to live in France, where they remained till their death. Hurly Burly was shut up and deserted for many years. Lately it has passed into new hands. The organ has been taken down and banished, and the room is a bed-chamber, more luxuriously furnished than any in the house. But no one sleeps in it twice.

Margaret Calderwood was carried to her grave the other day a very aged woman.

The Man of Science

JEROME K. JEROME

I met a man in the Strand one day that I knew very well, as I thought, though I had not seen him for years. We walked together to Charing Cross, and there we shook hands and parted. Next morning, I spoke of this meeting to a mutual friend, and then I learnt, for the first time, that the man had died six months before.

The natural inference was that I had mistaken one man for another, an error that, not having a good memory for faces, I frequently fall into. What was remarkable about the matter, however, was that throughout our walk I had conversed with the man under the impression that he was that other dead man, and, whether by coincidence or not, his replies had never once suggested to me my mistake.

As soon as I finished speaking, Jephson, who had been listening very thoughtfully, asked me if I believed in spiritualism 'to its fullest extent'.

'That is rather a large question,' I answered. 'What do you mean by "spiritualism to its fullest extent"?'

'Well, do you believe that the spirits of the dead have not only the power of revisiting this earth at their will, but that, when here, they have the power of action, or rather, of exciting to action. Let me put a definite case. A spiritualist friend of mine, a sensible and by no means imaginative man, once told me that a table, through the medium of which the spirit of a friend had been in the habit of communicating with him, came slowly across the room towards him, of its own accord, one night as he sat alone, and pinioned him against the wall. Now can any of you believe that, or can't you?'

'I could,' Brown took it upon himself to reply; 'but, before doing so, I should wish for an introduction to the friend who told you the story. Speaking generally,' he continued, 'it seems to me that the difference between what we call the natural and the supernatural is merely the difference between frequency and rarity of occurrence. Having regard to the phenomena we are compelled to admit, I think it illogical to disbelieve anything that we are not able to disprove.'

'For my part,' remarked MacShaugnassy, 'I can believe in the ability of our spirit friends to give the quaint entertainments credited to them much easier than I can in their desire to do so.'

'You mean,' added Jephson, 'that you cannot understand why a spirit, not compelled as we are by the exigencies of society, should care to spend its evenings carrying on a laboured and childish conversation with a room full of abnormally uninteresting people.'

'That is precisely what I cannot understand,' MacShaugnassy agreed.

'Nor I, either,' said Jephson. 'But I was thinking of something very different altogether. Suppose a man died with the dearest wish of his heart unfulfilled, do you believe that his spirit might have power to return to earth and complete the interrupted work?'

'Well,' answered MacShaugnassy, 'if one admits the possibility of spirits retaining any interest in the affairs of this world at all, it is certainly more reasonable to imagine them engaged upon a task such as you suggest, than to believe that they occupy themselves with the performance of mere drawing-room tricks. But what are you leading up to?'

'Why to this,' replied Jephson, seating himself straddle-legged across his chair, and leaning his arms upon the back. 'I was told a story this morning at the hospital by an old French doctor. The actual facts are few and simple; all that is known can be read in the Paris police records of forty-two years ago.

'The most important part of the case, however, is the part that is not known, and that never will be known.

'The story begins with a great wrong done by one man unto another man. What the wrong was I do not know. I am inclined to think, however, it was connected with a woman. I think that because he who had been wronged hated him who had wronged with a hate such as does not often burn in a man's brain unless it be fanned by the memory of a woman's breath.

'Still that is only conjecture, and the point is immaterial. The man who had done the wrong fled, and the other man followed him. It became a point to point race, the first man having the advantage of a day's start. The course was the whole world, and the stakes were the first man's life.

'Travellers were few and far between in those days, and this made the trail easy to follow. The first man, never knowing how far or how near the other was behind him, and hoping now and again that he might have baffled him, would rest for a while. The second man, knowing always just how far the first one was before him, never paused, and thus each day the man who was spurred by Hate drew nearer to the man who was spurred by Fear.

'At this town the answer to the never-varied question would be:

'"At seven o'clock last evening, M'sieur."

'"Seven—ah; eighteen hours. Give me something to eat, quick, while the horses are being put to."

'At the next the calculation would be sixteen hours.

'Passing a lonely châlet, Monsieur puts his head out of the window:

'"How long since a carriage passed this way, with a tall, fair man inside?"

'"Such a one passed early this morning, M'sieur."

'"Thanks, drive on, a hundred francs apiece if you are through the pass before daybreak."

'"And what for dead horses, M'sieur?"

'"Twice their value when living."

'One day the man who was ridden by Fear looked up, and saw before him the open door of a cathedral, and, passing in, knelt down and prayed. He prayed long and fervently, for men, when they are in sore straits, clutch eagerly at the straws of faith. He prayed that he might be forgiven his sin, and, more important still, that he might be pardoned the consequences of his sin, and be delivered from his adversary; and a few chairs from him, facing him, knelt his enemy, praying also.

'But the second man's prayer, being a thanksgiving merely, was short, so that when the first man raised his eyes, he saw the face of his enemy gazing at him across the chair tops, with a mocking smile upon it.

'He made no attempt to rise, bur remained kneeling, fascinated by the look of joy that shone out of the other man's eyes. And the other man moved the high-backed chairs one by one, and came towards him softly.

'Then, just as the man who had been wronged stood beside the man who had wronged him, full of gladness that his opportunity had come, there burst from the cathedral tower a sudden clash of bells, and the man whose opportunity had come broke his heart and fell back dead, with that mocking smile of his still playing round his mouth.

'And so he lay there.

'Then the man who had done the wrong rose up and passed out, praising God.

'What became of the body of the other man is not known. It was the body of a stranger who had died suddenly in the cathedral. There was none to identify it, none to claim it.

'Years passed away, and the survivor in the tragedy became a worthy and useful citizen, and a noted man of science.

'In his laboratory were many objects necessary to him in his researches, and prominent among them, stood in a certain corner, a human skeleton. It was a very old and much-mended skeleton, and one day the long-expected end arrived, and it tumbled to pieces.

'Thus it became necessary to purchase another.

'The man of science visited a dealer he well knew; a little parchment-faced old man who kept a dingy shop, where nothing was ever sold, within the shadow of the towers of Notre Dame.

'The little parchment-faced old man had just the very thing that Monsieur wanted—a singularly fine and well-proportioned "study". It should be sent round and set up in Monsieur's laboratory that very afternoon.

'The dealer was as good as his word. When Monsieur entered his laboratory that evening, the thing was in its place.

'Monsieur seated himself in his high-backed chair, and tried to collect his thoughts. But Monsieur's thoughts were unruly, and inclined to wander, and to wander always in one direction.

'Monsieur opened a large volume and commenced to read. He read of a man who had wronged another and fled from him, the other man following. Finding himself reading this, he closed the book angrily, and went and stood by the window and looked out. He saw before him the sun-pierced nave of a great cathedral, and on the stones lay a dead man with a mocking smile upon his face.

'Cursing himself for a fool, he turned away with a laugh. But his laugh was short-lived, for it seemed to him that something else in the room was laughing also. Struck suddenly still, with his feet glued to the ground, he stood listening for awhile: then sought with starting eyes the corner from where the sound had seemed to come. But the white thing standing there was only grinning.

'Monsieur wiped the damp sweat from his head and hands, and stole out.

'For a couple of days he did not enter the room again. On the third, telling himself that his fears were those of a hysterical girl, he opened the door and went in. To shame himself, he took his lamp in his hand, and crossing over to the far corner where the skeleton stood, examined it. A set of bones bought for a hundred francs. Was he a child, to be scared by such a bogey!

'He held his lamp up in front of the thing's grinning head. The flame of the lamp flickered as though a faint breath had passed over it.

'The man explained this to himself by saying that the walls of the house were old and cracked, and that the wind might creep in anywhere. He repeated this explanation to himself as he recrossed the room, walking backwards, with his eyes fixed on the thing. When he reached his desk, he sat down and gripped the arms of his chair till his fingers turned white.

'He tried to work, but the empty sockets in that grinning head seemed to be drawing him towards them. He rose and battled with his inclination to fly screaming from the room. Glancing fearfully about him, his eye fell upon a high screen, standing before the door. He dragged it forward, and placed it between himself and the thing, so that he could not see it—nor it see him. Then he sat down again to his work. For awhile he forced himself to look at the book in front of him, but at last, unable to control himself any longer, he suffered his eyes to follow their own bent.

'It may have been an hallucination. He may have accidentally placed the screen so as to favour such an illusion. But what he saw was a bony hand coming round the corner of the screen, and, with a cry, he fell to the floor in a swoon.

'The people of the house came running in, and lifting him up, carried him out, and laid him upon his bed. As soon as he recovered, his first question was, where had they found the thing—where was it when they entered the room? and when they told him they had seen it standing where it always stood, and had gone down into the room to look again, because of his frenzied entreaties, and returned trying to hide their smiles, he listened to their talk about overwork, and the necessity for change and rest, and said they might do with him as they would.

'So for many months the laboratory door remained locked. Then there came a chill autumn evening when the man of science opened it again, and closed it behind him.

'He lighted his lamp, and gathered his instruments and books around him, and sat down before them in his high-backed chair. And the old terror returned to him.

'But this time he meant to conquer himself. His nerves were stronger now, and his brain clearer; he would fight his unreasoning fear. He crossed to the door and locked himself in, and flung the key to the other end of the room, where it fell among jars and bottles with an echoing clatter.

'Later on, his old housekeeper, going her final round, tapped at his door and wished him good night, as was her custom. She received no response, at first, and, growing nervous, tapped louder and called again; and at length an answering "good night" came back to her.

'She thought little about it at the time, but afterwards she remembered that the voice that had replied to her had been strangely grating and mechanical. Trying to describe it, she likened it to such a voice as she would imagine coming from a statue.

'Next morning his door remained still locked. It was no unusual thing for him to work all night, and far into the next day, so no one thought to be surprised. When, however, evening came, and yet he did not appear, his servants gathered outside the room and whispered, remembering what had happened once before.

'They listened, but could hear no sound. They shook the door and called to him, then beat with their fists upon the wooden panels. But still no sound came from the room.

'Becoming alarmed, they decided to burst open the door, and, after many blows, it gave way and flew back, and they crowded in.

'He sat bolt upright in his high-backed chair. They thought at first he had died in his sleep. But when they drew nearer and the light fell upon him, they saw the livid marks of bony fingers round his throat; and in his eyes there was a terror such as is not often seen in human eyes.'

Brown was the first to break the silence that followed. He asked me if I had any brandy on board. He said he felt he should like just a nip of brandy before going to bed. That is one of the chief charms of Jephson's stories: they always make you feel you want a little brandy.

Canon Alberic's Scrap-book

M. R. JAMES

St Bertrand de Comminges is a decayed town on the spurs of the Pyrenees, not very far from Toulouse, and still nearer to Bagnères-de-Luchon. It was the site of a bishopric until the Revolution, and has a cathedral which is visited by a certain number of tourists. In the spring of 1883 an Englishman arrived at this old-world place—I can hardly dignify it with the name of city, for there are not a thousand inhabitants. He was a Cambridge man, who had come specially from Toulouse to see St Bertrand's Church, and had left two friends, who were less keen archaeologists than himself, in their hotel at Toulouse, under promise to join him on the following morning. Half an hour at the church would satisfy *them*, and all three could then pursue their journey in the direction of Auch. But our Englishman had come early on the day in question, and proposed to himself to fill a notebook and to use several dozens of plates in the process of describing and photographing every corner of the wonderful church that dominates the little hill of Comminges. In order to carry out this design satisfactorily, it was necessary to monopolize the verger of the church for the day. The verger or sacristan (I prefer the latter appellation, inaccurate as it may be) was accordingly sent for by the somewhat brusque lady who keeps the inn of the Chapeau Rouge; and when he came, the Englishman found him an unexpectedly interesting object of study. It was not in the personal appearance of the little, dry, wizened old man that the interest lay, for he was precisely like dozens of other church-guardians in France, but in a curious furtive, or rather hunted and oppressed, air which he had. He was perpetually half glancing behind him; the muscles of his back and shoulders seemed to be hunched in a continual nervous contraction, as if he were expecting every moment to find himself in the clutch of an enemy. The Englishman hardly knew whether to put him down as a man haunted by a fixed delusion, or as one oppressed by a guilty conscience, or as an unbearably henpecked husband. The probabilities, when reckoned up, certainly pointed to the last idea; but, still, the impression conveyed was that of a more formidable persecutor even than a termagant wife.

However, the Englishman (let us call him Dennistoun) was soon too deep in his notebook and too busy with his camera to give more than an occasional glance to the sacristan. Whenever he did look at him, he found him at no great distance, either huddling himself back against the wall or crouching in one of the gorgeous stalls. Dennistoun became rather fidgety after a time. Mingled suspicions that he was keeping the old man from his *déjeuner*, that he was regarded as likely to make away with St Bertrand's ivory crozier, or with the dusty stuffed crocodile that hangs over the font, began to torment him.

'Won't you go home?' he said at last; 'I'm quite well able to finish my notes alone; you can lock me in if you like. I shall want at least two hours more here, and it must be cold for you, isn't it?'

'Good heavens!' said the little man, whom the suggestion seemed to throw into a state of unaccountable terror, 'such a thing cannot be thought of for a moment. Leave monsieur alone in the church? No, no; two hours, three hours, all will be the same to me. I have breakfasted, I am not at all cold, with many thanks to monsieur.'

'Very well, my little man,' quoth Dennistoun to himself: 'you have been warned, and you must take the consequences.'

Before the expiration of the two hours, the stalls, the enormous dilapidated organ, the choir-screen of Bishop John de Mauléon, the remnants of glass and tapestry, and the objects in the treasure-chamber, had been well and truly examined; the sacristan still keeping at Dennistoun's heels, and every now and then whipping round as if he had been stung, when one or other of the strange noises that trouble a large empty building fell on his ear. Curious noises they were sometimes.

'Once,' Dennistoun said to me, 'I could have sworn I heard a thin metallic voice laughing high up in the tower. I darted an enquiring glance at my sacristan. He was white to the lips. "It is he—that is—it is no one; the door is locked," was all he said, and we looked at each other for a full minute.'

Another little incident puzzled Dennistoun a good deal. He was examining a large dark picture that hangs behind the altar, one of a series illustrating the miracles of St Bertrand. The composition of the picture is well-nigh indecipherable, but there is a Latin legend below, which runs thus:

'Qualiter S. Bertrandus liberavit hominem quem diabolus diu volebat strangulare.' (How St Bertrand delivered a man whom the Devil long sought to strangle.)

Dennistoun was turning to the sacristan with a smile and a jocular remark of some sort on his lips, but he was confounded to see the old man on his knees, gazing at the picture with the eye of a suppliant in agony, his hands tightly clasped, and a rain of tears on his cheeks. Dennistoun naturally pretended to have noticed nothing, but the question would not away from him, 'Why should a daub of this kind affect anyone so strongly?' He seemed to himself to be getting some sort of clue to the reason of the strange look that had been puzzling him all the day: the man must be a monomaniac; but what was his monomania?

It was nearly five o'clock; the short day was drawing in, and the church began to fill with shadows, while the curious noises—the muffled footfalls and distant talking voices that had been perceptible all day—seemed, no doubt because of the fading light and the consequently quickened sense of hearing, to become more frequent and insistent.

The sacristan began for the first time to show signs of hurry and impatience. He heaved a sigh of relief when camera and notebook were finally packed up and stowed away, and hurriedly beckoned Dennistoun to the western door of the church, under the tower. It was time to ring the Angelus. A few pulls at the reluctant rope, and the great bell Bertrande, high in the tower, began to speak, and swung her voice up among the pines and down to the valleys, loud with mountain-streams, calling the dwellers on those lonely hills to remember and repeat the salutation of the angel to her whom he called Blessed among women. With that a profound quiet seemed to fall for the first time that day upon the little town, and Dennistoun and the sacristan went out of the church.

On the doorstep they fell into conversation.

'Monsieur seemed to interest himself in the old choir-books in the sacristy.'

'Undoubtedly. I was going to ask you if there were a library in the town.'

'No, monsieur; perhaps there used to be one belonging to the Chapter, but it is now such a small place——' Here came a strange pause of irresolution, as it seemed; then, with a sort of plunge, he went on: 'But if monsieur is *amateur des vieux livres*, I have at home something that might interest him. It is not a hundred yards.'

At once all Dennistoun's cherished dreams of finding priceless manuscripts in untrodden corners of France flashed up, to die down again the next moment. It was probably a stupid missal of Plantin's printing, about 1580. Where was the likelihood that a place so near Toulouse would not have been ransacked long ago by collectors? However, it would be foolish not to go; he would reproach himself for ever after if he refused. So they set off. On the way the curious irresolution and sudden determination of the sacristan recurred to Dennistoun, and he wondered in a shamefaced way whether he was being decoyed into some purlieu to be made away with as a supposed rich Englishman. He contrived, therefore, to begin talking with his guide, and to drag in, in a rather clumsy fashion, the fact that he expected two friends to join him early the next morning. To his surprise, the announcement seemed to relieve the sacristan at once of some of the anxiety that oppressed him.

'That is well,' he said quite brightly—'that is very well. Monsieur will travel in company with his friends; they will be always near him. It is a good thing to travel thus in company—sometimes.'

The last word appeared to be added as an afterthought, and to bring with it a relapse into gloom for the poor little man.

They were soon at the house, which was one rather larger than its neighbours, stone-built, with a shield carved over the door, the shield of Alberic de Mauléon, a collateral descendant, Dennistoun tells me, of Bishop John de Mauléon. This Alberic was a Canon of Comminges from 1680 to 1701. The upper windows of the mansion were boarded up, and the whole place bore, as does the rest of Comminges, the aspect of decaying age.

Arrived on his doorstep, the sacristan paused a moment.

'Perhaps,' he said, 'perhaps, after all, monsieur has not the time?'

'Not at all—lots of time—nothing to do till tomorrow. Let us see what it is you have got.'

The door was opened at this point, and a face looked out, a face far younger than the sacristan's, but bearing something of the same distressing look: only here it seemed to be the mark, not so much of fear for personal safety as of acute anxiety on behalf of another. Plainly, the owner of the face was the sacristan's daughter; and, but for the expression I have described, she was a handsome girl enough. She brightened up considerably on seeing her father accompanied by an able-bodied stranger. A few remarks passed between father and daughter, of which Dennistoun only caught these words, said by the sacristan, 'He was laughing in the church', words which were answered only by a look of terror from the girl.

But in another minute they were in the sitting-room of the house, a small, high chamber with a stone floor, full of moving shadows cast by a wood-fire that flickered on a great hearth. Something of the character of an oratory was imparted to it by a tall crucifix, which reached almost to the ceiling on one side; the figure was painted of the natural colours, the cross was black. Under this stood a chest of some age and solidity, and when a lamp had been brought, and chairs set, the sacristan went to this chest, and produced therefrom, with growing excitement and nervousness, as Dennistoun thought, a large book, wrapped in a white cloth, on which cloth a cross was rudely embroidered in red thread. Even before the wrapping had been removed, Dennistoun began to be interested by the size and shape of the volume. 'Too large for a missal,' he thought, 'and not the shape of an antiphoner; perhaps it may be something good, after all.' The next moment the book was open, and Dennistoun felt that he had at last lit upon something better than good. Before him lay a large folio, bound, perhaps, late in the seventeenth century, with the arms of Canon Alberic de Mauléon stamped in gold on the sides. There may have been a hundred and fifty leaves of paper in the book, and on almost every one of them was fastened a leaf from an illuminated manuscript. Such a collection Dennistoun had hardly dreamed of in his wildest moments. Here were ten leaves from a copy of Genesis, illustrated with pictures, which could not be later than 700 AD. Further on was a complete set of pictures from a Psalter, of English execution, of the very finest kind that the thirteenth century could produce; and, perhaps best of all, there were twenty leaves of uncial writing in Latin, which, as a few words seen here and there told him at once, must belong to some very early unknown patristic treatise. Could it possibly be a fragment of the copy of Papias 'On the Words of Our Lord', which was known to have existed as late as the twelfth century at Nîmes?* In any case, his mind was made up; that book must return to Cambridge with him, even if he had to draw the whole of his balance from the bank and stay at St Bertrand till the money came. He glanced up at the sacristan to see if his face yielded any hint that the book was for sale. The sacristan was pale, and his lips were working.

'If monsieur will turn on to the end,' he said.

So monsieur turned on, meeting new treasures at every rise of a leaf; and at the end of the book he came upon two sheets of paper, of much more recent date than anything he had yet seen, which puzzled him considerably. They must be contemporary, he decided, with the unprincipled Canon Alberic, who had doubtless plundered the Chapter library of St Bertrand to form this priceless scrap-book. On the first of the paper sheets was a plan, carefully drawn and instantly recognizable by a person who knew the ground, of the south aisle and cloisters of St Bertrand's. There were curious signs looking like

* We now know that these leaves did contain a considerable fragment of that work, if not of that actual copy of it.

planetary symbols, and a few Hebrew words in the corners; and in the north-west angle of the cloister was a cross drawn in gold paint. Below the plan were some lines of writing in Latin, which ran thus:

'Responsa 12^{mi} Dec. 1694. Interrogatum est: Inveniamne? Responsum est: Invenies. Fiamne dives? Fies. Vivamne invidendus? Vives. Moriarne in lecto meo? Ita.' (Answers of the 12th of December, 1694. It was asked: Shall I find it? Answer: Thou shalt. Shall I become rich? Thou wilt. Shall I live an object of envy? Thou wilt. Shall I die in my bed? Thou wilt.)

'A good specimen of the treasure-hunter's record—quite reminds one of Mr Minor-Canon Quatremain in "Old St Paul's",' was Dennistoun's comment, and he turned the leaf.

What he then saw impressed him, as he has often told me, more than he could have conceived any drawing or picture capable of impressing him. And, though the drawing he saw is no longer in existence, there is a photograph of it (which I possess) which fully bears out that statement. The picture in question was a sepia drawing at the end of the seventeenth century, representing, one would say at first sight, a biblical scene; for the architecture (the picture represented an interior) and the figures had that semi-classical flavour about them which the artists of two hundred years ago thought appropriate to illustrations of the Bible. On the right was a King on his throne, the throne elevated on twelve steps, a canopy overhead, soldiers on either side—evidently King Solomon. He was bending forward with outstretched sceptre, in attitude of command; his face expressed horror and disgust, yet there was in it also the mark of imperious command and confident power. The left half of the picture was the strangest, however. The interest plainly centred there. On the pavement before the throne were grouped four soldiers, surrounding a crouching figure which must be described in a moment. A fifth soldier lay dead on the pavement, his neck distorted, and his eyeballs starting from his head. The four surrounding guards were looking at the King. In their faces the sentiment of horror was intensified; they seemed, in fact, only restrained from flight by their implicit trust in their master. All this terror was plainly excited by the being that crouched in their midst. I entirely despair of conveying by any words the impression which this figure makes upon anyone who looks at it. I recollect once showing the photograph of the drawing to a lecturer on morphology—a person of, I was going to say, abnormally sane and unimaginative habits of mind. He absolutely refused to be alone for the rest of that evening, and he told me afterwards that for many nights he had not dared to put out his light before going to sleep. However, the main traits of the figure I can at least indicate. At first you saw only a mass of coarse, matted black hair; presently it was seen that this covered a body of fearful thinness, almost a skeleton, but with the muscles standing out like wires. The hands were of a dusky pallor, covered, like the body, with long, coarse hairs, and hideously taloned. The eyes, touched in with a burning yellow, had intensely black pupils, and were fixed upon the throned King with a look of beast-like hate. Imagine one of the awful bird-catching spiders of South America translated into human form, and endowed with intelligence just less than human, and you will have some faint conception of the terror inspired by the appalling effigy. One remark is universally made by those to whom I have shown the picture: 'It was drawn from the life.'

As soon as the first shock of his irresistible fright had subsided, Dennistoun stole a look at his hosts. The sacristan's hands were pressed upon his eyes; his daughter, looking up at the cross on the wall, was telling her beads feverishly.

At last the question was asked, 'Is this book for sale?'

There was the same hesitation, the same plunge of determination, that he had noticed before, and then came the welcome answer, 'If monsieur pleases.'

'How much do you ask for it?'

'I will take two hundred and fifty francs.'

This was confounding. Even a collector's conscience is sometimes stirred, and Dennistoun's conscience was tenderer than a collector's.

'My good man!' he said again and again, 'your book is worth far more than two hundred and fifty francs, I assure you—far more.'

But the answer did not vary: 'I will take two hundred and fifty francs, not more.'

There was really no possibility of refusing such a chance. The money was paid, the receipt signed, a glass of wine drunk over the transaction, and then the sacristan seemed to become a new man. He stood upright, he ceased to throw those suspicious glances behind him, he actually laughed or tried to laugh. Dennistoun rose to go.

'I shall have the honour of accompanying monsieur to his hotel?' said the sacristan.

'Oh no, thanks! it isn't a hundred yards. I know the way perfectly, and there is a moon.'

The offer was pressed three or four times, and refused as often.

'Then, monsieur will summon me if—if he finds occasion; he will keep the middle of the road, the sides are so rough.'

'Certainly, certainly.' said Dennistoun, who was impatient to examine his prize by himself; and he stepped out into the passage with his book under his arm.

Here he was met by the daughter; she, it appeared, was anxious to do a little business on her own account; perhaps, like Gehazi, to 'take somewhat' from the foreigner whom her father had spared.

'A silver crucifix and chain for the neck; monsieur would perhaps be good enough to accept it?'

Well, really, Dennistoun hadn't much use for these things. What did mademoiselle want for it?

'Nothing—nothing in the world. Monsieur is more than welcome to it.'

The tone in which this and much more was said was unmistakably genuine, so that Dennistoun was reduced to profuse thanks, and submitted to have the chain put round his neck. It really seemed as if he had rendered the father and daughter some service which they hardly knew how to repay. As he set off with his book they stood at the door looking after him, and they were still looking when he waved them a last goodnight from the steps of the Chapeau Rouge.

Dinner was over, and Dennistoun was in his bedroom, shut up alone with his acquisition. The landlady had manifested a particular interest in him since he had told her that he had paid a visit to the sacristan and bought an old book from him. He thought, too, that he had heard a hurried dialogue between her and the said sacristan in the passage outside the *salle à manger*; some words to the effect that 'Pierre and Bertrand would be sleeping in the house' had closed the conversation.

At this time a growing feeling of discomfort had been creeping over him—nervous reaction, perhaps, after the delight of his discovery. Whatever it was, it resulted in a conviction that there was someone behind him, and that he was far more comfortable with his back to the wall. All this, of course, weighed light in the balance as against the obvious value of the collection he had acquired. And now, as I said, he was alone in his bedroom, taking stock of Canon Alberic's treasures, in which every moment revealed something more charming.

'Bless Canon Alberic!' said Dennistoun, who had an inveterate habit of talking to himself. 'I wonder where he is now? Dear me! I wish that landlady would learn to laugh in a more cheering manner; it makes one feel as if there was someone dead in the house. Half a pipe more, did you say? I think perhaps you are right. I wonder what that crucifix is that the young woman insisted on giving me? Last century, I suppose. Yes, probably. It is rather a nuisance of a thing to have round one's neck—just too heavy. Most likely her father has been wearing it for years. I think I might give it a clean up before I put it away.'

He had taken the crucifix off, and laid it on the table, when his attention was caught by an object lying on the red cloth just by his left elbow. Two or three ideas of what it might be flitted through his brain with their own incalculable quickness.

'A penwiper? No, no such thing in the house. A rat? No, too black. A large spider? I trust to goodness not—no. Good God! a hand like the hand in that picture!'

In another infinitesimal flash he had taken it in. Pale, dusky skin, covering nothing but bones and tendons of appalling strength; coarse black hairs, longer than ever grew on a human hand; nails rising from the ends of the fingers and curving sharply down and forward, grey, horny and wrinkled.

He flew out of his chair with deadly, inconceivable terror clutching at his heart. The shape, whose left hand rested on the table, was rising to a standing posture behind his seat, its right hand crooked above his scalp. There was black and tattered drapery about it; the coarse hair covered it as in the drawing. The lower jaw was thin—what can I call it?—shallow, like a beast's; teeth showed behind the black lips; there was no nose; the eyes, of a fiery yellow, against which the pupils showed black and intense, and the exulting hate and thirst to destroy life which shone there, were the most horrifying feature in the whole vision. There was intelligence of a kind in them—intelligence beyond that of a beast, below that of a man.

The feelings which this horror stirred in Dennistoun were the intensest physical fear and the most profound mental loathing. What did he do? What could he do? He has never been quite certain what words he said, but he knows that he spoke, that he grasped blindly at the silver crucifix, that he was conscious of a movement towards him on the part of the demon, and that he screamed with the voice of an animal in hideous pain.

Pierre and Bertrand, the two sturdy little serving-men, who rushed in, saw nothing, but felt themselves thrust aside by something that passed out between them, and found Dennistoun in a swoon. They sat up with him that night, and his two friends were at St Bertrand by nine o'clock next morning. He himself though still shaken and nervous, was almost himself by that time, and his story found credence with them, though not until they had seen the drawing and talked with the sacristan.

Almost at dawn the little man had come to the inn on some pretence, and had listened with the deepest interest to the story retailed by the landlady. He showed no surprise.

'It is he—it is he! I have seen him myself,' was his only comment;

and to all questionings but one reply was vouchsafed: 'Deux fois je l'ai vu; mille fois je l'ai senti.' He would tell them nothing of the provenance of the book, nor any details of his experiences. 'I shall soon sleep, and my rest will be sweet. Why should you trouble me?' he said.*

We shall never know what he or Canon Alberic de Mauléon suffered. At the back of that fateful drawing were some lines of writing which may be supposed to throw light on the situation:

> 'Contradictio Salomonis cum demonio nocturno.
> Albericus de Mauleone delineavit.
> V. Deus in adiutorium. Ps. Qui habitat.
> Sancte Bertrande, demoniorum effugator, intercede pro me miserrimo.
> Primum uidi nocte 12^mi Dec 1694: uidebo mox ultimum. Peccaui et passus sum, plura adhuc passurus. Dec. 29, 1701.'†

I have never quite understood what was Dennistoun's view of the events I have narrated. He quoted to me once a text from Ecclesiasticus: 'Some spirits there be that are created for vengeance, and in their fury lay on sore strokes.' On another occasion he said: 'Isaiah was a very sensible man; doesn't he say something about night monsters living in the ruins of Babylon? These things are rather beyond us at present.'

Another confidence of his impressed me rather, and I sympathized with it. We had been, last year, to Comminges, to see Canon Alberic's tomb. It is a great marble erection with an effigy of the Canon in a large wig and soutane, and an elaborate eulogy of his learning below. I saw Dennistoun talking for some time with the Vicar of St Bertrand's, and as we drove away he said to me: 'I hope it isn't wrong: you know I am a Presbyterian—but I—I believe there will be "saying of Mass and singing of dirges" for Alberic de Mauléon's rest.' Then he added, with

* He died that summer; his daughter married, and settled at St Papoul. She never understood the circumstances of her father's 'obsession'.

† i.e., The Dispute of Solomon with a demon of the night. Drawn by Alberic de Mauléon. *Versicle.* O Lord, make haste to help me. *Psalm.* Whoso dwelleth (xci.).

Saint Bertrand, who puttest devils to flight, pray for me most unhappy. I saw it first on the night of Dec. 12, 1694: soon I shall see it for the last time. I have sinned and suffered, and have more to suffer yet. Dec. 29, 1701.

The 'Gallia Christiana' gives the date of the Canon's death as December 31, 1701, 'in bed, of a sudden seizure'. Details of this kind are not common in the great work of the Sammarthani.

a touch of the Northern British in his tone, 'I had no notion they came so dear.'

The book is in the Wentworth Collection at Cambridge. The drawing was photographed and then burnt by Dennistoun on the day when he left Comminges on the occasion of his first visit.

Jerry Bundler

W. W. JACOBS

It wanted a few nights to Christmas, a festival for which the small market-town of Torchester was making extensive preparations. The narrow streets which had been thronged with people were now almost deserted; the cheap-jack from London, with the remnant of breath left him after his evening's exertions, was making feeble attempts to blow out his naphtha lamp, and the last shops open were rapidly closing for the night.

In the comfortable coffee-room of the old 'Boar's Head', half a dozen guests, principally commercial travellers, sat talking by the light of the fire. The talk had drifted from trade to politics, from politics to religion, and so by easy stages to the supernatural. Three ghost stories, never known to fail before, had fallen flat; there was too much noise outside, too much light within. The fourth story was told by an old hand with more success; the streets were quiet, and he had turned the gas out. In the flickering light of the fire, as it shone on the glasses and danced with shadows on the walls, the story proved so enthralling that George, the waiter, whose presence had been forgotten, created a very disagreeable sensation by suddenly starting up from a dark corner and gliding silently from the room.

'That's what I call a good story,' said one of the men, sipping his hot whisky. 'Of course it's an old idea that spirits like to get into the company of human beings. A man told me once that he travelled down the Great Western with a ghost and hadn't the slightest suspicion of it until the inspector came for tickets. My friend said the way that ghost tried to keep up appearances by feeling for it in all its pockets and looking on the floor was quite touching. Ultimately it gave it up and with a faint groan vanished through the ventilator.'

'That'll do, Hirst,' said another man.

'It's not a subject for jesting,' said a little old gentleman who had been an attentive listener. 'I've never seen an apparition myself, but I know people who have, and I consider that they form a very interesting link between us and the after-life. There's a ghost story connected with this house, you know.'

'Never heard of it,' said another speaker, 'and I've been here some years now.'

'It dates back a long time now,' said the old gentleman. 'You've heard about Jerry Bundler, George?'

'Well, I've just 'eard odds and ends, sir,' said the old waiter, 'but I never put much count to 'em. There was one chap 'ere what said 'e saw it, and the gov'ner sacked 'im prompt.'

'My father was a native of this town,' said the old gentleman, 'and knew the story well. He was a truthful man and a steady churchgoer, but I've heard him declare that once in his life he saw the appearance of Jerry Bundler in this house.'

'And who was this Bundler?' enquired a voice.

'A London thief, pickpocket, highwayman—anything he could turn his dishonest hand to,' replied the old gentleman; 'and he was run to earth in this house one Christmas week some eighty years ago. He took his last supper in this very room, and after he had gone up to bed a couple of Bow Street runners, who had followed him from London but lost the scent a bit, went upstairs with the landlord and tried the door. It was stout oak, and fast, so one went into the yard, and by means of a short ladder got on to the window-sill, while the other stayed outside the door. Those below in the yard saw the man crouching on the sill, and then there was a sudden smash of glass, and with a cry he fell in a heap on the stones at their feet. Then in the moonlight they saw the white face of the pickpocket peeping over the sill, and while some stayed in the yard, others ran into the house and helped the other man to break the door in. It was difficult to obtain an entrance even then, for it was barred with heavy furniture, but they got in at last, and the first thing that met their eyes was the body of Jerry dangling from the top of the bed by his own handkerchief.'

'Which bedroom was it?' asked two or three voices together.

The narrator shook his head. 'That I can't tell you; but the story goes that Jerry still haunts this house, and my father used to declare positively that the last time he slept here the ghost of Jerry Bundler lowered itself from the top of his bed and tried to strangle him.'

'That'll do,' said an uneasy voice. 'I wish you'd thought to ask your father which bedroom it was.'

'What for?' enquired the old gentleman.

'Well, I should take care not to sleep in it, that's all,' said the voice shortly.

'There's nothing to fear,' said the other. 'I don't believe for a moment that ghosts could really hurt one. In fact my father used to confess that it was only the unpleasantness of the thing that upset him, and that for all practical purposes Jerry's fingers might have been made of cotton-wool for all the harm they could do.'

'That's all very fine,' said the last speaker again; 'a ghost story is a ghost story, sir; but when a gentleman tells a tale of a ghost in the house in which one is going to sleep, I call it most ungentlemanly!'

'Pooh! nonsense!' said the old gentleman, rising; 'ghosts can't hurt you. For my own part; I should rather like to see one. Good night, gentlemen.'

'Good night,' said the others. "And I only hope Jerry'll pay you a visit," added the nervous man as the door closed.'

'Bring some more whisky, George,' said a stout commercial; 'I want keeping up when the talk turns this way.'

'Shall I light the gas, Mr Malcolm?' said George.

'No; the fire's very comfortable,' said the traveller. 'Now gentlemen, any of you know any more?'

'I think we've had enough,' said another man; 'we shall be thinking we see spirits next, and we're not all like the old gentleman who's just gone.'

'Old humbug!' said Hirst. 'I should like to put him to the test. Suppose I dress up as Jerry Bundler and go and give him a chance of displaying his courage?'

'Bravo!' said Malcolm huskily, drowning one or two faint 'Noes'. 'Just for the joke, gentlemen.'

'No, no! Drop it, Hirst,' said another man.

'Only for the joke,' said Hirst, somewhat eagerly. 'I've got some things upstairs in which I am going to play in the "Rivals"—knee-breeches, buckles, and all that sort of thing. It's a rare chance. If you'll wait a bit I'll give you a full dress rehearsal, entitled, "Jerry Bundler; or, The Nocturnal Strangler".'

'You won't frighten us,' said the commercial, with a husky laugh.

'I don't know that,' said Hirst sharply; 'it's a question of acting, that's all. I'm pretty good, ain't I, Somers?'

'Oh, you're all right—for an amateur,' said his friend, with a laugh.

'I'll bet you a level sov. you don't frighten me,' said the stout traveller.

'Done!' said Hirst. 'I'll take the bet to frighten you first and the old gentleman afterwards. These gentlemen shall be the judges.'

'You won't frighten us, sir,' said another man 'because we're prepared for you; but you'd better leave the old man alone. It's dangerous play.'

'Well, I'll try you first,' said Hirst, springing up. 'No gas, mind.'

He ran lightly upstairs to his room, leaving the others, most of whom had been drinking somewhat freely, to wrangle about his proceedings. It ended in two of them going to bed.

'He's crazy on acting,' said Somers, lighting his pipe. 'Thinks he's the equal of anybody almost. It doesn't matter with us, but I won't let him go to the old man. And he won't mind so long as he gets an opportunity of acting to us.'

'Well, I hope he'll hurry up,' said Malcolm, yawning; 'it's after twelve now.'

Nearly half an hour passed. Malcolm drew his watch from his pocket and was busy winding it, when George, the waiter, who had been sent on an errand to the bar, burst suddenly into the room and rushed towards them.

''E's comin', gentlemen,' he said breathlessly.

'Why, you're frightened, George,' said the stout commercial, with a chuckle.

'It was the suddenness of it,' said George sheepishly; 'and besides, I didn't look for seein' 'im in the bar. There's only a glimmer of light there, and 'e was sitting on the floor behind the bar. I nearly trod on 'im.'

'Oh, you'll never make a man, George,' said Malcolm.

'Well, it took me unawares,' said the waiter. 'Not that I'd have gone to the bar by myself if I'd known 'e was there, and I don't believe you would either sir.'

'Nonsense!' said Malcolm. 'I'll go and fetch him in.'

'You don't know what it's like, sir,' said George, catching him by the sleeve. 'It ain't fit to look at by yourself, it ain't, indeed. It's got the —— *What's that?*'

They all started at the sound of a smothered cry from the staircase and the sound of somebody running hurriedly along the passage. Before anybody could speak, the door flew open and a figure bursting into the room flung itself gasping and shivering upon them.

'What is it? What's the matter?' demanded Malcolm. 'Why, it's Mr Hirst.' He shook him roughly and then held some spirit to his lips. Hirst drank it greedily and with a sharp intake of his breath gripped him by the arm.

'Light the gas, George,' said Malcolm.

The waiter obeyed hastily. Hirst, a ludicrous but pitiable figure in knee-breeches and coat, a large wig all awry, and his face a mess of grease paint, clung to him, trembling.

'Now what's the matter?' asked Malcolm.

'I've seen it,' said Hirst, with a hysterical sob. 'O Lord, I'll never play the fcol again, never!'

'Seen what?' said the others.

'Him—it—the ghost—anything!' said Hirst wildly.

'Rot!' said Malcolm uneasily.

'I was coming down the stairs,' said Hirst. 'Just capering down—as I thought—it ought to do. I felt a tap——'

He broke off suddenly and peered nervously through the open door into the passage.

'I thought I saw it again,' he whispered. 'Look—at the foot of the stairs. Can you see anything?'

'No, there's nothing there,' said Malcolm, whose own voice shook a little. 'Go on. You felt a tap on your shoulder——'

'I turned round and saw it—a little wicked head and a white dead face. Pah!'

'That's what I saw in the bar,' said George. ''Orrid it was—devilish!'

Hirst shuddered, and, still retaining his nervous grip of Malcolm's sleeve, dropped into a chair.

'Well, it's a most unaccountable thing,' said the dumbfounded Malcolm, turning round to the others. 'It's the last time I come to this house.'

'I leave tomorrow,' said George. 'I wouldn't go down to that bar again by myself, no, not for fifty pounds!'

'It's talking about the thing that's caused it, I expect,' said one of the men; 'we've all been talking about this and having it in our minds. Practically we've been forming a spiritualistic circle without knowing it.'

'Hang the old gentleman!' said Malcolm heartily. 'Upon my soul, I'm half afraid to go to bed. It's odd they should both think they saw something.'

'I saw it as plain as I see you, sir,' said George solemnly. 'P'raps if you keep your eyes turned up the passage you'll see it for yourself.'

They followed the direction of his finger, but saw nothing, although one of them fancied that a head peeped round the corner of the wall.

'Who'll come down to the bar?' said Malcolm, looking round.

'You can go, if you like,' said one of the others, with a faint laugh; 'we'll wait here for you.'

The stout traveller walked towards the door and took a few steps up the passage. Then he stopped. All was quite silent, and he walked slowly to the end and looked down fearfully towards the glass partition which shut off the bar. Three times he made as though to go to it; then he turned back, and, glancing over his shoulder, came hurriedly back to the room.

'Did you see it, sir?' whispered George.

'Don't know,' said Malcolm shortly. 'I fancied I saw something, but it might have been fancy. I'm in the mood to see anything just now. How are you feeling now, sir?'

'Oh, I feel a bit better now,' said Hirst somewhat brusquely, as all eyes were turned upon him. 'I daresay you think I'm easily scared, but you didn't see it.'

'Not at all,' said Malcolm, smiling faintly despite himself.

'I'm going to bed,' said Hirst, noticing the smile and resenting it. 'Will you share my room with me, Somers?'

'I will with pleasure,' said his friend 'provided you don't mind sleeping with the gas on full all night.'

He rose from his seat, and bidding the company a friendly good night, left the room with his crestfallen friend. The others saw them to the foot of the stairs, and having heard their door close, returned to the coffee-room.

'Well, I suppose the bet's off?' said the stout commercial, poking the fire and then standing with his legs apart on the hearthrug; 'though, as far as I can see, I won it. I never saw a man so scared in all my life. Sort of poetic justice about it, isn't there?'

'Never mind about poetry or justice,' said one of his listeners; 'who's going sleep with me?'

'I will,' said Malcolm affably.

'And I suppose we share a room together, Mr Leek?' said the third man, turning to the fourth.

'No, thank you,' said the other briskly; 'I don't believe in ghosts. If anything comes into my room I shall shoot it.'

'That won't hurt a spirit, Leek,' said Malcolm decisively.

'Well the noise'll be like company to me,' said Leek, 'and it'll wake the house too. But if you're nervous, sir,' he added with a grin to the man who had suggested sharing his room, 'George'll be only too pleased to sleep on the door-mat inside your room, I know.'

'That I will, sir,' said George fervently; 'and if you gentlemen would only come down with me to the bar to put the gas out, I could never be sufficiently grateful.'

They went out in a body, with the exception of Leek, peering carefully before them as they went. George turned the light out in the bar and they returned unmolested to the coffee-room, and, avoiding the sardonic smile of Leek, prepared to separate for the night.

'Give me the candle while you put the gas out, George,' said the traveller.

The waiter handed it to him and extinguished the gas, and at the same moment all distinctly heard a step in the passage outside. It stopped at the door, and as they watched with bated breath, the door creaked and slowly opened. Malcolm fell back open-mouthed, as a white, leering face, with sunken eyeballs and close-cropped bullet head, appeared at the opening.

For a few seconds the creature stood regarding them, blinking in a strange fashion at the candle. Then, with a sidling movement, it came a little way into the room and stood there as if bewildered.

Not a man spoke or moved, but all watched with a horrible fascination as the creature removed its dirty neckcloth and its head rolled on its shoulder. For a minute it paused, and then holding the rag before it, moved towards Malcolm.

The candle went out suddenly with a flash and a bang. These was a smell of powder, and something writhing in the darkness on the floor. A faint, choking cough, and then silence. Malcolm was the first to speak. 'Matches,' he said in a strange voice. George struck one. Then he leapt at the gas and a burner flamed from the match. Malcolm touched the thing on the floor with his foot and found it soft. He looked at his companions. They mouthed enquiries at him, but he shook his head. He lit the candle, and, kneeling down, examined the silent thing on the floor. Then he rose swiftly, and dipping his handkerchief in the water jug, bent down again and grimly wiped the white face. Then he sprang back with a cry of incredulous horror, pointing at it. Leek's pistol fell to the floor and he shut out the sight with his hands, but the others, crowding forward, gazed spell-bound at the dead face of Hirst.

Before a word was spoken the door opened and Somers hastily entered the room. His eyes fell on the floor. 'Good God!' he cried. 'You didn't——'

Nobody spoke.

'I told him not to,' he said in a suffocating voice. 'I told him not to. I told him——'

He leaned against the wall, deathly sick, put his arms out feebly, and fell fainting into the traveller's arms.

An Eddy on the Floor

BERNARD CAPES

I had the pleasure of an invitation to one of those reunions or seances at the house, in a fashionable quarter, of my distant connection, Lady Barbara Grille, whereat it was my hostess's humour to gather together those many birds of alien feather and incongruous habit that will flock from the hedgerows to the least little flattering crumb of attention. And scarce one of them but thinks the simple feast is spread for him alone. And with so cheap a bait may a title lure.

That reference to so charming a personality should be in this place is a digression. She affects my narrative only inasmuch as I happened to meet at her house a gentleman who for a time exerted a considerable influence over my fortunes.

The next morning after the seance, my landlady entered with a card, which she presented to my consideration:

MAJOR JAMES SHRIKE,
H. M. PRISON, D

All astonishment, I bade my visitor up.

He entered briskly, fur-collared, hat in hand, and bowed as he stood on the threshold. He was a very short man—snub-nosed; rusty-whiskered; indubitably and unimpressively a cockney in appearance. He might have walked out of a Cruikshank etching.

I was beginning, 'May I enquire—' when the other took me up with a vehement frankness that I found engaging at once.

'This is a great intrusion. Will you pardon me? I heard some remarks of yours last night that deeply interested me. I obtained your name and address from our hostess, and took the liberty of——'

'Oh! pray be seated. Say no more. My kinswoman's introduction is all-sufficient. I am happy in having caught your attention in so motley a crowd.'

'She doesn't—forgive the impertinence—take herself seriously enough.'

'Lady Barbara? Then you've found her out?'

'Ah!—you're not offended?'

'Not in the least.'

'Good. It was a motley assemblage, as you say. Yet I'm inclined to think I found my pearl in the oyster. I'm afraid I interrupted—eh?'

'No, no, not at all. Only some idle scribbling. I'd finished.'

'You are a poet?'

'Only a lunatic. I haven't taken my degree.'

'Ah! it's a noble gift—the gift of song; precious through its rarity.'

I caught a note of emotion in my visitor's voice, and glanced at him curiously.

'Surely,' I thought, 'that vulgar, ruddy little face is transfigured.'

'But,' said the stranger, coming to earth, 'I am lingering beside the mark. I must try to justify my solecism in manners by a straight reference to the object of my visit. That is, in the first instance, a matter of business.'

'Business!'

'I am a man with a purpose, seeking the hopefullest means to an end. Plainly: if I could procure you the post of resident doctor at D—— gaol, would you be disposed to accept it?'

I looked my utter astonishment.

'I can affect no surprise at yours,' said the visitor. 'It is perfectly natural. Let me forestall some unnecessary expression of it. My offer seems unaccountable to you, seeing that we never met until last night. But I don't move entirely in the dark. I have ventured in the interval to inform myself as to the details of your career. I was entirely one with much of your expression of opinion as to the treatment of criminals, in which you controverted the crude and unpleasant scepticism of the lady you talked with. Combining the two, I come to the immediate conclusion that you are the man for my purpose.'

'You have dumbfounded me. I don't know what to answer. You have views, I know, as to prison treatment. Will you sketch them? Will you talk on, while I try to bring my scattered wits to a focus?'

'Certainly I will. Let me, in the first instance, recall to you a few words of your own. They ran somewhat in this fashion: Is not the man of practical genius the man who is most apt at solving the little problems of resourcefulness in life? Do you remember them?'

'Perhaps I do, in a cruder form.'

'They attracted me at once. It is upon such a postulate I base my practice. Their moral is this: To know the antidote the moment the snake bites. That is to have the intuition of divinity. We shall rise to it some day, no doubt, and climb the hither side of the new Olympus. Who knows? Over the crest the spirit of creation may be ours.'

I nodded, still at sea, and the other went on with a smile:

'I once knew a world-famous engineer with whom I used to

breakfast occasionally. He had a patent egg-boiler on the table, with a little double-sided ladle underneath to hold the spirit. He complained that his egg was always undercooked. I said, "Why not reverse the ladle so as to bring the deeper cut uppermost?" He was charmed with my perspicacity. The solution had never occurred to him. You remember, too, no doubt, the story of Coleridge and the horse collar. We aim too much at great developments. If we cultivate resourcefulness, the rest will follow. Shall I state my system *in nuce*? It is to encourage this spirit of resourcefulness.'

'Surely the habitual criminal has it in a marked degree?'

'Yes; but abnormally developed in a single direction. His one object is to out-manœuvre in a game of desperate and immoral chances. The tactical spirit in him has none of the higher ambition. It has felt itself in the degree only that stops at defiance.'

'That is perfectly true.'

'It is half self-conscious of an individuality that instinctively assumes the hopelessness of a recognition by duller intellects. Leaning to resentment through misguided vanity, it falls "all oblique". What is the cure for this? I answer, the teaching of a divine egotism. The subject must be led to a pure devotion to self. What he wishes to respect he must be taught to make beautiful and interesting. The policy of sacrifice to others has so long stunted his moral nature because it is a hypocritical policy. We are responsible to ourselves in the first instance; and to argue an eternal system of blind self-sacrifice is to undervalue the fine gift of individuality. In such he sees but an indefensible policy of force applied to the advantage of the community. He is told to be good—not that he may morally profit, but that others may not suffer inconvenience.'

I was beginning to grasp, through my confusion, a certain clue of meaning in my visitor's rapid utterance. The stranger spoke fluently, but in the dry, positive voice that characterizes men of will.

'Pray go on,' I said; 'I am digesting in silence.'

'We must endeavour to lead him to respect of self by showing him what his mind is capable of. I argue on no sectarian, no religious grounds even. Is it possible to make a man's self his most precious possession? Anyhow, I work to that end. A doctor purges before building up with a tonic. I eliminate cant and hypocrisy, and then introduce self-respect. It isn't enough to employ a man's hands only. Initiation in some labour that should prove wholesome and remunerative is a redeeming factor, but it isn't all. His mind must work also, and awaken to its capacities. If it rusts, the body reverts to inhuman instincts.'

'May I ask how you——?'

'By intercourse—in my own person or through my officials. I wish to have only those about me who are willing to contribute to my designs, and with whom I can work in absolute harmony. All my officers are chosen to that end. No doubt a dash of constitutional sentimentalism gives colour to my theories. I get it from a human trait in me that circumstances have obliged me to put a hoarding round.'

'I begin to gather daylight.'

'Quite so. My patients are invited to exchange views with their guardians in a spirit of perfect friendliness; to solve little problems of practical moment; to acquire the pride of self-reliance. We have competitions, such as certain newspapers open to their readers, in a simple form. I draw up the questions myself. The answers give me insight into the mental conditions of the competitors. Upon insight I proceed. I am fortunate in private means, and I am in a position to offer modest prizes to the winners. Whenever such a one is discharged, he finds awaiting him the tools most handy to his vocation. I bid him go forth in no pharisaical spirit, and invite him to communicate with me. I wish the shadow of the gaol to extend no further than the road whereon it lies. Henceforth, we are acquaintances with a common interest at heart. Isn't it monstrous that a state-fixed degree of misconduct should earn a man social ostracism? Parents are generally inclined to rule extra tenderness towards a child whose peccadilloes have brought him a whipping. For myself, I have no faith in police supervision. Give a culprit his term and have done with it. I find the majority who come back to me are ticket-of-leave men.

'Have I said enough? I offer you the reversion of the post. The present holder of it leaves in a month's time. Please to determine here and at once.'

'Very good. I have decided.'

'You will accept?'

'Yes.'

With my unexpected appointment as doctor to D—— gaol, I seemed to have put on the seven-league boots of success. No doubt it was an extraordinary degree of good fortune, even to one who had looked forward with a broad view of confidence; yet, I think, perhaps on account of the very casual nature of my promotion, I never took the post entirely seriously.

At the same time I was fully bent on justifying my little cockney patron's choice by a resolute subscription to his theories of prison management.

Major James Shrike inspired me with a curious conceit of impertinent respect. In person the very embodiment of that insignificant vulgarity, without extenuating circumstances, which is the type in caricature of the ultimate cockney, he possessed a force of mind and an earnestness of purpose that absolutely redeemed him on close acquaintanceship. I found him all he had stated himself to be, and something more.

He had a noble object always in view—the employment of sane and humanitarian methods in the treatment of redeemable criminals, and he strove towards it with completely untiring devotion. He was of those who never insist beyond the limits of their own understanding, clear-sighted in discipline, frank in relaxation, an altruist in the larger sense.

His undaunted persistence, as I learned, received ample illustration some few years prior to my acquaintance with him, when—his system being experimental rather than mature—a devastating epidemic of typhoid in the prison had for the time stultified his efforts. He stuck to his post; but so virulent was the outbreak that the prison commissioners judged a complete evacuation of the building and overhauling of the drainage to be necessary. As a consequence, for some eighteen months—during thirteen of which the Governor and his household remained sole inmates of the solitary pile (so sluggishly do we redeem our condemned social bog-lands)—the 'system' stood still for lack of material to mould. At the end of over a year of stagnation, a contract was accepted and workmen put in, and another five months saw the prison reordered for practical purposes.

The interval of forced inactivity must have sorely tried the patience of the Governor. Practical theorists condemned to rust too often eat out their own hearts. Major Shrike never referred to this period, and, indeed, laboriously snubbed any allusion to it.

He was, I have a shrewd notion, something of an officially petted reformer. Anyhow, to his abolition of the insensate barbarism of crank and treadmill in favour of civilizing methods no opposition was offered. Solitary confinement—a punishment outside all nature to a gregarious race—found no advocate in him. 'A man's own suffering mind,' he argued, 'must be, of all moral food, the most poisonous for him to feed on. Surround a scorpion with fire and he stings himself to death, they say. Throw a diseased soul entirely upon its own resources and moral suicide results.'

To sum up: his nature embodied humanity without sentimentalism, firmness without obstinacy, individuality without selfishness; his activity was boundless, his devotion to his system so real as to admit no utilitarian sophistries into his scheme of personal benevolence. Before I had been with him a week, I respected him as I had never respected man before.

One evening (it was during the second month of my appointment) we were sitting in his private study—a dark, comfortable room lined with books. It was an occasion on which a new characteristic of the man was offered to my inspection.

A prisoner of a somewhat unusual type had come in that day—a spiritualistic medium, convicted of imposture. To this person I casually referred.

'May I ask how you propose dealing with the newcomer?'

'On the familiar lines.'

'But, surely—here we have a man of superior education, of imagination even?'

'No, no, no! A hawker's opportuneness; that describes it. These fellows would make death itself a vulgarity.'

'You've no faith in their——'

'Not a tittle. Heaven forfend! A sheet and a turnip are poetry to their manifestations. It's as crude and sour soil for us to work on as any I know. We'll cart it wholesale.'

'I take you—excuse my saying so—for a supremely sceptical man.'

'As to what?'

'The supernatural.'

There was no answer during a considerable interval. Presently it came, with deliberate insistence:

'It is a principle with me to oppose bullying. We are here for a definite purpose—his duty plain to any man who *wills* to read it. There may be disembodied spirits who seek to distress or annoy where they can no longer control. If there are, mine, which is not yet divorced from its means to material action, declines to be influenced by any irresponsible whimsy, emanating from a place whose denizens appear to be actuated by a mere frivolous antagonism to all human order and progress.'

'But supposing you, a murderer, to be haunted by the presentment of your victim?'

'I will imagine that to be my case. Well, it makes no difference. My interest is with the great human system, in one of whose veins I am a circulating drop. It is my business to help to keep the system sound, to do my duty without fear or favour. If disease—say a fouled conscience—contaminates me, it is for me to throw off the incubus, not accept it, and transmit the poison. Whatever my lapses of nature, I owe it to the entire system to work for purity in my allotted sphere, and not to allow any microbe bugbear to ride me roughshod, to the detriment of my fellow drops.'

I laughed.

'It should be for you,' I said, 'to learn to shiver, like the boy in the fairy-tale.'

'I cannot,' he answered, with a peculiar quiet smile; 'and yet prisons, above all places, should be haunted.'

Very shortly after his arrival I was called to the cell of the medium, F——. He suffered, by his own statement, from severe pains in the head.

I found the man to be nervous, anaemic; his manner characterized by a sort of hysterical effrontery.

'Send me to the infirmary,' he begged. 'This isn't punishment, but torture.'

'What are your symptoms?'

'I see things; my case has no comparison with others. To a man of my super-sensitiveness close confinement is mere cruelty.'

I made a short examination. He was restless under my hands.

'You'll stay where you are,' I said.

He broke out into violent abuse, and I left him.

Later in the day I visited him again. He was then white and sullen; but under his mood I could read real excitement of some sort.

'Now, confess to me, my man,' I said, 'what do you see?'

He eyed me narrowly, with his lips a little shaky.

'Will you have me moved if I tell you?'

'I can give no promise till I know.'

He made up his mind after an interval of silence.

'There's something uncanny in my neighbourhood. Who's confined in the next cell—there, to the left?'

'To my knowledge it's empty.'

He shook his head incredulously.

'Very well,' I said, 'I don't mean to bandy words with you'; and I turned to go.

At that he came after me with a frightened choke.

'Doctor, your mission's a merciful one. I'm not trying to sauce you. For God's sake have me moved! I can see further than most, I tell you!'

The fellow's manner gave me pause. He was patently and beyond the pride of concealment terrified.

'What do you see?' I repeated stubbornly.

'It isn't that I see, but I know. The cell's *not* empty!'

I stared at him in considerable wonderment.

'I will make enquiries,' I said. 'You may take that for a promise. If the cell proves empty, you stop where you are.'

I noticed that he dropped his hands with a lost gesture as I left him. I was sufficiently moved to accost the warder who awaited me on the spot.

'Johnson,' I said, 'is that cell——'

'Empty, sir,' answered the man sharply and at once.

Before I could respond, F—— came suddenly to the door, which I still held open.

'You lying cur!' he shouted. 'You damned lying cur!'

The warder thrust the man back with violence.

'Now you, 49,' he said, 'dry up, and none of your sauce!' and he banged to the door with a sounding slap, and turned to me with a lowering face. The prisoner inside yelped and stormed at the studded panels.

'That cell's empty, sir,' repeated Johnson.

'Will you, as a matter of conscience, let me convince myself? I promised the man.'

'No, I can't.'

'You can't?'

'No, sir.'

'This is a piece of stupid discourtesy. You can have no reason, of course?'

'I can't open it—that's all.'

'Oh, Johnson! Then I must go to the fountainhead.'

'Very well, sir.'

Quite baffled by the man's obstinacy, I said no more, but walked off. If my anger was roused, my curiosity was piqued in proportion.

I had no opportunity of interviewing the Governor all day, but at night I visited him by invitation to play a game of piquet.

He was a man without 'incumbrances'—as a severe conservatism designates the *lares* of the cottage—and, at home, lived at his ease and indulged his amusements without comment.

I found him 'tasting' his books, with which the room was well lined, and drawing with relish at an excellent cigar in the intervals of the courses.

He nodded to me, and held out an open volume in his left hand.

'Listen to this fellow,' he said, tapping the page with his fingers:

'*"The most tolerable sort of Revenge, is for those wrongs which there is no Law to remedy. But then, let a man take heed, the Revenge be such, as there is no law to punish. Else, a man's Enemy, is still before hand, and it is two for one. Some, when they take Revenge, are Desirous the party should know, whence it cometh. This is the more Generous. For the Delight seemeth to be, not so much in doing the Hurt, as in making the Party repent: But Base and Crafty Cowards, are like the Arrow that flyeth in the Dark. Cosmus, Duke of Florence, had a Desperate Saying against Perfidious or Neglecting Friends, as if these wrongs were unpardonable. You shall read (saith he) that we are commanded to forgive our Enemies: But you never read, that we are commanded to forgive our Friends.*"'

'Is he not a rare fellow?'

'Who?' said I.

'Francis Bacon, who screwed his wit to his philosophy, like a hammer-head to its handle, and knocked a nail in at every blow. How many of our friends round about here would be picking oakum now if they had made a gospel of that quotation?'

'You mean they take no heed that the Law may punish for that for which it gives no remedy?'

'Precisely; and specifically as to revenge. The criminal, from the murderer to the petty pilferer, is actuated solely by the spirit of vengeance—vengeance blind and speechless—towards a system that forces him into a position quite outside his natural instincts.'

'As to that, we have left Nature in the thicket. It is hopeless hunting for her now.'

'We hear her breathing sometimes, my friend. Otherwise Her Majesty's prison locks would rust. But, I grant you, we have grown so unfamiliar with her that we call her simplest manifestations *super-natural* nowadays.'

'That reminds me. I visited F—— this afternoon. The man was in a queer way—not foxing, in my opinion. Hysteria, probably.'

'Oh! What was the matter with him?'

'The form it took was some absurd prejudice about the next cell—number 47. He swore it was not empty—was quite upset about it—said there was some infernal influence at work in his neigh-bourhood. Nerves, he finds, I suppose, may revenge themselves on one who has made a habit of playing tricks with them. To satisfy him, I asked Johnson to open the door of the next cell—'

'Well?'

'He refused.'

'It is closed by my orders.'

'That settles it, of course. The manner of Johnson's refusal was a bit uncivil, but——'

He had been looking at me intently all this time—so intently that I was conscious of a little embarrassment and confusion. His mouth was set like a dash between brackets, and his eyes glistened. Now his features relaxed, and he gave a short high neigh of a laugh.

'My dear fellow, you must make allowances for the rough old lurcher. He was a soldier. He is all cut and measured out to the regimental pattern. With him Major Shrike, like the king, can do no wrong. Did I ever tell you he served under me in India? He did; and, moreover, I saved his life there.'

'In an engagement?'

'Worse—from the bite of a snake. It was a mere question of will. I told him to wake and walk, and he did. They had thought him already in *rigor mortis*; and, as for him—well, his devotion to me since has been single to the last degree.'

'That's as it should be.'

'To be sure. And he's quite in my confidence. You must pass over the old beggar's churlishness.'

I laughed an assent. And then an odd thing happened. As I spoke, I had walked over to a bookcase on the opposite side of the room to that on which my host stood. Near this bookcase hung a mirror—an oblong affair, set in brass *repoussé* work—on the wall; and, happening to glance into it as I approached, I caught sight of the Major's reflection as he turned his face to follow my movement.

I say 'turned his face'—a formal description only. What met my startled gaze was an image of some nameless horror—of features grooved, and battered, and shapeless, as if they had been torn by a wild beast.

I gave a little indrawn gasp and turned about. There stood the Major, plainly himself, with a pleasant smile on his face.

'What's up?' said he.

He spoke abstractedly, pulling at his cigar; and I answered rudely, 'That's a damned bad looking-glass of yours!'

'I didn't know there was anything wrong with it,' he said, still abstracted and apart. And, indeed, when by sheer mental effort I forced myself to look again, there stood my companion as he stood in the room.

I gave a tremulous laugh, muttered something or nothing, and fell to examining the books in the case. But my fingers shook a trifle as I aimlessly pulled out one volume after another.

'Am *I* getting fanciful?' I thought—'I whose business it is to give practical account of every bugbear of the nerves. Bah! My liver must be out of order. A speck of bile in one's eye may look a flying dragon.'

I dismissed the folly from my mind, and set myself resolutely to inspecting the books marshalled before me. Roving amongst them, I pulled out, entirely at random, a thin, worn duodecimo, that was thrust well back at a shelf end, as if it shrank from comparison with its prosperous and portly neighbours. Nothing but chance impelled me to the choice; and I don't know to this day what the ragged volume was about. It opened naturally at a marker that lay in it—a folded slip of paper, yellow with age; and glancing at this, a printed name caught my eye.

With some stir of curiosity, I spread the slip out. It was a title-page to a volume, of poems, presumably; and the author was James Shrike.

I uttered an exclamation, and turned, book in hand.

'An author!' I said. '*You* an author, Major Shrike!'

To my surprise, he snapped round upon me with something like a glare of fury on his face. This the more startled me as I believed I had reason to regard him as a man whose principles of conduct had long disciplined a temper that was naturally hasty enough.

Before I could speak to explain, he had come hurriedly across the room and had rudely snatched the paper out of my hand.

'How did this get——' he began; then in a moment came to himself, and apologized for his ill manners.

'I thought every scrap of the stuff had been destroyed,' he said, and tore the page into fragments. 'It is an ancient effusion, doctor—perhaps the greatest folly of my life; but it's something of a sore subject with me, and I shall be obliged if you'll not refer to it again.'

He courted my forgiveness so frankly that the matter passed without embarrassment; and we had our game and spent a genial evening together. But memory of the queer little scene stuck in my mind, and I could not forbear pondering it fitfully.

Surely here was a new side-light that played upon my friend and superior a little fantastically.

Conscious of a certain vague wonder in my mind, I was traversing the prison, lost in thought, after my sociable evening with the Governor, when the fact that dim light was issuing from the open door of cell number 49 brought me to myself and to a pause in the corridor outside.

Then I saw that something was wrong with the cell's inmate, and that my services were required.

The medium was struggling on the floor, in what looked like an epileptic fit, and Johnson and another warder were holding him from doing an injury to himself.

The younger man welcomed my appearance with relief.

'Heard him guggling,' he said, 'and thought as something were up. You come timely, sir.'

More assistance was procured, and I ordered the prisoner's removal to the infirmary. For a minute, before following him, I was left alone with Johnson.

'It came to a climax, then?' I said, looking the man steadily in the face.

'He may be subject to 'em, sir,' he replied evasively.

I walked deliberately up to the closed door of the adjoining cell, which was the last on that side of the corridor. Huddled against the massive end wall, and half embedded in it, as it seemed, it lay in a certain shadow, and bore every sign of dust and disuse. Looking closely, I saw that the trap in the door was not only firmly bolted, but *screwed into its socket*.

I turned and said to the warder quietly—

'Is it long since this cell was in use?'

'You're very fond of asking questions,' he answered doggedly.

It was evident he would baffle me by impertinence rather than yield a confidence. A queer insistence had seized me—a strange desire to know more about this mysterious chamber. But, for all my curiosity, I flushed at the man's tone.

'You have your orders,' I said sternly, 'and do well to hold by them. I doubt, nevertheless, if they include impertinence to your superiors.'

'I look straight on my duty, sir,' he said, a little abashed. 'I don't wish to give offence.'

He did not, I feel sure. He followed his instinct to throw me off the scent, that was all.

I strode off in a fume, and after attending F—— in the infirmary, went promptly to my own quarters.

I was in an odd frame of mind, and for long tramped my sitting-room to and fro, too restless to go to bed, or, as an alternative, to settle down to a book. There was a welling up in my heart of some emotion that I could neither trace nor define. It seemed neighbour to terror, neighbour to an intense fainting pity, yet was not distinctly either of these. Indeed, where was cause for one, or the subject of the other? F—— might have endured mental sufferings which it was only human to help to end, yet F—— was a swindling rogue, who, once relieved, merited no further consideration.

It was not on him my sentiments were wasted. Who, then, was responsible for them?

There was a very plain line of demarcation between the legitimate spirit of enquiry and mere apish curiosity. I could recognize it, I have no doubt, as a rule, yet in my then mood, under the influence of a kind of morbid seizure, inquisitiveness took me by the throat. I could not whistle my mind from the chase of a certain graveyard will-o'-wisp; and on it went stumbling and floundering through bog and mire, until it fell into a state of collapse, and was useful for nothing else.

I went to bed and to sleep without difficulty, but I was conscious of myself all the time, and of a shadowless horror that seemed to come stealthily out of corners and to bend over and look at me, and to be nothing but a curtain or a hanging coat when I started and stared.

Over and over again this happened, and my temperature rose by leaps, and suddenly I saw that if I failed to assert myself, and promptly, fever would lap me in a consuming fire. Then in a moment I broke into a profuse perspiration, and sank exhausted into delicious unconsciousness.

Morning found me restored to vigour, but still with the maggot of curiosity in my brain. It worked there all day, and for many subsequent days, and at last it seemed as if my every faculty were honeycombed with its ramifications. Then 'this will not do', I thought, but still the tunnelling process went on.

At first I would not acknowledge to myself what all this mental to-do was about. I was ashamed of my new development, in fact, and nervous, too, in a degree of what it might reveal in the matter of moral degeneration; but gradually, as the curious devil mastered me, I grew into such harmony with it that I could shut my eyes no longer to the true purpose of its insistence. It was the *closed cell* about which my thoughts hovered like crows circling round carrion.

'In the dead waste and middle' of a certain night I awoke with a strange, quick recovery of consciousness. There was the passing of a single expiration, and I had been asleep and was awake. I had gone to bed with no sense of premonition or of resolve in a particular direction; I sat up a monomaniac. It was as if, swelling in the silent hours, the tumour of curiosity had come to a head, and in a moment it was necessary to operate upon it.

I make no excuse for my then condition. I am convinced I was the victim of some undistinguishable force, that I was an agent under the control of the supernatural, if you like. Some thought had been in my mind of late that in my position it was my duty to unriddle the mystery of the closed cell. This was a sop timidly held out to and rejected by my better reason. I sought—and I knew it in my heart—solution of the puzzle, because it was a puzzle with an atmosphere that vitiated my moral fibre. Now, suddenly, I knew I must act, or, by forcing self-control, imperil my mind's stability.

All strung to a sort of exaltation, I rose noiselessly and dressed myself with rapid, nervous hands. My every faculty was focused upon a solitary point. Without and around there was nothing but shadow and uncertainty. I seemed conscious only of a shaft of light, as it were, traversing the darkness and globing itself in a steady disc of radiance on a lonely door.

Slipping out into the great echoing vault of the prison in stockinged feet, I sped with no hesitation of purpose in the direction of the corridor that was my goal. Surely some resolute Providence guided and encompassed me, for no meeting with the night patrol occurred at any point to embarrass or deter me. Like a ghost myself, I flitted along the stone flags of the passages, hardly waking a murmur from them in my progress.

Without, I knew, a wild and stormy wind thundered on the walls of the prison. Within, where the very atmosphere was self-contained, a cold and solemn peace held like an irrevocable judgement.

I found myself as if in a dream before the sealed door that had for days harassed my waking thoughts. Dim light from a distant gas jet made a patch of yellow upon one of its panels; the rest was buttressed with shadow.

A sense of fear and constriction was upon me as I drew softly from my pocket a screwdriver I had brought with me. It never occurred to me, I swear, that the quest was no business of mine, and that even now I could withdraw from it, and no one be the wiser. But I was afraid—I was afraid. And there was not even the negative comfort of knowing that the neighbouring cell was tenanted. It gaped like a ghostly garret next door to a deserted house.

What reason had I to be there at all, or, being there, to fear? I can no more explain than tell how it was that I, an impartial follower of my vocation, had allowed myself to be tricked by that in the nerves I had made it my interest to study and combat in others.

My hand that held the tool was cold and wet. The stiff little shriek of the first screw, as it turned at first uneasily in its socket, sent a jarring thrill through me. But I persevered, and it came out readily by-and-by, as did the four or five others that held the trap secure.

Then I paused a moment; and, I confess, the quick pant of fear seemed to come grey from my lips. There were sounds about me—the deep breathing of imprisoned men; and I envied the sleepers their hard-wrung repose.

At last, in one access of determination, I put out my hand, and sliding back the bolt, hurriedly flung open the trap. An acrid whiff of dust assailed my nostrils as I stepped back a pace and stood expectant of anything—or nothing. What did I wish, or dread, or foresee? The complete absurdity of my behaviour was revealed to me in a moment. I could shake off the incubus here and now, and be a sane man again.

I giggled, with an actual ring of self-contempt in my voice, as I made a forward movement to close the aperture. I advanced my face to it, and inhaled the sluggish air that stole forth, and—God in heaven!

I had staggered back with that cry in my throat, when I felt fingers like iron clamps close on my arm and hold it. The grip, more than the face I turned to look upon in my surging terror, was forcibly human.

It was the warder Johnson who had seized me, and my heart bounded as I met the cold fury of his eyes.

'Prying!' he said, in a hoarse, savage whisper. 'So you will, will you? And now let the devil help you!'

It was not this fellow I feared, though his white face was set like a demon's; and in the thick of my terror I made a feeble attempt to assert my authority.

'Let me go!' I muttered. 'What! you dare?'

In his frenzy he shook my arm as a terrier shakes a rat, and, like a dog, he held on, daring me to release myself.

For the moment an instinct half-murderous leapt in me. It sank and was overwhelmed in a slough of some more secret emotion.

'Oh!' I whispered, collapsing, as it were, to the man's fury, even pitifully deprecating it. 'What is it? What's there? It drew me—something unnameable.'

He gave a snapping laugh like a cough. His rage waxed second by second. There was a maniacal suggestiveness in it; and not much longer, it was evident, could he have it under control. I saw it run and congest in his eyes; and, on the instant of its accumulation, he tore at me with a sudden wild strength, and drove me up against the very door of the secret cell.

The action, the necessity of self-defence, restored me to some measure of dignity and sanity.

'Let me go, you ruffian!' I cried, struggling to free myself from his grasp.

It was useless. He held me madly. There was no beating him off: and, so holding me, he managed to produce a single key from one of his pockets, and to slip it with a rusty clang into the lock of the door.

'You dirty, prying civilian!' he panted at me, as he swayed this way and that with the pull of my body. 'You shall have your wish, by G——! You want to see inside, do you? Look, then!'

He dashed open the door as he spoke, and pulled me violently into the opening. A great waft of the cold, dank air came at us, and with it—what?

The warder had jerked his dark lantern from his belt, and now—an arm of his still clasped about one of mine—snapped the slide open.

'Where is it?' he muttered, directing the disc of light round and about the floor of the cell. I ceased struggling. Some counter influence was raising an odd curiosity in me.

'Ah!' he cried, in a stifled voice, 'there you are, my friend!'

He was setting the light slowly travelling along the stone flags close by the wall over against us, and now, so guiding it, looked askance at me with a small, greedy smile.

'Follow the light, sir,' he whispered jeeringly.

I looked, and saw twirling on the floor, in the patch of radiance cast by the lamp, *a little eddy of dust*, it seemed. This eddy was never still, but went circling in that stagnant place without apparent cause or influence; and, as it circled, it moved slowly on by wall and corner, so that presently in its progress it must reach us where we stood.

Now, draughts will play queer freaks in quiet places, and of this trifling phenomenon I should have taken little note ordinarily. But, I must say at once, that as I gazed upon the odd moving thing my heart seemed to fall in upon itself like a drained artery.

'Johnson!' I cried, 'I must get out of this. I don't know what's the matter, or—Why do you hold me? D—— it! man, let me go; let me go, I say!'

As I grappled with him he dropped the lantern with a crash and flung his arms violently about me.

'You don't!' he panted, the muscles of his bent and rigid neck seeming actually to cut into my shoulder-blade. 'You don't, by G——! You came of your own accord, and now you shall take your bellyfull!'

It was a struggle for life or death, or, worse, for life and reason. But I was young and wiry, and held my own, if I could do little more. Yet there was something to combat beyond the mere brute strength of the man I struggled with, for I fought in an atmosphere of horror unexplainable, and I knew that inch by inch the *thing* on the floor was circling round in our direction.

Suddenly in the breathing darkness I felt it close upon us, gave one mortal yell of fear, and, with a last despairing fury, tore myself from the encircling arms, and sprang into the corridor without. As I plunged and leapt, the warder clutched at me, missed, caught a foot on the edge of the door, and, as the latter whirled to with a clap, fell heavily at my feet in a fit. Then, as I stood staring down upon him, steps sounded along the corridor and the voices of scared men hurrying up.

Ill and shaken, and, for the time, little in love with life, yet fearing death as I had never dreaded it before, I spent the rest of that horrible night huddled between my crumpled sheets, fearing to look forth, fearing to think, wild only to be far away, to be housed in some green and innocent hamlet, where I might forget the madness and the terror in learning to walk the unvext paths of placid souls. That unction I could lay to my heart, at least. I had done the manly part by the stricken warder, whom I had attended to his own home, in a row of little tenements that stood south of the prison walls. I had replied to all enquiries with some dignity and spirit, attributing my ruffled condition to an assault on the part of Johnson, when he was already under the shadow of his seizure. I had directed his removal, and grudged him no professional attention that it was in my power to bestow. But afterwards, locked into my room, my whole nervous system broke up like a trodden ant-hill, leaving me conscious of nothing but an aimless scurrying terror and the black swarm of thoughts, so that I verily fancied my reason would give under the strain.

Yet I had more to endure and to triumph over.

Near morning I fell into a troubled sleep, throughout which the drawn twitch of muscle seemed an accent on every word of ill-omen I had ever spelt out of the alphabet of fear. If my body rested, my brain was an open chamber for any toad of ugliness that listed to 'sit at squat' in.

Suddenly I woke to the fact that there was a knocking at my door—that there had been for some little time.

I cried, 'Come in!' finding a weak restorative in the mere sound of my own human voice; then, remembering the key was turned, bade the visitor wait until I could come to him.

Scrambling, feeling dazed and white-livered, out of bed, I opened the door, and met one of the warders on the threshold. The man looked scared, and his lips, I noticed, were set in a somewhat boding fashion.

'Can you come at once, sir?' he said. 'There's summat wrong with the Governor.'

'Wrong?' What's the matter with him?'

'Why'—he looked down, rubbed an imaginary protuberance smooth with his foot, and glanced up at me again with a quick, furtive expression—'he's got his face set in the grating of 47, and danged if a man Jack of us can get him to move or speak.'

I turned away, feeling sick. I hurriedly pulled on coat and trousers, and hurriedly went off with my summoner. Reason was all absorbed in a wildest phantasy of apprehension.

'Who found him?' I muttered, as we sped on.

'Vokins see him go down the corridor about half after eight, sir, and see him give a start like when he noticed the trap open. It's never been so before in my time. Johnson must ha' done it last night, before he were took.'

'Yes, yes.'

'The man said the Governor went to shut it, it seemed, and to draw his face to'ards the bars in so doin'. Then he see him a-lookin' through, as he thought; but nat'rally it weren't no business of his'n, and he went off about his work. But when he come anigh agen, fifteen minutes later, there were the Governor in the same position; and he got scared over it, and called out to one or two of us.'

'Why didn't one of you ask the Major if anything was wrong?'

'Bless you! we did; and no answer. And we pulled him, compatible with discipline, but——'

'But what?'

'He's stuck.'

'Stuck!'

'See for yourself, sir. That's all I ask.'

I did, a moment later. A little group was collected about the door of cell 47, and the members of it spoke together in whispers, as if they were frightened men. One young fellow, with a face white in patches, as if it had been floured, slid from them as I approached, and accosted me tremulously.

'Don't go anigh, sir. There's something wrong about the place.'

I pulled myself together, forcibly beating down the excitement reawakened by the associations of the spot. In the discomfiture of others' nerves I found my own restoration.

'Don't be an ass!' I said, in a determined voice. 'There's nothing here that can't be explained. Make way for me, please!

They parted and let me through, and I saw him. He stood, spruce, frock-coated, dapper, as he always was, with his face pressed against and *into* the grill, and either hand raised and clenched tightly round a bar of the trap. His posture was as of one caught and striving frantically to release himself; yet the narrowness of the interval between the rails precluded so extravagant an idea. He stood quite motionless—taut and on the strain, as it were—and nothing of his face was visible but the back ridges of his jawbones, showing white through a bush of red whiskers.

'Major Shrike!' I rapped out, and, allowing myself no hesitation, reached forth my hand and grasped his shoulder. The body vibrated under my touch, but he neither answered nor made sign of hearing me. Then I pulled at him forcibly, and ever with increasing strength. His fingers held like steel braces. He seemed glued to the trap, like Theseus to the rock.

Hastily I peered round, to see if I could get a glimpse of his face. I noticed enough to send me back with a little stagger.

'Has none of you got a key to this door?' I asked, reviewing the scared faces about me, than which my own was no less troubled, I feel sure.

'Only the Governor, sir,' said the warder who had fetched me. 'There's not a man but him amongst us that ever seen this opened.'

He was wrong there, I could have told him; but held my tongue, for obvious reasons.

'I want it opened. Will one of you feel in his pockets?'

Not a soul stirred. Even had not sense of discipline precluded, that of a certain inhuman atmosphere made fearful creatures of them all.

'Then,' said I, 'I must do it myself.'

I turned once more to the stiff-strung figure, had actually put hand on it, when an exclamation from Vokins arrested me.

'There's a key—there, sir!' he said—'stickin' out yonder between his feet.'

Sure enough there was—Johnson's, no doubt, that had been shot from its socket by the clapping to of the door, and afterwards kicked aside by the warder in his convulsive struggles.

I stooped, only too thankful for the respite, and drew it forth. I had seen it but once before, yet I recognized it at a glance.

Now, I confess, my heart felt ill as I slipped the key into the wards, and a sickness of resentment at the tyranny of Fate in making me its helpless minister surged up in my veins. Once, with my fingers on the iron loop, I paused, and ventured a fearful side glance at the figure whose crooked elbow almost touched my face; then, strung to the high pitch of inevitability, I shot the lock, pushed at the door, and in the act, made a back leap into the corridor.

Scarcely, in doing so, did I look for the totter and collapse outwards of the rigid form. I had expected to see it fall away, face down, into the cell, as its support swung from it. Yet it was, I swear, as if *something* from within had relaxed its grasp and given the fearful dead man a swingeing push outwards as the door opened.

It went on its back, with a dusty slap on the stone flags, and from all its spectators—me included—came a sudden drawn sound, like a wind in a keyhole.

What can I say, or how describe it? A dead thing it was—but the face!

Barred with livid scars where the grating rails had crossed it, the rest seemed to have been worked and kneaded into a mere featureless plate of yellow and expressionless flesh.

And it was this I had seen in the glass!

There was an interval following the experience above narrated, during which a certain personality that had once been mine was effaced or suspended, and I seemed a passive creature, innocent of the least desire of independence. It was not that I was actually ill or actually insane. A merciful Providence set my finer wits slumbering, that was all, leaving me a sufficiency of the grosser faculties that were necessary to the right ordering of my behaviour.

I kept to my room, it is true, and even lay a good deal in bed; but this was more to satisfy the busy scruples of a *locum tenens*—a practitioner of the neighbourhood, who came daily to the prison to officiate in my absence—than to cosset a complaint that in its inactivity was purely negative. I could review what had happened with a calmness as profound as if I had read of it in a book. I could have wished to continue my duties, indeed, had the power of insistence remained to me. But the saner medicus was acute where I had gone blunt, and bade me to the restful course. He was right. I was mentally stunned, and had I not slept off my lethargy, I should have gone mad in an hour—leapt at a bound, probably, from inertia to flaming lunacy.

I remembered everything, but through a fluffy atmosphere, so to speak. It was as if I looked on bygone pictures through ground glass that softened the ugly outlines.

Sometimes I referred to these to my substitute, who was wise to answer me according to my mood; for the truth left me unruffled, whereas an obvious evasion of it would have distressed me.

'Hammond,' I said one day, 'I have never yet asked you. How did I give my evidence at the inquest?'

'Like a doctor and a sane man.'

'That's good. But it was a difficult course to steer. You conducted the post-mortem. Did any peculiarity in the dead man's face strike you?'

'Nothing but this: that the excessive contraction of the bicipital muscles had brought the features into such forcible contact with the bars as to cause bruising and actual abrasion. He must have been dead some little time when you found him.'

'And nothing else? You noticed nothing else in his face—a sort of obliteration of what makes one human, I mean?'

'Oh, dear, no! nothing but the painful constriction that marks any ordinary fatal attack of angina pectoris.—There's a rum breach of promise case in the paper today. You should read it; it'll make you laugh.'

I had no more inclination to laugh than to sigh; but I accepted the change of subject with an equanimity now habitual to me.

One morning I sat up in bed, and knew that consciousness was wide awake in me once more. It had slept, and now rose refreshed, but trembling. Looking back, all in a flutter of new responsibility, along the misty path by way of which I had recently loitered, I shook with an awful thankfulness at sight of the pitfalls I had skirted and escaped—of the demons my witlessness had baffled.

The joy of life was in my heart again, but chastened and made pitiful by experience.

Hammond noticed the change in me directly he entered, and congratulated me upon it.

'Go slow at first, old man,' he said. 'You've fairly sloughed the old skin; but give the sun time to toughen the new one. Walk in it at present, and be content.'

I was, in great measure, and I followed his advice. I got leave of absence, and ran down for a month in the country to a certain house we wot of, where kindly ministration to my convalescence was only one of the many blisses to be put to an account of rosy days.

> Then did my love awake,
> Most like a lily-flower,
> And as the lovely queene of heaven,
> So shone shee in her bower.

Ah, me! ah, me! when was it? A year ago, or two-thirds of a lifetime? Alas! 'Age with stealing steps hath clawde me with his crowch.' And will the yews root in *my* heart, I wonder?

I was well, sane, recovered, when one morning, towards the end of my visit, I received a letter from Hammond, enclosing a packet addressed to me, and jealously sealed and fastened. My friend's communication ran as follows:

There died here yesterday afternoon a warder, Johnson—he who had that apoplectic seizure, you will remember, the night before poor Shrike's exit. I attended him to the end, and, being alone with him an hour before the finish, he took the enclosed from under his pillow, and a solemn oath from me that I would forward it direct to you sealed as you will find it, and permit no other soul to examine or even touch it. I acquit myself of the charge, but, my dear fellow, with an uneasy sense of the responsibility I incur in thus possibly suggesting to you a retrospect of events which you had much best consign to the limbo of the—not inexplainable, but not worth trying to explain. It was patent from what I have gathered that you were in an overstrung and excitable condition at that time, and that your temporary collapse was purely nervous in its character. It seems there was some nonsense abroad in the prison about a certain cell, and that there were fools who thought fit to associate Johnson's attack and the other's death with the opening of that cell's door. I have given the new Governor a tip, and he has stopped all that. We have examined the cell in company, and found it, as one might suppose, a very ordinary chamber. The two men died perfectly natural deaths, and there is the last to be said on the subject. I mention it only from the fear that the enclosed may contain some allusion to the rubbish, a perusal of which might check the wholesome convalescence of your thoughts. If you take my advice, you will throw the packet into the fire unread. At least, if you *do* examine it, postpone the duty till you feel yourself absolutely impervious to any mental trickery, and—bear in mind that you are a worthy member of a particularly matter-of-fact and unemotional profession.

I smiled at the last clause, for I was now in a condition to feel a rather warm shame over my erst weak-knee'd collapse before a sheet and an illuminated turnip. I took the packet to my bedroom, shut the door, and sat myself down by the open window. The garden lay below me, and the dewy meadows beyond. In the one, bees were busy ruffling the ruddy gillyflowers and April stocks; in the other, the hedge twigs were all frosted with Mary buds, as if Spring had brushed them with the fleece of her wings in passing.

I fetched a sigh of content as I broke the seal of the packet and brought out the enclosure. Somewhere in the garden a little sardonic laugh was clipt to silence. It came from groom or maid, no doubt; yet it thrilled me with an odd feeling of uncanniness, and I shivered slightly.

'Bah!' I said to myself determinedly. 'There is a shrewd nip in the wind, for all the show of sunlight'; and I rose, pulled down the window, and resumed my seat.

Then in the closed room, that had become deathly quiet by contrast, I opened and read the dead man's letter.

SIR, I hope you will read what I here put down. I lay it on you as a solemn injunction, for I am a dying man, and I know it. And to who is my death due, and the Governor's death, if not to you, for your pryin' and curiosity, as surely as if you had drove a nife through our harts? Therefore, I say, Read this, and take my burden from me, for it has been a burden; and now it is right that you that interfered should have it on your own mortal shoulders. The Major is dead and I am dying, and in the first of my fit it went on in my head like cimbells that the trap was left open, and that if he passed he would look in and *it* would get him. For he knew not fear, neither would he submit to bullying by God or devil.

Now I will tell you the truth, and Heaven quit you of your responsibility in our destruction.

There wasn't another man to me like the Governor in all the countries of the world. Once he brought me to life after doctors had given me up for dead; but he willed it, and I lived; and ever afterwards I loved him as a dog loves it master. That was in the Punjab; and I came home to England with him, and was his servant when he got his appointment to the jail here. I tell you he was a proud and fierce man, but under control and tender to those he favoured; and I will tell you also a strange thing about him. Though he was a soldier and an officer, and strict in discipline as made men fear and admire him, his hart at bottom was all for books, and literature, and such-like gentle crafts. I had his confidence, as a man gives his confidence to his dog, and before others. In this way I learnt the bitter sorrow of his life. He had once hoped to be a poet, acknowledged as such before the world. He was by natur' an idelist, as they call it, and God knows what it meant to him to come out of the woods, so to speak, and swet in the dust of cities; but he did it, for his will was of tempered steel. He buried his dreams in the clouds and came down to earth greatly resolved, but with one undying hate. It is not good to hate as he could, and worse to be hated by such as him; and I will tell you the story, and what it led to.

It was when he was a subaltern that he made up his mind to the plunge. For years he had placed all his hopes and confidents in a book of verses he had wrote, and added to, and improved during that time. A little encouragement, a little word of praise, was all he looked for, and then he was redy to buckle to again, profitin' by advice, and do better. He put all the love and beauty of his hart into that book, and at last, after doubt, and anguish, and much diffidents, he published it, and give it to the world. Sir, it fell what they call still-born from the press. It was like a green leaf flutterin' down in a dead wood. To a proud and hopeful man, bubblin' with music, the pain of neglect, when he come to relize it, was terrible. But nothing was said, and there was nothing to say. In silence he had to endure and suffer.

But one day, during manoovers, there came to the camp a grey-faced man, a newspaper correspondent, and young Shrike nocked up a friendship with him. Now how it come about I cannot tell, but so it did that this skip-kennel wormed the lad's sorrow out of him, and his confidents, swore he'd been damnabilly used, and that when he got back he'd crack up the book himself in his own paper. He was a fool for his pains, and a serpent in his croolty. The notice come out as promised, and, my God! the author was laughed and mocked at from beginning to end. Even confidentses he had given to the creature was twisted to his ridicule, and his very appearance joked over. And the mess got wind of it, and made a rare story for the dog days.

He bore it like a soldier and that he became hart and liver from the moment. But he put something to the account of the grey-faced man and locked it up in his breast.

He come across him again years afterwards in India, and told him very politely that he hadn't forgotten him, and didn't intend to. But he was anigh losin' sight of him there for ever and a day, for the creature took cholera, or what looked like it, and rubbed shoulders with death and the devil before he pulled through. And he come across him again over here, and that was the last of him, as you shall see presently.

Once, after I knew the Major (he were Captain then), I was a-brushin' his coat, and he stood a long while before the glass. Then he twisted upon me, with a smile on his mouth, and says he—

'The dog was right, Johnson: this isn't the face of a poet. I was a presumtious ass, and born to cast up figgers with a pen behind my ear.'

'Captain,' I says, 'if you was skinned, you'd look like any other man without his. The quality of a soul isn't expressed by a coat.'

'Well,' he answers, 'my soul's pretty clean-swept, I think, save for one Bluebeard chamber in it that's been kep' locked ever so many years. It's nice and dirty by this time, I expect,' he says. Then the grin comes on his mouth again. 'I'll open it some day,' he says, 'and look. There's something in it about comparing me to a dancing dervish, with the wind in my petticuts. Perhaps I'll get the chance to set somebody else dancing by-and-by.'

He did, and took it, and the Bluebeard chamber come to be opened in this very jail.

It was when the system was lying fallow, so to speak, and the prison was deserted. Nobody was there but him and me and the echoes from the empty courts. The contract for restoration hadn't been signed, and for months, and more than a year, we lay idle, nothing bein' done.

Near the beginnin' of this period, one day comes, for the third time of the Major's seein' him, the grey-faced man. 'Let bygones be bygones,' he says. 'I was a good friend to you, though you didn't know it; and now, I expect, you're in the way to thank me.'

'I am,' says the Major.

'Of course,' he answers. ' Where would be your fame and reputation as one of the leadin' prison reformers of the day if you had kep' on in that riming nonsense?'

'Have you come for my thanks?' says the Governor.

'I've come,' says the grey-faced man, 'to examine and report upon your system.'

'For your paper?'

'Possibly; but to satisfy myself of its efficacy, in the first instance.'

'You aren't commissioned, then?'

'No; I come on my own responsibility.'

'Without consultation with any one?'

'Absolutely without. I haven't even a wife to advise me,' he says, with a yellow grin. What once passed for cholera had set the bile on his skin like paint, and he had caught a manner of coughing behind his hand like a toast-master.

'I know,' says the Major, looking him steady in the face, 'that what you say about me and my affairs is sure to be actuated by conscientious motives.'

'Ah,' he answers. 'You're sore about that review still, I see.'

'Not at all,' says the Major; 'and, in proof, I invite you to be my guest for the night, and tomorrow I'll show you over the prison and explain my system.'

The creature cried, 'Done!' and they set to and discussed jail matters in great earnestness. I couldn't guess the Governor's intentions, but, somehow, his manner troubled me. And yet I can remember only one point of his talk. He were always dead against making public show of his birds. 'They're there for reformation, not ignominy,' he'd say. Prisons in the old days were often, with the asylum and the work'us, made the holiday show-places of towns. I've heard of one Justice of the Peace, up North, who, to save himself trouble, used to sign a lot of blank orders for leave to view, so that applicants needn't bother him when they wanted to go over. They've changed all that, and the Governor were instrumental in the change.

'It's against my rule,' he said that night, 'to exhibit to a stranger without a Government permit; but, seein' the place is empty, and for old remembrance' sake, I'll make an exception in your favour, and you shall learn all I can show you of the inside of a prison.'

Now this was natural enough; but I was uneasy.

He treated his guest royly; so much that when we assembled the next mornin' for the inspection, the grey-faced man were shaky as a wet dog. But the Major were all set prim and dry, like the soldier he was.

We went straight away down corridor B, and at cell 47 we stopped.

'We will begin our inspection here,' said the Governor. 'Johnson, open the door.'

I had the keys of the row; fitted in the right one, and pushed open the door.

'After you, sir,' said the Major; and the creature walked in, and he shut the door on him.

I think he smelt a rat at once, for he began beating on the wood and calling out to us. But the Major only turned round to me with his face like a stone.

'Take that key from the bunch,' he said, 'and give it to me.'

I obeyed, all in a tremble, and he took and put it in his pocket.

'My God, Major.' I whispered, 'what are you going to do with him?'

'Silence, sir!' he said; 'How dare you question your superior officer!'

And the noise inside grew louder.

The Governor, he listened to it a moment like music; then he unbolted and flung open the trap, and the creature's face came at it like a wild beast's.

'Sir,' said the Major to it, 'you can't better understand my system than by experiencing it. What an article for your paper you could write already—almost as pungint a one as that in which you ruined the hopes and prospects of a young cockney poet.'

The man mouthed at the bars. He was half-mad, I think, in that one minute.

'Let me out!' he screamed. 'This is a hidius joke! Let me out!'

'When you are quite quiet—deathly quiet,' said the Major, 'you shall come out. Not before'; and he shut the trap in its face very softly.

'Come, Johnson, march!' he said, and took the lead, and we walked out of the prison.

I was like to faint, but I dared not disobey, and the man's screeching followed us all down the empty corridors and halls, until we shut the first great door on it.

It may have gone on for hours, alone in that awful emptiness. The creature was a reptile, but the thought sickened my heart.

And from that hour till his death, five months later, he rotted and maddened in his dreadful tomb.'

There was more, but I pushed the ghastly confession from me at this point in uncontrollable loathing and terror. Was it possible—possible, that injured vanity could so falsify its victim's every tradition of decency?

'Oh!' I muttered, 'what a disease is ambition! Who takes one step towards it puts his foot on Alsirat!'

It was minutes before my shocked nerves were equal to a resumption of the task; but at last I took it up again, with a groan.

I don't think at first I realized the full mischief the Governor intended to do. At least, I hoped he only meant to give the man a good fright and then let him go. I might have known better. How could he ever release him without ruining himself?

The next morning he summoned me to attend him. There was a strange new look of triumph in his face, and in his hand he held a heavy hunting-crop. I pray to God he acted in madness, but my duty and obedience was to him.

'There is sport towards, Johnson,' he said. 'My dervish has got to dance.'

I followed him quiet. We listened when I opened the jail door, but the place was silent as the grave. But from the cell, when we reached it, came a low, whispering sound.

The Governor slipped the trap and looked through.

'All right,' he said, and put the key in the door and flung it open.

He were sittin' crouched on the ground, and he looked up at us vacant-like. His face were all fallen down, as it were, and his mouth never ceased to shake and whisper.

The Major shut the door and posted me in a corner. Then he moved to the creature with his whip.

'Up!' he cried. 'Up, you dervish, and dance to us!' and he brought the thong with a smack across his shoulders.

The creature leapt under the blow, and then to his feet with a cry, and the Major whipped him till he danced. All round the cell he drove him, lashing and cutting—and again, and many times again, until the poor thing rolled on the floor whimpering and sobbing. I shall have to give an account of this some day. I shall have to whip my master with a red-hot serpent round the blazing furnace of the pit, and I shall do it with agony, because here my love and my obedience was to him.

When it was finished, he bade me put down food and drink that I had brought with me, and come away with him; and we went, leaving him rolling on the floor of the cell, and shut him alone in the empty prison until we should come again at the same time tomorrow.

So day by day this went on, and the dancing three or four times a week, until at last the whip could be left behind, for the man would scream and begin to dance at the mere turning of the key in the lock. And he danced for four months, but not the fifth.

Nobody official came near us all this time. The prison stood lonely as a deserted ruin where dark things have been done.

Once, with fear and trembling, I asked my master how he would account for the inmate of 47 if he was suddenly called upon by authority to open the cell; and he answered, smiling—

'I should say it was my mad brother. By his own account, he showed me a brother's love, you know. It would be thought a liberty; but the authorities, I think, would stretch a point for me. But if I got sufficient notice, I should clear out the cell.'

I asked him how, with my eyes rather than my lips, and he answered me only with a look.

And all this time he was, outside the prison, living the life of a good man—helping the needy, ministering to the poor. He even entertained occasionally, and had more than one noisy party in his house.

But the fifth month the creature danced no more. He was a dumb, silent animal then, with matted hair and beard; and when one entered he would only look up at one pitifully, as if he said, 'My long punishment is nearly ended.' How it came that no enquiry was ever made about him I know not, but none ever was. Perhaps he was one of the wandering gentry that nobody ever knows where they are next. He was unmarried, and had apparently not told of his intended journey to a soul.

And at the last he died in the night. We found him lying stiff and stark in the morning, and scratched with a piece of black crust on a stone of the wall these strange words: 'An Eddy on the Floor.' Just that—nothing else.

Then the Governor came and looked down, and was silent. Suddenly he caught me by the shoulder.

'Johnson,' he cried, 'if it was to do again, I would do it! I repent of nothing. But he has paid the penalty, and we call quits. May he rest in peace!'

'Amen!' I answered low. Yet I knew our turn must come for this.

We buried him in quicklime under the wall where the murderers lie, and I made the cell trim and rubbed out the writing, and the Governor locked all up and took away the key. But he locked in more than he bargained for.

For months the place was left to itself, and neither of us went anigh 47. Then one day the workmen was to be put in, and the Major he took me round with him for a last examination of the place before they come.

He hesitated a bit outside a particular cell; but at last he drove in the key and kicked open the door.

'My God!' he says, 'he's dancing still!'

My heart was thumpin', I tell you, as I looked over his shoulder. What did we see? What you well understand, sir; but, for all it was no more than that, we knew as well as if it was shouted in our ears that it was him, dancin'. It went round by the walls and drew towards us, and as it stole near I screamed out, 'An Eddy on the Floor!' and seized and dragged the Major out and clapped to the door behind us.

'Oh!' I said, 'in another moment it would have had us.'

He looked at me gloomily.

'Johnson,' he said, 'I'm not to be frightened or coerced. He may dance, but he shall dance alone. Get a screwdriver and some screws and fasten up this trap. No one from this time looks into this cell.'

I did as he bid me, swetin'; and I swear all the time I wrought I dreaded a hand would come through the trap and clutch mine.

On one pretex' or another, from that day till the night you meddled with it, he kep' that cell as close shut as a tomb. And he went his ways, discardin' the past from that time forth. Now and again a over-sensitive prisoner in the next cell would complain of feelin' uncomfortable. If possible, he would be removed to another; if not, he was dam'd for his fancies. And so it might be goin' on to now, if you hadn't pried and interfered. I don't blame you at this moment, sir. Likely you were an instrument in the hands of Providence; only, as the instrument, you must now take the burden of the truth on your own shoulders. I am a dying man, but I cannot die till I have confessed. Per'aps you may find it in your hart some day to give up a prayer for me—but it must be for the Major as well.

Your obedient servant,
J. JOHNSON

What comment of my own can I append to this wild narrative? Professionally, and apart from personal experiences, I should rule it the composition of an epileptic. That a noted journalist, nameless as he was and is to me, however nomadic in habit, could disappear from human ken, and his fellows rest content to leave him unaccounted for, seems a tax upon credulity so stupendous that I cannot seriously endorse the statement.

Yet, also—there *is* that little matter of my personal experience.

The Tomb of Sarah

F. G. LORING

My father was the head of a celebrated firm of church restorers and decorators about sixty years ago. He took a keen interest in his work, and made an especial study of any old legends or family histories that came under his observation. He was necessarily very well read and thoroughly well posted in all questions of folklore and medieval legend. As he kept a careful record of every case he investigated the manuscripts he left at his death have a special interest. From amongst them I have selected the following, as being a particularly weird and extraordinary experience. In presenting it to the public I feel it is superfluous to apologize for its supernatural character.

MY FATHER'S DIARY

1841.—*June 17th*. Received a commission from my old friend Peter Grant to enlarge and restore the chancel of his church at Hagarstone, in the wilds of the West Country.

July 5th. Went down to Hagarstone with my head man, Somers. A very long and tiring journey.

July 7th. Got the work well started. The old church is one of special interest to the antiquarian, and I shall endeavour while restoring it to alter the existing arrangements as little as possible. One large tomb, however, must be moved bodily ten feet at least to the southward. Curiously enough, there is a somewhat forbidding inscription upon it in Latin, and I am sorry that this particular tomb should have to be moved. It stands amongst the graves of the Kenyons, an old family which has been extinct in these parts for centuries. The inscription on it runs thus:

SARAH.
1630.

FOR THE SAKE OF THE DEAD AND THE WELFARE
OF THE LIVING, LET THIS SEPULCHRE REMAIN
UNTOUCHED AND ITS OCCUPANT UNDISTURBED TILL
THE COMING OF CHRIST.
IN THE NAME OF THE FATHER, THE SON, AND
THE HOLY GHOST.

July 8th. Took counsel with Grant concerning the 'Sarah Tomb'. We are both very loth to disturb it, but the ground has sunk so beneath it that the safety of the church is in danger; thus we have no choice. However, the work shall be done as reverently as possible under our own direction.

Grant says there is a legend in the neighbourhood that it is the tomb of the last of the Kenyons, the evil Countess Sarah, who was murdered in 1630. She lived quite alone in the old castle, whose ruins still stand three miles from here on the road to Bristol. Her reputation was an evil one even for those days. She was a witch or were-woman, the only companion of her solitude being a familiar in the shape of a huge Asiatic wolf. This creature was reputed to seize upon children, or failing these, sheep and other small animals, and convey them to the castle, where the Countess used to suck their blood. It was popularly supposed that she could never be killed. This, however, proved a fallacy, since she was strangled one day by a mad peasant woman who had lost two children, she declaring that they had both been seized and carried off by the Countess's familiar. This is a very interesting story, since it points to a local superstition very similar to that of the Vampire, existing in Slavonic and Hungarian Europe.

The tomb is built of black marble, surmounted by an enormous slab of the same material. On the slab is a magnificent group of figures. A young and handsome woman reclines upon a couch; round her neck is a piece of rope, the end of which she holds in her hand. At her side is a gigantic dog with bared fangs and lolling tongue. The face of the reclining figure is a cruel one: the corners of the mouth are curiously lifted, showing the sharp points of long canine or dog teeth. The whole group, though magnificently executed, leaves a most unpleasant sensation.

If we move the tomb it will have to be done in two pieces, the covering slab first and then the tomb proper. We have decided to remove the covering slab tomorrow.

July 9th. 6 p.m. A very strange day.

By noon everything was ready for lifting off the covering stone, and after the men's dinner we started the jacks and pulleys. The slab lifted easily enough, though if fitted closely into its seat and was further secured by some sort of mortar or putty, which must have kept the interior perfectly air-tight.

None of us were prepared for the horrible rush of foul, mouldy air that escaped as the cover lifted clear of its seating. And the contents that gradually came into view were more startling still. There lay the fully dressed body of a woman, wizened and shrunk and ghastly pale as if from starvation. Round her neck was a loose cord, and, judging by the scars still visible, the story of death of strangulation was true enough.

The most horrible part, however, was the extraordinary freshness of the body. Except for the appearance of starvation, life might have been only just extinct. The flesh was soft and white, the eyes were wide open and seemed to stare at us with a fearful understanding in them. The body itself lay on mould, without any pretence to coffin or shell.

For several moments we gazed with horrible curiosity, and then it became too much for my workmen, who implored us to replace the covering slab. That, of course, we would not do; but I set the carpenters to work at once to make a temporary cover while we moved the tomb to its new position. This is a long job, and will take two or three days at least.

July 9th. 9 p.m. Just at sunset we were startled by the howling of, seemingly, every dog in the village. It lasted for ten minutes or a quarter of an hour, and then ceased as suddenly as it began. This, and a curious mist that has risen round the church, makes me feel rather anxious about the 'Sarah Tomb'. According to the best-established traditions of the Vampire-haunted countries, the disturbance of dogs or wolves at sunset is supposed to indicate the presence of one of these fiends, and local fog is always considered to be a certain sign. The Vampire has the power of producing it for the purpose of concealing its movements near its hiding-place at any time.

I dare not mention or even hint my fears to the Rector, for he is, not unnaturally perhaps, a rank disbeliever in many things that I know, from experience, are not only possible but even probable. I must work this out alone at first, and get his aid without his knowing in what direction he is helping me. I shall now watch till midnight at least.

10.15 p.m. As I feared and half expected. Just before ten there was another outburst of the hideous howling. It was commenced most distinctly by a particularly horrible and blood-curdling wail from the vicinity of the churchyard. The chorus lasted only a few minutes, however, and at the end of it I saw a large dark shape, like a huge dog, emerge from the fog and lope away at a rapid canter towards the open country. Assuming this to be what I fear, I shall see it return soon after midnight.

12.30 p.m. I was right. Almost as midnight struck I saw the beast returning. It stopped at the spot where the fog seemed to commence, and lifting up its head, gave tongue to that particularly horrible long-drawn wail that I had noticed as preceding the outburst earlier in the evening.

Tomorrow I shall tell the Rector what I have seen; and if, as I expect, we hear of some neighbouring sheepfold having been raided, I shall get him to watch with me for this nocturnal marauder. I shall also examine the 'Sarah Tomb' for something which he may notice without any previous hint from me.

July 10th. I found the workmen this morning much disturbed in mind about the howling of the dogs. 'We doan't like it, zur,' one of them said to me—'we doan't like it; there was summat abroad last night that was unholy.' They were still more uncomfortable when the news came round that a large dog had made a raid upon a flock of sheep, scattering them far and wide, and leaving three of them dead with torn throats in the field.

When I told the Rector of what I had seen and what was being said in the village, he immediately decided that we must try and catch or at least identify the beast I had seen. 'Of course,' said he, 'it is some dog lately imported into the neighbourhood, for I know of nothing about here nearly as large as the animal you describe, though its size may be due to the deceptive moonlight.'

This afternoon I asked the Rector, as a favour, to assist me in lifting the temporary cover that was on the tomb, giving as an excuse the reason that I wished to obtain a portion of the curious mortar with which it had been sealed. After a slight demur he consented, and we raised the lid. If the sight that met our eyes gave me a shock, at least it appalled Grant.

'Great God!' he exclaimed; 'the woman is alive!'

And so it seemed for a moment. The corpse had lost much of its starved appearance and looked hideously fresh and alive. It was still wrinkled and shrunken, but the lips were firm, and of the rich red hue of health. The eyes, if possible, were more appalling than ever, though fixed and staring. At one corner of the mouth I thought I noticed a slight dark-coloured froth, but I said nothing about it then.

'Take your piece of mortar, Harry,' gasped Grant, 'and let us shut the tomb again. God help me! Parson though I am, such dead faces frighten me!'

Nor was I sorry to hide that terrible face again; but I got my bit of mortar, and I have advanced a step towards the solution of the mystery.

This afternoon the tomb was moved several feet towards its new position, but it will be two or three days yet before we shall be ready to replace the slab.

10.15 p.m. Again the same howling at sunset, the same fog enveloping the church, and at ten o'clock the same great beast slipping silently out into the open country. I must get the Rector's help and watch for its return. But precautions we must take, for if things are as I believe, we take our lives in our hands when we venture out into the night to waylay the—Vampire. Why not admit it at once? For that the beast I have seen is the Vampire of that evil thing in the tomb I can have no reasonable doubt.

Not yet come to its full strength, thank Heaven! after the starvation of nearly two centuries, for at present it can only maraud as wolf apparently. But, in a day or two, when full power returns, that dreadful woman in new strength and beauty will be able to leave her refuge. Then it would not be sheep merely that would satisfy her disgusting lust for blood, but victims that would yield their life-blood without a murmur to her caressing touch—victims that, dying of her foul embrace, themselves must become Vampires in their turn to prey on others.

Mercifully my knowledge gives me a safeguard; for that little piece of mortar that I rescued today from the tomb contains a portion of the Sacred Host, and who holds it, humbly and firmly believing in its virtue, may pass safely through such an ordeal as I intend to submit myself and the Rector to tonight.

12.30 p.m. Our adventure is over for the present, and we are back safe.

After writing the last entry recorded above, I went off to find Grant and tell him that the marauder was out on the prowl again. 'But, Grant,' I said, 'before we start out tonight I must insist that you will let me prosecute this affair in my own way; you must promise to put yourself completely under my orders, without asking any questions as to the why and wherefore.'

After a little demur, and some excusable chaff on his part at the serious view I was taking of what he called a 'dog hunt', he gave me his promise. I then told him that we were to watch tonight and try and track the mysterious beast, but not to interfere with it in any way. I think, in spite of his jests, that I impressed him with the fact that there might be, after all, good reason for my precautions.

It was just after eleven when we stepped out into the still night.

Our first move was to try and penetrate the dense fog round the church, but there was something so chilly about it, and a faint smell so disgustingly rank and loathsome, that neither our nerves nor our stomachs were proof against it. Instead, we stationed ourselves in the dark shadow of a yew tree that commanded a good view of the wicket entrance to the churchyard.

At midnight the howling of the dogs began again, and in a few minutes we saw a large grey shape, with green eyes shining like lamps, shamble swiftly down the path towards us.

The Rector started forward, but I laid a firm hand upon his arm and whispered a warning 'Remember!' Then we both stood very still and watched as the great beast cantered swiftly by. It was real enough, for we could hear the clicking of its nails on the stone flags. It passed within a few yards of us, and seemed to be nothing more nor less than a great grey wolf, thin and gaunt, with bristling hair and dripping jaws. It stopped where the mist commenced, and turned round. It was truly a horrible sight, and made one's blood run cold. The eyes burnt like fires, the upper lip was snarling and raised, showing the great canine teeth, while round the mouth clung and dripped a dark-coloured froth.

It raised its head and gave tongue to its long wailing howl, which was answered from afar by the village dogs. After standing for a few moments it turned and disappeared into the thickest part of the fog.

Very shortly afterwards the atmosphere began to clear, and within ten minutes the mist was all gone, the dogs in the village were silent, and the night seemed to reassume its normal aspect. We examined the spot where the beast had been standing and found, plainly enough upon the stone flags, dark spots of froth and saliva.

'Well, Rector,' I said, 'will you admit now, in view of the things you have seen today, in consideration of the legend, the woman in the tomb, the fog, the howling dogs, and, last but not least, the mysterious beast you have seen so close, that there is something not quite normal in it all? Will you put yourself unreservedly in my hands and help me, *whatever I may do*, to first make assurance doubly sure, and finally take the necessary steps for putting an end to this horror of the night?' I saw that the uncanny influence of the night was strong upon him, and wished to impress it as much as possible.

'Needs must,' he replied, 'when the Devil drives: and in the face of what I have seen I must believe that some unholy forces are at work. Yet, how can they work in the sacred precincts of a church? Shall we not call rather upon Heaven to assist us in our need.'

'Grant,' I said solemnly, 'that we must do, each in his own way. God helps those who help themselves, and by His help and the light of my knowledge we must fight this battle for Him and the poor lost soul within.'

We then returned to the rectory and to our rooms, though I have sat up to write this account while the scene is fresh in my mind.

July 11th. Found the workmen again very much disturbed in their minds, and full of a strange dog that had been seen during the night by several people, who had hunted it. Farmer Stotman, who had been watching his sheep (the same flock that had been raided the night before), had surprised it over a fresh carcass and tried to drive it off, but its size and fierceness so alarmed him that he had beaten a hasty retreat for a gun. When he returned the animal was gone, though he found that three more sheep from his flock were dead and torn.

The 'Sarah Tomb' was moved today to its new position; but it was a long, heavy business, and there was not time to replace the covering slab. For this I was glad, as in the prosaic light of day the Rector almost disbelieves the events of the night, and is prepared to think everything to have been magnified and distorted by our imagination.

As, however, I could not possibly proceed with my war of extermination against this foul thing without assistance, and as there is nobody else I can rely upon, I appealed to him for one more night—to convince him that it was no delusion, but a ghastly, horrible truth, which must be fought and conquered for our own sakes, as well as that of all those living in the neighbourhood.

'Put yourself in my hands, Rector,' I said, 'for tonight at least. Let us take those precautions which my study of the subject tells me are the right ones. Tonight you and I must watch in the church; and I feel assured that tomorrow you will be as convinced as I am, and be equally prepared to take those awful steps which I know to be proper, and I must warn you that we shall find a more startling change in the body lying there than you noticed yesterday.'

My words came true; for on raising the wooden cover once more the rank stench of a slaughterhouse arose, making us feel positively sick. There lay the Vampire, but how changed from the starved and shrunken corpse we saw two days ago for the first time! The wrinkles had almost disappeared, the flesh was firm and full, the crimson lips grinned horribly over the long pointed teeth, and a distinct smear of blood had trickled down one corner of the mouth. We set our teeth, however, and hardened our hearts. Then we replaced the cover and put what we had collected into a safe place in the vestry. Yet even now Grant could not believe that there was any real or pressing danger concealed in that awful tomb, as he raised strenuous objections to any apparent desecration of the body without further proof. This he shall have tonight. God grant that I am not taking too much on myself! If there is any truth in old legends it would be easy enough to destroy the Vampire now; but Grant will not have it.

I hope for the best of this night's work, but the danger in waiting is very great.

6 p.m. I have prepared everything: the sharp knives, the pointed stake, fresh garlic, and the wild dog-roses. All these I have taken and concealed in the vestry, where we can get at them when our solemn vigil commences.

If either or both of us die with our fearful task undone, let those reading my record see that this is done. I lay it upon them as a solemn obligation. 'That the Vampire be pierced through the heart with the stake, then let the Burial Service be read over the poor clay at last released from its doom. Thus shall the Vampire cease to be, and a lost soul rest.'

July 12th. All is over. After the most terrible night of watching and horror one Vampire at least will trouble the world no more. But how thankful should we be to a merciful Providence that that awful tomb was not disturbed by anyone not having the knowledge necessary to deal with its dreadful occupant! I write this with no feelings of self-complacency, but simply with a great gratitude for the years of study I have been able to devote to this special subject.

And now to my tale.

Just before sunset last night the Rector and I locked ourselves into the church, and took up our position in the pulpit. It was one of those pulpits, to be found in some churches, which is entered from the vestry, the preacher appearing at a good height through an arched opening in the wall. This gave us a sense of security (which we felt we needed), a good view of the interior, and direct access to the implements which I had concealed in the vestry.

The sun set and the twilight gradually deepened and faded. There was, so far, no sign of the usual fog, nor any howling of the dogs. At nine o'clock the moon rose, and her pale light gradually flooded the aisles, and still no sign of any kind from the 'Sarah Tomb'. The Rector had asked me several times what he might expect, but I was determined that no words or thought of mine should influence him, and that he should be convinced by his own senses alone.

By half-past ten we were both getting very tired, and I began to think that perhaps after all we should see nothing that night. However, soon after eleven we observed a light mist rising from the 'Sarah Tomb'. It seemed to scintillate and sparkle as it rose, and curled in a sort of pillar or spiral.

I said nothing, but I heard the Rector give a sort of gasp as he clutched my arm feverishly. 'Great Heaven!' he whispered, 'it is taking shape.'

And, true enough, in a very few moments we saw standing erect by the tomb the ghastly figure of the Countess Sarah!

She looked thin and haggard still, and her face was deadly white; but the crimson lips looked like a hideous gash in the pale cheeks, and her eyes glared like red coals in the gloom of the church.

It was a fearful thing to watch as she stepped unsteadily down the aisle, staggering a little as if from weakness and exhaustion. This was perhaps natural, as her body must have suffered much physically from her long incarceration, in spite of the unholy forces which kept it fresh and well.

We watched her to the door, and wondered what would happen; but it appeared to present no difficulty, for she melted through it and disappeared.

'Now, Grant,' I said, 'do you believe?'

'Yes,' he replied, 'I must. Everything is in your hands, and I will obey your commands to the letter, if you can only instruct me how to rid my poor people of this unnameable terror.'

'By God's help I will,' said I; 'but you shall be yet more convinced first, for we have a terrible work to do, and much to answer for in the future, before we leave the church again this morning. And now to work, for in its present weak state the Vampire will not wander far, but may return at any time, and must not find us unprepared.'

We stepped down from the pulpit and, taking dog-roses and garlic from the vestry, proceeded to the tomb. I arrived first and, throwing off the wooden cover, cried, 'Look! it is empty!' There was nothing there! Nothing except the impress of the body in the loose damp mould!

I took the flowers and laid them in a circle round the tomb, for legend teaches us that Vampires will not pass over these particular blossoms if they can avoid it.

Then, eight or ten feet away, I made a circle on the stone pavement, large enough for the Rector and myself to stand in, and within the circle I placed the implements that I had brought into the church with me.

'Now,' I said, 'from this circle, which nothing unholy can step across, you shall see the Vampire face to face, and see her afraid to cross that other circle of garlic and dog-roses to regain her unholy refuge. But on no account step beyond the holy place you stand in, for the Vampire has a fearful strength not her own, and, like a snake, can draw her victim willingly to his own destruction.'

Now so far my work was done, and, calling the Rector, we stepped into the Holy Circle to await the Vampire's return.

Nor was this long delayed. Presently a damp, cold odour seemed to pervade the church, which made our hair bristle and flesh to creep. And then, down the aisle with noiseless feet came That which we watched for.

I heard the Rector mutter a prayer, and I held him tightly by the arm, for he was shivering violently.

Long before we could distinguish the features we saw the glowing eyes and the crimson sensual mouth. She went straight to her tomb, but stopped short when she encountered my flowers. She walked right round the tomb seeking a place to enter, and as she walked she saw us. A spasm of diabolical hate and fury passed over her face; but it quickly vanished, and a smile of love, more devilish still, took its place. She stretched out her arms towards us. Then we saw that round her mouth gathered a bloody froth, and from under her lips long pointed teeth gleamed and champed.

She spoke: a soft soothing voice, a voice that carried a spell with it, and affected us both strangely, particularly the Rector. I wished to test as far as possible, without endangering our lives, the Vampire's power.

Her voice had a soporific effect, which I resisted easily enough, but which seemed to throw the Rector into a sort of trance. More than this: it seemed to compel him to her in spite of his efforts to resist.

'Come!' she said—'come! I give sleep and peace—sleep and peace—sleep and peace.'

She advanced a little towards us; but not far, for I noted that the Sacred Circle seemed to keep her back like an iron hand.

My companion seemed to become demoralized and spellbound. He tried to step forward and, finding me detain him, whispered, 'Harry, let go! I must go! She is calling me! I must! I must! Oh, help me! help me!' And he began to struggle.

It was time to finish.

'Grant!' I cried, in a loud, firm voice, 'in the name of all that you hold sacred, have done and play the man!' He shuddered violently and gasped, 'Where am I?' Then he remembered, and clung to me convulsively for a moment.

At this a look of damnable hate changed the smiling face before us, and with a sort of shriek she staggered back.

'Back!' I cried: 'back to your unholy tomb! No longer shall you molest the suffering world! Your end is near.'

It was fear that now showed itself in her beautiful face (for it was beautiful in spite of its horror) as she shrank back, back and over the circlet of flowers, shivering as she did so. At last, with a low mournful cry, she appeared to melt back again into her tomb.

As she did so the first gleams of the rising sun lit up the world, and I knew all danger was over for the day.

Taking Grant by the arm, I drew him with me out of the circle and led him to the tomb. There lay the Vampire once more, still in her living death as we had a moment before seen her in her devilish life. But in the eyes remained that awful expression of hate, and cringing, appalling fear.

Grant was pulling himself together.

'Now,' I said, 'will you dare the last terrible act and rid the world for ever of this horror?'

'By God!' he said solemnly, 'I will. Tell me what to do.'

'Help me to lift her out of her tomb. She can harm us no more,' I replied.

With averted faces we set to our terrible task, and laid her out upon the flags.

'Now,' I said, 'read the Burial Service over the poor body, and then let us give it its release from this living hell that holds it.'

Reverently the Rector read the beautiful words, and reverently I made the necessary responses. When it was over I took the stake and, without giving myself time to think, plunged it with all my strength through the heart.

As though really alive, the body for a moment writhed and kicked convulsively, and an awful heart-rending shriek woke the silent church; then all was still.

Then we lifted the poor body back; and, thank God! the consolation that legend tells is never denied to those who have to do such awful work as ours came at last. Over the face stole a great and solemn peace; the lips lost their crimson hue, the prominent sharp teeth sank back into the mouth, and for a moment we saw before us the calm, pale face of a most beautiful woman, who smiled as she slept. A few minutes more, and she faded away to dust before our eyes as we watched. We set to work and cleaned up every trace of our work, and then departed for the rectory. Most thankful were we to step out of the church, with its horrible associations, into the rosy warmth of the summer morning.

With the above end the notes in my father's diary, though a few days later this further entry occurs:

July 15th. Since the 12th everything has been quiet and as usual. We replaced and sealed up the 'Sarah Tomb' this morning. The workmen were surprised to find the body had disappeared, but took it to be the natural result of exposing it to the air.

One odd thing came to my ears today. It appears that the child of one of the villagers strayed from home the night of the 11th inst., and was found asleep in a coppice near the church, very pale and quite exhausted. There were two small marks on her throat, which have since disappeared.

What does this mean? I have, however, kept it to myself, as, now that the Vampire is no more, no further danger either to that child or any other is to be apprehended. It is only those who die of the Vampire's embrace that become Vampires at death in their turn.

The Case of Vincent Pyrwhit

BARRY PAIN

The death of Vincent Pyrwhit, JP, of Ellerdon House, in the county of Buckinghamshire, would in the ordinary way have received no more attention than the death of any other simple country gentleman. The circumstances of his death, however, though now long since forgotten, were sensational, and attracted some notice at the time. It was one of those cases which is easily forgotten within a year, except just in the locality where it occurred. The most sensational circumstances of the case never came before the public at all. I give them here simply and plainly. The psychical people may make what they like of them.

Pyrwhit himself was a very ordinary country gentleman, a good fellow, but in no way brilliant. He was devoted to his wife, who was some fifteen years younger than himself, and remarkably beautiful. She was quite a good woman, but she had her faults. She was fond of admiration, and she was an abominable flirt. She misled men very cleverly, and was then sincerely angry with them for having been misled. Her husband never troubled his head about these flirtations, being assured quite rightly that she was a good woman. He was not jealous; she, on the other hand was possessed of a jealousy amounting almost to insanity. This might have caused trouble if he had ever provided her with the slightest basis on which her jealousy could work, but he never did. With the exception of his wife, women bored him. I believe she did once or twice try to make a scene for some preposterous reason which was no reason at all; but nothing serious came of it, and there was never a real quarrel between them.

On the death of his wife, after a prolonged illness, Pyrwhit wrote and asked me to come down to Ellerdon for the funeral, and to remain at least a few days with him. He would be quite alone, and I was his oldest friend. I hate attending funerals, but I *was* his oldest friend, and I was, moreover, a distant relation of his wife. I had no choice and I went down.

There were many visitors in the house for the funeral, which took place in the village churchyard, but they left immediately afterwards. The air of heavy gloom which had hung over the house seemed to lift a little. The servants (servants are always emotional) continued to break down at intervals, noticeably Pyrwhit's man, Williams, but Pyrwhit himself was self-possessed. He spoke of his wife with great affection and regret, but still he could speak of her and not unsteadily. At dinner he also spoke of one or two other subjects, of politics and of his duties as a magistrate, and of course he made the requisite fuss about his gratitude to me for coming down to Ellerdon at that time. After dinner we sat in the library, a room well and expensively furnished, but without the least attempt at taste. There were a few oil paintings on the walls, a presentation portrait of himself, and a landscape or two—all more or less bad, as far as I remember. He had eaten next to nothing at dinner, but he had drunk a good deal; the wine, however, did not seem to have the least effect upon him. I had got the conversation definitely off the subject of his wife when I made a blunder. I noticed an Erichsen's extension standing on his writing-table. I said:

'I didn't know that telephones had penetrated into the villages yet.'

'Yes,' he said, 'I believe they are common enough now. I had that one fitted up during my wife's illness to communicate with her bedroom on the floor above us on the other side of the house.'

At that moment the bell of the telephone rang sharply.

We both looked at each other. I said with the stupid affectation of calmness one always puts on when one is a little bit frightened:

'Probably a servant in that room wishes to speak to you.'

He got up, walked over to the machine, and swung the green cord towards me. The end of it was loose.

'I had it disconnected this morning,' he said; 'also the door of that room is locked, and no one can possibly be in it.'

He had turned the colour of grey blotting-paper; so probably had I.

The bell rang again—a prolonged, rattling ring.

'Are you going to answer it?' I said.

'I am not ,' he answered firmly.

'Then,' I said, 'I shall answer it myself. It is some stupid trick, a joke not in the best of taste, for which you will probably have to sack one or other of your domestics.'

'My servants,' he answered, 'would not have done that. Besides, don't you see it is impossible? The instrument is disconnected.'

'The bell rang all the same. I shall try it.'

I picked up the receiver.

'Are you there?' I called.

The voice which answered me was unmistakably the rather high staccato voice of Mrs Pyrwhit.

'I want you,' it said, 'to tell my husband that he will be with me tomorrow.'

I still listened. Nothing more was said.

I repeated, 'Are you there?' and still there was no answer.

I turned to Pyrwhit.

'There is no one there,' I said. 'Possibly there is thunder in the air affecting the bell in some mysterious way. There must be some simple explanation, and I'll find it all out tomorrow.'

He went to bed early that night. All the following day I was with him. We rode together, and I expected an accident every minute, but none happened. All the evening I expected him to turn suddenly faint and ill, but that also did not happen. When at about ten o'clock he excused himself and said goodnight, I felt distinctly relieved. He went up to his room and rang for Williams.

The rest is, of course, well known. The servant's reason had broken down, possibly the immediate cause being the death of Mrs Pyrwhit. On entering his master's bedroom, without the least hesitation, he raised a loaded revolver which he carried in his hand, and shot Pyrwhit through the heart. I believe the case is mentioned in some of the textbooks on homicidal mania.

The Shadows on the Wall

MARY E. WILKINS

'Henry had words with Edward in the study the night before Edward died,' said Caroline Glynn.

She was elderly, tall, and harshly thin, with a hard colourlessness of face. She spoke not with acrimony, but with grave severity. Rebecca Ann Glynn, younger, stouter, and rosy of face between her crinkling puffs of grey hair, gasped, by way of assent. She sat in a wide flounce of black silk in the corner of the sofa, and rolled terrified eyes from her sister Caroline to her sister Mrs Stephen Brigham, who had been Emma Glynn, the one beauty of the family. She was beautiful still, with a large, splendid, full-blown beauty; she filled a great rocking-chair with her superb bulk of femininity, and swayed gently back and forth, her black silks whispering and her black frills fluttering. Even the shock of death (for her brother Edward lay dead in the house), could not disturb her outward serenity of demeanour. She was grieved over the loss of her brother: he had been the youngest, and she had been fond of him, but never had Emma Brigham lost sight of her own importance amidst the waters of tribulation. She was always awake to the consciousness of her own stability in the midst of vicissitudes and the splendour of her permanent bearing.

But even her expression of masterly placidity changed before her sister Caroline's announcement and her sister Rebecca Ann's gasp of terror and distress in response.

'I think Henry might have controlled his temper, when poor Edward was so near his end,' said she with an asperity which disturbed slightly the roseate curves of her beautiful mouth.

'Of course he did not *know*,' murmured Rebecca Ann in a faint tone strangely out of keeping with her appearance.

One involuntarily looked again to be sure that such a feeble pipe came from that full-swelling chest.

'Of course he did not know it,' said Caroline quickly. She turned on her sister with a strange sharp look of suspicion. 'How could he have known it?' said she. Then she shrank as if from the other's possible answer. 'Of course you and I both know he could not,' said she conclusively, but her pale face was paler than it had been before.

Rebecca gasped again. The married sister, Mrs Emma Brigham, was now sitting up straight in her chair; she had ceased rocking, and was eyeing them both intently with a sudden accentuation of family likeness in her face. Given one common intensity of emotion and similar lines showed forth, and the three sisters of one race were evident.

'What do you mean?' said she impartially to them both. Then she, too, seemed to shrink before a possible answer. She even laughed an evasive sort of laugh. 'I guess you don't mean anything,' said she, but her face wore still the expression of shrinking horror.

'Nobody means anything,' said Caroline firmly. She rose and crossed the room toward the door with grim decisiveness.

'Where are you going?' asked Mrs Brigham.

'I have something to see to,' replied Caroline, and the others at once knew by her tone that she had some solemn and sad duty to perform in the chamber of death.

'Oh,' said Mrs Brigham.

After the door had closed behind Caroline, she turned to Rebecca.

'Did Henry have many words with him?' she asked.

'They were talking very loud,' replied Rebecca evasively, yet with an answering gleam of ready response to the other's curiosity in the quick lift of her soft blue eyes.

Mrs Brigham looked at her. She had not resumed rocking. She still sat up straight with a slight knitting of intensity on her fair forehead, between the pretty rippling curves of her auburn hair.

'Did you—hear anything?' she asked in a low voice with a glance toward the door.

'I was just across the hall in the south parlour, and that door was open and this door ajar,' replied Rebecca with a slight flush.

'Then you must have——'

'I couldn't help it.'

'Everything?'

'Most of it.'

'What was it?'

'The old story.'

'I suppose Henry was mad, as he always was, because Edward was living on here for nothing, when he had wasted all the money father left him.'

Rebecca nodded with a fearful glance at the door.

When Emma spoke again her voice was still more hushed. 'I know how he felt,' said she. 'He had always been so prudent himself, and worked hard at his profession, and there Edward had never done anything but spend, and it must have looked to him as if Edward was living at his expense, but he wasn't.'

'No, he wasn't.'

'It was the way father left the property—that all the children should have a home here—and he left money enough to buy the food and all if we had all come home.'

'Yes.'

'And Edward had a right here according to the terms of father's will, and Henry ought to have remembered it.'

'Yes, he ought.'

'Did he say hard things?'

'Pretty hard from what I heard.'

'What?'

'I heard him tell Edward that he had no business here at all, and he thought he had better go away.'

'What did Edward say?'

'That he would stay here as long as he lived and afterward, too, if he was a mind to, and he would like to see Henry get him out; and then——'

'What?'

'Then he laughed.'

'What did Henry say.'

'I didn't hear him say anything, but——'

'But what?'

'I saw him when he came out of this room.'

'He looked mad?'

'You've seen him when he looked so.'

Emma nodded; the expression of horror on her face had deepened.

'Do you remember that time he killed the cat because she had scratched him?'

'Yes. Don't!'

Then Caroline re-entered the room. She went up to the stove in which a wood fire was burning—it was a cold, gloomy day of fall—and she warmed her hands, which were reddened from recent washing in cold water.

Mrs Brigham looked at her and hesitated. She glanced at the door, which was still ajar, as it did not easily shut, being still swollen with the damp weather of the summer. She rose and pushed it together with a sharp thud which jarred the house. Rebecca started painfully with a half exclamation. Caroline looked at her disapprovingly.

'It is time you controlled your nerves, Rebecca,' said she.

'I can't help it,' replied Rebecca with almost a wail. 'I am nervous. There's enough to make me so, the Lord knows.'

'What do you mean by that?' asked Caroline with her old air of sharp suspicion, and something between challenge and dread of its being met.

Rebecca shrank.

'Nothing,' said she.

'Then I wouldn't keep speaking in such a fashion.'

Emma, returning from the closed door, said imperiously that it ought to be fixed, it shut so hard.

'It will shrink enough after we have had the fire a few days,' replied Caroline. 'If anything is done to it, it will be too small; there will be a crack at the sill.'

'I think Henry ought to be ashamed of himself for talking as he did to Edward,' said Mrs Brigham abruptly, but in an almost inaudible voice.

'Hush!' said Caroline, with a glance of actual fear at the closed door.

'Nobody can hear with the door shut.'

'He must have heard it shut, and——'

'Well, I can say what I want to before he comes down, and I am not afraid of him.'

'I don't know who is afraid of him! What reason is there for anybody to be afraid of Henry?' demanded Caroline.

Mrs Brigham trembled before her sister's look. Rebecca gasped again. 'There isn't any reason, of course. Why should there be?'

'I wouldn't speak so, then. Somebody might overhear you and think it was queer. Miranda Joy is in the south parlour sewing, you know.'

'I thought she went upstairs to stitch on the machine.'

'She did, but she has come down again.'

'Well, she can't hear.'

'I say again I think Henry ought to be ashamed of himself. I shouldn't think he'd ever get over it, having words with poor Edward the very night before he died. Edward was enough sight better disposition than Henry, with all his faults. I always thought a great deal of poor Edward, myself.'

Mrs Brigham passed a large fluff of handkerchief across her eyes; Rebecca sobbed outright.

'Rebecca,' said Caroline admonishingly, keeping her mouth stiff and swallowing determinately.

'I never heard him speak a cross word, unless he spoke cross to Henry that last night. I don't know, but he did from what Rebecca overheard,' said Emma.

'Not so much cross as sort of soft, and sweet, and aggravating,' sniffled Rebecca.

'He never raised his voice,' said Caroline; 'but he had his way.'

'He had a right to in this case.'

'Yes, he did.'

'He had as much of a right here as Henry,' sobbed Rebecca, 'and now he's gone, and he will never be in this home that poor father left him and the rest of us again.'

'What do you really think ailed Edward?' asked Emma in hardly more than a whisper. She did not look at her sister.

Caroline sat down in a nearby armchair, and clutched the arms convulsively until her thin knuckles whitened.

'I told you,' said she.

Rebecca held her handkerchief over her mouth, and looked at them above it with terrified, streaming eyes.

'I know you said that he had terrible pains in his stomach, and had spasms, but what do you think made him have them?'

'Henry called it gastric trouble. You know Edward has always had dyspepsia.'

Mrs Brigham hesitated a moment. 'Was there any talk of an—examination?' said she.

Then Caroline turned on her fiercely.

'No,' said she in a terrible voice. 'No.'

The three sisters' souls seemed to meet on one common ground of terrified understanding through their eyes. The old-fashioned latch of the door was heard to rattle, and a push from without made the door shake ineffectually. 'It's Henry,' Rebecca sighed rather than whispered. Mrs Brigham settled herself after a noiseless rush across the floor into her rocking-chair again, and was swaying back and forth with her head comfortably leaning back, when the door at last yielded and Henry Glynn entered. He cast a covertly sharp, comprehensive glance at Mrs Brigham with her elaborate calm; at Rebecca quietly huddled in the corner of the sofa with her handkerchief to her face and only one small reddened ear as attentive as a dog's uncovered and revealing her alertness for his presence; at Caroline sitting with a strained composure in her armchair by the stove. She met his eyes quite firmly with a look of inscrutable fear, and defiance of the fear and of him.

Henry Glynn looked more like this sister than the others. Both had the same hard delicacy of form and feature, both were tall and almost emaciated, both had a sparse growth of grey blond hair far back from high intellectual foreheads, both had an almost noble aquilinity of feature. They confronted each other with the pitiless immovability of two statues in whose marble lineaments emotions were fixed for all eternity.

Then Henry Glynn smiled and the smile transformed his face. He looked suddenly years younger, and an almost boyish recklessness and irresolution appeared in his face. He flung himself into a chair with a gesture which was bewildering from its incongruity with his general appearance. He leaned his head back, flung one leg over the other, and looked laughingly at Mrs Brigham.

'I declare, Emma, you grow younger every year,' he said.

She flushed a little, and her placid mouth widened at the corners. She was susceptible to praise.

'Our thoughts today ought to belong to the one of us who will *never* grow older,' said Caroline in a hard voice.

Henry looked at her, still smiling. 'Of course, we none of us forget that,' said he, in a deep, gentle voice, 'but we have to speak to the living, Caroline, and I have not seen Emma for a long time, and the living are as dear as the dead.'

'Not to me,' said Caroline.

She rose, and went abruptly out of the room again. Rebecca also rose and hurried after her, sobbing loudly.

Henry looked slowly after them.

'Caroline is completely unstrung,' said he.

Mrs Brigham rocked. A confidence in him inspired by his manner was stealing over her. Out of that confidence she spoke quite easily and naturally.

'His death was very sudden,' said she.

Henry's eyelids quivered slightly but his gaze was unswerving.

'Yes,' said he; 'it was very sudden. He was sick only a few hours.'

'What did you call it?'

'Gastric.'

'You did not think of an examination?'

'There was no need. I am perfectly certain as to the cause of his death.'

Suddenly Mrs Brigham felt a creep as of some live horror over her very soul. Her flesh prickled with cold, before an inflection of his voice. She rose, tottering on weak knees.

'Where are you going?' asked Henry in a strange, breathless voice.

Mrs Brigham said something incoherent about some sewing which she had to do, some black for the funeral, and was out of the room. She went up to the front chamber which she occupied. Caroline was there. She went close to her and took her hands, and the two sisters looked at each other.

'Don't speak, don't, I won't have it!' said Caroline finally in an awful whisper.

'I won't,' replied Emma.

That afternoon the three sisters were in the study, the large front room on the ground floor across the hall from the south parlour, when the dusk deepened.

Mrs Brigham was hemming some black material. She sat close to the west window for the waning light. At last she laid her work on her lap.

'It's no use, I cannot see to sew another stitch until we have a light,' said she.

Caroline, who was writing some letters at the table, turned to Rebecca, in her usual place on the sofa.

'Rebecca, you had better get a lamp,' she said.

Rebecca started up; even in the dusk her face showed her agitation.

'It doesn't seem to me that we need a lamp quite yet,' she said in a piteous, pleading voice like a child's.

'Yes, we do,' returned Mrs Brigham peremptorily. 'We must have a light. I must finish this tonight or I can't go to the funeral, and I can't see to sew another stitch.'

'Caroline can see to write letters, and she is further from the window than you are,' said Rebecca.

'Are you trying to save kerosene or are you lazy, Rebecca Glynn?' cried Mrs Brigham. 'I can go and get the light myself, but I have this work all in my lap.'

Caroline's pen stopped scratching.

'Rebecca, we must have the light,' said she.

'Had we better have it in here?' asked Rebecca weakly.

'Of course! Why not?' cried Caroline sternly.

'I am sure I don't want to take my sewing into the other room, when it is all cleaned up for tomorrow,' said Mrs Brigham.

'Why, I never heard such a to-do about lighting a lamp.'

Rebecca rose and left the room. Presently she entered with a lamp—a large one with a white porcelain shade. She set it on a table, an old-fashioned card-table which was placed against the opposite wall from the window. That wall was clear of bookcases and books, which were only on three sides of the room. That opposite wall was taken up with three doors, the one small space being occupied by the table. Above the table on the old-fashioned paper, of a white satin gloss, traversed by an indeterminate green scroll, hung quite high a small gilt and black-framed ivory miniature taken in her girlhood of the mother of the family. When the lamp was set on the table beneath it, the tiny pretty face painted on the ivory seemed to gleam out with a look of intelligence.

'What have you put that lamp over there for?' asked Mrs Brigham, with more of impatience than her voice usually revealed. 'Why didn't you set it in the hall and have done with it. Neither Caroline nor I can see if it is on that table.'

'I thought perhaps you would move,' replied Rebecca hoarsely.

'If I do move, we can't both sit at that table. Caroline has her paper all spread around. Why don't you set the lamp on the study table in the middle of the room, then we can both see?'

Rebecca hesitated. Her face was very pale. She looked with an appeal that was fairly agonizing at her sister Caroline.

'Why don't you put the lamp on this table, as she says?' asked Caroline, almost fiercely. 'Why do you act so, Rebecca?'

'I should think you *would* ask her that,' said Mrs Brigham. 'She doesn't act like herself at all.'

Rebecca took the lamp and set it on the table in the middle of the room without another word. Then she turned her back upon it quickly and seated herself on the sofa, and placed a hand over her eyes as if to shade them, and remained so.

'Does the light hurt your eyes, and is that the reason why you didn't want the lamp?' asked Mrs Brigham kindly.

'I always like to sit in the dark,' replied Rebecca chokingly. Then she snatched her handkerchief hastily from her pocket and began to weep. Caroline continued to write, Mrs Brigham to sew.

Suddenly Mrs Brigham as she sewed glanced at the opposite wall. The glance became a steady stare. She looked intently, her work suspended in her hands. Then she looked away again and took a few more stitches, then she looked again, and again turned to her task. At last she laid her work in her lap and stared concentratedly. She looked from the wall around the room, taking note of the various objects; she looked at the wall long and intently. Then she turned to her sisters.

'What *is* that?' said she.

'What?' asked Caroline harshly; her pen scratched loudly across the paper.

Rebecca gave one of her convulsive gasps.

'That strange shadow on the wall,' replied Mrs Brigham.

Rebecca sat with her face hidden: Caroline dipped her pen in the inkstand.

'Why don't you turn around and look?' asked Mrs Brigham in a wondering and somewhat aggrieved way.

'I am in a hurry to finish this letter, if Mrs Wilson Ebbit is going to get word in time to come to the funeral,' replied Caroline shortly.

Mrs Brigham rose, her work slipping to the floor, and she began walking around the room, moving various articles of furniture, with her eyes on the shadow.

Then suddenly she shrieked out:

'Look at this awful shadow! What is it? Caroline, look, look! Rebecca, look! *What is it?*'

All Mrs Brigham's triumphant placidity was gone. Her handsome face was livid with horror. She stood stiffly pointing at the shadow.

'Look!' said she, pointing her finger at it. 'Look! What is it?'

Then Rebecca burst out in a wild wail after a shuddering glance at the wall:

'Oh, Caroline, there it is again! There it is again!'

'Caroline Glynn, you look!' said Mrs Brigham. 'Look! What is that dreadful shadow?'

Caroline rose, turned, and stood confronting the wall.

'How should I know?' she said.

'It has been there every night since he died,' cried Rebecca.

'Every night?'

'Yes. He died Thursday and this is Saturday; that makes three nights,' said Caroline rigidly. she stood as if holding herself calm with a vise of concentrated will.

'It—it looks like—like——' stammered Mrs Brigham in a tone of intense horror.

'I know what it looks like well enough,' said Caroline. 'I've got eyes in my head.'

'It looks like Edward,' burst out Rebecca in a sort of frenzy of fear. 'Only——'

'Yes, it does,' assented Mrs Brigham, whose horror-stricken tone matched her sister's, 'only—— Oh, it is awful! What is it, Caroline?'

'I ask you again, how should I know?' replied Caroline. 'I see it there like you. How should I know any more than you?'

'It *must* be something in the room,' said Mrs Brigham, staring wildly around.

'We moved everything in the room the first night it came,' said Rebecca; 'it is not anything in the room.'

Caroline turned upon her with a sort of fury. 'Of course it is something in the room,' said she. 'How you act! What do you mean by talking so? Of course it is something in the room.'

'Of course, it is,' agreed Mrs Brigham, looking at Caroline suspiciously. 'Of course it must be. It is only a coincidence. It just happens so. Perhaps it is that fold of the window curtain that makes it. It must be something in the room.'

'It is not anything in the room,' repeated Rebecca with obstinate horror.

The door opened suddenly and Henry Glynn entered. He began to speak, then his eyes followed the direction of the others'. He stood stock still staring at the shadow on the wall. It was life size and stretched across the white parallelogram of a door, half across the wall space on which the picture hung.

'What is that?' he demanded in a strange voice.

'It must be due to something in the room,' Mrs Brigham said faintly.

'It is not due to anything in the room,' said Rebecca again with the shrill insistency of terror.

'How you act, Rebecca Glynn,' said Caroline.

Henry Glynn stood and stared a moment longer. His face showed a gamut of emotions—horror, conviction, then furious incredulity. Suddenly he began hastening hither and thither about the room. He moved the furniture with fierce jerks, turning ever to see the effect upon the shadow on the wall. Not a line of its terrible outlines wavered.

'It must be something in the room!' he declared in a voice which seemed to snap like a lash.

His face changed. The inmost secrecy of his nature seemed evident until one almost lost sight of his lineaments. Rebecca stood close to her sofa, regarding him with woeful, fascinated eyes. Mrs Brigham clutched Caroline's hand. They both stood in a corner out of his way. For a few moments he raged about the room like a caged wild animal. He moved every piece of furniture; when the moving of a piece did not affect the shadow, he flung it to the floor, the sisters watching.

Then suddenly he desisted. He laughed and began straightening the furniture which he had flung down.

'What an absurdity,' he said easily. 'Such a to-do about a shadow.'

'That's so,' assented Mrs Brigham, in a scared voice which she tried to make natural. As she spoke she lifted a chair near her.

'I think you have broken the chair that Edward was so fond of,' said Caroline.

Terror and wrath were struggling for expression on her face. Her mouth was set, her eyes shrinking. Henry lifted the chair with a show of anxiety.

'Just as good as ever,' he said pleasantly. He laughed again, looking at his sisters. 'Did I scare you?' he said. 'I should think you might be used to me by this time. You know my way of wanting to leap to the bottom of a mystery, and that shadow does look—queer, like—and I thought if there was any way of accounting for it I would like to without any delay.'

'You don't seem to have succeeded,' remarked Caroline dryly, with a slight glance at the wall.

Henry's eyes followed hers and he quivered perceptibly.

'Oh, there is no accounting for shadows,' he said, and he laughed again. 'A man is a fool to try to account for shadows.'

Then the supper bell rang, and they all left the room, but Henry kept his back to the wall, as did, indeed, the others.

Mrs Brigham pressed close to Caroline as she crossed the hall. 'He looked like a demon!' she breathed in her ear.

Henry led the way with an alert motion like a boy; Rebecca brought up the rear; she could scarcely walk, her knees trembled so.

'I can't sit in that room again this evening,' she whispered to Caroline after supper.

'Very well, we will sit in the south room,' replied Caroline. 'I think we will sit in the south parlour,' she said aloud; 'it isn't as damp as the study, and I have a cold.'

So they all sat in the south room with their sewing. Henry read the newspaper, his chair drawn close to the lamp on the table. About nine o'clock he rose abruptly and crossed the hall to the study. The three sisters looked at one another. Mrs Brigham rose, folded her rustling skirts compactly around her, and began tiptoeing toward the door.

'What are you going to do?' enquired Rebecca agitatedly.

'I am going to see what he is about,' replied Mrs Brigham cautiously.

She pointed as she spoke to the study door across the hall; it was ajar. Henry had striven to pull it together behind him, but it had somehow swollen beyond the limit with curious speed. It was still ajar and a streak of light showed from top to bottom. The hall lamp was not lit.

'You had better stay where you are,' said Caroline with guarded sharpness.

'I am going to see,' repeated Mrs Brigham firmly.

Then she folded her skirts so tightly that her bulk with its swelling curves was revealed in a black silk sheath, and she went with a slow toddle across the hall to the study door. She stood there, her eye at the crack.

In the south room Rebecca stopped sewing and sat watching with dilated eyes. Caroline sewed steadily. What Mrs Brigham, standing at the crack in the study door, saw was this:

Henry Glynn, evidently reasoning that the source of the strange shadow must be between the table on which the lamp stood and the wall, was making systematic passes and thrusts all over and through the intervening space with an old sword which had belonged to his father.

Not an inch was left unpierced. He seemed to have divided the space into mathematical sections. He brandished the sword with a sort of cold fury and calculation; the blade gave out flashes of light, the shadow remained unmoved. Mrs Brigham, watching, felt herself cold with horror.

Finally Henry ceased and stood with the sword in hand and raised as if to strike, surveying the shadow on the wall threateningly. Mrs Brigham toddled back across the hall and shut the south room door behind her before she related what she had seen.

'He looked like a demon!' she said again. 'Have you got any of that old wine in the house, Caroline? I don't feel as if I could stand much more.'

Indeed, she looked overcome. Her handsome placid face was worn and strained and pale.

'Yes, there's plenty,' said Caroline; 'you can have some when you go to bed.'

'I think we had all better take some,' said Mrs Brigham. 'Oh, my God, Caroline, what——'

'Don't ask and don't speak,' said Caroline.

'No, I am not going to,' replied Mrs Brigham; 'but——'

Rebecca moaned aloud.

'What are you doing that for?' asked Caroline harshly.

'Poor Edward,' returned Rebecca.

'That is all you have to groan for,' said Caroline. 'There is nothing else.'

'I am going to bed,' said Mrs Brigham. 'I sha'n't be able to be at the funeral if I don't.'

Soon the three sisters went to their chambers and the south parlour was deserted. Caroline called to Henry in the study to put out the light before he came upstairs. They had been gone about an hour when he came into the room bringing the lamp which had stood in the study. He set it on the table and waited a few minutes, pacing up and down. His face was terrible, his fair complexion showed livid; his blue eyes seemed dark blanks of awful reflections.

Then he took the lamp up and returned to the library. He set the lamp on the centre table, and the shadow sprang out on the wall. Again he studied the furniture and moved it about, but deliberately, with none of his former frenzy. Nothing affected the shadow. Then he returned to the south room with the lamp and again waited. Again he returned to the study and placed the lamp on the table, and the shadow sprang out upon the wall. It was midnight before he went up stairs. Mrs Brigham and the other sisters, who could not sleep, heard him.

The next day was the funeral. That evening the family sat in the south room. Some relatives were with them. Nobody entered the study until Henry carried a lamp in there after the others had retired for the night. He saw again the shadow on the wall leap to an awful life before the light.

The next morning at breakfast Henry Glynn announced that he had to go to the city for three days. The sisters looked at him with surprise. He very seldom left home, and just now his practice had been neglected on account of Edward's death. He was a physician.

'How can you leave your patients now?' asked Mrs Brigham wonderingly.

'I don't know how to, but there is no other way,' replied Henry easily. 'I have had a telegram from Doctor Mitford.'

'Consultation?' enquired Mrs Brigham.

'I have business,' replied Henry.

Doctor Mitford was an old classmate of his who lived in a neighbouring city and who occasionally called upon him in the case of a consultation.

After he had gone Mrs Brigham said to Caroline that after all Henry had not said that he was going to consult with Doctor Mitford, and she thought it very strange.

'Everything is very strange,' said Rebecca with a shudder.

'What do you mean?' enquired Caroline sharply.

'Nothing,' replied Rebecca.

Nobody entered the library that day, nor the next, nor the next. The third day Henry was expected home, but he did not arrive and the last train from the city had come.

'I call it pretty queer work,' said Mrs Brigham. 'The idea of a doctor leaving his patients for three days anyhow, at such a time as this, and I know he has some very sick ones; he said so. And the idea of a consultation lasting three days! There is no sense in it, and *now* he has not come. I don't understand it, for my part.'

'I don't either,' said Rebecca.

They were all in the south parlour. There was no light in the study opposite, and the door was ajar.

Presently Mrs Brigham rose—she could not have told why; something seemed to impel her, some will outside her own. She went out of the room, again wrapping her rustling skirts around that she might pass noiselessly, and began pushing at the swollen door of the study.

'She has not got any lamp,' said Rebecca in a shaking voice.

Caroline, who was writing letters, rose again, took a lamp (there were two in the room) and followed her sister. Rebecca had risen, but she stood trembling, not venturing to follow.

The doorbell rang, but the others did not hear it; it was on the south door on the other side of the house from the study. Rebecca, after hesitating until the bell rang the second time, went to the door; she remembered that the servant was out.

Caroline and her sister Emma entered the study. Caroline set the lamp on the table. They looked at the wall. 'Oh, my God,' gasped Mrs Brigham, 'there are—there are *two*—shadows.' The sisters stood clutching each other, staring at the awful things on the wall. Then Rebecca came in, staggering, with a telegram in her hand. 'Here is—a telegram,' she gasped. 'Henry is—dead.'

Father Macclesfield's Tale

R. H. BENSON

Monsignor Maxwell announced next day at dinner that he had already arranged for the evening's entertainment. A priest, whose acquaintance he had made on the Palatine, was leaving for England the next morning; and it was our only chance therefore of hearing his story. That he had a story had come to the Canon's knowledge in the course of a conversation on the previous afternoon.

'He told me the outline of it,' he said, 'I think it very remarkable. But I had a great deal of difficulty in persuading him to repeat it to the company this evening. But he promised at last. I trust, gentlemen, you do not think I have presumed in begging him to do so.'

Father Macclesfield arrived at supper.

He was a little unimposing dry man, with a hooked nose, and grey hair. He was rather silent at supper; but there was no trace of shyness in his manner as he took his seat upstairs, and without glancing round once, began in an even and dispassionate voice:

'I once knew a Catholic girl that married an old Protestant three times her own age. I entreated her not to do so; but it was useless. And when the disillusionment came she used to write to me piteous letters, telling me that her husband had in reality no religion at all. He was a convinced infidel; and scouted even the idea of the soul's immortality.

'After two years of married life the old man died. He was about sixty years old; but very hale and hearty till the end.

'Well, when he took to his bed, the wife sent for me; and I had half-a-dozen interviews with him; but it was useless. He told me plainly that he wanted to believe—in fact he said that the thought of annihilation was intolerable to him. If he had had a child he would not have hated death so much; if his flesh and blood in any manner survived him, he could have fancied that he had a sort of vicarious life left; but as it was there was no kith or kin of his alive; and he could not bear that.'

Father Macclesfield sniffed cynically, and folded his hands.

'I may say that his death-bed was extremely unpleasant. He was a coarse old fellow, with plenty of strength in him; and he used to make remarks about the churchyard—and—and in fact the worms, that used to send his poor child of a wife half fainting out of the room. He had lived an immoral life too, I gathered.

'Just at the last it was—well—disgusting. He had no consideration (God knows why she married him!). The agony was a very long one; he caught at the curtains round the bed; calling out; and all his words were about death, and the dark. It seemed to me that he caught hold of the curtains as if to hold himself into this world. And at the very end he raised himself clean up in bed, and stared horribly out of the window that was open just opposite.

'I must tell you that straight away beneath the window lay a long walk, between sheets of dead leaves with laurels on either side, and the branches meeting overhead, so that it was very dark there even in summer; and at the end of the walk away from the house was the churchyard gate.'

Father Macclesfield paused and blew his nose. Then he went on still without looking at us.

'Well the old man died; and he was carried along this laurel path, and buried.

'His wife was in such a state that I simply dared not go away. She was frightened to death; and, indeed, the whole affair of her husband's dying was horrible. But she would not leave the house. She had a fancy that it would be cruel to him. She used to go down twice a day to pray at the grave; but she never went along the laurel walk. She would go round by the garden and in at a lower gate, and come back the same way, or by the upper garden.

'This went on for three or four days. The man had died on a Saturday, and was buried on Monday; it was in July; and he had died about eight o'clock.

'I made up my mind to go on the Saturday after the funeral. My curate had managed along very well for a few days; but I did not like to leave him for a second Sunday.

'Then on the Friday at lunch—her sister had come down, by the way, and was still in the house—on the Friday the widow said something about never daring to sleep in the room where the old man had died. I told her it was nonsense, and so on; but you must remember she was in a dreadful state of nerves, and she persisted. So I said I would sleep in the room myself. I had no patience with such ideas then.

'Of course she said all sorts of things, but I had my way; and my things were moved in on Friday evening.

'I went to my new room about a quarter before eight to put on my cassock for dinner. The room was very much as it had been—rather dark because of the trees at the end of the walk outside. There was the four-poster there with the damask curtains; the table and chairs, the cupboard where his clothes were kept, and so on.

'When I had put my cassock on, I went to the window to look out. To right and left were the gardens, with the sunlight just off them, but still very bright and gay, with the geraniums, and exactly opposite was the laurel walk, like a long green shady tunnel, dividing the upper and lower lawns.

'I could see straight down it to the churchyard gate, which was about a hundred yards away, I suppose. There were limes overhead, and laurels, as I said, on each side.

'Well—I saw someone coming up the walk; but it seemed to me at first that he was drunk. He staggered several times as I watched; I suppose he would be fifty yards away—and once I saw him catch hold of one of the trees and cling against it as if he were afraid of falling. Then he left it, and came on again slowly, going from side to side, with his hands out. He seemed desperately keen to get to the house.

'I could see his dress; and it astonished me that a man dressed so should be drunk; for he was quite plainly a gentleman. He wore a white top hat, and a grey cut-away coat, and grey trousers, and I could make out his white spats.

'Then it struck me he might be ill; and I looked harder than ever, wondering whether I ought to go down.

'When he was about twenty yards away he lifted his face; and, it struck me as very odd, but it seemed to me he was extraordinarily like the old man we had buried on Monday; but it was darkish where he was, and the next moment he dropped his face, threw up his hands and fell flat on his back.

'Well of course I was startled at that, and I leaned out of the window and called out something. He was moving his hands I could see, as if he were in convulsions; and I could hear the dry leaves rustling.

'Well, then I turned and ran out and downstairs.'

Father Macclesfield stopped a moment.

'Gentlemen,' he said abruptly, 'when I got there, there was not a sign of the old man. I could see that the leaves had been disturbed, but that was all.'

There was an odd silence in the room as he paused; but before any of us had time to speak he went on.

'Of course I did not say a word of what I had seen. We dined as usual; I smoked for an hour or so by myself after prayers; and then I went up to bed. I cannot say I was perfectly comfortable, for I was not; but neither was I frightened.

'When I got to my room I lit all my candles, and then went to a big cupboard I had noticed, and pulled out some of the drawers. In the bottom of the third drawer I found a grey cut-away coat and grey trousers; I found several pairs of white spats in the top drawer; and a white hat on the shelf above. That is the first incident.'

'Did you sleep there, Father?' said a voice softly.

'I did,' said the priest; 'there was no reason why I should not. I did not fall asleep for two or three hours; but I was not disturbed in any way; and came to breakfast as usual.

'Well, I thought about it all a bit; and finally I sent a wire to my curate telling him I was detained. I did not like to leave the house just then.'

Father Macclesfield settled himself again in his chair and went on, in the same dry uninterested voice.

'On Sunday we drove over to the Catholic Church, six miles off, and I said Mass. Nothing more happened till the Monday evening.

'That evening I went to the window again about a quarter before eight, as I had done both on the Saturday and Sunday. Everything was perfectly quiet, till I heard the churchyard gate unlatch; and I saw a man come through.

'But I saw almost at once that it was not the same man I had seen before; it looked to me like a keeper, for he had a gun across his arm; then I saw him hold the gate open an instant, and a dog came through and began to trot up the path towards the house with his master following.

'When the dog was about fifty yards away he stopped dead, and pointed.

'I saw the keeper throw his gun forward and come up softly; and as he came the dog began to slink backwards. I watched very closely, clean forgetting why I was there; and the next instant something—it was too shadowy under the trees to see exactly what it was—but something about the size of a hare burst out of the laurels and made straight up the path, dodging from side to side, but coming like the wind.

'The beast could not have been more than twenty yards from me, when the keeper fired, and the creature went over and over in the dry leaves, and lay struggling and screaming. It was horrible! But what astonished me was that the dog did not come up. I heard the keeper snap out something, and then I saw the dog making off down the avenue in the direction of the churchyard as hard as he could go.

'The keeper was running now towards me; but the screaming of the hare, or of whatever it was, had stopped; and I was astonished to see the man come right up to where the beast was struggling and kicking, and then stop as if he was puzzled.

'I leaned out of the window and called to him.

'"Right in front of you, man," I said. "For God's sake kill the brute."

'He looked up at me, and then down again.

'"Where is it, sir," he said. "I can't see it anywhere."

'And there lay the beast clear before him all the while, not a yard away, still kicking.

'Well, I went out of the room and downstairs and out to the avenue.

'The man was standing there still, looking terribly puzzled, but the hare was gone. There was not a sign of it. Only the leaves were disturbed, and the wet earth showed beneath.

'The keeper said that it had been a great hare; he could have sworn to it; and that he had orders to kill all hares and rabbits in the garden enclosure. Then he looked rather odd.

'"Did you see it plainly, sir," he asked.

'I told him, not very plainly; but I thought it a hare too.

'"Yes, sir," he said, "it was a hare, sure enough; but do you know, sir, I thought it to be a kind of silver grey with white feet. I never saw one like that before!"

'The odd thing was that not a dog would come near, his own dog was gone; but I fetched the yard dog—a retriever, out of his kennel in the kitchen yard; and if ever I saw a frightened dog it was this one. When we dragged him up at last, all whining and pulling back, he began to snap at us so fiercely that we let go, and he went back like the wind to his kennel. It was the same with the terrier.

'Well, the bell had gone, and I had to go in and explain why I was late; but I didn't say anything about the colour of the hare. That was the second incident.'

Father Macclesfield stopped again, smiling reminiscently to himself. I was very much impressed by his quiet air and composure. I think it helped his story a good deal.

Again, before we had time to comment or question he went on.

'The third incident was so slight that I should not have mentioned it, or thought anything of it, if it had not been for the others; but it seemed to me there was a kind of diminishing gradation of energy, which explained. Well, now you shall hear.

'On the other nights of that week I was at my window again; but nothing happened till the Friday. I had arranged to go for certain next day; the widow was much better and more reasonable, and even talked of going abroad herself in the following week.

'On that Friday evening I dressed a little earlier, and went down to the avenue this time, instead of staying at my window, at about twenty minutes to eight.

'It was rather a heavy depressing evening, without a breath of wind; and it was darker than it had been for some days.

'I walked slowly down the avenue to the gate and back again; and, I suppose it was fancy, but I felt more uncomfortable than I had felt at all up to then. I was rather relieved to see the widow come out of the house and stand looking down the avenue. I came out myself then and went towards her. She started rather when she saw me and then smiled.

'"I thought it was someone else," she said. "Father, I have made up my mind to go. I shall go to town tomorrow, and start on Monday. My sister will come with me."

'I congratulated her; and then we turned and began to walk back to the lime avenue. She stopped at the entrance, and seemed unwilling to come any further.

'"Come down to the end," I said, "and back again. There will be time before dinner."

'She said nothing; but came with me; and we went straight down to the gate and then turned to come back.

'I don't think either of us spoke a word; I was very uncomfortable indeed by now; and yet I had to go on.

'We were half way back I suppose when I heard a sound like a gate rattling; and I whisked round in an instant, expecting to see someone at the gate. But there was no one.

'Then there came a rustling overhead in the leaves; it had been dead still before. Then I don't know why, but I took my friend suddenly by the arm and drew her to one side out of the path, so that we stood on the right hand, not a foot from the laurels.

'She said nothing, and I said nothing; but I think we were both looking this way and that, as if we expected to see something.

'The breeze died, and then sprang up again; but it was only a breath. I could hear the living leaves rustling overhead, and the dead leaves underfoot; and it was blowing gently from the churchyard.

'Then I saw a thing that one often sees; but I could not take my eyes off it, nor could she. It was a little column of leaves, twisting and turning and dropping and picking up again in the wind, coming slowly up the path. It was a capricious sort of draught, for the little scurry of leaves went this way and that, to and fro across the path. It came up to us, and I could feel the breeze on my hands and face. One leaf struck me softly on the cheek, and I can only say that I shuddered as if it had been a toad. Then it passed on.

'You understand, gentlemen, it was pretty dark; but it seemed to me that the breeze died and the column of leaves—it was no more than a little twist of them—sank down at the end of the avenue.

'We stood there perfectly still for a moment or two; and when I turned, she was staring straight at me, but neither of us said one word.

'We did not go up the avenue to the house. We pushed our way through the laurels, and came back by the upper garden.

'Nothing else happened; and the next morning we all went off by the eleven o'clock train.

'That is all, gentlemen.'

Thurnley Abbey

PERCEVAL LANDON

Three years ago I was on my way out to the East, and as an extra day in London was of some importance, I took the Friday evening mail-train to Brindisi instead of the usual Thursday morning Marseilles express. Many people shrink from the long forty-eight-hour train journey through Europe, and the subsequent rush across the Mediterranean on the nineteen-knot *Isis* or *Osiris*; but there is really very little discomfort on either the train or the mail-boat, and unless there is actually nothing for me to do, I always like to save the extra day and a half in London before I say goodbye to her for one of my longer tramps. This time—it was early, I remember, in the shipping season, probably about the beginning of September—there were few passengers, and I had a compartment in the P. & O. Indian express to myself all the way from Calais. All Sunday I watched the blue waves dimpling the Adriatic, and the pale rosemary along the cuttings; the plain white towns, with their flat roofs and their bold 'duomos', and the grey-green gnarled olive orchards of Apulia. The journey was just like any other. We ate in the dining-car as often and as long as we decently could. We slept after luncheon; we dawdled the afternoon away with yellow-backed novels; sometimes we exchanged platitudes in the smoking-room, and it was there that I met Alastair Colvin.

Colvin was a man of middle height, with a resolute, well-cut jaw; his hair was turning grey; his moustache was sun-whitened, otherwise he was clean-shaven—obviously a gentleman, and obviously also a pre-occupied man. He had no great wit. When spoken to, he made the usual remarks in the right way, and I dare say he refrained from banalities only because he spoke less than the rest of us; most of the time he buried himself in the Wagon-lit Company's time-table, but seemed unable to concentrate his attention on any one page of it. He found that I had been over the Siberian railway, and for a quarter of an hour he discussed it with me. Then he lost interest in it, and rose to go to his compartment. But he came back again very soon, and seemed glad to pick up the conversation again.

Of course this did not seem to me to be of any importance. Most travellers by train become a trifle infirm of purpose after thirty-six hours' rattling. But Colvin's restless way I noticed in somewhat marked contrast with the man's personal importance and dignity; especially ill suited was it to his finely made large hand with strong, broad, regular nails and its few lines. As I looked at his hand I noticed a long, deep, and recent scar of ragged shape. However, it is absurd to pretend that I thought anything was unusual. I went off at five o'clock on Sunday afternoon to sleep away the hour or two that had still to be got through before we arrived at Brindisi.

Once there, we few passengers transhipped our hand baggage, verified our berths—there were only a score of us in all—and then, after an aimless ramble of half an hour in Brindisi, we returned to dinner at the Hôtel International, not wholly surprised that the town had been the death of Virgil. If I remember rightly, there is a gaily painted hall at the International—I do not wish to advertise anything, but there is no other place in Brindisi at which to await the coming of the mails—and after dinner I was looking with awe at a trellis overgrown with blue vines, when Colvin moved across the room to my table. He picked up *Il Secolo*, but almost immediately gave up the pretence of reading it. He turned squarely to me and said:

'Would you do me a favour?'

One doesn't do favours to stray acquaintances on Continental expresses without knowing something more of them than I knew of Colvin. But I smiled in a noncommittal way, and asked him what he wanted. I wasn't wrong in part of my estimate of him; he said bluntly:

'Will you let me sleep in your cabin on the *Osiris*?' And he coloured a little as he said it.

Now, there is nothing more tiresome than having to put up with a stable-companion at sea, and I asked him rather pointedly:

'Surely there is room for all of us?' I thought that perhaps he had been partnered off with some mangy Levantine, and wanted to escape from him at all hazards.

Colvin, still somewhat confused, said: 'Yes; I am in a cabin by myself. But you would do me the greatest favour if you would allow me to share yours.'

This was all very well, but, besides the fact that I always sleep better when alone, there had been some recent thefts on board English liners, and I hesitated, frank and honest and self-conscious as Colvin was. Just then the mail-train came in with a clatter and a rush of escaping steam, and I asked him to see me again about it on the boat when we started. He answered me curtly—I suppose he saw the mistrust in my manner—'I am a member of White's.' I smiled to myself as he said it, but I remembered in a moment that the man—if he were really what he claimed to be, and I make no doubt that he was—must have been sorely put to it before he urged the fact as a guarantee of his respectability to a total stranger at a Brindisi hotel.

That evening, as we cleared the red and green harbour-lights of Brindisi, Colvin explained. This is his story in his own words.

'When I was travelling in India some years ago, I made the acquaintance of a youngish man in the Woods and Forests. We camped out together for a week, and I found him a pleasant companion. John Broughton was a light-hearted soul when off duty, but a steady and capable man in any of the small emergencies that continually arise in that department. He was liked and trusted by the natives, and though a trifle over-pleased with himself when he escaped to civilization at Simla or Calcutta, Broughton's future was well assured in Government service, when a fair-sized estate was unexpectedly left to him, and he joyfully shook the dust of the Indian plains from his feet and returned to England. For five years he drifted about London. I saw him now and then. We dined together about every eighteen months, and I could trace pretty exactly the gradual sickening of Broughton with a merely idle life. He then set out on a couple of long voyages, returned as restless as before, and at last told me that he had decided to marry and settle down at his place, Thurnley Abbey, which had long been empty. He spoke about looking after the property and standing for his constituency in the usual way. Vivien Wilde, his fiancée, had, I suppose, begun to take him in hand. She was a pretty girl with a deal of fair hair and rather an exclusive manner; deeply religious in a narrow school, she was still kindly and high-spirited, and I thought that Broughton was in luck. He was quite happy and full of information about his future.

'Among other things, I asked him about Thurnley Abbey. He confessed that he hardly knew the place. The last tenant, a man called Clarke, had lived in one wing for fifteen years and seen no one. He had been a miser and a hermit. It was the rarest thing for a light to be seen at the Abbey after dark. Only the barest necessities of life were ordered, and the tenant himself received them at the side-door. His one half-caste manservant, after a month's stay in the house, had abruptly left without warning, and had returned to the Southern States. One thing Broughton complained bitterly about: Clarke had wilfully spread the rumour among the villagers that the Abbey was haunted, and had even condescended to play childish tricks with spirit-lamps and salt in order to scare trespassers away at night. He had been detected in the act of this tomfoolery, but the story spread, and no one, said Broughton, would venture near the house except in broad daylight. The hauntedness of Thurnley Abbey was now, he said with a grin, part of the gospel of the countryside, but he and his young wife were going to change all that. Would I propose myself any time I liked? I, of course, said I would, and equally, of course, intended to do nothing of the sort without a definite invitation.

'The house was put in thorough repair, though not a stick of the old furniture and tapestry were removed. Floors and ceilings were relaid: the roof was made watertight again, and the dust of half a century was scoured out. He showed me some photographs of the place. It was called an Abbey, though as a matter of fact it had been only the infirmary of the long-vanished Abbey of Closter some five miles away. The larger part of this building remained as it had been in pre-Reformation days, but a wing had been added in Jacobean times, and that part of the house had been kept in something like repair by Mr Clarke. He had in both the ground and first floors set a heavy timber door, strongly barred with iron, in the passage between the earlier and the Jacobean parts of the house, and had entirely neglected the former. So there had been a good deal of work to be done.

'Broughton, whom I saw in London two or three times about this period, made a deal of fun over the positive refusal of the workmen to remain after sundown. Even after the electric light had been put into every room, nothing would induce them to remain, though, as Broughton observed, electric light was death on ghosts. The legend of the Abbey's ghosts had gone far and wide, and the men would take no risks. They went home in batches of five and six, and even during the daylight hours there was an inordinate amount of talking between one and another, if either happened to be out of sight of his companion. On the whole, though nothing of any sort or kind had been conjured up even by their heated imaginations during their five months' work upon the Abbey, the belief in the ghosts was rather strengthened than otherwise in Thurnley because of the men's confessed nervousness, and local tradition declared itself in favour of the ghost of an immured nun.

'"Good old nun!" said Broughton.

'I asked him whether in general he believed in the possibility of ghosts, and, rather to my surprise, he said that he couldn't say he entirely disbelieved in them. A man in India had told him one morning in camp that he believed that his mother was dead in England, as her vision had come to his tent the night before. He had not been alarmed, but had said nothing, and the figure vanished again. As a matter of fact, the next possible dak-walla brought on a telegram announcing the mother's death. "There the thing was," said Broughton. But at Thurnley he was practical enough. He roundly cursed the idiotic selfishness of Clarke, whose silly antics had caused all the inconvenience. At the same time, he couldn't refuse to sympathize to some extent with the ignorant workmen. "My own idea," said he, "is that if a ghost ever does come in one's way, one ought to speak to it."

'I agreed. Little as I knew of the ghost world and its conventions, I had always remembered that a spook was in honour bound to wait to be spoken to. It didn't seem much to do, and I felt that the sound of one's own voice would at any rate reassure oneself as to one's wakefulness. But there are few ghosts outside Europe—few, that is, that a white man can see—and I had never been troubled with any. However, as I have said, I told Broughton that I agreed.

'So the wedding took place, and I went to it in a tall hat which I bought for the occasion, and the new Mrs Broughton smiled very nicely at me afterwards. As it had to happen, I took the Orient Express that evening and was not in England again for nearly six months. Just before I came back I got a letter from Broughton. He asked if I could see him in London or come to Thurnley, as he thought I should be better able to help him than anyone else he knew. His wife sent a nice message to me at the end, so I was reassured about at least one thing. I wrote from Budapest that I would come and see him at Thurnley two days after my arrival in London, and as I sauntered out of the Pannonia into the Kerepesi Utcza to post my letters, I wondered of what earthly service I could be to Broughton. I had been out with him after tiger on foot, and I could imagine few men better able at a pinch to manage their own business. However, I had nothing to do, so after dealing with some small accumulations of business during my absence, I packed a kit-bag and departed to Euston.

'I was met by Broughton's great limousine at Thurnley Road station, and after a drive of nearly seven miles we echoed through the sleepy streets of Thurnley village, into which the main gates of the park thrust themselves, splendid with pillars and spreadeagles and tom-cats rampant atop of them. I never was a herald, but I know that the Broughtons have the right to supporters—Heaven knows why! From the gates a quadruple avenue of beech-trees led inwards for a quarter of a mile. Beneath them a neat strip of fine turf edged the road and ran back until the poison of the dead beech-leaves killed it under the trees. There were many wheel-tracks on the road, and a comfortable little pony trap jogged past me laden with a country parson and his wife and daughter. Evidently there was some garden party going on at the Abbey. The road dropped away to the right at the end of the avenue, and I could see the Abbey across a wide pasturage and a broad lawn thickly dotted with guests.

'The end of the building was plain. It must have been almost mercilessly austere when it was first built, but time had crumbled the edges and toned the stone down to an orange-lichened grey wherever it showed behind its curtain of magnolia, jasmine, and ivy. Further on was the three-storied Jacobean house, tall and handsome. There had not been the slightest attempt to adapt the one to the other, but the kindly ivy had glossed over the touching-point. There was a tall flèche in the middle of the building, surmounting a small bell tower. Behind the house there rose the mountainous verdure of Spanish chestnuts all the way up the hill.

'Broughton had seen me coming from afar, and walked across from his other guests to welcome me before turning me over to the butler's care. This man was sandy-haired and rather inclined to be talkative. He could, however, answer hardly any questions about the house; he had, he said, only been there three weeks. Mindful of what Broughton had told me, I made no enquiries about ghosts, though the room into which I was shown might have justified anything. It was a very large low room with oak beams projecting from the white ceiling. Every inch of the walls, including the doors, was covered with tapestry, and a remarkably fine Italian fourpost bedstead, heavily draped, added to the darkness and dignity of the place. All the furniture was old, well made, and dark. Underfoot there was a plain green pile carpet, the only new thing about the room except the electric light fittings and the jugs and basins. Even the looking-glass on the dressing-table was an old pyramidal Venetian glass set in heavy repoussé frame of tarnished silver.

'After a few minutes' cleaning up, I went downstairs and out upon the lawn, where I greeted my hostess. The people gathered there were of the usual country type, all anxious to be pleased and roundly curious as to the new master of the Abbey. Rather to my surprise, and quite to my pleasure, I rediscovered Glenham, whom I had known well in old days in Barotseland: he lived quite close, as, he remarked with a grin, I ought to have known. "But," he added, "I don't live in a place like this." He swept his hand to the long, low lines of the Abbey in obvious admiration, and then, to my intense interest, muttered beneath his breath, "Thank God!" He saw that I had overheard him, and turning to me said decidedly, "Yes, 'thank God' I said, and I meant it. I wouldn't live at the Abbey for all Broughton's money."

'"But surely," I demurred, "you know that old Clarke was discovered in the very act of setting light to his bug-a-boos?"

'Glenham shrugged his shoulders. "Yes, I know about that. But there is something wrong with the place still. All I can say is that Broughton is a different man since he has lived here. I don't believe that he will remain much longer. But—you're staying here?—well, you'll hear all about it tonight. There's a big dinner, I understand." The conversation turned off to old reminiscences, and Glenham soon after had to go.

'Before I went to dress that evening I had twenty minutes' talk with Broughton in his library. There was no doubt that the man was altered, gravely altered. He was nervous and fidgety, and I found him looking at me only when my eye was off him. I naturally asked him what he wanted of me. I told him I would do anything I could, but that I couldn't conceive what he lacked that I could provide. He said with a lustreless smile that there was, however, something, and that he would tell me the following morning. It struck me that he was somehow ashamed of himself. and perhaps ashamed of the part he was asking me to play. However, I dismissed the subject from my mind and went up to dress in my palatial room. As I shut the door a draught blew out the Queen of Sheba from the wall, and I noticed that the tapestries were not fastened to the wall at the bottom. I have always held very practical views about spooks, and it has often seemed to me that the slow waving in firelight of loose tapestry upon a wall would account for ninety-nine per cent of the stories one hears. Certainly the dignified undulation of this lady with her attendants and huntsmen—one of whom was untidily cutting the throat of a fallow deer upon the very steps on which King Solomon, a grey-faced Flemish nobleman with the order of the Golden Fleece, awaited his fair visitor—gave colour to my hypothesis.

'Nothing much happened at dinner. The people were very much like those of the garden party. A young woman next to me seemed anxious to know what was being read in London. As she was far more familiar than I with the most recent magazines and literary supplements, I found salvation in being myself instructed in the tendencies of modern fiction. All true art, she said, was shot through and through with melancholy. How vulgar were the attempts at wit that marked so many modern books! From the beginning of literature it had always been tragedy that embodied the highest attainment of every age. To call such works morbid merely begged the question. No thoughtful man—she looked sternly at me through the steel rim of her glasses—could fail to agree with me. Of course, as one would, I immediately and properly said that I slept with Pett Ridge and Jacobs under my pillow at night, and that if *Jorrocks* weren't quite so large and cornery, I would add him to the company. She hadn't read any of them, so I was saved—for a time. But I remember grimly that she said that the dearest wish of her life was to be in some awful and soul-freezing situation of horror, and I remember that she dealt hardly with the hero of Nat Paynter's vampire story, between nibbles at her brown-bread ice. She was a cheerless soul, and I couldn't help thinking that if there were many such in the neighbourhood, it was not surprising that old Glenham had been stuffed with some nonsense or other about the Abbey. Yet nothing could well have been less creepy than the glitter of silver and glass, and the subdued lights and cackle of conversation all round the dinner-table.

'After the ladies had gone I found myself talking to the rural dean. He was a thin, earnest man, who at once turned the conversation to old Clarke's buffooneries. But, he said, Mr Broughton had introduced such a new and cheerful spirit, not only into the Abbey, but, he might say, into the whole neighbourhood, that he had great hopes that the ignorant superstitions of the past were from henceforth destined to oblivion. Thereupon his other neighbour, a portly gentleman of independent means and position, audibly remarked "Amen", which damped the rural dean, and we talked of partridges past, partridges present, and pheasants to come. At the other end of the table Broughton sat with a couple of his friends, red-faced hunting men. Once I noticed that they were discussing me, but I paid no attention to it at the time. I remembered it a few hours later.

'By eleven all the guests were gone, and Broughton, his wife, and I were alone together under the fine plaster ceiling of the Jacobean drawing-room. Mrs Broughton talked about one or two of the neighbours, and then, with a smile, said that she knew I would excuse her, shook hands with me, and went off to bed. I am not very good at analysing things, but I felt that she talked a little uncomfortably and with a suspicion of effort, smiled rather conventionally, and was obviously glad to go. These things seem trifling enough to repeat, but I had throughout the faint feeling that everything was not square. Under the circumstances, this was enough to set me wondering what on earth the service could be that I was to render—wondering also whether the whole business were not some ill-advised jest in order to make me come down from London for a mere shooting-party.

'Broughton said little after she had gone. But he was evidently labouring to bring the conversation round to the so-called haunting of the Abbey. As soon as I saw this, of course I asked him directly about it. He then seemed at once to lose interest in the matter. There was no doubt about it: Broughton was somehow a changed man, and to my mind he had changed in no way for the better. Mrs Broughton seemed no sufficient cause. He was clearly very fond of her, and she of him. I reminded him that he was going to tell me what I could do for him in the morning, pleaded my journey, lighted a candle, and went upstairs with him. At the end of the passage leading into the old house he grinned weakly and said, "Mind, if you see a ghost, do talk to it; you said you would." He stood irresolutely a moment and then turned away. At the door of his dressing-room he paused once more: "I'm here," he called out, "if you should want anything. Good night," and he shut his door.

'I went along the passage to my room, undressed, switched on a lamp beside my bed, read a few pages of *The Jungle Book*, and then, more than ready for sleep, turned the light off and went fast asleep.

'Three hours later I woke up. There was not a breath of wind outside. There was not even a flicker of light from the fireplace. As I lay there, an ash tinkled slightly as it cooled, but there was hardly a gleam of the dullest red in the grate. An owl cried among the silent Spanish chestnuts on the slope outside. I idly reviewed the events of the day, hoping that I should fall off to sleep again before I reached dinner. But at the end I seemed as wakeful as ever. There was no help for it. I must read my *Jungle Book* again till I felt ready to go off, so I fumbled for the pear at the end of the cord that hung down inside the bed, and I switched on the bedside lamp. The sudden glory dazzled me for a moment. I felt under my pillow for my book with half-shut eyes. Then, growing used to the light, I happened to look down to the foot of my bed.

'I can never tell you really what happened then. Nothing I could ever confess in the most abject words could even faintly picture to you what I felt. I know that my heart stopped dead, and my throat shut automatically. In one instinctive movement I crouched back up against the head-boards of the bed, staring at the horror. The movement set my heart going again, and the sweat dripped from every pore. I am not a particularly religious man, but I had always believed that God would never allow any supernatural appearance to present itself to man in such a guise and in such circumstances that harm, either bodily or mental, could result to him. I can only tell you that at that moment both my life and my reason rocked unsteadily on their seats.'

The other *Osiris* passengers had gone to bed. Only he and I remained leaning over the starboard railing, which rattled uneasily now and then under the fierce vibration of the over-engined mail-boat. Far over, there were the lights of a few fishing-smacks riding out the night, and a great rush of white combing and seething water fell out and away from us overside.

At last Colvin went on:

'Leaning over the foot of my bed, looking at me, was a figure swathed in a rotten and tattered veiling. This shroud passed over the head, but left both eyes and the right side of the face bare. It then followed the line of the arm down to where the hand grasped the bed-end. The face was not entirely that of a skull, though the eyes and the flesh of the face were totally gone. There was a thin, dry skin drawn tightly over the features, and there was some skin left on the hand. One wisp of hair crossed the forehead. It was perfectly still. I looked at it, and it looked at me, and my brains turned dry and hot in my head. I had still got the pear of the electric lamp in my hand, and I played idly with it; only I dared not turn the light out again. I shut my eyes, only to open them in a hideous terror the same second. The thing had not moved. My heart was thumping, and the sweat cooled me as it evaporated. Another cinder tinkled in the grate, and a panel creaked in the wall.

'My reason failed me. For twenty minutes, or twenty seconds, I was able to think of nothing else but this awful figure, till there came, hurtling through the empty channels of my senses, the remembrance that Broughton and his friends had discussed me furtively at dinner. The dim possibility of its being a hoax stole gratefully into my unhappy mind, and once there, one's pluck came creeping back along a thousand tiny veins. My first sensation was one of blind unreasoning thankfulness that my brain was going to stand the trial. I am not a timid man, but the best of us needs some human handle to steady him in time of extremity, and in this faint but growing hope that after all it might be only a brutal hoax, I found the fulcrum that I needed. At last I moved.

'How I managed to do it I cannot tell you, but with one spring towards the foot of the bed I got within arm's-length and struck out one fearful blow with my fist at the thing. It crumbled under it, and my hand was cut to the bone. With a sickening revulsion after my terror, I dropped half-fainting across the end of the bed. So it was merely a foul trick after all. No doubt the trick had been played many a time before: no doubt Broughton and his friends had had some large bet among themselves as to what I should do when I discovered the gruesome thing. From my state of abject terror I found myself transported into an insensate anger. I shouted curses upon Broughton. I dived rather than climbed over the bed-end on to the sofa. I tore at the robed skeleton—how well the whole thing had been carried out, I thought—I broke the skull against the floor, and stamped upon its dry bones. I flung the head away under the bed, and rent the brittle bones of the trunk in pieces. I snapped the thin thigh-bones across my knee, and flung them in different directions. The shin-bones I set up against a stool and broke with my heel. I raged like a Berserker against the loathly thing, and stripped the ribs from the backbone and slung the breastbone against the cupboard. My fury increased as the work of destruction went on. I tore the frail rotten veil into twenty pieces, and the dust went up over everything, over the clean blotting-paper and the silver inkstand. At last my work was done. There was but a raffle of broken bones and strips of parchment and crumbling wool. Then, picking up a piece of the skull—it was the cheek and temple bone of the right side, I remember—I opened the door and went down the passage to Broughton's dressing-room. I remember still how my sweat-dripping pyjamas clung to me as I walked. At the door I kicked and entered.

'Broughton was in bed. He had already turned the light on and seemed shrunken and horrified. For a moment he could hardly pull himself together. Then I spoke. I don't know what I said. Only I know that from a heart full and over-full with hatred and contempt, spurred on by shame of my own recent cowardice, I let my tongue run on. He answered nothing. I was amazed at my own fluency. My hair still clung lankily to my wet temples, my hand was bleeding profusely, and I must have looked a strange sight. Broughton huddled himself up at the head of the bed just as I had. Still he made no answer, no defence. He seemed preoccupied with something besides my reproaches, and once or twice moistened his lips with his tongue. But he could say nothing though he moved his hands now and then, just as a baby who cannot speak moves its hands.

'At last the door into Mrs Broughton's room opened and she came in, white and terrified. "What is it? What is it? Oh, in God's name! what is it?" she cried again and again, and then she went up to her husband and sat on the bed in her night-dress, and the two faced me. I told her what the matter was. I spared her husband not a word for her presence there. Yet he seemed hardly to understand. I told the pair that I had spoiled their cowardly joke for them. Broughton looked up.

'"I have smashed the foul thing into a hundred pieces," I said. Broughton licked his lips again and his mouth worked. "By God!" I shouted, "it would serve you right if I thrashed you within an inch of your life. I will take care that not a decent man or woman of my acquaintance ever speaks to you again. And there," I added, throwing the broken piece of the skull upon the floor beside his bed, "there is a souvenir for you, of your damned work tonight!"

'Broughton saw the bone, and in a moment it was his turn to frighten me. He squealed like a hare caught in a trap. He screamed and screamed till Mrs Broughton, almost as bewildered as myself, held on to him and coaxed him like a child to be quiet. But Broughton—and as he moved I thought that ten minutes ago I perhaps looked as terribly ill as he did—thrust her from him, and scrambled out of the bed on to the floor, and still screaming put out his hand to the bone. It had blood on it from my hand. He paid no attention to me whatever. In truth I said nothing. This was a new turn indeed to the horrors of the evening. He rose from the floor with the bone in his hand and stood silent. He seemed to be listening. "Time, time, perhaps," he muttered, and almost at the same moment fell at full length on the carpet, cutting his head against the fender. The bone flew from his hand and came to rest near the door. I picked Broughton up, haggard and broken, with blood over his face. He whispered hoarsely and quickly, "Listen, listen!" We listened.

'After ten seconds' utter quiet, I seemed to hear something. I could not be sure, but at last there was no doubt. There was a quiet sound as of one moving along the passage. Little regular steps came towards us over the hard oak flooring. Broughton moved to where his wife sat, white and speechless, on the bed, and pressed her face into his shoulder.

'Then, the last thing that I could see as he turned the light out, he fell forward with his own head pressed into the pillow of the bed. Something in their company, something in their cowardice, helped me, and I faced the open doorway of the room, which was outlined fairly clearly against the dimly lighted passage. I put out one hand and touched Mrs Broughton's shoulder in the darkness. But at the last moment I too failed. I sank on my knees and put my face in the bed. Only we all heard. The footsteps came to the door, and there they stopped. The piece of bone was lying a yard inside the door. There was a rustle of moving stuff, and the thing was in the room. Mrs Broughton was silent: I could hear Broughton's voice praying, muffled in the pillow: I was cursing my own cowardice. Then the steps moved out again on the oak boards of the passage, and I heard the sounds dying away. In a flash of remorse I went to the door and looked out. At the end of the corridor I thought I saw something that moved away. A moment later the passage was empty. I stood with my forehead against the jamb of the door almost physically sick.

'"You can turn the light on," I said, and there was an answering flare. There was no bone at my feet. Mrs Broughton had fainted. Broughton was almost useless, and it took me ten minutes to bring her to. Broughton only said one thing worth remembering. For the most part he went on muttering prayers. But I was glad afterwards to recollect that he had said that thing. He said in a colourless voice, half as a question, half as a reproach, "You didn't speak to her."

'We spent the remainder of the night together. Mrs Broughton actually fell off into in a kind of sleep before dawn, but she suffered so horribly in her dreams that I shook her into consciousness again. Never was dawn so long in coming. Three or four times Broughton spoke to himself. Mrs Broughton would then just tighten her hold on his arm, but she could say nothing. As for me, I can honestly say that I grew worse as the hours passed and the light strengthened. The two violent reactions had battered down my steadiness of view, and I felt that the foundations of my life had been built upon the sand. I said nothing, and after binding up my hand with a towel, I did not move. It was better so. They helped me and I helped them, and we all three knew that our reason had gone very near to ruin that night. At last, when the light came in pretty strongly, and the birds outside were chattering and singing, we felt that we must do something. Yet we never moved. You might have thought that we should particularly dislike being found as we were by the servants: yet nothing of that kind mattered a straw, and an overpowering listlessness bound us as we sat, until Chapman, Broughton's man, actually knocked and opened the door. None of us moved. Broughton, speaking hardly and stiffly, said, "Chapman you can come back in five minutes." Chapman, was a discreet man, but it would have made no difference to us if he had carried his news to the "room" at once.

'We looked at each other and I said I must go back. I meant to wait outside till Chapman returned. I simply dared not re-enter my bedroom alone. Broughton roused himself and said that he would come with me. Mrs Broughton agreed to remain in her own room for five minutes if the blinds were drawn up and all the doors left open.

'So Broughton and I, leaning stiffly one against the other, went down to my room. By the morning light that filtered past the blinds we could see our way, and I released the blinds. There was nothing wrong in the room from end to end, except smears of my own blood on the end of the bed, on the sofa, and on the carpet where I had torn the thing to pieces.'

Colvin had finished his story. There was nothing to say. Seven bells stuttered out from the fo'c'sle, and the answering cry wailed through the darkness. I took him downstairs.

'Of course I am much better now, but it is a kindness of you to let me sleep in your cabin.'

The Kit-bag

ALGERNON BLACKWOOD

When the words 'Not Guilty' sounded through the crowded courtroom that dark December afternoon, Arthur Wilbraham, the great criminal KC, and leader for the triumphant defence, was represented by his junior; but Johnson, his private secretary, carried the verdict across to his chambers like lightning.

'It's what we expected, I think,' said the barrister, without emotion; 'and, personally, I am glad the case is over.' There was no particular sign of pleasure that his defence of John Turk, the murderer, on a plea of insanity, had been successful, for no doubt he felt, as everybody who had watched the case felt, that no man had ever better deserved the gallows.

'I'm glad too,' said Johnson. He had sat in the court for ten days watching the face of the man who had carried out with callous detail one of the most brutal and cold-blooded murders of recent years.

The counsel glanced up at his secretary. They were more than employer and employed; for family and other reasons, they were friends. 'Ah, I remember; yes,' he said with a kind smile, 'and you want to get away for Christmas? You're going to skate and ski in the Alps, aren't you? If I was your age I'd come with you.'

Johnson laughed shortly. He was a young man of twenty-six, with a delicate face like a girl's. 'I can catch the morning boat now,' he said; 'but that's not the reason I'm glad the trial is over. I'm glad it's over because I've seen the last of that man's dreadful face. It positively haunted me. That white skin, with the black hair brushed low over the forehead, is a thing I shall never forget, and the description of the way the dismembered body was crammed and packed with lime into that——'

'Don't dwell on it, my dear fellow,' interrupted the other, looking at him curiously out of his keen eyes, 'don't think about it. Such pictures have a trick of coming back when one least wants them.' He paused a moment. 'Now go,' he added presently, 'and enjoy your holiday. I shall want all your energy for my Parliamentary work when you get back. And don't break your neck skiing.'

Johnson shook hands and took his leave. At the door he turned suddenly.

'I knew there was something I wanted to ask you,' he said. 'Would you mind lending me one of your kit-bags? It's too late to get one tonight, and I leave in the morning before the shops are open.'

'Of course; I'll send Henry over with it to your rooms. You shall have it the moment I get home.'

'I promise to take great care of it,' said Johnson gratefully, delighted to think that within thirty hours he would be nearing the brilliant sunshine of the high Alps in winter. The thought of that criminal court was like an evil dream in his mind.

He dined at his club and went on to Bloomsbury, where he occupied the top floor in one of those old, gaunt houses in which the rooms are large and lofty. The floor below his own was vacant and unfurnished, and below that were other lodgers whom he did not know. It was cheerless, and he looked forward heartily to a change. The night was even more cheerless: it was miserable, and few people were about. A cold, sleety rain was driving down the streets before the keenest east wind he had ever felt. It howled dismally among the big, gloomy houses of the great squares, and when he reached his rooms he heard it whistling and shouting over the world of black roofs beyond his windows.

In the hall he met his landlady, shading a candle from the draughts with her thin hand. 'This come by a man from Mr Wilbr'im's, sir.'

She pointed to what was evidently the kit-bag, and Johnson thanked her and took it upstairs with him. 'I shall be going abroad in the morning for ten days, Mrs Monks,' he said. 'I'll leave an address for letters.'

'And I hope you'll 'ave a merry Christmas, sir,' she said, in a raucous, wheezy voice that suggested spirits, 'and better weather than this.'

'I hope so too,' replied her lodger, shuddering a little as the wind went roaring down the street outside.

When he got upstairs he heard the sleet volleying against the window panes. He put his kettle on to make a cup of hot coffee, and then set about putting a few things in order for his absence. 'And now I must pack—such as my packing is,' he laughed to himself, and set to work at once.

He liked the packing, for it brought the snow mountains so vividly before him, and made him forget the unpleasant scenes of the past ten days. Besides, it was not elaborate in nature. His friend had lent him the very thing—a stout canvas kit-bag, sack-shaped, with holes round the neck for the brass bar and padlock. It was a bit shapeless, true, and not much to look at, but its capacity was unlimited, and there was no need to pack carefully. He shoved in his waterproof coat, his fur cap and gloves, his skates and climbing boots, his sweaters, snow-boots, and ear-caps; and then on the top of these he piled his woollen shirts and underwear, his thick socks, puttees, and knickerbockers. The dress suit came next, in case the hotel people dressed for dinner, and then, thinking of the best way to pack his white shirts, he paused a moment to reflect. 'That's the worst of these kit-bags,' he mused vaguely, standing in the centre of the sitting-room, where he had come to fetch some string.

It was after ten o'clock. A furious gust of wind rattled the windows as though to hurry him up, and he thought with pity of the poor Londoners whose Christmas would be spent in such a climate, whilst he was skimming over snowy slopes in bright sunshine, and dancing in the evening with rosy-cheeked girls——Ah! that reminded him; he must put in his dancing-pumps and evening socks. He crossed over from his sitting-room to the cupboard on the landing where he kept his linen.

And as he did so he heard someone coming softly up the stairs.

He stood still a moment on the landing to listen. It was Mrs Monks's step, he thought; she must be coming up with the last post. But then the steps ceased suddenly, and he heard no more. They were at least two flights down, and he came to the conclusion they were too heavy to be those of his bibulous landlady. No doubt they belonged to a late lodger who had mistaken his floor. He went into his bedroom and packed his pumps and dress-shirts as best he could.

The kit-bag by this time was two-thirds full, and stood upright on its own base like a sack of flour. For the first time he noticed that it was old and dirty, the canvas faded and worn, and that it had obviously been subjected to rather rough treatment. It was not a very nice bag to have sent him—certainly not a new one, or one that his chief valued. He gave the matter a passing thought, and went on with his packing. Once or twice, however, he caught himself wondering who it could have been wandering down below, for Mrs Monks had not come up with letters, and the floor was empty and unfurnished. From time to time, moreover, he was almost certain he heard a soft tread of someone padding about over the bare boards—cautiously, stealthily, as silently as possible—and, further, that the sounds had been lately coming distinctly nearer.

For the first time in his life he began to feel a little creepy. Then, as though to emphasize this feeling, an odd thing happened: as he left the bedroom, having just packed his recalcitrant white shirts, he noticed that the top of the kit-bag lopped over towards him with an extraordinary resemblance to a human face. The canvas fell into a fold like a nose and forehead, and the brass rings for the padlock just filled the position of the eyes. A shadow—or was it a travel stain? for he could not tell exactly—looked like hair. It gave him rather a turn, for it was so absurdly, so outrageously, like the face of John Turk, the murderer.

He laughed, and went into the front room, where the light was stronger.

'That horrid case has got on my mind,' he thought; 'I shall be glad of a change of scene and air.' In the sitting-room, however, he was not pleased to hear again that stealthy tread upon the stairs, and to realize that it was much closer than before, as well as unmistakably real. And this time he got up and went out to see who it could be creeping about on the upper staircase at so late an hour.

But the sound ceased; there was no one visible on the stairs. He went to the floor below, not without trepidation, and turned on the electric light to make sure that no one was hiding in the empty rooms of the unoccupied suite. There was not a stick of furniture large enough to hide a dog. Then he called over the banisters to Mrs Monks, but there was no answer, and his voice echoed down into the dark vault of the house, and was lost in the roar of the gale that howled outside. Everyone was in bed and asleep—everyone except himself and the owner of this soft and stealthy tread.

'My absurd imagination, I suppose,' he thought. 'It must have been the wind after all, although—it seemed so *very* real and close, I thought.' He went back to his packing. It was by this time getting on towards midnight. He drank his coffee up and lit another pipe—the last before turning in.

It is difficult to say exactly at what point fear begins, when the causes of that fear are not plainly before the eyes. Impressions gather on the surface of the mind, film by film, as ice gathers upon the surface of still water, but often so lightly that they claim no definite recognition from the consciousness. Then a point is reached where the accumulated impressions become a definite emotion, and the mind realizes that something has happened. With something of a start, Johnson suddenly recognized that he felt nervous—oddly nervous; also, that for some time past the causes of this feeling had been gathering slowly in his mind, but that he had only just reached the point where he was forced to acknowledge them.

It was a singular and curious malaise that had come over him, and he hardly knew what to make of it. He felt as though he were doing something that was strongly objected to by another person, another person, moreover, who had some right to object. It was a most disturbing and disagreeable feeling, not unlike the persistent promptings of conscience: almost, in fact, as if he were doing something he knew to be wrong. Yet, though he searched vigorously and honestly in his mind, he could nowhere lay his finger upon the secret of this growing uneasiness, and it perplexed him. More, it distressed and frightened him.

'Pure nerves, I suppose,' he said aloud with a forced laugh. 'Mountain air will cure all that! Ah,' he added, still speaking to himself, 'and that reminds me—my snow-glasses.'

He was standing by the door of the bedroom during this brief soliloquy, and as he passed quickly towards the sitting-room to fetch them from the cupboard he saw out of the corner of his eye the indistinct outline of a figure standing on the stairs, a few feet from the top. It was someone in a stooping position, with one hand on the banisters, and the face peering up towards the landing. And at the same moment he heard a shuffling footstep. The person who had been creeping about below all this time had at last come up to his own floor. Who in the world could it be? And what in the name of Heaven did he want?

Johnson caught his breath sharply and stood stock still. Then, after a few seconds' hesitation, he found his courage, and turned to investigate. The stairs, he saw to his utter amazement, were empty; there was no one. He felt a series of cold shivers run over him, and something about the muscles of his legs gave a little and grew weak. For the space of several minutes he peered steadily into the shadows that congregated about the top of the staircase where he had seen the figure, and then he walked fast—almost ran, in fact—into the light of the front room; but hardly had he passed inside the doorway when he heard someone come up the stairs behind him with a quick bound and go swiftly into his bedroom. It was a heavy, but at the same time a stealthy footstep—the tread of somebody who did not wish to be seen. And it was at this precise moment that the nervousness he had hitherto experienced leaped the boundary line, and entered the state of fear, almost of acute, unreasoning fear. Before it turned into terror there was a further boundary to cross, and beyond that again lay the region of pure horror. Johnson's position was an unenviable one.

'By Jove! That *was* someone on the stairs, then,' he muttered, his flesh crawling all over; 'and whoever it was has now gone into my bedroom.' His delicate, pale face turned absolutely white, and for some minutes he hardly knew what to think or do. Then he realized intuitively that delay only set a premium upon fear; and he crossed the landing boldly and went straight into the other room, where, a few seconds before, the steps had disappeared.

'Who's there? Is that you, Mrs Monks?' he called aloud, as he went, and heard the first half of his words echo down the empty stairs, while the second half fell dead against the curtains in a room that apparently held no other human figure than his own.

'Who's there?' he called again, in a voice unnecessarily loud and that only just held firm. 'What do you want here?'

The curtains swayed very slightly, and, as he saw it, his heart felt as if it almost missed a beat; yet he dashed forward and drew them aside with a rush. A window, streaming with rain, was all that met his gaze. He continued his search, but in vain; the cupboards held nothing but rows of clothes, hanging motionless; and under the bed there was no sign of anyone hiding. He stepped backwards into the middle of the room, and, as he did so, something all but tripped him up. Turning with a sudden spring of alarm he saw—the kit-bag.

'Odd!' he thought. 'That's not where I left it!' A few moments before it had surely been on his right, between the bed and the bath; he did not remember having moved it. It was very curious. What in the world was the matter with everything? Were all his senses gone queer? A terrific gust of wind tore at the windows, dashing the sleet against the glass with the force of small gunshot, and then fled away howling dismally over the waste of Bloomsbury roofs. A sudden vision of the Channel next day rose in his mind and recalled him sharply to realities.

There's no one here at any rate; that's quite clear!' he exclaimed aloud. Yet at the time he uttered them he knew perfectly well that his words were not true and that he did not believe them himself. He felt exactly as though someone was hiding close about him, watching all his movements, trying to hinder his packing in some way. 'And two of my senses,' he added, keeping up the pretence, 'have played me the most absurd tricks: the steps I heard and the figure I saw were both entirely imaginary.'

He went back to the front room, poked the fire into a blaze, and sat down before it to think. What impressed him more than anything else was the fact that the kit-bag was no longer where he had left it. It had been dragged nearer to the door.

What happened afterwards that night happened, of course, to a man already excited by fear, and was perceived by a mind that had not the full and proper control, therefore, of the senses. Outwardly, Johnson remained calm and master of himself to the end, pretending to the very last that everything he witnessed had a natural explanation, or was merely delusions of his tired nerves. But inwardly, in his very heart, he knew all along that someone had been hiding downstairs in the empty suite when he came in, that this person had watched his opportunity and then stealthily made his way up to the bedroom, and that all he saw and heard afterwards, from the moving of the kit-bag to—well, to the other things this story has to tell—were caused directly by the presence of this invisible person.

And it was here, just when he most desired to keep his mind and thoughts controlled, that the vivid pictures received day after day upon the mental plates exposed in the courtroom of the Old Bailey, came strongly to light and developed themselves in the dark room of his inner vision. Unpleasant, haunting memories have a way of coming to life again just when the mind least desires them—in the silent watches of the night, on sleepless pillows, during the lonely hours spent by sick and dying beds. And so now, in the same way, Johnson saw nothing but the dreadful face of John Turk, the murderer, lowering at him from every corner of his mental field of vision; the white skin, the evil eyes, and the fringe of black hair low over the forehead. All the pictures of those ten days in court crowded back into his mind unbidden, and very vivid.

'This is all rubbish and nerves,' he exclaimed at length, springing with sudden energy from his chair. 'I shall finish my packing and go to bed. I'm overwrought, overtired. No doubt, at this rate I shall hear steps and things all night!'

But his face was deadly white all the same. He snatched up his field-glasses and walked across to the bedroom, humming a music-hall song as he went—a trifle too loud to be natural; and the instant he crossed the threshold and stood within the room something turned cold about his heart, and he felt that every hair on his head stood up.

The kit-bag lay close in front of him, several feet nearer to the door than he had left it, and just over its crumpled top he saw a head and face slowly sinking down out of sight as though someone were crouching behind it to hide, and at the same moment a sound like a long-drawn sigh was distinctly audible in the still air about him between the gusts of the storm outside.

Johnson had more courage and will-power than the girlish indecision of his face indicated; but at first such a wave of terror came over him that for some seconds he could do nothing but stand and stare. A violent trembling ran down his back and legs, and he was conscious of a foolish, almost a hysterical, impulse to scream aloud.

That sigh seemed in his very ear, and the air still quivered with it. It was unmistakably a human sigh.

'Who's there?' he said at length, finding his voice; but though he meant to speak with loud decision, the tones came out instead in a faint whisper, for he had partly lost the control of his tongue and lips.

He stepped forward, so that he could see all round and over the kit-bag. Of course there was nothing there, nothing but the faded carpet and the bulging canvas sides. He put out his hands and threw open the mouth of the sack where it had fallen over, being only three parts full, and then he saw for the first time that round the inside, some six inches from the top, there ran a broad smear of dull crimson. It was an old and faded blood stain. He uttered a scream, and drew back his hands as if they had been burnt. At the same moment the kit-bag gave a faint, but unmistakable, lurch forward towards the door.

Johnson collapsed backwards, searching with his hands for the support of something solid, and the door, being further behind him than he realized, received his weight just in time to prevent his falling, and shut to with a resounding bang. At the same moment the swinging of his left arm accidentally touched the electric switch, and the light in the room went out.

It was an awkward and disagreeable predicament, and if Johnson had not been possessed of real pluck he might have done all manner of foolish things. As it was, however, he pulled himself together, and groped furiously for the little brass knob to turn the light on again. But the rapid closing of the door had set the coats hanging on it a-swinging, and his fingers became entangled in a confusion of sleeves and pockets, so that it was some moments before he found the switch. And in those few moments of bewilderment and terror two things happened that sent him beyond recall over the boundary into the region of genuine horror—he distinctly heard the kit-bag shuffling heavily across the floor in jerks, and close in front of his face sounded once again the sigh of a human being.

In his anguished efforts to find the brass button on the wall he nearly scraped the nails from his fingers, but even then, in those frenzied moments of alarm—so swift and alert are the impressions of a mind keyed-up by a vivid emotion—he had time to realize that he dreaded the return of the light, and that it might be better for him to stay hidden in the merciful screen of darkness. It was but the impulse of a moment, however, and before he had time to act upon it he had yielded automatically to the original desire, and the room was flooded again with light.

But the second instinct had been right. It would have been better for him to have stayed in the shelter of the kind darkness. For there, close before him, bending over the half-packed kit-bag, clear as life in the merciless glare of the electric light, stood the figure of John Turk, the murderer. Not three feet from him the man stood, the fringe of black hair marked plainly against the pallor of the forehead, the whole horrible presentment of the scoundrel, as vivid as he had seen him day after day in the Old Bailey, when he stood there in the dock, cynical and callous, under the very shadow of the gallows.

In a flash Johnson realized what it all meant: the dirty and much-used bag; the smear of crimson within the top; the dreadful stretched condition of the bulging sides. He remembered how the victim's body had been stuffed into a canvas bag for burial, the ghastly, dismembered fragments forced with lime into this very bag; and the bag itself produced as evidence—it all came back to him as clear as day . . .

Very softly and stealthily his hand groped behind him for the handle of the door, but before he could actually turn it the very thing that he most of all dreaded came about, and John Turk lifted his devil's face and looked at him. At the same moment that heavy sigh passed through the air of the room, formulated somehow into words: 'It's my bag. And I want it.'

Johnson just remembered clawing the door open, and then falling in a heap upon the floor of the landing, as he tried frantically to make his way into the front room.

He remained unconscious for a long time, and it was still dark when he opened his eyes and realized that he was lying, stiff and bruised, on the cold boards. Then the memory of what he had seen rushed back into his mind, and he promptly fainted again. When he woke the second time the wintry dawn was just beginning to peep in at the windows, painting the stairs a cheerless, dismal grey, and he managed to crawl into the front room, and cover himself with an overcoat in the armchair, where at length he fell asleep.

A great clamour woke him. He recognized Mrs Monks's voice, loud and voluble.

'What! You ain't been to bed, sir! Are you ill, or has anything 'appened? And there's an urgent gentleman to see you, though it ain't seven o'clock yet, and——'

'Who is it?' he stammered. 'I'm all right, thanks. Fell asleep in my chair, I suppose.'

'Someone from Mr Wilb'rim's, and he says he ought to see you quick before you go abroad, and I told him——'

'Show him up, please, at once,' said Johnson, whose head was whirling, and his mind was still full of dreadful visions.

Mr Wilbraham's man came in with many apologies, and explained briefly and quickly that an absurd mistake had been made, and that the wrong kit-bag had been sent over the night before.

'Henry somehow got hold of the one that came over from the courtoom, and Mr Wilbraham only discovered it when he saw his own lying in his room, and asked why it had not gone to you,' the man said.

'Oh!' said Johnson stupidly.

'And he must have brought you the one from the murder case instead, sir, I'm afraid,' the man continued, without the ghost of an expression on his face. 'The one John Turk packed the dead body in. Mr Wilbraham's awful upset about it, sir, and told me to come over first thing this morning with the right one, as you were leaving by the boat.'

He pointed to a clean-looking kit-bag on the floor, which he had just brought. 'And I was to bring the other one back, sir,' he added casually.

For some minutes Johnson could not find his voice. At last he pointed in the direction of his bedroom. 'Perhaps you would kindly unpack it for me. Just empty the things out on the floor.'

The man disappeared into the other room, and was gone for five minutes. Johnson heard the shifting to and fro of the bag, and the rattle of the skates and boots being unpacked.

'Thank you, sir,' the man said, returning with the bag folded over his arm. 'And can I do anything more to help you, sir?'

'What is it?' asked Johnson, seeing that he still had something he wished to say.

The man shuffled and looked mysterious. 'Beg pardon, sir, but knowing your interest in the Turk case, I thought you'd maybe like to know what's happened——'

'Yes.'

'John Turk killed hisself last night with poison immediately on getting his release, and he left a note for Mr Wilbraham saying as he'd be much obliged if they'd have him put away, same as the woman he murdered, in the old kit-bag.'

'What time—did he do it?' asked Johnson.

'Ten o'clock last night, sir, the warder says.'

SOURCES

THE stories have been arranged in chronological order of publication. Usually this means a story's first appearance in a magazine; but where this information is either not known to the present editors or is not applicable, the date of first publication in book form is given. Place of publication is London unless otherwise stated.

'The Old Nurse's Story' by Elizabeth Gaskell (1810–65). First published in *Household Words* (Christmas Number, 1852); reprinted in *Lizzie Leigh, and other tales* (Chapman & Hall, 1855).

'An Account of Some Strange Disturbances in Aungier Street' by J[oseph] S[heridan] Le Fanu (1814–73). First published in the *Dublin University Magazine* (Dec. 1853); first reprinted in *Madam Crowl's Ghost*, ed. M. R. James (Bell & Co., 1923).

'The Miniature' by John Yonge Akerman (1803–76). From *Legends of Old London* (Arthur Hall, Virtue & Co., 1853).

'The Last House in C—— Street' by Dinah Mulock (Mrs Craik, 1828–87). First published in *Fraser's Magazine* (Aug. 1856); reprinted in *Nothing New. Tales* (Hurst & Blackett, 2 vols., 1857).

'To be Taken with a Grain of Salt' by Charles Dickens (1812–70). First published in *All the Year Round* (Christmas Number, 1865) as a companion to Rosa Mulholland's 'Not to be Taken at Bed-time'.

'The Botathen Ghost' by R. S. Hawker (1803–75). First published in *All the Year Round* (18 May 1867); reprinted in *The Prose Works of Rev. R. S. Hawker*, ed. J. S. Godwin (Edinburgh and London: Blackwood & Sons, 1893).

'The Truth, the Whole Truth, and Nothing but the Truth' by Rhoda Broughton (1840–1920). First published in *Temple Bar* (Feb. 1868); reprinted in *Tales for Christmas Eve* (Leipzig: Tauchnitz, 1872; London: Bentley, 1873).

'The Romance of Certain Old Clothes' by Henry James (1843–1916). First published in the *Atlantic Monthly* (Feb. 1868); reprinted in *A Passionate Pilgrim, and other tales* (Boston: James R. Osgood & Co., 1875).

'Pichon & Sons, of the Croix Rousse', Anon. *A Stable for Nightmares*, the Christmas Number of *Tinsleys' Magazine* for 1868.

'Reality or Delusion?' by Mrs Henry Wood (née Ellen Price, 1814–87). First published in *The Argosy*, owned and edited by Mrs Henry Wood (Dec. 1868); reprinted in *Johnny Ludlow*, First Series (Bentley, 3 vols., 1874), with minor textual variations. The text followed here is 1874.

'Uncle Cornelius His Story' by George MacDonald (1824–1905). First published in *St Paul's Magazine* (Jan. 1869); reprinted in *Works of Fancy and Imagination* (Strahan & Co., 10 vols., 1871), vol. x.

'The Shadow of a Shade' by Tom Hood (1835–74). First published in *Frozen In, a series of stories related in a snow-storm, Bow Bells Annual* (Christmas 1869).

'At Chrighton Abbey' by Mary Elizabeth Braddon (1835–1915). First published in *Belgravia* (May 1871); reprinted in *Milly Darrell, and other tales* (John Maxwell, 3 vols., 1873).

'No Living Voice' by Thomas Street Millington (1821–1906?). Published anonymously in *Temple Bar* (Apr. 1872). The story is ascribed to Millington, a clergyman whose output included adventure stories for boys, in the *Wellesley Index to Victorian Periodicals*.

'Miss Jéromette and the Clergyman' by Wilkie Collins (1824–89). First published (as 'The Clergyman's Confession') in the *Canadian Monthly* (Aug.–Sept. 1875); reprinted in *Little Novels* (Chatto & Windus, 3 vols., 1887).

'The Story of Clifford House', Anon. Published in *The Mistletoe Bough* (Christmas 1878), edited by Mary Elizabeth Braddon. It is tempting to suppose the story was written by Miss Braddon herself, but there is no evidence to support the ascription.

'Was It An Illusion?' by Amelia B[lanford] Edwards (1831–92). From *Arrowsmith's Christmas Annual* (1881).

'The Open Door' by Charlotte Elizabeth [Mrs J. H.] Riddell (1832–1906). From *Weird Stories* (Hogg, 1882).

'The Captain of the "Pole-star"' by Sir Arthur Conan Doyle (1859–1930). First published in *Temple Bar* (Jan. 1883); reprinted in *The Captain of the 'Pole-star', and other tales* (London and New York: Longman's, Green, 1890).

'The Body-Snatcher' by Robert Louis Stevenson (1850–94). First published in the *Pall Mall Magazine* (Christmas Number, 1884); issued in book form by the Merriam Company (New York, 1895); reprinted in *Tales and Fantasies* (Chatto & Windus, 1905).

'The Story of the Rippling Train' by Mary Louisa Molesworth (1839–1921). First published in *Longman's Magazine* (Oct. 1887); reprinted in *Four Ghost Stories* (Macmillan, 1888).

'At the End of the Passage' by Rudyard Kipling (1865–1936). First published in *Lippincott's Magazine* (Aug. 1890); reprinted in *Life's Handicap* (Macmillan, 1891).

'"To Let"' by B[ithia] M[ary] Croker (1849–1920). First published in *London Society* (Christmas Number, 1890); reprinted in '*To Let*' (Chatto & Windus, 1893).

'John Charrington's Wedding' by E[dith] Nesbit (1858–1924). First published in *Temple Bar* (Sept. 1891); reprinted in *Grim Tales* (A. D. Innes, 1893).

'The Haunted Organist of Hurly Burly' by Rosa Mulholland (1841–1921). From *The Haunted Organist of Hurly Burly* (Hutchinson, n.d. [1891]).

'The Man of Science' by Jerome K[lapka] Jerome (1859–1927). First published in *The Idler* (edited by Jerome and Robert Barr, Sept. 1892). The story was told within part vi of the serialized *Novel Notes* (issued in book form by the Leadenhall Press, 1893). The title is the present editors'.

'Canon Alberic's Scrap-book', by M[ontague] R[hodes] James (1862–1936). First published in the *National Review* (Mar. 1895); reprinted in *Ghost Stories of an Antiquary* (Edward Arnold, 1904).

'Jerry Bundler' by W. W. Jacobs (1863–1943). First published in the *Windsor Magazine* (Dec. 1897); reprinted in *Light Freights* (Methuen, 1901).

'An Eddy on the Floor' by Bernard Capes (1854–1918). From *At a Winter's Fire* (C. A. Pearson, 1899).

'The Tomb of Sarah' by F. G. Loring (1869–1951). *Pall Mall Magazine* (Dec. 1900).

'The Case of Vincent Pyrwhit' by Barry Pain (pseudonym of Eric Odell, 1865–1928). From *Stories in the Dark* (Grant Richards, 1901).

'The Shadows on the Wall' by Mary E[leanor] Wilkins (also known as Wilkins-Freeman, 1852–1930). First published in *Everbody's Magazine*, vol. viii (1902); reprinted in *The Wind in the Rose-bush* (John Murray, 1903).

'Father Macclesfield's Tale' by R. H. Benson (1871–1914). From *A Mirror of Shalott, composed of tales told at a symposium* (Sir I. Pitman & Sons, 1907).

'Thurnley Abbey' by Perceval Landon (1869–1927). From *Raw Edges* (Heinemann, 1908).

'The Kit-bag' by Algernon Blackwood (1869–1951). *Pall Mall Magazine* (Dec. 1908).

SELECT CHRONOLOGICAL CONSPECTUS OF GHOST STORIES 1840–1910

This conspectus contains details of the main ghost-story collections published between 1840 and 1910, and also of fiction collections that contain individual ghost stories.

1840–9

Catherine Crowe *The Night-side of Nature; or, Ghosts and ghost-seers* (T. C. Newby, 2 vols., 1848).

1850–9

J. S. Le Fanu, *Ghost Stories and Tales of Mystery* (Dublin and London: James McGlashan, William S. Orr & Co., 1851); with four illustrations by 'Phiz'. Includes 'The Watcher' [reprinted in later collections as 'The Familiar'], 'Schalken the Painter'.

Elizabeth Gaskell, *Lizzie Leigh, and other tales* (Chapman & Hall, 1855). Includes 'The Old Nurse's Story'.

Catherine Crowe, *Ghosts and Family Legends. A volume for Christmas* (T. C. Newby, 1859 for 1858).

Dinah Maria Mulock [Mrs Craik], *Nothing New. Tales* (Hurst & Blackett, 2 vols., 1857). Includes 'The Last House in C—— Street'.

1860–9

Amelia B. Edwards, *Miss Carew* (Hurst & Blackett, 3 vols., 1865). Includes (vol. i) 'The Eleventh of March', 'Number Three' [i.e. 'How the Third Floor Knew the Potteries']; (vol. iii) 'My Brother's Ghost Story', 'The North Mail' [i.e. 'The Phantom Coach'].

M. E. Braddon, *Ralph the Bailiff, and other tales* (Ward, Lock and Tyler, n.d. [1867].) Includes 'The Cold Embrace', 'Eveline's Visitant'.

1870–9

J. S. Le Fanu, *Chronicles of Golden Friars* (Bentley, 3 vols., 1871). Includes (vol. i) 'Madam Crowl's Ghost' (incorporated into 'A Strange Adventure in the Life of Miss Laura Mildmay').

—— *In a Glass Darkly* (Bentley, 3 vols., 1872). Includes (vol. i) 'Green Tea', 'The Familiar' [i.e. 'The Watcher'], 'Mr Justice Harbottle'; (vol. iii) 'Carmilla'.

Rhoda Broughton, *Tales for Christmas Eve* (Bentley, 1873). Includes 'The Truth, the Whole Truth, and Nothing but the Truth'. 'The Man with the Nose'.

Amelia B. Edwards, *Monsieur Maurice* (Hurst and Blackett, 3 vols., 1873). Includes (vol. i) 'An Engineer's Story'; (vol. ii) 'The New Pass', 'A Night on the Borders of the Black Forest', 'The Story of Salome', 'In the Confessional'; (vol. iii) 'The Four-fifteen Express'.

M. E. Braddon, *Milly Darrell, and other tales* (John Maxwell, 1873). Includes 'At Chrighton Abbey'.

Amelia B. Edwards, *A Night on the Borders of the Black Forest* (Leipzig: Tauchnitz, 1874). Contains title story, 'Sister Johanna's Story'.

Mrs J. H. Riddell, *Frank Sinclair's Wife, and other stories* (Tinsley, 3 vols., 1874). Includes 'Forewarned, Forearmed', 'Hertford O'Donnell's Warning' [later reprinted as 'The Banshee's Warning' in the volume of that name, 1894].

Mrs Henry Wood, *Johnny Ludlow*, 1st series (Bentley, 3 vols., 1874). Includes 'Reality or Delusion?'

Henry James, *A Passionate Pilgrim* (Boston: James Osgood, 1875). Includes 'The Last of the Valerii', 'The Romance of Certain Old Clothes'.

M. E. Braddon, *Weavers and Weft, and other tales* (John Maxwell, 3 vols., 1877). Includes 'John Granger', 'Her Last Appearance'.

1880–9

J. S. Le Fanu, *The Purcell Papers* (Bentley, 3 vols., 1880). Includes 'The Ghost and the Bone-setter'.

Mrs Henry Wood, *Johnny Ludlow*, 2nd series (Bentley, 1880). Includes 'Seen in the Moonlight'.

M. E. Braddon, *Flower and Weed, and other tales* (John and Robert Maxwell, 1882). Includes 'The Shadow in the Corner'.

Mrs J. H. Riddell, *Weird Stories* (Hogg, 1882). 'Walnut-tree House', 'The Open Door', 'Nut-bush Farm', 'The Old House in Vauxhall Walk', 'Sandy the Tinker', 'Old Mrs Jones'.

Florence Marryat, *The Ghost of Charlotte Cray* (Leipzig: Tauchnitz, 1883). Includes title story, 'The Invisible Tenants of Rushmere', 'Little White Souls'.

Margaret Oliphant, *Two Stories of the Seen and the Unseen* (Edinburgh and London: Blackwood, 1885). 'The Open Door', 'Old Lady Mary'.

Wilkie Collins, *Little Novels* (Chatto & Windus, 3 vols., 1887). Includes (vol. i) 'Mrs Zant and the Ghost'; (vol. ii) 'Miss Jéromette and the Clergyman'.

Mrs J. H. Riddell, *Idle Tales* (Ward & Downey, 1887). Includes 'The Last of Squire Ennismore'.

Mary Louisa Molesworth, *Four Ghost Stories* (Macmillan, 1888). 'Lady

Farquhar's Old Lady', 'Witnessed by Two', 'Unexplained', 'The Story of the Rippling Train'.

Sir Arthur Conan Doyle, *Mysteries and Adventures* (Walter Scott, 1889). Includes 'The Silver Hatchet'.

Mrs J. H. Riddell, *Princess Sunshine, and other stories* (Ward & Downey, 2 vols., 1889). Includes 'A Terrible Vengeance'.

1890–9

Sir Arthur Conan Doyle, *The Captain of the 'Pole-star', and other tales* (London and New York: Longman's, Green, 1890). Includes title story, 'The Great Kleinplatz Experiment', 'John Barrington Cowles'.

Rudyard Kipling, *The Phantom 'Rickshaw, and other tales* (first English edition, Sampson Low, Marston, Searle & Rivington, 1890). Includes title story, 'My Own True Ghost Story'.

Vernon Lee [Violet Paget], *Hauntings. Fantastic stories* (Heinemann, 1890). 'Amour Dure', 'Dionea', 'Oke of Okehurst', 'A Wicked Voice'.

Mrs Henry Wood, *Johnny Ludlow*, 4th series (Bentley, 1890). Includes 'Sandstone Torr', 'A Curious Experience'.

Rudyard Kipling, *Life's Handicap* (Macmillan, 1891). Includes 'At the End of the Passage'.

Rosa Mulholland, *The Haunted Organist of Hurly Burly* (Hutchinson, n.d. [1891]). Includes title story, 'The Ghost at the Rath', 'The Ghost of Wildwood Chase'.

Robert Barr, *In a Steamer Chair, and other shipboard tales* (Chatto & Windus, 1892). Includes 'Share and Share Alike', 'The Man Who Was Not on the Passenger List'.

Henry James, *The Lesson of the Master* (New York and London: Macmillan, 1892). Includes 'Sir Edmund Orme'.

A. Quiller-Couch, *'I Saw Three Ships', and other winter's tales* (Cassell, 1892). Includes 'A Blue Pantomime', 'The Haunted Dragoon'.

B. M. Croker, *'To Let'* (Chatto & Windus, 1893).

Henry James, *The Real Thing* (New York and London: Macmillan, 1893). Includes 'Nona Vincent', 'Sir Dominick Ferrand'.

E. Nesbit, *Grim Tales* (A. D. Innes, 1893). Includes 'Man-Size in Marble', 'John Charrington's Wedding', 'Uncle Abraham's Romance'.

—— *Something Wrong* (A. D. Innes, 1893). Includes 'Hurst of Hurstcote'.

Sir Arthur Conan Doyle, *Round the Red Lamp* (Methuen, 1894). Includes 'Lot No. 249'.

F. Marion Crawford, *The Upper Berth* (T. Fisher Unwin, 1894). Title story.

J. S. Le Fanu, *The Watcher, and other weird stories* (Downey & Co., 1894). Illustrated by Brinsley Le Fanu. Reprinted stories.

F. Murray Gilchrist, *The Stone Dragon* (Methuen, 1894). Includes 'Midsummer Madness', 'The Return', 'The Pageant of Ghosts'.

Louisa [Mrs Alfred] Baldwin, *The Shadow on the Blind, and other ghost stories* (Dent, 1895). Includes title story, 'How He Left the Hotel'. The volume was dedicated to Louisa Baldwin's nephew, Rudyard Kipling.

Ralph Adams Cram, *Black Spirits and White. A book of ghost stories* (Chicago: Stone and Kimball, 1895). Includes 'In Kropfsberg Keep'.

A. Quiller-Couch, *Wandering Heath. Stories, studies, and sketches* (Cassell, 1895). Includes 'The Roll-call of the Reef'.

Henry James, *Embarrassments* (Heinemann, 1896). Includes 'The Friends of the Friends' [under the title 'The Way It Came'].

Mary Louisa Molesworth, *Uncanny Tales* (Hutchinson, 1896). Includes 'Half-way Between the Stiles', 'At the Dip of the Road'.

The Countess of Munster [Wilhemina Fitzclarence], *Ghostly Tales* (Hutchinson, 1896).

Vincent O'Sullivan, *A Book of Bargains* (Leonard Smithers, 1896). Includes 'The Business of Madame Jahn', 'When I Was Dead'.

H. G. Wells, *The Plattner Story, and others* (Methuen, 1897). Includes 'The Red Room'.

Ella D'Arcy, *Modern Instances* (London and New York: John Lane, 1898). Includes 'The Villa Lucienne'.

Henry James, *The Two Magics* (1898). Includes 'The Turn of the Screw'.

Bernard Capes, *At a Winter's Fire* (Pearson, 1899). Includes 'An Eddy on the Floor', 'The Black Reaper'.

Dick Donovan [pseudonym of J. E. Preston-Muddock], *Tales of Terror* (Chatto & Windus, 1899).

E. and H. Heron [Hesketh and Kate Prichard], *Ghosts. Being the experiences of Flaxman Low* (Pearson, 1899). Includes 'The Story of "The Spaniards", Hammersmith', 'The Story of Medhans Lea'.

Mrs J. H. Riddell, *Handsome Phil, and other tales* (F. V. White, 1899). Includes 'Conn Kilrea'.

1900–10

Henry James, *The Soft Side* (Methuen, 1900). Includes 'The Great Good Place', 'Maud-Evelyn', 'The Real Right Thing', 'The Third Person'.

A. Quiller-Couch, *Old Fires and Profitable Ghosts* (Cassell, 1900). Includes 'The Seventh Man', 'A Pair of Hands'.

Richard Marsh, *The Seen and the Unseen* (Methuen, 1900). Includes 'The Violin', 'The Fifteenth Man'.

W. W. Jacobs, *Light Freights* (Methuen, 1901). Includes 'Jerry Bundler'.

Barry Pain, *Stories in the Dark* (Grant Richards, 1901). Includes 'The Case of Vincent Pyrwhit', 'The Undying Thing'.

Bernard Capes, *Plots* (Methuen, 1902). Includes 'The Face on the Sheet'.

W. W. Jacobs, *The Lady of the Barge* (Methuen, 1902). Includes 'The Monkey's Paw', 'The Well'.

Richard Marsh, *Between the Dark and the Daylight* (Digby, Long & Co., 1902). Includes 'The Haunted Chair'.

Mrs Baillie Reynolds [G. M. Robins], *The Relations and What They Related. A series of weird tales* (Hutchinson, 1902). Includes 'The Blathwaytes', 'A Twilight Experience', 'A Chance of Travel'.

Mary E. Wilkins, *The Wind in the Rose-bush* (John Murray, 1903). Title story, 'The Shadows on the Wall', 'Luella Miller', 'The Southwest Chamber', 'The Vacant Lot', 'The Lost Ghost'.

S. Baring-Gould, *A Book of Ghosts* (Methuen, 1904). Includes 'The Leaden Ring', 'A Dead Finger', 'The 9.30 Up-train'.

M. R. James, *Ghost Stories of an Antiquary* (Edward Arnold, 1904). 'Canon Alberic's Scrap-book', 'Lost Hearts', 'The Mezzotint', 'The Ash-tree', 'Number 13', 'Count Magnus', '"Oh, Whistle, and I'll Come To You, My Lad"', 'The Treasure of Abbot Thomas'.

Rudyard Kipling, *Traffics and Discoveries* (Macmillan, 1904). Includes 'They' (issued separately in a special illustrated edition in 1905).

Gertrude Atherton, *The Bell in the Fog, and other stories* (Macmillan, 1905).

R. L. Stevenson, *Tales and Fantasies* (Chatto & Windus, 1905). Includes 'The Body-Snatcher'.

Algernon Blackwood, *The Empty House, and other ghost stories* (Eveleigh Nash, 1906). Includes title story, 'A Case of Eavesdropping', 'Keeping His Promise'.

R. H. Benson, *A Mirror of Shalott. Composed of tales told at a symposium* (Sir I. Pitman, 1907). Includes 'Father Macclesfield's Tale'.

Algernon Blackwood, *The Listener, and other stories* (Eveleigh Nash, 1907). Includes title story, 'The Willows', The Woman's Ghost Story'.

Sir Arthur Conan Doyle, *Round the Fire Stories* (Smith, Elder, 1908). Includes 'The Leather Funnel', 'Playing With Fire', 'The Brown Hand'.

Algernon Blackwood, *John Silence, Physician Extraordinary* (Eveleigh Nash, 1908). Includes 'Ancient Sorceries', 'The Nemesis of Fire'.

Perceval Landon, *Raw Edges* (Heinemann, 1908). Includes 'Thurnley Abbey'.

W. W. Jacobs, *Sailors' Knots* (Methuen, 1909). Includes 'The Toll-house'.

W. F. Harvey, *Midnight House, and other tales* (Dent, 1910).

E. Nesbit, *Fear* (Stanley Paul, 1910). Includes 'The Shadow', 'The Violet Car'.

Edith Wharton, *Tales of Men and Ghosts* (Macmillan, 1910). Includes 'The Eyes'.